W9-AHH-240

FIFTH EDITION

SOILS AND FOUNDATIONS

CHENG LIU
JACK B. EVETT
The University of North Carolina at Charlotte

Upper Saddle River, New Jersey
Columbus, Ohio

To Kimmie, Jonathan, and Michele Liu
and
Linda, Susan, Scott, Sarah, Sallie, and Kayla Evett

Library of Congress Cataloging-in-Publication Data
Liu, Cheng
Soils and foundations / Cheng Liu, Jack B. Evett.—5th ed.
p. cm.
ISBN 0-13-025517-3
1. Soil mechanics. 2. Foundations. I. Title.
TA710 .L548 2001
624.1′5136—dc21 00-039195

Vice President and Publisher: Dave Garza
Editor in Chief: Stephen Helba
Executive Editor: Ed Francis
Production Editor: Christine M. Buckendahl
Production Coordination: Terry Routley, Carlisle Publishers Services
Design Coordinator: Robin G. Chukes
Cover Designer: Jeff Vanik
Cover photo: International Stock
Production Manager: Matthew Ottenweller
Marketing Manager: Jamie Van Voorhis

This book was set in Giovanni by Carlisle Communications, Ltd., and was printed and bound by R. R. Donnelley & Sons Company. The cover was printed by Phoenix Color Corp.

10 9 8 7 6 5 4 3 2 1
ISBN 0-13-025517-3

CONTENTS

PREFACE

We have attempted to prepare an introductory, practical textbook for soil mechanics and foundations, which emphasizes design and practical applications that are supported by basic theory. Written in a simple and direct style that should make it very easy to read and understand the subject matter, this book contains an abundance of both example problems within each chapter and work problems at the end of each chapter. In addition, there are ample diagrams, charts, and illustrations throughout to help better explain the subject matter. In summary, we have tried to extract the salient and essential aspects of soils and foundations and to present them in a simple and straightforward manner.

The preceding paragraph, slightly modified, began the preface of the first four editions of *Soils and Foundations,* and we think that it aptly relates our basic philosophy in preparing the fifth edition. We have, however, updated material where applicable and added substantial amounts of new and essential material to the fifth edition. We believe the result is a much stronger, more comprehensive, and therefore better book.

We urge students using this book to review each illustration as it is cited and to study each example problem very carefully. Believing that example problems are an extremely effective means of learning a subject such as soils and foundations, we have included an abundance of these problems, and we believe that they will be very useful in mastering the material in the book.

We want to express our sincere appreciation to Carlos G. Bell, formerly of The University of North Carolina at Charlotte, and to W. Kenneth Humphries, former Dean of Engineering at the University of South Carolina, who read our original manuscript and offered many helpful suggestions. Also, we would like to acknowledge the late Donald Steila of the Department of Geography and Earth Science at The University of North Carolina at Charlotte, who reviewed Chapter 1. We also thank Alan Stadler of the Department of Civil Engineering at The University of North Carolina at Charlotte for reviewing the new material on soil stabilization in Chapter 4. Finally, we thank the other reviewers of this edition for their helpful comments and suggestions: Mathew A. Dettman, Western Kentucky University; Gene R. Francisco, Northern Wisconsin Technical College; Charles A. Matrosic, Ferris State University; and Mohammad Naiafi, Missouri Western State College.

We hope that you will enjoy using the book. We would be pleased to receive your comments, suggestions, and/or criticisms.

Cheng Liu
Jack B. Evett
Charlotte, North Carolina

1

Formation of Natural Soil Deposits

1–1 INTRODUCTION

Soil is more or less taken for granted by the average person. It makes up the ground on which we live, it is for growing crops, and it makes us dirty. Beyond these observations, most people are not overly concerned with soil. There are, however, some people who *are* deeply concerned. These include certain engineers as well as geologists, contractors, hydrologists, farmers, agronomists, soil chemists, and others.

Most structures of all types rest either directly or indirectly upon soil, and proper analysis of the soil and design of the structure's foundation are necessary to ensure a safe structure free of undue settling and/or collapse. A comprehensive knowledge of the soil in a specific location is also important in many other contexts. Thus, study of soils should be an important component in the education of civil engineers.

Chapter 1 relates the formation of natural soil deposits; it describes the sources of soil. Chapter 2 introduces and defines various engineering properties of soil. Subsequent chapters deal with evaluation of these properties and with essential interrelationships of soil with structures of various types.

1–2 ROCKS—THE SOURCES OF SOILS

Soil is composed of particles, large and small, and it may be necessary to include as "soil" not only solid matter but also air and water. Normally, the particles are the result of weathering (disintegration and decomposition) of rocks and decay of vegetation. Some soil particles may, over a period of time, become consolidated under the weight of overlying material and become rock. In fact, cycles of rock distintegrating to form soil, soil becoming consolidated under great pressure and heat to form rock, rock disintegrating to form soil, and so on, have occurred repeatedly throughout geologic time. The differentiation between soil and rock is not sharp; but from an engineering perspective, if material can be removed without blasting, it is usually considered to be "soil," whereas if blasting is required, it might be regarded as "rock."

Rocks can be classified into three basic groups that reflect their origin and/or method of formation: *igneous, sedimentary,* and *metamorphic.*

Igneous Rocks

Igneous rocks form when magma (molten matter) such as that produced by erupting volcanoes cools sufficiently to solidify. Volcanic action, normally referred to as *volcanism,* can occur beneath or upon the earth's surface. Volcanoes probably produced the minority of earth's igneous rocks, however. During the earth's formative stages, its surface may well have been largely molten, thus not requiring magma to move to the surface from great depths. It is likely that great amounts of Precambrian rock formed in this fashion.

Igneous rocks can be coarse-grained or fine-grained, depending on whether cooling occurred slowly or rapidly. Relatively slow cooling occurs when magma is trapped in the crust below the earth's surface (such as at the core of a mountain range), whereas more rapid cooling occurs if the magma reaches the surface while molten (e.g., lava flow).

Of coarse-grained igneous rocks, the most common is *granite,* a hard rock rich in quartz, widely used as a construction material and for monuments. Others are *syenites, diorites,* and *gabbros.* Most common of the fine-grained igneous rocks is *basalt,* a hard, dark-colored rock rich in ferromagnesian minerals and often used in road construction. Others are *rhyolites* and *andesites.*

Being generally hard, dense, and durable, igneous rocks often make good construction materials. Also, they typically have high bearing capacities and therefore make good foundation material.

Sedimentary Rocks

Sedimentary rocks compose the great majority of rocks found on the earth's surface. They are formed when mineral particles, fragmented rock particles, and remains of certain organisms are transported by wind, water, and ice (with water being the predominant transporting agent) and deposited, typically in layers, to form sediments. Over a period of time as layers accumulate at a site, pressure on lower layers resulting from the weight of overlying strata hardens the deposits, forming sedimentary rocks. In addition, deposits may be solidified and cemented by certain minerals (e.g., silica, iron oxides, calcium carbonate). Sedimentary rocks can be identified easily when their layered appearance is observable. The most common sedimentary rocks are *shale, sandstone, limestone,* and *dolomite.*

Shale, the most abundant of the sedimentary rocks, is formed by consolidation of clays or silts. Organic matter or lime may also be present. Shales have a laminated structure and often exhibit a tendency to split along laminations. They can become soft and revert to clayey or silty material if soaked in water for a period of time. Shales vary in strength from soft (may be scratched with a fingernail and easily excavated) to hard (requiring explosives to excavate). Shales are sometimes referred to as *claystone* or *siltstone,* depending on whether they were formed from clays or silts, respectively.

Sandstone, consisting primarily of quartz, is formed by pressure and the cementing action of silica (SiO_2), calcite (calcium carbonate, $CaCO_3$), iron oxide, or clay. Strength and durability of sandstones vary widely depending on the kind of cementing material and degree of cementation as well as the amount of pressure involved.

Limestone is sedimentary rock composed primarily of calcium carbonate hardened underwater by cementing action (rather than pressure); it may contain some clays or organic materials within fissures or cavities. Like the strength of shales and sandstones, that of limestones varies considerably from soft to hard (and therefore durable), with actual strength depending largely on the rock's texture and degree of cementation. (A porous texture means lower strength.) Limestones occasionally have thin layers of sandstone and often contain fissures, cavities, and caverns, which may be empty or partly or fully filled with clay.

Dolomites are similar in grain structure and color to limestones and are, in fact, limestones in which the calcite ($CaCO_3$) interbonded with magnesium. Hence, the principal ingredient of dolomites is calcium magnesium carbonate [$CaMg(CO_3)_2$]. Dolomites and limestones can be differentiated by placing a drop of dilute hydrochloric acid on the rock. A quick reaction forming small white bubbles is indicative of limestone; no reaction, or a very slow one, means that the rock is dolomite.

As indicated, the degrees of strength and hardness of sedimentary rocks are variable, and engineering use of such rocks varies accordingly. Relatively hard shale makes a good foundation material. Sandstones are generally good construction materials. Limestone and dolomite, if strong, can be both good foundation and construction materials.

Metamorphic Rocks

Metamorphic rocks are much less common at the earth's surface than are sedimentary rocks. They are produced when sedimentary or igneous rocks literally change their texture and structure as well as mineral and chemical composition, as a result of heat, pressure, and shear. Granite metamorphoses to *gneiss,* a coarse-grained, banded rock. *Schist,* a medium- to coarse-grained rock, results from high-grade metamorphism of both basalt and shale. Low-grade metamorphism of shale produces *slate,* a fine-textured rock that splits into sheets. Sandstone is transformed to *quartzite,* a highly weather-resistant rock; limestone and dolomite change to *marble,* a hard rock capable of being highly polished. Gneiss, schist, and slate are *foliated* (layered); quartzite and marble are *nonfoliated.*

Metamorphic rocks can be hard and strong if unweathered. They can be good construction materials—marble is often used for buildings and monuments—but foliated metamorphic rocks often contain planes of weakness that can diminish strength. Metamorphic rocks sometimes contain weak layers between very hard layers.

1–3 ROCK WEATHERING AND SOIL FORMATION

As related in the preceding section, soil particles are the result of weathering of rocks and organic decomposition. Weathering is achieved by *mechanical* (*physical*) and *chemical* means.

Mechanical weathering disintegrates rocks into small particles by temperature changes, frost action, rainfall, running water, wind, ice, abrasion, and other physical phenomena. These cause rock distintegration by breaking, grinding, crushing, and so on. The effect of temperature change is especially important. Rocks subjected to large temperature variations expand and contract like other materials, possibly causing structural deterioration and eventual breakdown of rock material. When temperatures drop below the freezing point, water trapped in rock crevices freezes, expands, and can thereby break rock apart. Smaller particles produced by mechanical weathering maintain the same chemical composition as the original rock.

Chemical weathering causes chemical decomposition of rock, which can drastically change its physical and chemical characteristics. This type of weathering results from reactions of rock minerals with oxygen, water, acids, salts, and so on. It may include such processes as oxidation, solution (strictly speaking, solution is a physical process), carbonation, leaching, and hydrolysis. These cause chemical weathering actions that can (1) increase the volume of material, thereby causing subsequent material breakdown, (2) dissolve parts of rock matter, yielding voids that make remaining matter more susceptible to breaking, and (3) react with the cementing material, thereby loosening particles.

The type of soil produced by rock weathering is largely dependent on rock type. Of igneous rocks, granites tend to decompose to silty sands and sandy silts with some clays. Basalts and other rocks containing ferromagnesian minerals (but little or no silica) decompose primarily to clayey soils. With regard to sedimentary rocks, decomposed shales produce clays and silts, whereas sandstones again become sandy soils. Weathered limestones can produce a variety of soil types, with fine-grained ones being common. Of metamorphic rocks, gneiss and schist generally decompose to form silt–sand soils, whereas slate tends more to clayey soils. Weathered marble often produces fine-grained soils; quartzite decomposes to more coarse-grained soils, including both sands and gravels.

1–4 SOIL DEPOSITS

Soils produced by rock weathering can be categorized according to where they are ultimately deposited relative to the location of the parent rock. Some soils remain where they were formed, simply overlying the rock from which they came. These are known as *residual soils.* Others are transported from their place of origin and deposited elsewhere. They are called *transported soils.*

Residual soils have general characteristics that depend in part on the type of rock from which they came. Particle sizes, shapes, and composition can vary widely, as do depths of residual soil deposits—all depending on the amount and type of weathering. The actual depth of a residual soil deposit depends on the rate at which rock weathering has occurred at the location and the presence or absence of any erosive agents that would have carried soil away.

Transported soils are formed when rock weathers at one site and the particles are moved to another location. Some common transporting agents for particles are (1) gravity, (2) running water, (3) glaciers, and (4) wind. Transported soils can therefore be categorized with regard to these agents as *gravity deposits, alluvial deposits, glacial deposits,* and *wind deposits.*

Gravity Deposits

Gravity deposits are soil deposits transported by the effect of gravity. A common example is the landslide. Gravity deposits, which are not generally carried very far, tend to be loosely compacted and otherwise exhibit little change in the general character of soil material as a result of being transported.

Alluvial Deposits

Alluvial deposits, having been transported by moving water, are found in the vicinity of rivers. Rainwater falling on land areas runs overland, eroding and transporting soil and rock particles as it goes, and eventually enters a creek or river. Continuously moving water can carry particles and deposit them a considerable distance from their former location. All soils carried and deposited by flowing water are called *alluvial deposits.* Lack of vegetation may allow enormous amounts of erosion leading to vast alluvial deposits (e.g., the Mississippi Delta).

Rivers are capable of transporting particles of all sizes, ranging from very fine silts in suspension to, in some cases, large boulders. The greater the velocity of river flow, the larger will be the size of particles that can be carried. Hence, a sluggish creek may carry only fine-grained sediment, whereas a flooding river transports all particle sizes, including large rocks. The relationship between river velocity and size of particle carried also affects the manner in which particles are deposited. As river velocity decreases, relatively larger particles settle and are deposited first. If the velocity decreases further, the next-larger-size particles settle out.

Alluvial deposits are often composed of various soil types because different types of soil tend to mix as they are carried downstream. They do, however, tend to be layered because settling rates are proportional to particle size.

The nature of soil can be greatly influenced by past alluvial transport and deposits. For example, at a location where a river's velocity decreases, such as when the channel widens significantly or its slope decreases substantially, coarser soil particles settle, forming submerged, flat, triangular deposits known as *alluvial fans.* When flooding rivers, which normally carry a heavy sediment load, overflow their banks, the overflowing water experiences a decrease in velocity. Larger particles, such as sands and gravels, tend to settle more quickly; their deposits can form *natural levees*

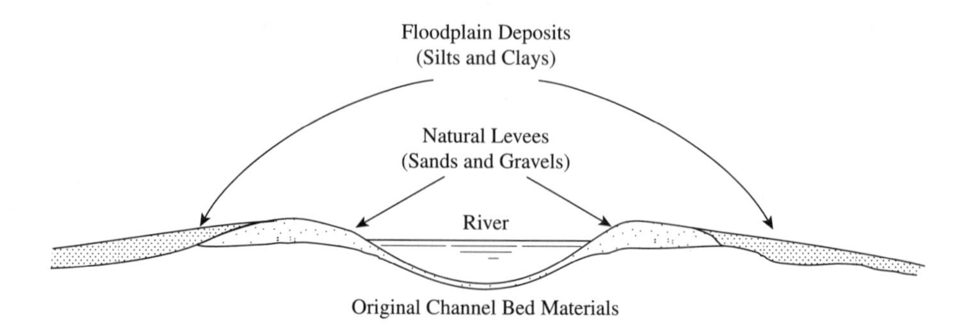

FIGURE 1–1 Natural levees and floodplain deposits.

along riverbanks (see Figure 1–1). (These natural levees may someday be washed away by a more severe flood.) Smaller particles, such as silts and clays, settle less quickly, forming *floodplain deposits* in areas beyond the levees (Figure 1–1). (However, smaller rivers can have floodplain deposits without forming levees.)

Another type of alluvial deposit occurs when rivers meander (i.e., follow a winding and turning course). As water moves through a channel bend, velocity along the inside edge decreases, whereas that along the outer one increases. Consequently, particle erosion may occur along the outer edge with deposition along the inner edge. This action can, over a period of time, increase the amount of bend and significantly alter the river channel and adjacent land area. Eventually, the river may cut across a large bend, as shown in Figure 1–2, leaving the old channel bend isolated. Water remaining in the isolated bend forms an *oxbow lake,* which can eventually fill in with floodplain deposits (usually silty and organic materials). Ultimately, the entire filled-in oxbow lake may be covered by additional floodplain deposits, leaving a hidden deposit of undesirable, high plastic, and/or organic silt, silty clay, and peat.

Sediments deposited at the mouths of creeks and rivers flowing into lakes, bays, or seas are known as *deltas.* Those deposited in lakes and seas are called *lacustrine* and *marine deposits,* respectively. These deposits tend to be loose and compressible and may contain organic material. They are therefore generally undesirable from an engineering point of view.

Glacial Deposits

Glacial deposits result, of course, from the action of glaciers. Many years ago (over 10,000), glaciers, unimaginably enormous sheets of ice, moved southward across much of the northern United States (as well as Europe and other areas). As they progressed, virtually everything in their paths, including soils and rocks ranging in size from the finest clays to huge boulders, was picked up and transported. As they were being carried by glaciers, soils and rocks were mixed together, thrashed about, bro-

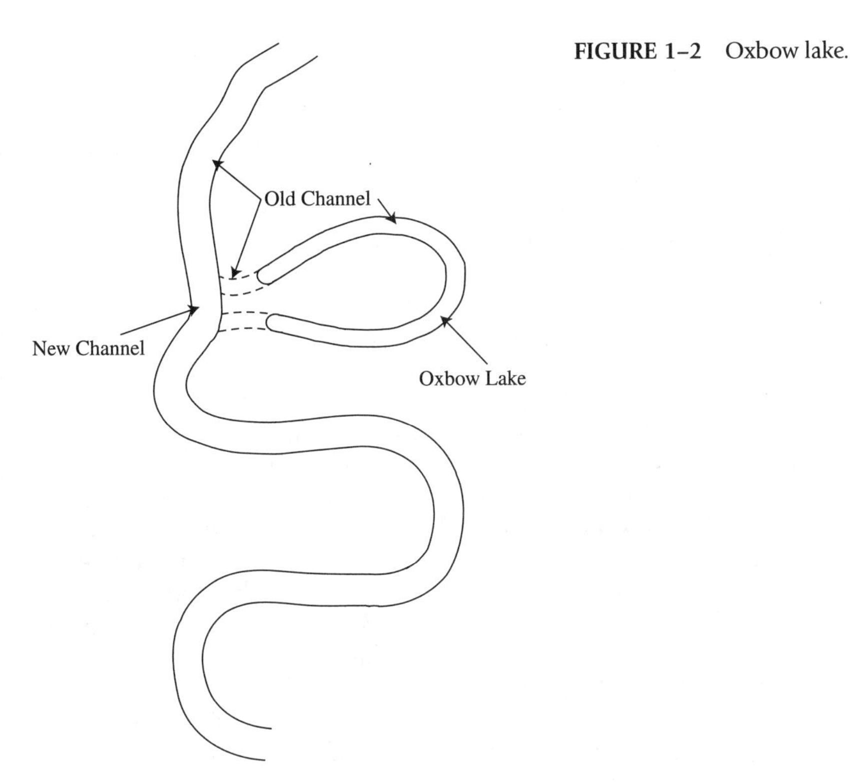

FIGURE 1–2 Oxbow lake.

ken, crushed, and so on, by enormous internal glacial pressures. Consequently, glacial deposits can contain all types of soils.

Some soil particles were directly deposited by moving glaciers; others were taken from glaciers by water flowing from the ice to be deposited in lakes or transported in rivers flowing away from the ice; still others were deposited *en masse* when glaciers ultimately melted and disappeared. Direct glacial deposits, known as *moraines,* are heterogeneous mixtures composed of all sizes of particles (from boulders to clay) that the ice accumulated as it traveled. *Eskers* are ridges or mounds of boulders, gravel, and sand formed when such materials flowing in streams on, within, or beneath glaciers were deposited as the stream's bed load.

The quality of soils in glacial deposits as foundations and construction materials is somewhat variable because of the different types of soils found in such deposits. Often these soils make good materials because of the intense compaction they have undergone, although those containing mostly clays are not as strong, are often compressible, and may therefore cause problems if used for foundations or construction materials.

Wind Deposits

Wind deposits (also known as *aeolian deposits*) obviously have wind as the transporting agent. Wind is a very important agent in certain areas and has the potential to move soil particles over large distances.

Winds can move sandy soil particles by rolling them along the ground as well as sending them short distances through the air. Wind-deposited sands are known as *dunes,* and they tend to occur in sandy desert areas and along sandy beaches on the downwind side. Sands from dunes can be used for certain construction purposes.

Fine-grained soils can be airborne over long distances by winds. Silty soils are more amenable than clayey soils to wind transport, however, because a clayey soil's bonding or cohesion reduces its wind erosion. A wind-deposited silt is known as *loess,* significant deposits of which are found in the general vicinity of the Mississippi and Missouri Rivers in the United States, and in Europe and Asia (especially northern China). Loess is generally a hard and stable soil when unsaturated because of cementation from calcium carbonate and iron oxide. It tends to lose its cementation when wetted, however, and to become soft and mushy. Loessial deposits typically have a yellow–brown (buff) color, low density, and relatively uniform grain size. These deposits are generally able to stand on vertical cuts and exhibit high vertical permeability. Because of their low strength when wet, however, special care must be taken during the design and construction of foundations over such deposits.

Ashes from erupting volcanoes can also produce wind deposits. Consisting of fine-sized igneous rock fragments, volcanic ash is light and porous, and deposits tend to decompose quickly, often changing into plastic clays. The great Mt. St. Helens eruption produced not lava but ash.

It should be noted in concluding this section that soil deposits seldom occur in nature in neat "packages"—that is, a soil of exactly the same type at all depths throughout a construction site. An area with "original" glacial deposits may subsequently have been overlain by alluvial deposits possessing different characteristics. Even if all the soil at a given job site is of the same deposit, its properties may vary from place to place throughout the site.

For these reasons, subsurface investigation of an area is extremely important. One cannot just look at the surface and know what is beneath. Using quantitative results obtained from subsurface investigation together with qualitative knowledge of the origins of the soil(s) at the site, soils engineers can produce an adequate foundation design to ensure against failure or undue settling of a structure. (Subsurface investigation is covered in Chapter 3.)

2

ENGINEERING PROPERTIES OF SOILS

2–1 SOIL TYPES

Soils may be separated into three very broad categories: *cohesionless, cohesive,* and *organic* soils. In the case of cohesionless soils, the soil particles do not tend to stick together. Cohesive soils are characterized by very small particle size where surface chemical effects predominate. The particles do tend to stick together—the result of water–particle interaction and attractive forces between particles. Cohesive soils are therefore both sticky and plastic. Organic soils are typically spongy, crumbly, and compressible. They are undesirable for use in supporting structures.

Three common types of cohesionless soils are *gravel, sand,* and *silt.* Gravel has particle sizes greater than about 2 millimeters (mm), whereas particle sizes for sand range from about 0.1 to 2 mm. Both gravel and sand may be further divided into "fine" (as fine sand) and "coarse" (as coarse sand). Gravel and sand can be classified according to particle size by sieve analysis. Silt has particle sizes that range from about 0.005 to 0.1 mm.

The common type of cohesive soil is clay, which has particle sizes less than about 0.005 mm. Clayey soils cannot be separated by sieve analysis into size categories because no practical sieve can be made with openings so small; instead, particle sizes may be determined by observing settling velocities of the particles in a water mixture.

Soils can also be categorized strictly in terms of grain size. Two such categories are *coarse-grained* and *fine-grained.* Gravel and sand, with soil grains coarser than 0.075 mm, or a No. 200 sieve size, are coarse-grained (also referred to as *granular* soils); silt and clay, with soil grains finer than 0.075 mm, are fine-grained.

Engineering properties of granular soils are affected by their grain sizes and shapes as well as by their grain-size distributions and their compactness (see Section 2–9). Granular soils, except for loose sand, generally possess excellent engineering properties. Exhibiting large bearing capacities and experiencing relatively small settlements, they make outstanding foundation materials for supporting roads and structures. Granular soils also make excellent backfill materials for retaining walls because they are easily compacted and easily drained, and because they exert small lateral pressures. In addition, as a result of high shear strengths and ease of compaction, granular soils make superior embankment material. One drawback, however, is that the high permeabilities of granular soils make them poor, or even unacceptable, for use alone as earthen dikes or dams.

Cohesive soils (mostly clays, but also silty clays and clay–sand mixtures with clay being predominant) exhibit generally undesirable engineering properties compared with those of granular soils. They tend to have lower shear strengths and to lose shear strength further upon wetting or other physical disturbances. They can be plastic and compressible, and they expand when wetted and shrink when dried. Some types expand and shrink greatly upon wetting and drying—a very undesirable feature. Cohesive soils can *creep* (deform plastically) over time under constant load, especially when the shear stress is approaching its shear strength, making them prone to landslides. They develop large lateral pressures and have low permeabilities. For these reasons, cohesive soils—unlike granular soils—are generally poor materials for retaining-wall backfills. Being impervious, however, they make better core materials for earthen dams and dikes.

Silty soils are on the border between clayey and sandy soils. They are fine-grained like clays but cohesionless like sands. Silty soils possess undesirable engineering properties. They exhibit high capillarity and susceptibility to frost action, yet they have low permeabilities and low relative densities.

Any soil containing a sufficient amount of organic matter to affect its engineering properties is called *organic soil.* As mentioned previously, organic soils are typically spongy, crumbly, and compressible. In addition, they possess low shear strengths and may contain harmful materials. Organic soils are essentially unacceptable for supporting foundations.

More precise classifications of these soil types by particle size according to two systems—the American Association of State Highway and Transportation Officials (AASHTO) system and the Unified Soil Classification System (USCS)—are given in Table 2–1. It is clear from variations between these classifications that boundaries between soil types are more or less arbitrary.

In most applications in this book, soils are categorized as cohesionless or cohesive, with cohesionless generally implying a sandy soil and cohesive, a clayey soil. Some soils encountered in practice are mixtures of both types and therefore exhibit characteristics of both.

2–2 GRAIN-SIZE ANALYSIS AND ATTERBERG LIMITS

Never will a natural soil be encountered in which all particles are exactly the same size and shape. Both cohesionless and cohesive soils, as well as mixtures of the two, will always contain particles of varying sizes. Properties of a soil are greatly influenced by the sizes of its particles and distribution of grain sizes throughout the soil mass. Hence, in many engineering applications, it is not sufficient to know only that a given soil is clay, sand, rock, gravel, or silt. It is also necessary to know something about the distribution of grain sizes of the soil.

TABLE 2–1
Soil Classification Based on Grain Size[1]

Agency	Coarse-Grained: Gravel	Coarse-Grained: Coarse Sand	Coarse-Grained: Fine Sand	Fine-Grained: Silt	Fine-Grained: Clay
AASHTO	75–2.00 (3-in.–No. 10 sieves)	2.00–0.425 (No. 10–No. 40 sieves)	0.425–0.075 (No. 40–No. 200 sieves)	0.075–0.002	<0.002
USCS	Coarse: 75–19.0 (3-in.–3/4-in. sieves) Fine: 19.0–4.75 (3/4-in.–No. 4 sieves)	4.75–2.00 (No. 4–No. 10 sieves) Medium sand: 2.00–0.425 (No. 10–No. 40 sieves)	0.425–0.075 (No. 40–No. 200 sieves)	Fines <0.075 (silt or clay)	

[1] All grain sizes are in millimeters.

TABLE 2–2
U.S. Standard Sieve Numbers and Their Sieve Openings

U.S. Standard Sieve Number	Sieve Opening (mm)
4	4.75
10	2.00
20	0.850
40	0.425
60	0.250
100	0.150
200	0.075

In the case of most cohesionless soils, distribution of grain size can be determined by sieve analysis. A sieve is similar to a cook's flour sifter. It is an apparatus containing a wire mesh with openings the same size and shape. When soil is passed through a sieve, soil particles smaller than the opening size of the sieve will pass through, whereas those larger than the opening size will be retained. Certain sieve-size openings between 4.75 and 0.075 mm are designated by U.S. Standard Sieve Numbers, as given in Table 2–2. Thus, grain sizes within this range can be classified according to U.S. Standard Sieve Numbers.

In practice, sieves of different opening sizes are stacked, with the largest opening size at the top and a pan at the bottom. Soil is poured in at the top, and soil particles pass downward through the sieves until they are retained on a particular sieve (see Figure 2–1). The stack of sieves is mechanically agitated during this procedure. At the end of the procedure, the soil particles retained on each sieve can be weighed and the results presented graphically in the form of a grain-size distribution curve. This is normally a semilog plot with grain size (diameter) along the abscissa on a logarithmic scale and percentage passing that grain size along the ordinate on an arithmetic scale. Example 2–1 illustrates the analysis of the results of a sieve test, including the preparation of a grain-size distribution curve.

EXAMPLE 2–1

Given

An air-dry soil sample weighing 2000 grams (g) is brought to the soils laboratory for mechanical grain-size analysis. The laboratory data are as follows:

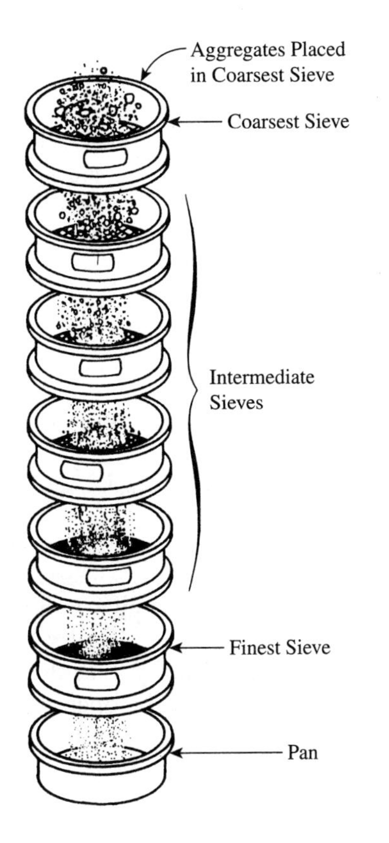

FIGURE 2–1 Sieve analysis [1].*

*Numbers in brackets refer to the references at the end of each chapter.

U.S. Sieve Size	Size Opening (mm)	Weight Retained (g)
¾ in.	19.0	0
⅜ in.	9.50	158
No. 4	4.75	308
No. 10	2.00	608
No. 40	0.425	652
No. 100	0.150	224
No. 200	0.075	42
Pan	—	8

Required

A grain-size distribution curve for this soil sample.

TABLE 2–3
Sieve Analysis Data for Example 2–1

(1) Sieve Number	(2) Sieve Opening (mm)	(3) Weight Retained (g)	(4) Percentage Retained	(5) Cumulative Percentage Retained	(6) Percentage Passing
¾ in.	19.0	0	0	0	100.0
⅜ in.	9.50	158	7.9	7.9	92.1
No. 4	4.75	308	15.4	23.3	76.7
No. 10	2.00	608	30.4	53.7	46.3
No. 40	0.425	652	32.6	86.3	13.7
No. 100	0.150	224	11.2	97.5	2.5
No. 200	0.075	42	2.1	99.6	0.4
Pan	—	8	0.4	100.0	—

Solution

To plot the gradation curve, one must first calculate the percentage retained on each sieve, the cumulative percentage retained, and the percentage passing through each sieve, then tabulate the results, as shown in Table 2–3.

$$\text{Total sample weight} = 2000 \text{ g}$$

1. The percentage retained on each sieve is obtained by dividing the weight retained on each sieve by the total sample weight. Thus,

$$\text{Percentage retained on 3/4-in. sieve} = \frac{0 \text{ g}}{2000 \text{ g}} \times 100\% = 0\%$$

$$\text{Percentage retained on 3/8-in. sieve} = \frac{158 \text{ g}}{2000 \text{ g}} \times 100\% = 7.9\%$$

$$\text{Percentage retained on No. 4 sieve} = \frac{308 \text{ g}}{2000 \text{ g}} \times 100\% = 15.4\% \qquad \text{etc.}$$

Therefore,

$$\text{Column (4)} = \frac{\text{Column (3)}}{\text{Total sample weight}} \times 100\%$$

2. The cumulative percentage retained on each sieve is obtained by summing the percentage retained on all coarser sieves. Thus,

$$\begin{aligned}
\text{Cumulative percentage retained on 3/4-in. sieve} &= 0\% \\
\text{Cumulative percentage retained on 3/8-in. sieve} &= 0\% + 7.9\% = 7.9\% \\
\text{Cumulative percentage retained on No. 4 sieve} &= 7.9\% + 15.4\% \\
&= 23.3\% \\
\text{Cumulative percentage retained on No. 10 sieve} &= 23.3\% + 30.4\% \\
&= 53.7\% \qquad \text{etc.}
\end{aligned}$$

3. The percentage passing through each sieve is obtained by subtracting from 100% the cumulative percentage retained on the sieves. Thus,

Percentage passing through 3/4-in. sieve = 100% − 0% = 100%
Percentage passing through 3/8-in. sieve = 100% − 7.9% = 92.1%
Percentage passing through No. 4 sieve = 100% − 23.3%
= 76.7% etc.

Therefore, column (6) = 100% − column (5).

4. Upon completion of these calculations, the grain-size distribution curve is obtained by plotting column (2), sieve opening (mm), versus column (6), percentage passing through, on semilog paper. The percentage passing is always plotted as the ordinate on the arithmetic scale and the sieve opening as the abscissa on the log scale (see Figure 2–2).

Several useful parameters can be determined from grain-size distribution curves. The diameter of soil particles at which 50% passes (i.e., 50% of the soil by weight is finer than this size) is known as the *median size* and is denoted by D_{50}. The diameter at which 10% passes is called the *effective size* and is denoted by D_{10}. Two coefficients used only in the Unified Soil Classification System (see Section 2–3) are the *coefficient of uniformity* (C_u) and the *coefficient of curvature* (C_c), which are defined as follows:

$$C_u = \frac{D_{60}}{D_{10}} \tag{2–1}$$

$$C_c = \frac{(D_{30})^2}{D_{60}D_{10}} \tag{2–2}$$

where D_{60} and D_{30} are the soil particle diameters corresponding to 60 and 30%, respectively, passing on the cumulative grain-size distribution curve.

Median size gives an "average" particle size for a given soil sample; other parameters offer some indication of the particle size range. Effective size gives the maximum particle diameter of the smallest 10% of soil particles. It is this size to which permeability and capillarity are related. C_u and C_c have little or no meaning when more than 5% of the soil is finer than a No. 200 sieve opening (0.075 mm).

In the case of cohesive soils, distribution of grain size is not determined by sieve analysis because the particles are too small. Particle sizes may be determined by the hydrometer method, which is a process for indirectly observing the settling velocities of the particles in a soil–water mixture. Another valuable technique for analyzing cohesive soils is by use of *Atterberg limits,* which is described in the remainder of this section.

Atterberg [2, 3] defined four states of consistency for cohesive soils. (Consistency refers to their degree of firmness.) These states are *liquid, plastic, semisolid,* and *solid* (see Figure 2–3). The dividing line between liquid and plastic states is the *liquid limit;* the dividing line between plastic and semisolid states is the *plastic limit;* and the dividing line between semisolid and solid states is the *shrinkage limit* (Fig. 2–3). If a soil in the liquid state is gradually dried out, it will pass through the liquid limit, plastic state, plastic limit, semisolid state, and shrinkage limit, and will

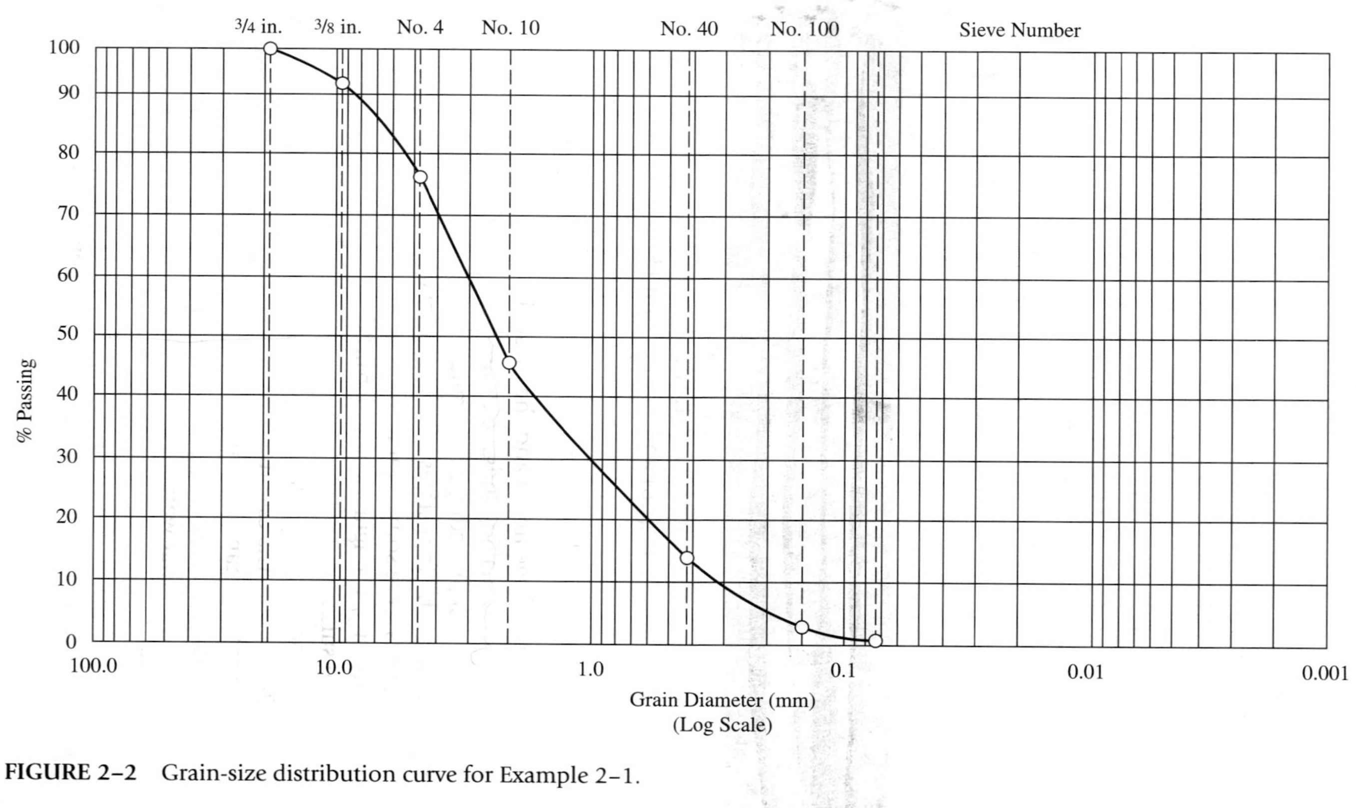

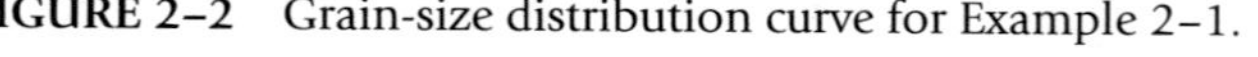

FIGURE 2–2 Grain-size distribution curve for Example 2–1.

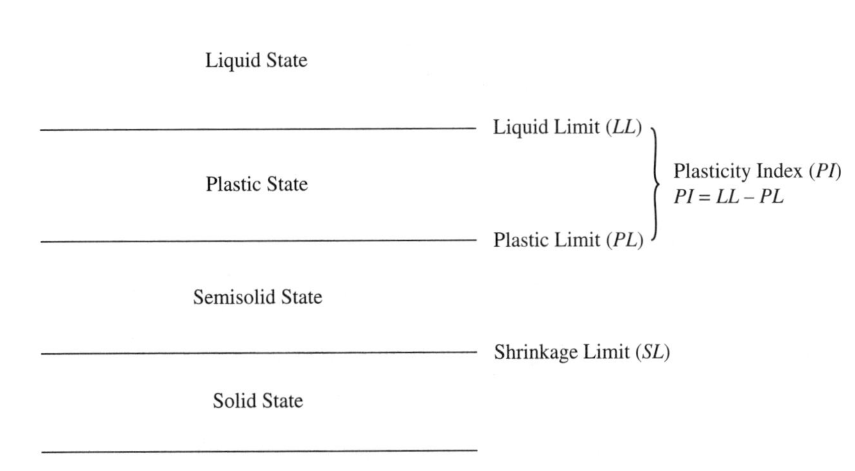

FIGURE 2–3 Atterberg limits [3].

reach the solid state. The liquid, plastic, and shrinkage limits are quantified, therefore, in terms of water content. For example, the liquid limit is reported in terms of the water content at which soil changes from the liquid state to the plastic state. The difference between the liquid limit (LL) and plastic limit (PL) is the *plasticity index* (PI); that is,

$$PI = LL - PL \tag{2-3}$$

The liquid, plastic, and shrinkage limits and the plasticity index are useful parameters in classifying soils and in making judgments as to their applications.

Standard laboratory test procedures are available to determine Atterberg limits. Although Atterberg defined the four states of consistency for cohesive soils, his original consistency limit tests were somewhat arbitrary and did not yield entirely consistent results. Subsequently, Casagrande standardized the tests, thereby increasing reproducibility of test results.

Casagrande developed a *liquid limit device* for use in determining liquid limits. As shown in Figure 2–4, it consists essentially of a "cup" that is raised and dropped 10 mm by a manually rotated handle. In performing a liquid limit test, a standard groove is cut in a remolded soil sample in the cup using a standard grooving tool. The liquid limit is defined as that water content at which the standard groove will close a distance of 1/2 in. (12.7 mm) along the bottom of the groove at exactly 25 blows (drops) of the cup. Because it is difficult to mix the soil with the precise water content at which the groove will close 1/2 in. at exactly 25 blows, tests are usually run on samples with differing water contents, and a straight-line plot of water content versus the logarithm of the number of blows required to close the groove 1/2 in. is prepared. From this plot, which is known as a *flow curve,* the particular water content corresponding to 25 blows is read and reported as the liquid limit.

FIGURE 2–4 Liquid limit device.
Source: Courtesy of Soiltest, Inc.

The plastic limit is evaluated quantitatively in the laboratory by finding the water content at which a thread of soil begins to crumble when it is manually rolled out on a glass plate to a diameter of 1/8 in. and breaks up into segments about 1/8 to 3/8 in. (3 to 10 mm) in length. If threads can be rolled to smaller diameters, the soil is too wet (i.e., it is above the plastic limit). If threads crumble before reaching the 1/8-in. diameter, the soil is too dry and the plastic limit has been surpassed.

The shrinkage limit, the dividing line between the semisolid and solid states, is quantified for a given soil as a specific water content, and from a physical standpoint it is the water content that is just sufficient to fill the voids when the soil is at the minimum volume it will attain on drying. In other words, the smallest water content at which a soil can be completely saturated is called the *shrinkage limit.* Below the shrinkage limit, any water content change will *not* result in volume change; above the shrinkage limit, any water content change *will* result in an accompanying volume change (see Figure 2–5).

The general procedure for determining the shrinkage limit is begun by placing a sample in an evaporating dish and mixing it with enough distilled water to fill the soil voids completely. After a shrinkage dish is coated with petroleum jelly, wet soil is taken from the evaporating dish with a spatula and placed in the shrinkage dish. The placement should be done in three parts, with steps taken each time to drive all air out of the soil. After the shrinkage dish and wet soil are weighed, the soil is set aside to dry in air. It is then oven dried overnight, after which the shrinkage dish and dry soil are weighed. After the oven-dried soil pat is removed from the shrinkage dish, its volume can be determined by mercury displacement. In addition, the weight and volume of the empty shrinkage dish must be determined. The latter (i.e., the volume of the shrinkage dish) is also obtained by mercury displacement, and it is the same as the volume of the wet soil pat. With these data known, the shrinkage limit can be

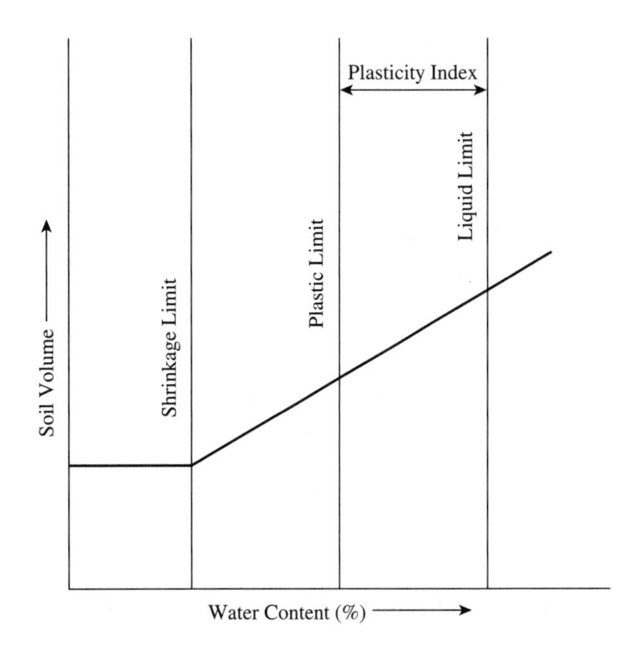

FIGURE 2–5 Definition of the shrinkage limit.

determined by the following equation:

$$SL = w - \left[\frac{(V - V_o)\rho_w}{M_o}\right] \times 100 \qquad (2\text{–}4)$$

where SL = shrinkage limit (expressed as a percentage)
w = water content of wet soil in the shrinkage dish, %
V = volume of wet soil pat (same as volume of shrinkage dish), cm^3
V_o = volume of oven-dried soil pat, cm^3
ρ_w = approximate density of water equal to 1.0 g/cm^3
M_o = mass of oven-dried soil pat, g

The mass of the oven-dried soil pat, M_o, is determined by subtracting the mass of the dish coated with petroleum jelly from the mass of the dish coated with petroleum jelly plus the oven-dried soil. V and V_o should be expressed in cubic centimeters and M_o, in grams.

Detailed procedures for laboratory determinations of liquid, plastic, and shrinkage limits are given in *Soil Properties: Testing, Measurement, and Evaluation,* 4th edition, by Liu and Evett [4].

2–3 SOIL CLASSIFICATION SYSTEMS

In order to be able to describe, in general, a specific soil without listing values of its many soil parameters, it would be convenient to have some kind of generalized classification system. In practice, a number of different classification systems have evolved, most of which were developed to meet specific needs of the particular group that developed a given system. Today, however, only two such systems—the American Association of State Highway and Transportation Officials (AASHTO) system and the Unified Soil Classification System (USCS)—are widely used in engineering practice.

The AASHTO system is widely used in highway work and is followed by nearly all state departments of highways and/or transportation in the United States. Most federal agencies (such as the U.S. Army Corps of Engineers and the U.S. Department of the Interior, Bureau of Reclamation) use the Unified Soil Classification System; it is also utilized by many engineering consulting companies and soil-testing laboratories in the United States. Both of these classification systems are presented in detail later in this section.

Some years ago, the Federal Aviation Administration (FAA) had its own soil classification system, known appropriately as the FAA classification system, for designing airport pavements. Now, however, the FAA uses the Unified Soil Classification System. If one needs information about the FAA classification system, it can be found in the first two editions of this book.

AASHTO Classification System [5]

In the late 1920s, the U.S. Bureau of Public Roads (currently the Federal Highway Administration) developed a soil classification system known as the BPR classification system. It was subsequently revised several times, with the 1945 revision becoming what is basically today's AASHTO system.

Required parameters for classification by the AASHTO system are grain-size analysis, liquid limit, and plasticity index. With values of these parameters known, one enters the first (left) column of Table 2–4 and determines whether or not known parameters meet the limiting values in that column. If they do, then the soil classification is that given at the top of the column (A-1-a, if known parameters meet the limiting values in the first column). If they do not, one enters the next column (to the right) and determines whether or not known parameters meet the limiting values in that column. The procedure is repeated until the *first* column is reached in which known parameters meet the limiting values in that column. The soil classification for the given soil is indicated at the top of that particular column.

Once a soil has been classified using Table 2–4, it can be further described using a *group index*. This index utilizes the percent of soil passing a No. 200 sieve, the liquid limit, and the plasticity index. With known values of these parameters, the group index is computed from the following equation:

$$\text{Group index} = (F - 35)[0.2 + 0.005(LL - 40)] + 0.01(F - 15)(PI - 10) \quad \textbf{(2–5)}$$

where F = percentage of soil passing a No. 200 sieve
LL = liquid limit
PI = plasticity index

The group index computed from Eq. (2–5) is rounded off to the nearest whole number and appended in parentheses to the group designation determined from Table 2–4. If the computed group index is either zero or negative, the number zero is used

TABLE 2–4
Classification of Soils and Soil-Aggregate Mixtures by AASHTO Classification System [5]

General Classification	Granular Materials (35% or less passing 0.075 mm)							Silt–Clay Materials (more than 35% passing 0.075 mm)			
	A-1			*A-2*							*A-7*
Group Classification	*A-1-a*	*A-1-b*	*A-3*	*A-2-4*	*A-2-5*	*A-2-6*	*A-2-7*	*A-4*	*A-5*	*A-6*	*A-7-5, A-7-6*
Sieve analysis: Percent passing:											
2.00 mm (No. 10)	50 max.	—	—	—	—	—	—	—	—	—	—
0.425 mm (No. 40)	30 max.	50 max.	51 min.	—	—	—	—	—	—	—	—
0.075 mm (No. 200)	15 max.	25 max.	10 max.	35 max.	35 max.	35 max.	35 max.	36 min.	36 min.	36 min.	36 min.
Characteristics of fraction passing 0.425 mm (No. 40):											
Liquid limit	—		—	40 max.	41 min.	40 max.	41 min.	40 max.	41 min.	40 max.	41 min.
Plasticity index	6 max.		NP	10 max.	10 max.	11 min.	11 min.	10 max.	10 max.	11 min.	11 min.[1]
Usual types of significant constituent materials	Stone fragments, gravel, and sand		Fine sand	Silty or clayey gravel and sand				Silty soils		Clayey soils	
General ratings as subgrade	Excellent to good						Fair to poor				

[1]Plasticity index of A-7-5 subgroup is equal to or less than *LL* minus 30. Plasticity index of A-7-6 subgroup is greater than *LL* minus 30.

as the group index and should be appended to the group designation. If preferred, Figure 2–6 may be used instead of Eq. (2–5) to determine the group index.

As a general rule, the value of soil as a subgrade material is an inverse ratio to its group index (i.e., the lower the index, the better the material). Table 2–5 gives some general descriptions of the various classification groups according to the AASHTO system.

EXAMPLE 2–2

Given

A sample of soil was tested in the laboratory, and results of the laboratory tests were as follows:

1. Liquid limit = 42.3%.
2. Plastic limit = 15.8%.
3. The following sieve analysis data:

U.S. Sieve Size	Percentage Passing
No. 4	100.0
No. 10	93.2
No. 40	81.0
No. 200	60.2

Required

Classify the soil sample by the AASHTO classification system.

Solution

By the AASHTO classification system:

$$\text{Plasticity index } (PI) = \text{Liquid limit } (LL) - \text{Plastic limit } (PL)$$
$$PI = 42.3\% - 15.8\% = 26.5\%$$

From Table 2–4, the sample is classified as A-7. According to the AASHTO classification system, the plasticity index of the A-7-5 subgroup is equal to or less than the liquid limit minus 30, and the plasticity index of the A-7-6 subgroup is greater than the liquid limit minus 30 (see footnote under Table 2–4).

$$LL - 30\% = 42.3\% - 30\% = 12.3\%$$
$$[PI = 26.5\%] > [LL - 30\% = 12.3\%]$$

Hence, this is A-7-6 material.

From Figure 2–6 (group index chart), with $LL = 42.3\%$ and the percentage passing the No. 200 sieve = 60.2%, the partial group index for $LL = 5.3$. With $PI = 26.5\%$

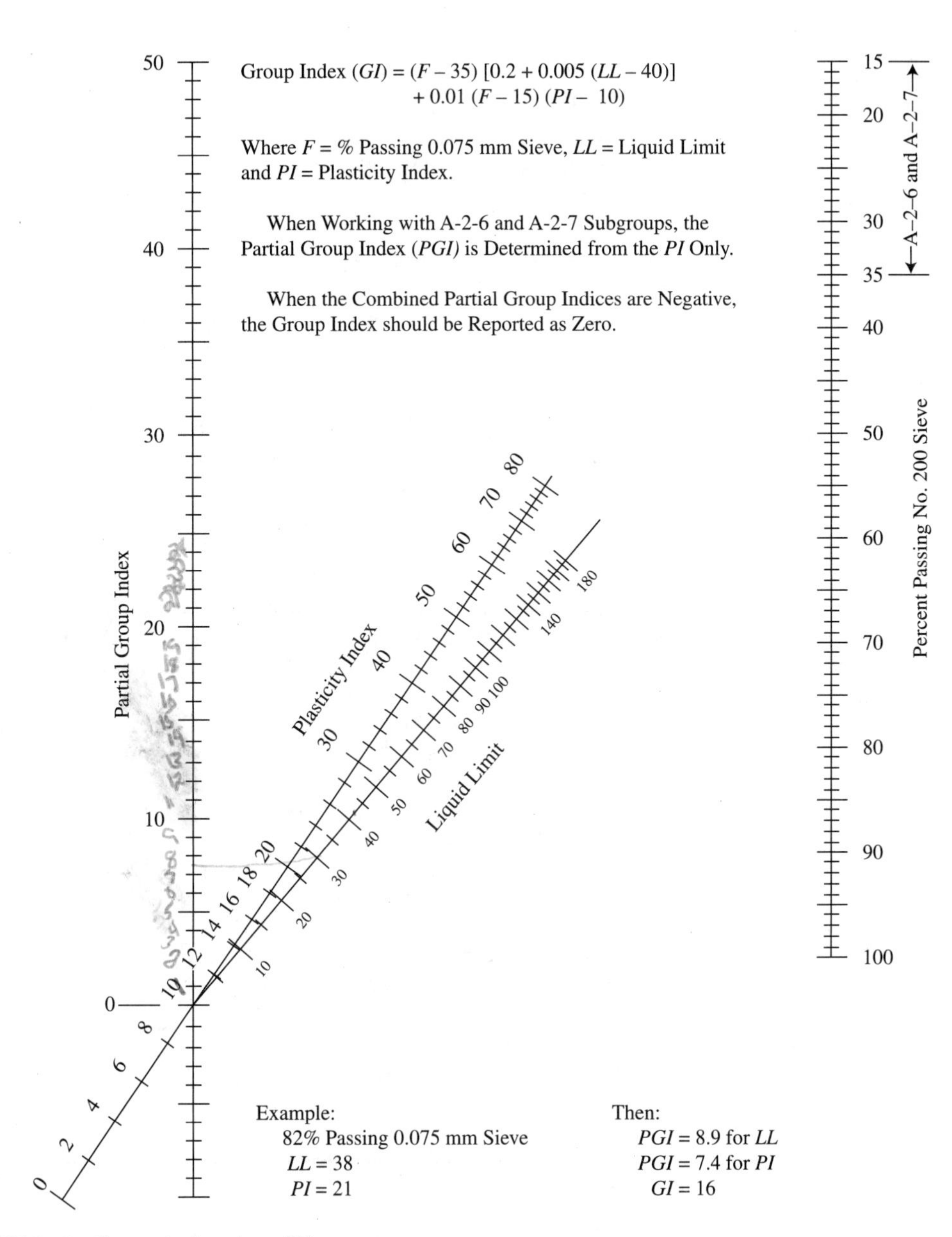

FIGURE 2–6 Group index chart [5].

and the percentage passing the No. 200 sieve = 60.2%, the partial group index for *PI* = 7.5. Hence,

$$\text{Total group index} = 5.3 + 7.5 = 12.8$$

Hence, the soil is A-7-6 (13), according to the AASHTO classification system.

TABLE 2–5
Descriptions of AASHTO Classification Groups [5]

(1) *Granular Materials.* Containing 35% or less passing 0.075 mm (No. 200) sieve, Note 1.

(1.1) *Group A-1:* The typical material of this group is a well-graded mixture of stone fragments or gravel, coarse sand, fine sand and a nonplastic or feebly plastic soil binder. However, this group includes also stone fragments, gravel, coarse sand, volcanic cinders, etc. without soil binder.

(1.1.1) Subgroup A-1-a includes those materials consisting predominantly of stone fragments or gravel, either with or without a well-graded binder of fine material.

(1.1.2) Subgroup A-1-b includes those materials consisting predominantly of coarse sand either with or without a well-graded soil binder.

(1.2) *Group A-3:* The typical material of this group is fine beach sand or fine desert blow sand without silty or clay fines or with a very small amount of nonplastic silt. The group includes also stream-deposited mixtures of poorly graded fine sand and limited amounts of coarse sand and gravel.

(1.3) *Group A-2:* This group includes a wide variety of "granular" materials which are border-line between the materials falling in Groups A-1 and A-3 and silt–clay materials of Group A-4, A-5, A-6, and A-7. It includes all materials containing 35% or less passing the 0.075-mm sieve which cannot be classified as A-1 or A-3, due to fines content or plasticity or both, in excess of the limitations for those groups.

(1.3.1) Subgroups A-2-4 and A-2-5 include various granular materials containing 35% or less passing the 0.075 mm sieve and with a minus 0.425-mm (No. 40) portion having the characteristics of the A-4 and A-5 groups. These groups include such materials as gravel and coarse sand with silt contents or plasticity indexes in excess of the limitations of Group A-1, and fine sand with nonplastic silt content in excess of the limitations of Group A-3.

(1.3.2) Subgroups A-2-6 and A-2-7 include materials similar to those described under Subgroups A-2-4 and A-2-5 except that the fine portion contains plastic clay having the characteristics of the A-6 or A-7 group.

Note 1: Classification of materials in the various groups applies only to the fraction passing the 75 mm sieve. Therefore, any specification regarding the use of A-1, A-2, or A-3 materials in construction should state whether boulders (retained on 3-in. sieve) are permitted.

Unified Soil Classification System [6–8]

The Unified Soil Classification System was originally developed by Casagrande [6] and is utilized by the Corps of Engineers. In this system, soils fall within one of three major categories: coarse-grained, fine-grained, and highly organic soils. These categories are further subdivided into 15 basic soil groups. The following group symbols are used in the Unified System:

TABLE 2–5 *continued*

(2) *Silt–Clay Materials.* Containing more than 35percent passing the 0.075 mm sieve.

(2.1) *Group A-4:* The typical material of this group is a nonplastic or moderately plastic silty soil usually having 75% or more passing the 0.075-mm sieve. The group includes also mixtures of fine silty soil and up to 64% of sand and gravel retained on 0.075 mm sieve.

(2.2) *Group A-5:* The typical material of this group is similar to that described under Group A-4, except that it is usually of diatomaceous or micaceous character and may be highly elastic as indicated by the high liquid limit.

(2.3) *Group A-6:* The typical material of this group is a plastic clay soil usually having 75% or more passing the 0.075 mm sieve. The group includes also mixtures of fine clayey soil and up to 64% of sand and gravel retained on the 0.075 mm sieve. Materials of this group usually have high volume change between wet and dry states.

(2.4) *Group A-7:* The typical material of this group is similar to that described under Group A-6, except that it has the high liquid limits characteristic of the A-5 group and may be elastic as well as subject to high volume change.

(2.4.1) Subgroup A-7-5 includes those materials with moderate plasticity indexes in relation to liquid limit and which may be highly elastic as well as subject to considerable volume change.

(2.4.2) Subgroup A-7-6 includes those materials with high plasticity indexes in relation to liquid limit and which are subject to extremely high volume change.

Note 2: Highly organic soils (peak or muck) may be classified as an A-8 group. Classification of these materials is based on visual inspection, and is not dependent on percentage passing the 0.075 mm (No. 200) sieve, liquid limit or plasticity index. The material is composed primarily of partially decayed organic matter, generally has a fibrous texture, dark brown or black color and odor of decay. These organic materials are unsuitable for use in embankments and subgrades. They are highly compressible and have low strength.

G	Gravel
S	Sand
M	Silt
C	Clay
O	Organic
PT	Peat
W	Well graded
P	Poorly graded

Normally, two group symbols are used to classify soils. For example, SW indicates well-graded sand. Table 2–6 lists the 15 soil groups, including each one's name and symbol, as well as specific details for classifying soils by this system.

In order to classify a given soil by the Unified System, its grain-size distribution, liquid limit, and plasticity index must first be determined. With these values known,

TABLE 2–6

Soil Classification Chart by Unified Soil Classification System [8]

This table 1st

Criteria for Assigning Group Symbols and Group Names Using Laboratory Tests[A]				Soil Classification	
				Group Symbol	*Group Name*[B]
Coarse-grained soils: More than 50% retained on No. 200 sieve	*Gravels:* More than 50% of coarse fraction retained on No. 4 sieve	*Clean gravels:* Less than 5% fines[C]	$C_u \geq 4$ and $1 \leq C_c \leq 3$[E]	GW	Well-graded gravel[E]
			$C_u < 4$ and/or $1 > C_c > 3$[E]	GP	Poorly graded gravel[E]
		Gravels with fines: More than 12% fines[C]	Fines classify as ML or MH	GM	Silty gravel[F,G,H]
			Fines classify as CL or CH	GC	Clayey gravel[F,G,H]
	Sands: 50% or more of coarse fraction passes No. 4 sieve	*Clean sands:* Less than 5% fines[D]	$C_u \geq 6$ and $1 \leq C_c \leq 3$[E]	SW	Well-graded sand[I]
			$C_u < 6$ and/or $1 > C_c > 3$[E]	SP	Poorly graded sand[I]
		Sands with fines: More than 12% fines[D]	Fines classify as ML or MH	SM	Silty sand[G,H,I]
			Fines classify as CL or CH	SC	Clayey sand[G,H,I]
Fine-grained soils: 50% or more passes the No. 200 sieve	*Silts and clays:* Liquid limit less than 50	Inorganic	$PI > 7$ and plots on or above "A" line[J]	CL	Lean clay[K,L,M]
			$PI < 4$ or plots below "A" line[J]	ML	Silt[K,L,M]
		Organic	$\frac{\text{Liquid limit—oven dried}}{\text{Liquid limit—not dried}} < 0.75$	OL	Organic clay[K,L,M,N] Organic silt[K,L,M,O]
	Liquid limit 50 or more	Inorganic	PI plots on or above "A" line	CH	Fat clay[K,L,M]
			PI plots below "A" line	MH	Elastic silt[K,L,M]
		Organic	$\frac{\text{Liquid limit—oven dried}}{\text{Liquid limit—not dried}} < 0.75$	OH	Organic clay[K,L,M,P] Organic silt[K,L,M,Q]
Highly organic soils	Primarily organic matter, dark in color, and organic color			PT	Peat

[A]Based on the material passing the 3-in. (75-mm) sieve.
[B]If field sample contained cobbles or boulders, or both, add "with cobbles or boulders, or both" to group name.
[C]Gravels with 5 to 12% fines require dual symbols:
GW-GM, well-graded gravel with silt
GW-GC, well-graded gravel with clay
GP-GM, poorly graded gravel with silt
GP-GC, poorly graded gravel with clay
[D]Sands with 5 to 12% fines require dual symbols:
SW-SM, well-graded sand with silt
SW-SC, well-graded sand with clay
SP-SM, poorly graded sand with silt
SP-SC, poorly graded sand with clay
[E]$C_u = D_{60}/D_{10}$, $C_c = (D_{30})^2/(D_{10} \times D_{60})$.
[F]If soil contains ≥ 15% sand, add "with sand" to group name.
[G]If fines classify as CL-ML, use dual symbol GC-GM or SC-SM.
[H]If fines are organic, add "with organic fines" to group name.
[I]If soil contains ≥ 15% gravel, add "with gravel" to group name.
[J]If Atterberg limits plot in hatched area, soil is a CL-ML silty clay.
[K]If soil contains 15 to 29% plus No. 200, add "with sand" or "with gravel," whichever is predominant.
[L]If soil contains ≥ 30% plus No. 200, predominantly sand, add "sandy" to group name.
[M]If soil contains ≥ 30% plus No. 200, predominantly gravel, add "gravelly" to group name.
[N]$PI \geq 4$ and plots on or above "A" line.
[O]$PI < 4$ or plots below "A" line.
[P]PI plots on or above "A" line.
[Q]PI plots below "A" line.

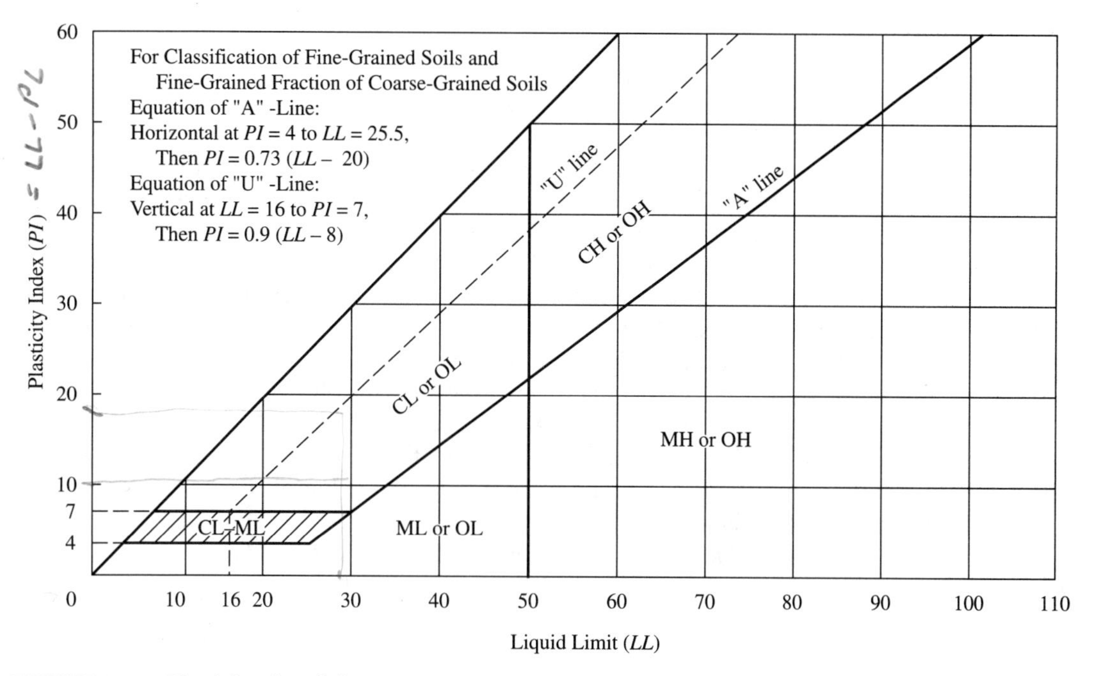

FIGURE 2–7 Plasticity chart [8].

the soil can be classified by using Table 2–6 and Figure 2–7. The Unified Soil Classification System is published as ASTM D 2487.

EXAMPLE 2–3

Given

A sample of soil was tested in the laboratory with the following results:

1. Liquid limit = 30.0%.
2. Plastic limit = 12.0%.
3. Sieve analysis data:

U.S. Sieve Size	Percentage Passing
⅜ in.	100.0
No. 4	76.5
No. 10	60.0
No. 40	39.7
No. 200	15.2

Required

Classify the soil by the Unified Soil Classification System.

Solution

Because the percentage retained on the No. 200 sieve (100 − 15.2, or 84.8%) is more than 50%, go to the block labeled "Coarse-grained soils" in Table 2–6. The sample consists of 100 − 15.2, or 84.8%, coarse-grain sizes, and 100 − 76.5, or 23.5%, was retained on the No. 4 sieve. Thus, the percentage of coarse fraction retained on the No. 4 sieve is (23.5/84.8) (100), or 27.7%, and the percentage of coarse fraction that passed the No. 4 sieve is 72.3%. Because 72.3% is greater than 50%, go to the block labeled "Sands" in Table 2–6. The soil is evidently a sand. Because the sample contains 15.2% passing the No. 200 sieve, which is greater than 12% fines, go to the block labeled "Sands with fines: More than 12% fines." Refer next to the plasticity chart (Figure 2–7). With a liquid limit of 30.0% and plasticity index of 18.0% (recall that the plasticity index is the difference between the liquid and plastic limits, or 30.0 − 12.0), the sample is located above the "A" line, and the fines are classified as CL. Return to Table 2–6, and go to the block labeled "SC." Thus, this soil is classified SC according to the Unified Soil Classification System.

EXAMPLE 2–4

Given

A sample of soil was tested in the laboratory with the following results:

1. Liquid limit = NP (nonplastic).
2. Plastic limit = NP (nonplastic).
3. Sieve analysis data:

U.S. Sieve Size	Percentage Passing
1 in.	100
¾ in.	85
½ in.	70
⅜ in.	60
No. 4	48
No. 10	30
No. 40	16
No. 100	10
No. 200	2

Required

Classify the soil by the USCS.

Solution

Because the percentage retained on the No. 200 sieve (100 − 2, or 98%) is more than 50%, go to the block labeled "Coarse-grained soils" in Table 2–6. The sample consists of 100 − 2, or 98%, coarse-grain sizes, and 100 − 48, or 52%, was retained on the No. 4 sieve. Thus, the percentage of coarse fraction retained on the No. 4 sieve is 52/98, or 53.1%. Because 53.1% is greater than 50%, go to the block labeled "Gravels" in Table 2–6. The soil is evidently a gravel. Because the sample contains 2% passing the No. 200 sieve, which is less than 5% fines, go to the block labeled "Clean gravels: Less than 5% fines." The next block indicates that the coefficients of uniformity (C_u) and curvature (C_c) must be evaluated.

$$C_u = \frac{D_{60}}{D_{10}} \tag{2–1}$$

$$C_c = \frac{(D_{30})^2}{D_{60}D_{10}} \tag{2–2}$$

Values of D_{60}, D_{30}, and D_{10} are determined from the grain-size distribution curve (see Figure 2–8) to be 9.5, 2.00, and 0.150 mm, respectively. Hence,

$$C_u = \frac{9.5 \text{ mm}}{0.150 \text{ mm}} = 63.3$$

$$C_c = \frac{(2.00 \text{ mm})^2}{(9.5 \text{ mm})(0.150 \text{ mm})} = 2.8$$

Because C_u (63.3) is greater than 4, and C_c (2.8) is between 1 and 3, this sample meets both criteria for a well-graded gravel. Hence, from Table 2–6 the soil is classified GW (i.e., well-graded gravel) according to the Unified Soil Classification System.

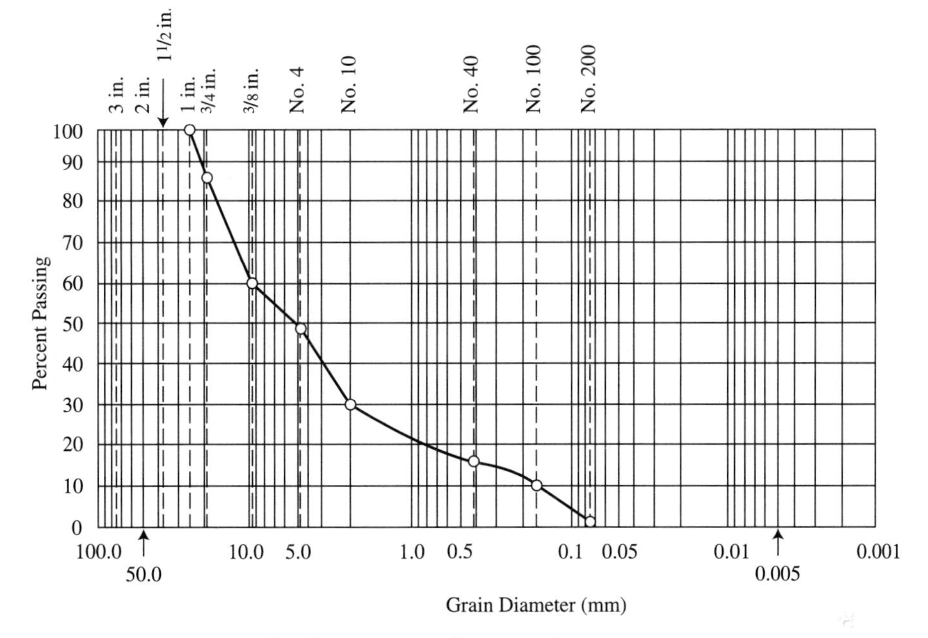

FIGURE 2–8 Grain-size distribution curve for Example 2–4.

EXAMPLE 2–5

Given

A sample of inorganic soil was tested in the laboratory with the following results:

1. Liquid limit = 42.3%.
2. Plastic limit = 15.8%.
3. Sieve analysis data:

U.S. Sieve Size	Percentage Passing
No. 4	100.0
No. 10	93.2
No. 40	81.0
No. 200	60.2

Required

Classify the soil sample by the Unified Soil Classification System.

Solution

Because the percentage passing the No. 200 sieve is 60.2%, which is greater than 50%, go to the lower block (labeled "Fine-grained soils") in Table 2–6. The liquid limit is 42.3%, which is less than 50%, so go to the block labeled "Silts and clays: Liquid limit less than 50." Now, because the sample is an inorganic soil, and the plasticity index is 42.3 − 15.8, or 26.5%, which is greater than 7, refer next to the plasticity chart (Figure 2–7). With a liquid limit of 42.3% and plasticity index of 26.5%, the sample is located above the "A" line. Return to Table 2–6 and go to the block labeled "CL." Thus, the soil is classified CL according to the Unified Soil Classification System.

2–4 COMPONENTS OF SOILS

Soils contain three components, which may be characterized as solid, liquid, and gas. The solid components of soils are weathered rock and (sometimes) decayed vegetation. The liquid component of soils is almost always water (often with dissolved matter), and the gas component is air. The volume of water and air combined is referred to as the *void*.

Figure 2–9 gives a block diagram showing the components of a soil. These components may be considered in terms of both their volumes and their weights/masses. In Figure 2–9, terms V, V_a, V_w, V_s, and V_v represent total volume and volume of air, water, solid matter, and voids, respectively. Terms W, W_a, W_w, and W_s stand for total weight and weight of air, water, and solid matter, respectively. Similarly, terms M, M_a, M_w, and M_s denote total mass and mass of air, water, and solid matter, respectively. The weight and mass of air (W_a and M_a) are both virtually zero.

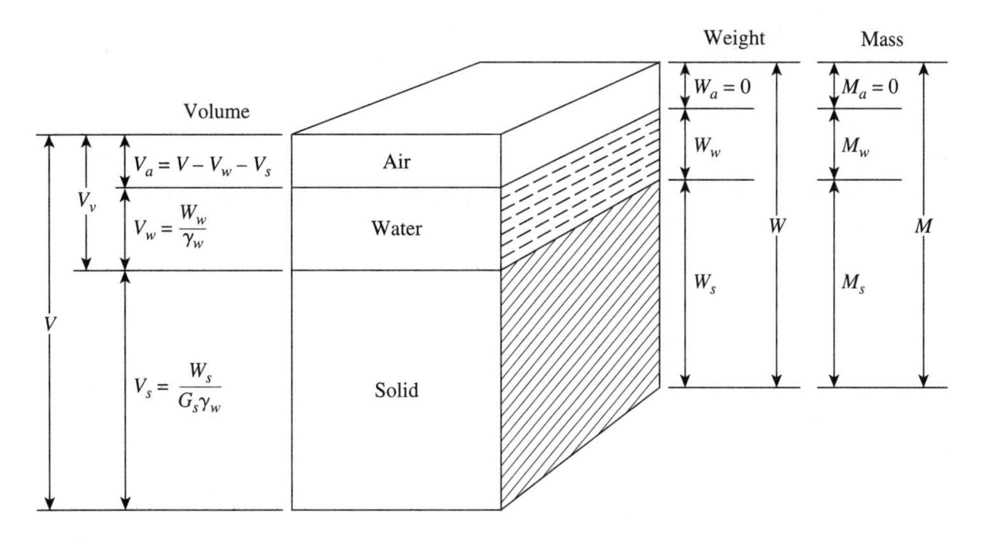

FIGURE 2–9 Block diagram showing components of soil.

2–5 WEIGHT/MASS AND VOLUME RELATIONSHIPS

A number of important relationships exist among the components of soil in terms of both weight/mass and volume. These relationships define new parameters that are useful in working with soils.

In terms of volume, the following new parameters are important—*void ratio, porosity,* and *degree of saturation.* Void ratio (e) is the ratio (expressed as a decimal fraction) of volume of voids to volume of solids.

$$e = \frac{V_v}{V_s} \tag{2-6}$$

Porosity (n) is the ratio (expressed as a percentage) of volume of voids to total volume.

$$n = \frac{V_v}{V} \times 100\% \tag{2-7}$$

Degree of saturation (S) is the ratio (expressed as a percentage) of volume of water to volume of voids.

$$S = \frac{V_w}{V_v} \times 100\% \tag{2-8}$$

In terms of weight/mass, the new parameters are *water content, unit weight, dry unit weight, unit mass* (or *density*), *dry unit mass* (or *dry density*), and *specific gravity of solids.* (*Note:* The terms *unit weight* and *unit mass* imply *wet unit weight* and *wet unit mass.* If *dry unit weight* or *dry unit mass* is intended, the adjective *dry* is indicated explicitly.) Water content (w) is the ratio (expressed as a percentage) of weight of water to weight of solids or the ratio of mass of water to mass of solids.

$$w = \frac{W_w}{W_s} \times 100\% = \frac{M_w}{M_s} \times 100\% \qquad (2\text{–}9)$$

Unit weight (γ) is total weight (weight of solid plus weight of water) divided by total volume (volume of solid plus volume of water plus volume of air).

$$\gamma = \frac{W}{V} \qquad (2\text{–}10)$$

Dry unit weight (γ_d) is weight of solids divided by total volume.

$$\gamma_d = \frac{W_s}{V} \qquad (2\text{–}11)$$

Unit mass (ρ) is total mass divided by total volume.

$$\rho = \frac{M}{V} \qquad (2\text{–}12)$$

Dry unit mass (ρ_d) is mass of solids divided by total volume.

$$\rho_d = \frac{M_s}{V} \qquad (2\text{–}13)$$

Specific gravity of solids (G_s) is the ratio of unit weight of solids (weight of solids divided by volume of solids) to unit weight of water or of unit mass of solids (mass of solids divided by volume of solids) to unit mass of water.

$$G_s = \frac{W_s/V_s}{\gamma_w} = \frac{W_s}{V_s\gamma_w} \qquad (2\text{–}14)$$

$$G_s = \frac{M_s/V_s}{\rho_w} = \frac{M_s}{V_s\rho_w} \qquad (2\text{–}15)$$

where γ_w and ρ_w are the unit weight and unit mass of water, respectively.

The unit weight of water varies slightly with temperature, but at normal temperatures, it has a value of around 62.4 pounds per cubic foot (lb/ft^3) or 9.81 kilonewtons per cubic meter (kN/m^3). The unit mass (density) of water is 1000 kilograms per cubic meter (kg/m^3) or 1 gram per cubic centimeter (g/cm^3). A useful conversion factor is as follows: 1 lb/ft^3 = 0.1571 kN/m^3, or 1 kN/m^3 = 6.366 lb/ft^3.

Soils engineers must be proficient in determining these parameters based on laboratory evaluations of weight/mass and volume of the components of a soil. Use of a block diagram (as shown in Figure 2–9) is recommended to help obtain answers more quickly and accurately. Five example problems follow.

EXAMPLE 2–6

Given

1. The weight of a chunk of moist soil sample is 45.6 lb.
2. The volume of the soil chunk measured before drying is 0.40 ft^3.
3. After the sample is dried out in an oven, its weight is 37.8 lb.
4. The specific gravity of solids is 2.65.

Required

1. Water content.
2. Unit weight of moist soil.
3. Void ratio.
4. Porosity.
5. Degree of saturation.

Solution

See Figure 2–10. (Boldface data on the figure indicate given information. Other data are calculated in the solution of the problem.)

1. Water content $(w) = \dfrac{W_w}{W_s} \times 100\% = \dfrac{45.6 \text{ lb} - 37.8 \text{ lb}}{37.8 \text{ lb}} \times 100\%$
   $= 20.6\%$

2. Unit weight of moist soil $(\gamma) = \dfrac{W}{V} = \dfrac{45.6 \text{ lb}}{0.40 \text{ ft}^3} = 114.0 \text{ lb/ft}^3$

3. $V_w = \dfrac{W_w}{\gamma_w} = \dfrac{45.6 \text{ lb} - 37.8 \text{ lb}}{62.4 \text{ lb/ft}^3} = 0.13 \text{ ft}^3$

   $V_s = \dfrac{W_s}{G_s\gamma_w} = \dfrac{37.8 \text{ lb}}{(2.65)(62.4 \text{ lb/ft}^3)} = 0.23 \text{ ft}^3$

   $V_a = V - V_w - V_s = 0.40 \text{ ft}^3 - 0.13 \text{ ft}^3 - 0.23 \text{ ft}^3 = 0.04 \text{ ft}^3$

   $V_v = V - V_s = 0.40 \text{ ft}^3 - 0.23 \text{ ft}^3 = 0.17 \text{ ft}^3$

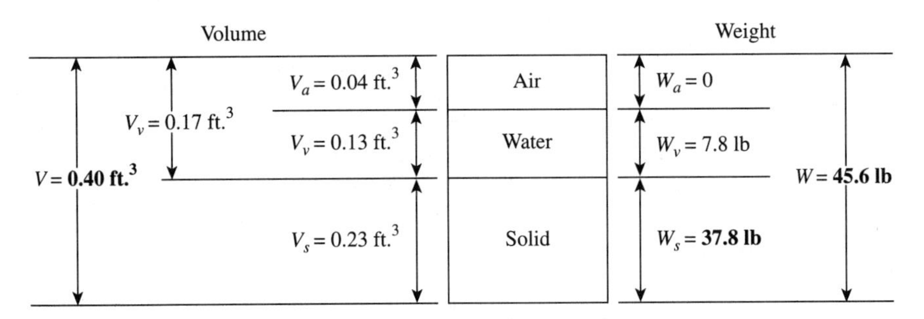

FIGURE 2–10 Block diagram showing components of soil for Example 2–6.

or

$$V_v = V_a + V_w = 0.04\ \text{ft}^3 + 0.13\ \text{ft}^3 = 0.17\ \text{ft}^3$$

$$\text{Void ratio } (e) = \frac{V_v}{V_s} = \frac{0.17\ \text{ft}^3}{0.23\ \text{ft}^3} = 0.74$$

4. $\text{Porosity } (n) = \dfrac{V_v}{V} \times 100\% = \dfrac{0.17\ \text{ft}^3}{0.40\ \text{ft}^3} \times 100\% = 42.5\%$

5. $\text{Degree of saturation } (S) = \dfrac{V_w}{V_v} \times 100\% = \dfrac{0.13\ \text{ft}^3}{0.17\ \text{ft}^3} \times 100\% = 76.5\%$

EXAMPLE 2–7

Given

1. The moist mass of a soil specimen is 20.7 kg.
2. The specimen's volume measured before drying is 0.011 m^3.
3. The specimen's dried mass is 16.3 kg.
4. The specific gravity of solids is 2.68.

Required

1. Void ratio.
2. Degree of saturation.
3. Wet unit mass.
4. Dry unit mass.
5. Wet unit weight.
6. Dry unit weight.

Solution

See Figure 2–11.

1. $V_s = \dfrac{M_s}{G_s\rho_w} = \dfrac{16.3\ \text{kg}}{(2.68)(1000\ \text{kg/m}^3)} = 0.0061\ \text{m}^3$

$$V_w = \frac{M_w}{\rho_w} = \frac{20.7\ \text{kg} - 16.3\ \text{kg}}{1000\ \text{kg/m}^3} = 0.0044\ \text{m}^3$$

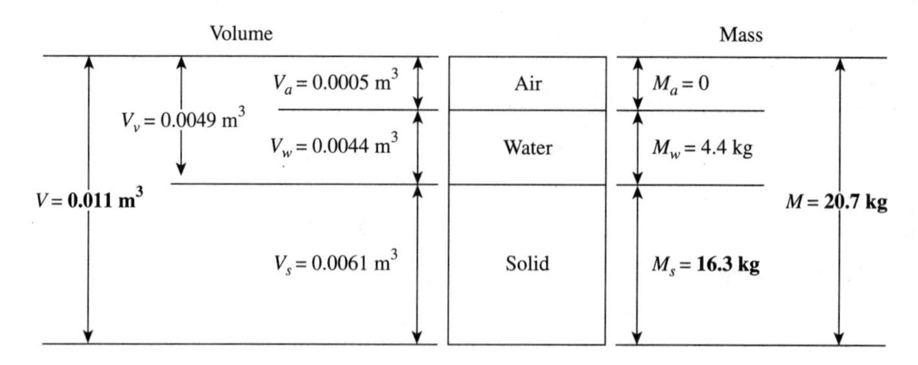

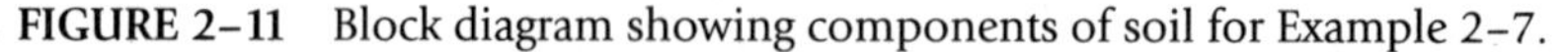

FIGURE 2–11 Block diagram showing components of soil for Example 2–7.

$$V_a = V - V_w - V_s = 0.011\ \text{m}^3 - 0.0044\ \text{m}^3 - 0.0061\ \text{m}^3 = 0.0005\ \text{m}^3$$

$$V_v = V - V_s = 0.011\ \text{m}^3 - 0.0061\ \text{m}^3 = 0.0049\ \text{m}^3$$

or

$$V_v = V_a + V_w = 0.0005\ \text{m}^3 + 0.0044\ \text{m}^3 = 0.0049\ \text{m}^3$$

$$\text{Void ratio } (e) = \frac{V_v}{V_s} = \frac{0.0049\ \text{m}^3}{0.0061\ \text{m}^3} = 0.80$$

2. Degree of saturation $(S) = \dfrac{V_w}{V_v} \times 100\% = \dfrac{0.0044\ \text{m}^3}{0.0049\ \text{m}^3} \times 100\% = 89.8\%$

3. Wet unit mass $(\rho) = \dfrac{M}{V} = \dfrac{20.7\ \text{kg}}{0.011\ \text{m}^3} = 1882\ \text{kg/m}^3$

4. Dry unit mass $(\rho_d) = \dfrac{M_s}{V} = \dfrac{16.3\ \text{kg}}{0.011\ \text{m}^3} = 1482\ \text{kg/m}^3$

5. Wet unit weight $(\gamma) = \rho g = (1882\ \text{kg/m}^3)(9.81\ \text{m/s}^2)$

$$= 18{,}460\ \frac{\text{kg} \cdot \text{m}}{\text{s}^2} \Big/ \text{m}^3 = 18{,}460\ \text{N/m}^3$$

$$= 18.46\ \text{kN/m}^3$$

6. Dry unit weight $(\gamma_d) = \rho_d g = (1482\ \text{kg/m}^3)(9.81\ \text{m/s}^2)$

$$= 14{,}540\ \frac{\text{kg} \cdot \text{m}}{\text{s}^2} \Big/ \text{m}^3 = 14{,}540\ \text{N/m}^3$$

$$= 14.54\ \text{kN/m}^3$$

EXAMPLE 2–8

Given

An undisturbed soil sample has the following data:

1. Void ratio = 0.78.
2. Water content = 12%.
3. Specific gravity of solids = 2.68.

Required

1. Wet unit weight.
2. Dry unit weight.
3. Degree of saturation.
4. Porosity.

Solution

See Figure 2–12. Because the void ratio $(e) = 0.78$,

$$\frac{V_v}{V_s} = 0.78; \qquad V_v = 0.78V_s \qquad \textbf{(A)}$$

$$V_v + V_s = V = 1\ \text{m}^3 \qquad \textbf{(B)}$$

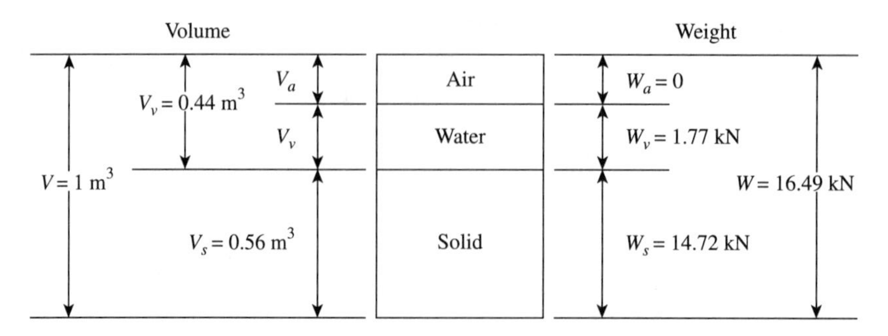

FIGURE 2–12 Block diagram showing components of soil for Example 2–8.

(A volume of 1 m^3 is assumed.)
Substitute Eq. (A) into Eq. (B).

$$0.78V_s + V_s = 1\ \text{m}^3$$

$$V_s = 0.56\ \text{m}^3$$
$$V_v = 1\ \text{m}^3 - 0.56\ \text{m}^3 = 0.44\ \text{m}^3$$
$$V_s = \frac{W_s}{G_s\gamma_w}; \qquad 0.56\ \text{m}^3 = \frac{W_s}{(2.68)(9.81\ \text{kN/m}^3)}$$
$$W_s = 14.72\ \text{kN}$$

From the given water content, $W_w/W_s = 0.12$,

$$W_w = 0.12W_s = (0.12)(14.72\ \text{kN}) = 1.77\ \text{kN}$$

1. Wet unit weight (γ) $= \dfrac{W}{V} = \dfrac{W_w + W_s}{V} = \dfrac{1.77\ \text{kN} + 14.72\ \text{kN}}{1\ \text{m}^3}$
   $= 16.49\ \text{kN/m}^3$
2. Dry unit weight (γ_d) $= \dfrac{W_s}{V} = \dfrac{14.72\ \text{kN}}{1\ \text{m}^3} = 14.72\ \text{kN/m}^3$
3. $V_w = \dfrac{W_w}{\gamma_w} = \dfrac{1.77\ \text{kN}}{9.81\ \text{kN/m}^3} = 0.18\ \text{m}^3$

   Degree of saturation (S) $= \dfrac{V_w}{V_v} \times 100\% = \dfrac{0.18\ \text{m}^3}{0.44\ \text{m}^3} \times 100\% = 40.9\%$
4. Porosity (n) $= \dfrac{V_v}{V} \times 100\% = \dfrac{0.44\ \text{m}^3}{1\ \text{m}^3} \times 100\% = 44.0\%$

EXAMPLE 2–9

Given

1. A 100% saturated soil has a wet unit weight of 120 lb/ft^3.
2. The water content of this saturated soil was determined to be 36%.

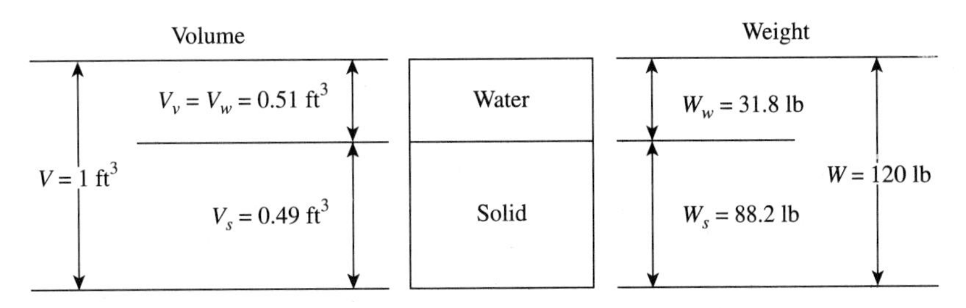

FIGURE 2–13 Block diagram showing components of soil for Example 2–9.

Required

1. Void ratio.
2. Specific gravity of solids.

Solution

See Figure 2–13.

$$W_w + W_s = 120 \text{ lb} \tag{A}$$

$$\frac{W_w}{W_s} = 0.36 \tag{B}$$

From Eq. (B), $W_w = 0.36W_s$; substitute into Eq. (A).

$$0.36W_s + W_s = 120 \text{ lb}$$
$$W_s = 88.2 \text{ lb}$$
$$W_w = 0.36W_s = (0.36)(88.2 \text{ lb}) = 31.8 \text{ lb}$$

1. $V_w = \dfrac{W_w}{\gamma_w} = \dfrac{31.8 \text{ lb}}{62.4 \text{ lb/ft}^3} = 0.51 \text{ ft}^3$

   $V_s = V - V_w = 1 \text{ ft}^3 - 0.51 \text{ ft}^3 = 0.49 \text{ ft}^3$

   $e = \dfrac{V_v}{V_s} = \dfrac{V_w}{V_s} = \dfrac{0.51 \text{ ft}^3}{0.49 \text{ ft}^3} = 1.04$

   Note: In this problem, because the soil is 100% saturated, $V_v = V_w$.

2. $V_s = \dfrac{W_s}{G_s\gamma_w}$; $\quad 0.49 \text{ ft}^3 = \dfrac{88.2 \text{ lb}}{(G_s)(62.4 \text{ lb/ft}^3)}$

   $G_s = 2.88$

EXAMPLE 2–10

Given

A soil sample has the following data:

1. Void ratio = 0.94.
2. Degree of saturation = 35%.
3. Specific gravity of solids = 2.71.

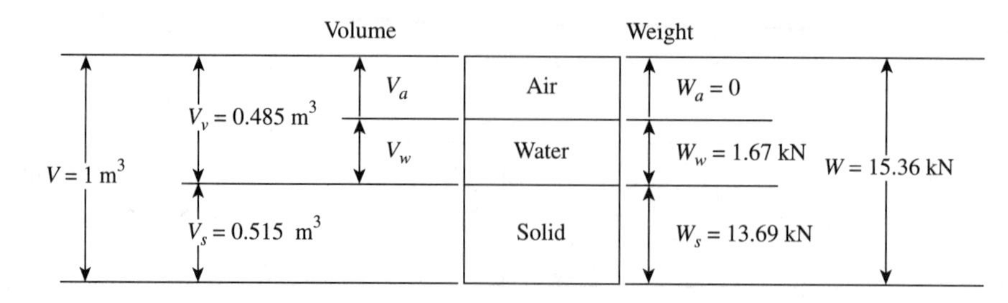

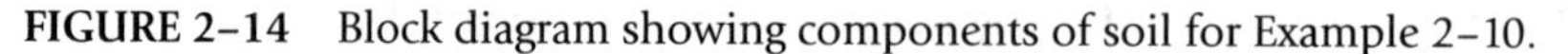
FIGURE 2–14 Block diagram showing components of soil for Example 2–10.

Required

1. Water content.
2. Unit weight.

Solution

See Figure 2–14. From the given void ratio, $e = V_v/V_s = 0.94$,

$$V_v = 0.94V_s \quad \text{(A)}$$
$$V_v + V_s = 1 \text{ m}^3 \quad \text{(B)}$$

Substitute Eq. (A) into Eq. (B).

$$0.94V_s + V_s = 1 \text{ m}^3$$
$$V_s = 0.515 \text{ m}^3$$
$$V_v = 0.485 \text{ m}^3$$

From the given degree of saturation, $S = V_w/V_v = 0.35$,

$$V_w = 0.35V_v$$
$$V_w = (0.35)(0.485 \text{ m}^3) = 0.170 \text{ m}^3$$
$$W_w = (0.170 \text{ m}^3)(9.81 \text{ kN/m}^3) = 1.67 \text{ kN}$$
$$W_s = (V_s)(G_s)(\gamma_w) = (0.515 \text{ m}^3)(2.71)(9.81 \text{ kN/m}^3) = 13.69 \text{ kN}$$

1. Water content $(w) = \dfrac{W_w}{W_s} \times 100\% = \dfrac{1.67 \text{ kN}}{13.69 \text{ kN}} \times 100\% = 12.2\%$
2. Unit weight $(\gamma) = \dfrac{W}{V} = \dfrac{W_w + W_s}{V} = \dfrac{1.67 \text{ kN} + 13.69 \text{ kN}}{1 \text{ m}^3} = 15.36 \text{ kN/m}^3$

2–6 PERMEABILITY, CAPILLARITY, AND FROST HEAVE

As indicated in Section 2–4, water is a component of soil, and its presence in a given soil may range from virtually none to saturation—the latter case occurring when the soil's void space is completely filled with water. Soil properties and characteristics are influenced by changes in water content. This section introduces three phenomena that are directly related to water in soil: permeability, capillarity, and frost heave. These as well as other factors pertaining to water in soil are discussed in more detail in Chapter 5.

Permeability refers to the movement of water within soil. Actual water movement is through the voids, which might be thought of as small, interconnected, irregular conduits. Because the water moves through the voids, it follows that soils with large voids (such as sands) are generally more permeable than those with smaller voids (such as clays). Additionally, because soils with large voids generally have large void ratios, it may be generalized that permeability tends to increase as the void ratio increases. Because water movement can have profound effects on soil properties and characteristics, it is an important consideration in certain engineering applications. Construction procedures, as well as the behavior of completed structures, can be significantly influenced by water movement within soil. For example, the rate of consolidation of soil and related settlement of structures on soil are highly dependent on how permeable a given soil is. Also, the amount of leakage through and under dams and hydrostatic uplift on dams (and other structures) are influenced by soil permeability. Additional examples where permeability is a factor in geotechnical engineering are infiltration into excavations and dewatering therefrom, stability of slopes and embankments, and erosion. The type, manner, and practical effects of water movement are discussed in Chapter 5. The flow of water through soil is governed by Darcy's law, which is also covered in Chapter 5.

Capillarity refers to the rise of water (or other liquids) in a small-diameter tube inserted into the water, the rise being caused by both cohesion of the water's molecules (surface tension) and adhesion of the water to the tube's wall. The amount of rise of water in the tube above the water level surrounding the tube is inversely proportional to the tube's diameter. With soils, capillarity occurs at the water table (see Section 3–4) when water rises from saturated soil below into dry or partially saturated soil above the water table. The "capillary tubes" through which water rises in soils are actually the void spaces among soil particles. Because the voids interconnect in varying directions (not just vertically) and are irregular in size and shape, accurate calculation of the height of capillary rise is virtually impossible. It is known, however, that the height of capillary rise is associated with the mean diameter of a soil's voids, which is in turn related to average grain size. In general, the smaller the grain size, the smaller the void space, and consequently the greater will be the capillary rise. Thus, clayey soils, with the smallest grain size, should theoretically experience the greatest capillary rise, although the rate of rise may be very slow because of the characteristically low permeability of such soils. In fact, the largest capillary rise for any particular length of time generally occurs in soils of medium grain sizes (such as silts and very fine sands).

It is well known from physics that water expands when it is cooled and freezes. When the temperature in a soil mass drops below water's freezing point, water in the voids freezes and therefore expands, causing the soil mass to move upward. This vertical expansion of soil caused by freezing water within is known as *frost heave.* Serious damage may result from frost heave when structures such as pavements and building foundations supported by soil are lifted. Because the amount of frost heave (i.e., upward soil movement) is not necessarily uniform in a horizontal direction, cracking of pavements and building walls and/or floors may occur. When the temperature rises above the freezing point, frozen soil thaws from the top downward. Because resulting melted water near the surface cannot drain through underlying frozen soil, an increase in water content of the upper soil, a decrease in its strength, and subsequent settlement of structures occur. Clearly, alternate lifting and settling of pavements and structures as a result of frost heave are undesirable, may cause serious structural damage, and should be avoided or at least minimized.

2–7 COMPRESSIBILITY

When soil is compressed, its volume is decreased. This decrease in volume results from reduction in voids within the soil and consequently can be expressed as a reduction in void ratio (e). Soil compression, which results from loading and causes reduction in the volume of voids (or decrease in void ratio), is usually brought on by the extruding of water and/or air from the soil. If saturated soil is subjected to the weight of a building and water is subsequently squeezed out or otherwise lost, resulting soil compression can cause undue building settlement. If water is added to the soil, soil expansion may occur, causing building uplift.

Settlement resulting from the compressibility of soil varies depending on whether a soil is cohesionless or cohesive. Cohesionless soils (such as sands and gravels) generally compress relatively quickly. In most cases, most of the settlement a structure built on cohesionless soil will undergo takes place during the construction phase. Additionally, compression of cohesionless soils can be induced by vibration more easily and more quickly than of cohesive soils.

Compressibility is more pronounced in the case of cohesive soils (such as clays), where soil moisture plays a part in particle interaction. Because of lower permeabilities, cohesive soils compress much more slowly because the expulsion of water from the small soil pores is so slow. Hence, the ultimate volume decrease of a cohesive soil and associated settlement of a structure built on this soil may not occur until some time after the structure is loaded.

It is helpful to consider total settlement as a two-phase process—*immediate settlement* and *consolidation settlement*. Immediate settlement occurs very rapidly—within days or even hours after a structure is loaded. Consolidation settlement occurs over an extended period of time (months or years) and is characteristic of cohesive soils. Consolidation settlement can be further divided into *primary consolidation* and *secondary consolidation* (sometimes called *creep*). Primary consolidation occurs first; it occurs faster and is generally larger, easier to predict, and more important than secondary consolidation. Secondary consolidation occurs subsequent to primary consolidation. It is thought to occur less due to extrusion of water from the voids and more as a result of some type of plastic deformation of the soil.

The preceding discussion of compressibility of soil is presented here to give a brief introduction to this subject because the purpose of this chapter is to introduce various engineering properties of soils. A more comprehensive treatment of compressibility is given in Chapter 7.

2–8 SHEAR STRENGTH [9]*

Shear strength of soil refers to its ability to resist shear stresses. Shear stresses exist in a sloping hillside or result from filled land, weight of footings, and so on. If a given soil does not have sufficient shear strength to resist such shear stresses, failures in the forms of landslides and footing failures will occur.

Shear strength results from frictional resistance to sliding, interlocking between adjacent solid particles in the soil, and cohesion and adhesion between adjacent soil particles. Because the ability of soil to support an imposed load is determined by its shear strength, the shear strength of soil is of great importance in foundation design (e.g., in determining a soil's bearing capacity), lateral earth pressure calculation (e.g., for retaining wall and sheet piling designs), slope stability analysis (for earth cuts, dams, embankments, etc.), pile design, and many other considerations. As a matter of fact, shear strength of soil is a factor in most soil problems.

The shear strength of a given soil may be approximately described by the Coulomb equation:

$$s = c + \bar{\sigma} \tan \phi \tag{2-16}$$

*Wayne C. Teng, *Foundation Design*, © 1962. Reprinted by permission of Prentice-Hall, Inc. (This footnote applies to all succeeding citations to this reference in this book.)

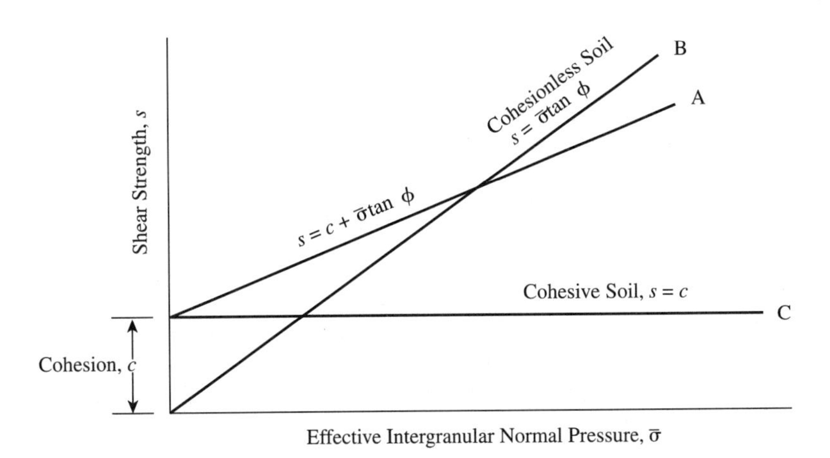

FIGURE 2–15 Shear strength diagram [9].

where s = shear strength
c = cohesion
$\bar{\sigma}$ = effective intergranular normal (perpendicular to the shear plane) pressure
ϕ = angle of internal friction
$\tan \phi$ = coefficient of friction

This equation is represented graphically by line A in Figure 2–15.

In the case of cohesionless soil (such as sand), there is virtually no cohesion ($c = 0$) and Eq. (2–16) reverts to $s = \bar{\sigma} \tan \phi$. This is represented graphically by line B in Figure 2–15. With cohesive soil (such as clay), the angle of internal friction (ϕ) can be taken to be zero for many foundation design problems. If ϕ is zero, Eq. (2–16) reverts to $s = c$. This is represented graphically by line C in Figure 2–15.

The preceding discussion of shear strength of soil is presented here to give an introduction to this subject. A more comprehensive treatment of shear strength of both cohesionless and cohesive soils, including certain long-term effects on shear strength of cohesive soil, is given in Chapter 8.

The shear strength parameters, c and ϕ, in Eq. (2–16) can be determined directly or indirectly by standard field or laboratory tests (see Chapter 8).

2–9 COMPACTNESS—RELATIVE DENSITY

In granular soils, compressibility and shear strength (covered in the two preceding sections) are related to the compactness of the soil grains. For a soil in its densest condition, its void ratio is the lowest, and it exhibits the highest shear strength and the greatest resistance to compression. Conversely, in its loosest condition, its void ratio is the highest, and its shear strength and resistance to compression are the lowest. Soils in a natural state generally exhibit characteristics somewhere between these two extremes. *Compactness* refers to the relative condition of a given soil between these two extremes.

To evaluate the relative condition of a given granular soil, the *in situ* void ratio can be determined and compared to the void ratio when the soil is in its densest condition and when it is in its loosest condition. Then, the *relative density* (D_r) can be evaluated by the equation:

$$D_r = \frac{e_{max} - e_0}{e_{max} - e_{min}} \times 100\% \qquad \textbf{(2–17)}$$

where e_{max} = highest void ratio possible for a given soil (void ratio of the soil in its loosest condition)
e_0 = void ratio of the soil in-place
e_{min} = lowest void ratio possible for the soil (void ratio of the soil in its densest condition)

Relative density can also be evaluated in terms of maximum, minimum, and in-place dry unit weights (γ_{max}, γ_{min}, and γ, respectively) by the equation:

$$D_r = \frac{\gamma_{max}(\gamma - \gamma_{min})}{\gamma(\gamma_{max} - \gamma_{min})} \times 100\% \qquad (2\text{–}18)$$

This equation is generally more convenient to use because it is easier to evaluate dry unit weights than void ratios.

Values of γ_{min} or e_{max} for a given soil can be determined by performing standard laboratory tests on a quantity of the soil that has been dried, pulverized, and poured slowly from a small height through a funnel into a container. Values of γ_{max} or e_{min} can be found in the laboratory by prolonged vibration of the soil under a vertical load.

Clearly, the relative density of any soil varies between 0 and 100%. Soils having relative densities less than 15% are considered to be "very loose," whereas those with values between 15 and 35% are "loose." "Medium dense" soils have relative densities between 35 and 65%, whereas "dense" soils have values between 65 and 85%. Soils with relative densities greater than 85% are considered to be "very dense."

Relative density may be used as an indicator of the degree of compactness of *in situ* soils and/or of compacted fills. In the latter case, a required relative density might be a specification requirement. Relative density may also be used as a rough indicator of soil stability. A low value of relative density indicates a "loose" soil, which would tend to be relatively unstable, whereas a soil with a high value of relative density would tend to be more stable.

EXAMPLE 2–11

Given

1. A fine, dry sand with an in-place unit weight of 18.28 kN/m^3.
2. The specific gravity of solids is 2.67.
3. The void ratio at its densest condition is 0.361.
4. The void ratio at its loosest condition is 0.940.

Required

Relative density of the sand.

Solution

From Eq. (2–14),

$$V_s = \frac{W_s}{G_s\gamma_w} = \frac{18.28 \text{ kN}}{(2.67)(9.81 \text{ kN/m}^3)} = 0.6979 \text{ m}^3$$

$$V_v = V - V_s = 1 \text{ m}^3 - 0.6979 \text{ m}^3 = 0.3021 \text{ m}^3$$

$$e_0 = \frac{V_v}{V_s} = \frac{0.3021 \text{ m}^3}{0.6979 \text{ m}^3} = 0.433$$

From Eq. (2–17),

$$D_r = \frac{e_{max} - e_0}{e_{max} - e_{min}} \times 100\% \qquad (2\text{–}17)$$

$$D_r = \frac{0.940 - 0.433}{0.940 - 0.361} \times 100\% = 87.6\%$$

2–10 PROBLEMS

2–1. Draw a gradation curve and find the median size, effective size, and coefficients of uniformity and of curvature for a soil sample that has the following test data for mechanical grain-size analysis:

U.S. Sieve Size	Size Opening (mm)	Weight Retained (g)
⅜ in.	9.50	0
No. 4	4.75	42
No. 10	2.00	146
No. 40	0.425	458
No. 100	0.150	218
No. 200	0.075	73
Pan	—	63

2–2. A sample of soil was tested in the laboratory, and the test results were listed as follows. Classify the soil by both the AASHTO system and the Unified Soil Classification System.

1. Liquid limit = 29%.
2. Plastic limit = 19%.
3. Mechanical grain-size analysis:

U.S. Sieve Size	Percentage Passing
1 in.	100
¾ in.	90
⅜ in.	82
No. 4	70
No. 10	65
No. 40	54
No. 200	25

2–3. An undisturbed chunk of soil has a wet weight of 62 lb and a volume of 0.56 ft^3. When dried out in an oven, the soil weighs 50 lb. If the specific gravity of solids is found to be 2.64, determine the water content, wet unit weight of soil, dry unit weight of soil, void ratio, porosity, and degree of saturation.

2–4. A 72-cm^3 sample of moist soil weighs 141.5 g. When it is dried out in an oven, it weighs 122.7 g. The specific gravity of solids is found to be 2.66. Find the water content, void ratio, porosity, degree of saturation, and wet and dry unit weights.

2–5. A soil specimen has a water content of 18% and a wet unit weight of 118.5 lb/ft^3. The specific gravity of solids is found to be 2.72. Find the dry unit weight, void ratio, and degree of saturation.

2–6. An undisturbed soil sample has a void ratio of 0.56, water content of 15%, and specific gravity of solids of 2.64. Find the wet and dry unit weights in lb/ft^3, porosity, and degree of saturation.

2–7. A 100% saturated soil has a wet unit weight of 112.8 lb/ft^3, and its water content is 42%. Find the void ratio and specific gravity of solids.

2–8. A 100% saturated soil has a void ratio of 1.33 and a water content of 48%. Find the unit weight of soil in lb/ft^3 and specific gravity of solids.

2–9. The water content of a 100% saturated soil is 35%, and the specific gravity of solids is 2.70. Determine the void ratio and unit weight in lb/ft^3.

2–10. A soil sample has the following data:

1. Degree of saturation = 42%.
2. Void ratio = 0.85.
3. Specific gravity of solids = 2.74.

Find its water content and unit weight in lb/ft^3.

2–11. A 0.082-m^3 sample of soil weighs 1.445 kN. When it is dried out in an oven, it weighs 1.301 kN. The specific gravity of solids is found to be 2.65. Find the water content, void ratio, porosity, degree of saturation, and wet and dry unit weights.

2–12. The wet unit weight of a soil sample is 18.55 kN/m^3. Its specific gravity of solids and water content are 2.72 and 12.3%, respectively. Find the dry unit weight, void ratio, and degree of saturation.

2–13. A fine sand has an in-place unit weight of 18.85 kN/m^3 and a water content of 5.2%. The specific gravity of solids is 2.66. Void ratios at densest and loosest conditions are 0.38 and 0.92, respectively. Find the relative density.

2–14. Derive an expression for $e = f(n)$, where e is void ratio and n is porosity.

2–15. Derive an expression for $n = f(e)$, where n is porosity and e is void ratio.

2–16. A sand sample has a porosity of 38% and specific gravity of solids of 2.66. Find the void ratio and wet unit weight in lb/ft^3 if the degree of saturation is 35%.

2–17. A proposed earthen dam will contain 5,000,000 m^3 of earth. Soil to be taken from a borrow pit will be compacted to a void ratio of 0.78. The void ratio of soil in the borrow pit is 1.12. Estimate the volume of soil that must be excavated from the borrow pit.

2–18. A soil sample with a water content of 14.5% and unit weight of 128.2 lb/ft^3 was dried to a unit weight of 118.8 lb/ft^3 without changing its void ratio. What is its new water content?

2–19. The unit weight, relative density, water content, and specific gravity of solids of a given sand are 17.98 kN/m^3, 62%, 7.6%, and 2.65, respectively.

1. If the minimum void ratio for this soil is 0.35, what would be its maximum void ratio?
2. What is its unit weight in the loosest condition?

2–20. A soil sample has a degree of saturation of 30.4% and void ratio of 0.85. How much water must be added per cubic foot of soil to increase the degree of saturation to 100%?

2–21. A soil sample has the following properties:

1. $e_{max} = 0.95$.
2. $e_{min} = 0.38$.
3. $D_r = 47\%$.
4. $G_s = 2.65$.

Find dry and saturated unit weights in both lb/ft^3 and kN/m^3.

References

[1] *The Asphalt Handbook,* Manual Series No. 4 [MS-4], Asphalt Institute, College Park, Md., 1989.
[2] A. Atterberg, Various papers published in the *Int. Mitt. Bodenkd,* 1911, 1912.
[3] B. K. Hough, *Basic Soils Engineering,* 2nd ed., The Ronald Press Company, New York, 1969. Copyright © 1969, John Wiley & Sons, Inc. Reprinted by permission of John Wiley & Sons, Inc.
[4] Cheng Liu and Jack B. Evett, *Soil Properties: Testing, Measurement, and Evaluation,* 4th ed., Prentice Hall, Upper Saddle River, N.J., 2000.
[5] *Standard Specifications for Transportation Materials and Methods of Sampling and Testing,* Part I, *Specifications,* 13th ed., AASHTO, 1982.
[6] A. Casagrande, "Classification and Identification of Soils," *Trans. ASCE,* **113,** 901 (1948).
[7] *The Unified Soil Classification System,* Waterways Exp. Sta. Tech. Mem. 3-357, U.S. Army Corps of Engineers, Vicksburg, Miss., 1953.
[8] *1989 Annual Book of ASTM Standards,* ASTM, Philadelphia, 1989. Copyright, American Society for Testing and Materials, 1916 Race Street, Philadelphia, PA 19103. Reprinted with permission.
[9] Wayne C. Teng, *Foundation Design,* © 1962. Reprinted by permission of Prentice-Hall, Inc.

3

SOIL EXPLORATION

3–1 INTRODUCTION

In Chapter 2, various engineering properties of soils were presented. An evaluation of these properties is absolutely necessary in any rational design of structures resting on, in, or against soil. To evaluate these properties, it is imperative that soils engineers visit proposed construction sites and collect and test soil samples, in order to evaluate and record results in a useful and meaningful form.

Chapter 3 deals with evaluation of soil properties, including reconnaissance, steps of soil exploration (boring, sampling, and testing), and the record of field exploration. Although different types of soil tests are discussed in this chapter, detailed test methods are outside the scope of this book. For specific step-by-step procedures, the reader is referred to *Soil Properties: Testing, Measurement, and Evaluation,* 4th edition, by Liu and Evett (Prentice Hall, 2000).

3-2 RECONNAISSANCE

A reconnaissance is a preliminary examination or survey of a job site. Usually, some useful information on the area (e.g., maps or aerial photographs) will already be available, and an astute person can learn much about surface conditions and get a general idea of subsurface conditions by simply visiting the site, observing thoroughly and carefully, and properly interpreting what is seen.

The first step in the preliminary soil survey of an area should be to collect and study any pertinent information that is already available. This could include general geologic and topographical information available in the form of geologic and topographic maps, obtainable from federal, state, and local governmental agencies (e.g., U.S. Geological Survey, Soil Conservation Service of the U.S. Department of Agriculture, and various state geologic surveys).

Aerial photographs can provide geologic information over large areas. Proper interpretation of these photographs may reveal land patterns, sinkhole cavities, landslides, surface drainage patterns, and the like. Such information can usually be obtained on a more widespread and thorough basis by aerial photography than by visiting the project site. Specific details on this subject are, however, beyond the scope of this book. For more information, the reader is referred to the many books available on aerial photo interpretation.

After carefully collecting and studying available pertinent information, the soils engineer should then visit the site in person, observe thoroughly and carefully, and interpret what is seen. The ability to do this successfully requires considerable practice and experience; however, a few generalizations are given next.

To begin with, significant details on surface conditions and general information about subsurface conditions in an area may be obtained by observing general topographical characteristics at the proposed job site and at nearby locations where soil was cut or eroded (such as railroad and highway cuts, ditch and stream erosion, and quarries), thereby exposing subsurface soil strata.

The general topographical characteristics of an area can be of significance. Any unusual conditions (e.g., swampy areas or dump areas, such as sanitary landfills) deserve particular attention in soil exploration.

Because the presence of water is often a major consideration in working with soil and associated structures, several observations regarding water may be made during reconnaissance. Groundwater tables may be noted by observing existing wells. Historical high watermarks may be recorded on buildings, trees, and so on.

Often, valuable information can be obtained by talking with local inhabitants of an area. Such information could include the flooding history, erosion patterns, mud slides, soil conditions, depths of overburden, groundwater levels, and the like.

One final consideration is that the reconnoiterer should take numerous photographs of the proposed construction site, exposed subsurface strata, adjacent structures, and so on. These can be invaluable in subsequent analysis and design processes and in later comparisons of conditions before and after construction.

The authors hope the preceding discussion in this section has made the reader aware of the importance of reconnaissance with regard to soil exploration at a proposed construction site. In addition to providing important information, the results of reconnaissance help determine the necessary scope of subsequent soil exploration.

At some point prior to beginning any subsurface exploration (Section 3–3), it is important that underground utilities (water mains, sewer lines, etc.) be located to assist the soils engineer in planning and carrying out subsequent subsurface exploration.

3–3 STEPS OF SOIL EXPLORATION

After all possible preliminary information is obtained as indicated in the preceding section, the next step is the actual subsurface soil exploration. It should be done by experienced personnel, using appropriate equipment. Much of soil mechanics practice can be successful only if one has long experience with which to compare each new problem.

Soil exploration may be thought of as consisting of three steps—boring, sampling, and testing. *Boring* refers to drilling or advancing a hole in the ground; *sampling* refers to removing soil from the hole; and *testing* refers to determining characteristics or properties of the soil. These three steps appear simple in concept but are quite difficult in good practice and are discussed in detail in the remainder of this section.

Boring

Some of the more common types of borings are *auger borings, wash borings, test pits,* and *core borings.*

An *auger* (see Figure 3–1) is a screwlike tool used to bore a hole. Some augers are operated by hand; others are power operated. As the hole is bored a short distance, the auger may be lifted to remove soil. Removed soil can be used for field classification and laboratory testing, but it must not be considered as an undisturbed soil sample. It is difficult to use augers in either very soft clay or coarse sand because the hole tends to refill when the auger is removed. Also, it may be difficult or impossible to use an auger below the water table because most saturated soils will not cling sufficiently to the auger for lifting. Hand augers may be used for boring to a depth of about 20 ft (6 m); power augers may be used to bore much deeper and quicker.

Wash borings (see Figure 3–2) consist of simultaneous drilling and jetting action. To begin with, a casing is usually driven into the ground. A chopping bit attached to the end of a drilling rod (or wash pipe) is driven by hammer, thereby breaking up the soil in the casing. Jetting action is accomplished by pumping water downward through the drilling bit. Water emerges at the chopping bit and further serves to break up the soil. Returning water transports soil to the ground surface, where samples can be collected for examination and classification. Such samples are, of course, disturbed samples whose water content has been increased.

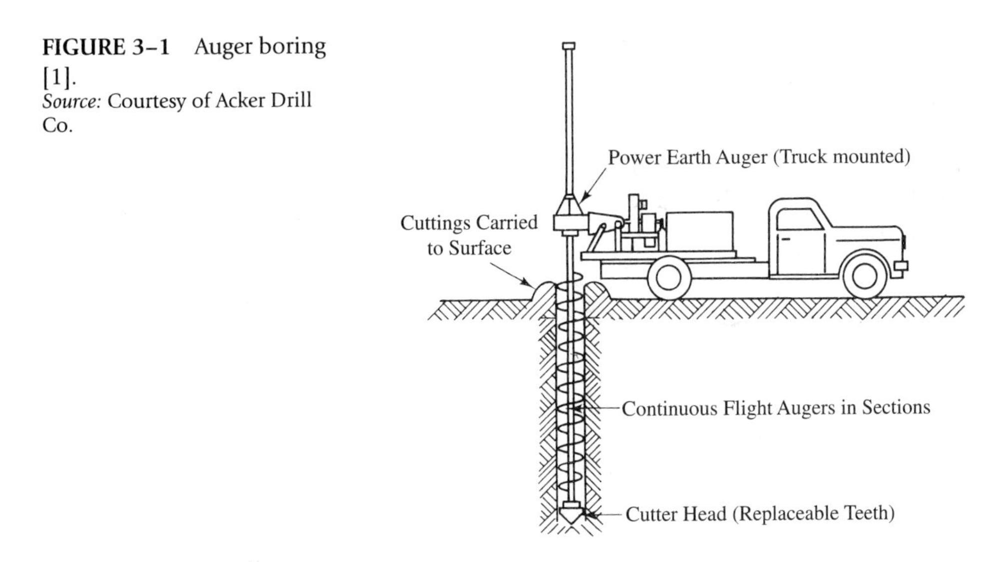

FIGURE 3–1 Auger boring [1].
Source: Courtesy of Acker Drill Co.

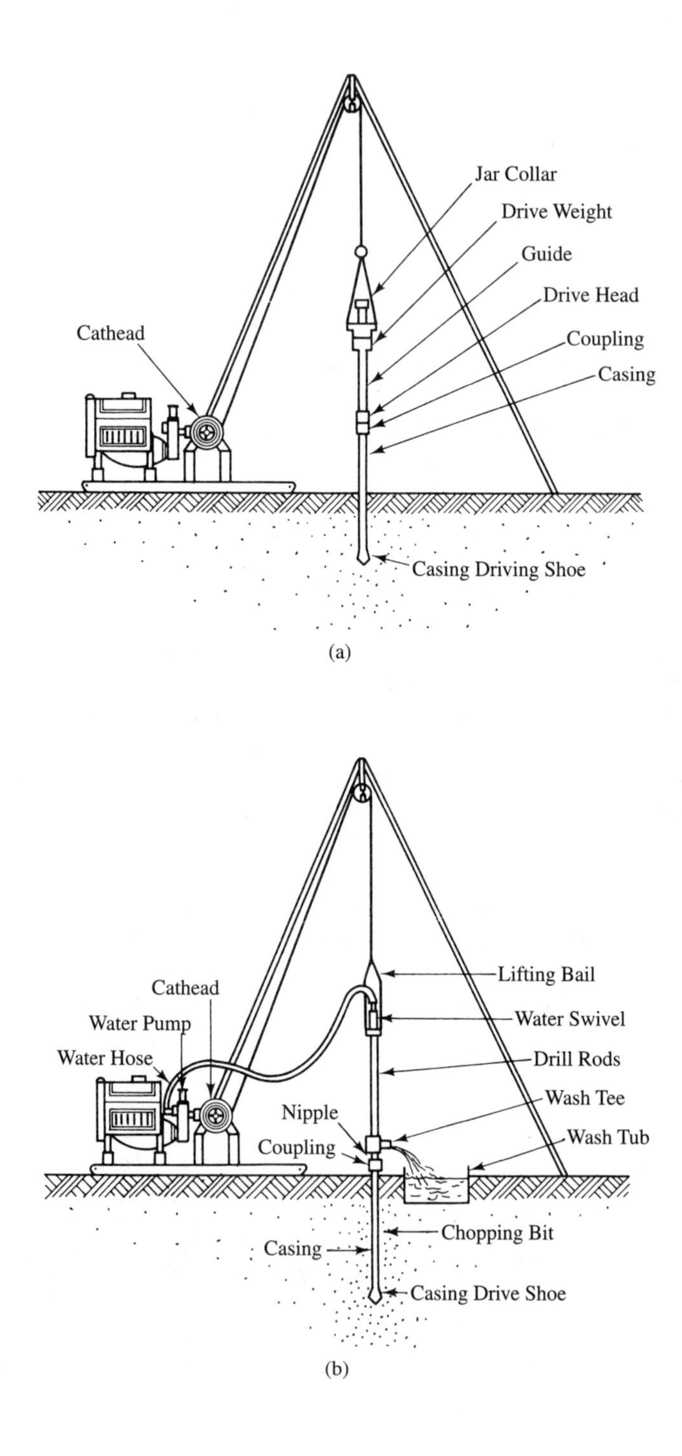

FIGURE 3–2 Typical setup for wash boring: (a) driving casing; (b) chopping and jetting [1].
Source: Courtesy of Acker Drill Co.

Test pits are excavations into the earth that permit a direct, visual inspection of the soil along the sides of the pit. As depicted in Figure 3–3, they may be large enough to allow a person to enter them and make inspections by viewing the exposed walls, taking color photographs of the soil in its natural condition, testing *in situ,* and taking undisturbed samples. Clearly, the soil strata (including thicknesses and stiffnesses of strata), texture and grain size of the soil along with visual classification of soils, soil moisture content, detection of fissures or cracks in the soil, and location of groundwater, among others, can be easily and accurately determined throughout the depth of the test pit. Soil samples can be obtained by carving an undisturbed sample from the pit's sides or bottom or by pushing a thin-walled steel tube into the pit's sides or bottom and extracting a sample by pulling the tube out. (Undisturbed samples should be preserved with wax to prevent moisture loss while the samples are transported to the laboratory.)

Test pits are excavated either manually or by power equipment, such as a backhoe or bulldozer (see Figure 3–4). For deeper pits, the excavation may need to be shored to protect persons entering the pits.

Soil inspection using test pits has several advantages. They are relatively rapid and inexpensive, and they provide a clear picture of the variation in soil properties with increasing depth. They also permit easy and reliable *in situ* testing and sampling. Another advantage of test pits is that they allow the detection and removal of larger soil particles (gravel or rocks, for example) for identification and testing; this may not be possible with boring samplers. On the other hand, test pits are generally limited by practical considerations as to depth; they generally do not extend deeper than 10 to 15 ft, whereas auger boring samplers can extend to much greater depths. Also, a high water table may preclude or limit the use of test pits.

Oftentimes, the presence of subsurface rock at a construction site can be important. Many times, construction projects have been delayed at considerable expense upon encountering unexpected rock in an excavation area. On the other hand, the presence of rock may be desirable if it can be used to support the load of an overlying structure. For these and other reasons, an investigation of subsurface rock in a project area is an important part of soil investigation.

Core borings are commonly used to drill into and through rock formations. Because rock is invariably harder than sandy and clayey soils, the sampling tools used for drilling in soil are usually not adequate for investigating subsurface rock. Core borings are performed using a core barrel, a hardened steel or steel alloy tube with a hard cutting bit containing tungsten carbide or commercial diamond chips (see Figure 3–5). Core barrels are typically 5 to 10 cm (2 to 4 in.) in diameter and 60 to 300 cm (2 to 10 ft) long.

Core borings are performed by attaching the core barrel and cutting bit to rods and rotating them with a drill, while water or air, serving as a coolant, is pushed (pumped) through the rods and barrel, emerging at the bit. The core remains in the core barrel and may be removed for examination by bringing the barrel to the surface. The rock specimen can be removed from the barrel, placed in the core box (see Figure 3–6), and sent to the soils laboratory for testing and analysis. The (empty) core barrel can then be used for another boring.

FIGURE 3–3 Test pit [2].

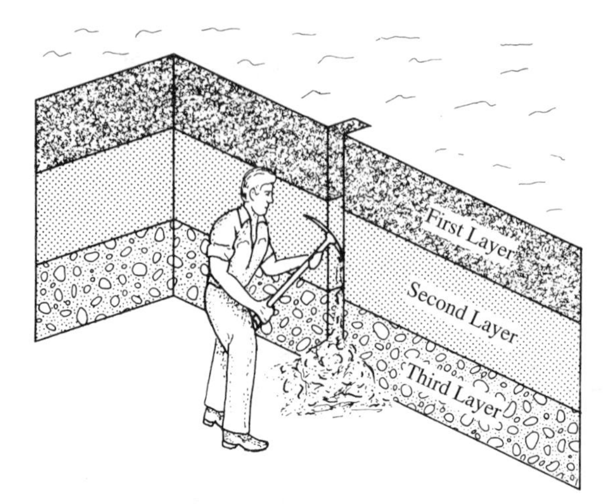

FIGURE 3–4 Backhoe (Courtesy of Caterpillar, Inc.)

FIGURE 3–5 Cutting bit for rock coring.

FIGURE 3–6 Core box containing rock core samples.

A wealth of information can be obtained from the laboratory testing and analysis of a rock core boring. The type of rock (such as granite, sandstone), its texture (coarse-grained or fine-grained, or some mixture of the two), degree of stratification (such as laminations), orientation of rock formation (bedding planes vertical, horizontal, or in between), and the presence of weathering, fractures, fissures, faults, or seams can be observed. Also, compression tests can be performed on core samples to determine the rock's compressive strength, and permeability tests can be done to see how underground water flow might be affected. All of the foregoing information can be invaluable in the design process and to prevent costly "surprises" that may be encountered during excavations.

Core recovery is the length of core obtained divided by the distance drilled. For example, a laminated shale stratum with a number of clay seams would likely exhibit a relatively small percentage of core recovery because the clayey soil originally located between laminations may have been washed or blown away by the water or air, respectively, during the drilling process. On the other hand, a larger percentage of core recovery would be expected in the case of granite.

Preceding paragraphs have discussed some of the more common types of borings. Once a means of boring has been decided upon, the question arises as to how many borings should be made. Obviously, the more borings made, the better the analysis of subsurface conditions should be. Borings are expensive, however, and a balance must be made between the cost of additional borings and the value of information gained from them.

As a rough guide for initial spacing of borings, the following are offered: for multistory buildings, 50 to 100 ft (15 to 30 m); for one-story buildings, earthen dams, and borrow pits, 100 to 200 ft (30 to 60 m); and for highways (subgrade), 500 to 1000 ft (150 to 300 m). These spacings may be increased if soil conditions are found to be relatively uniform and must be decreased if found to be nonuniform.

Once the means of boring and the spacing have been decided upon, a final question arises as to how deep the borings should be. In general, borings should extend through any unsuitable foundation strata (unconsolidated fill, organic soils, compressible layers such as soft, fine-grained soils, etc.) until soil of acceptable bearing capacity (hard or compact soil) is reached. If soil of acceptable bearing capacity is encountered at shallow depths, one or more borings should extend to a sufficient depth to ensure that an underlying weaker layer, if found, will have a negligible effect on surface stability and settlement. In compressible fine-grained strata, borings should extend to a depth at which stress from the superimposed load is so small that surface settlement is negligible. In the case of very heavy structures, including tall buildings, borings in most cases should extend to bedrock. In all cases, it is advisable to investigate drilling at least one boring to bedrock.

The preceding discussion presented some general considerations regarding boring depths. A more definitive criterion for determining required minimum depths of test borings in cohesive soils is to carry borings to a depth where the increase in stress due to foundation loading (i.e., weight of the structure) is less than 10% of the effective overburden pressure. Figures 3–7, 3–8, and 3–9 were developed [3] to determine minimum depths of borings based on the 10% increase in stress criterion for cohesive soils. Figure 3–7 is for a continuous footing (such as a wall footing). Figure 3–8 is for a square footing with a design pressure between 1000 and 9000 lb/ft^2, and Figure 3–9 is for a square footing with a design pressure between 100 and 1000 lb/ft^2. If the groundwater table is at the footing's base, the buoyant weight (submerged unit weight) of the soil should be used in these figures. If the groundwater table is lower than distance B below the footing (B is the footing's width), the wet unit weight should be used. For intermediate conditions, an interpolation can be

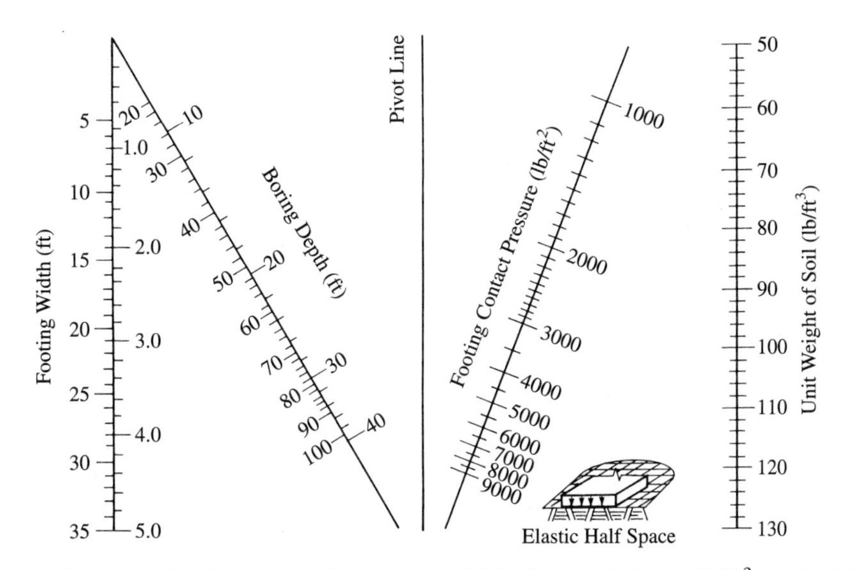

FIGURE 3–7 Infinite strip loading—Boussinesq-type solid (1 ft = 0.3048 m; 1 lb/ft^2 = 47.88 N/m^2; 1 lb/ft^3 = 0.1571 kn/m^3) [3].

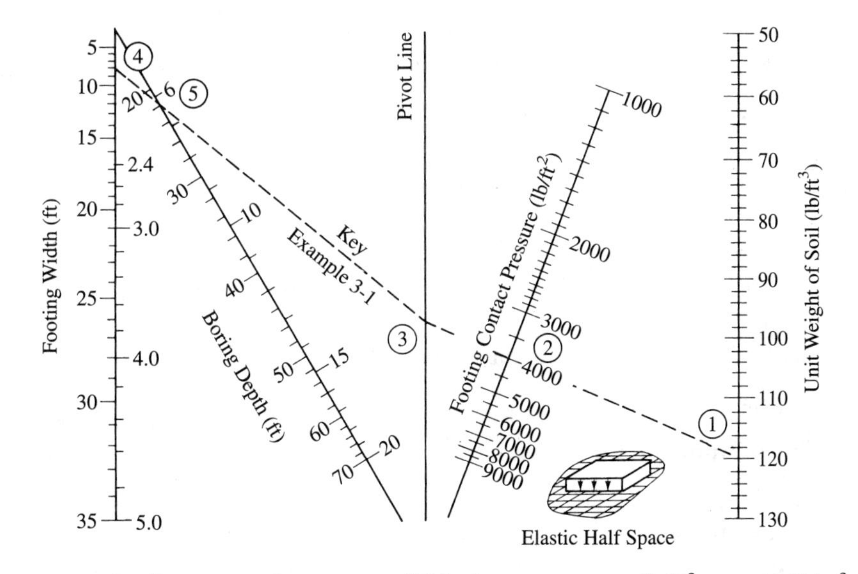

FIGURE 3–8 Square loading—Boussinesq-type solid (1 ft = 0.3048 m; 1 lb/ft^2 = 47.88 N/m^2; 1 lb/ft^3 = 0.1571 kN/m^3) [3].

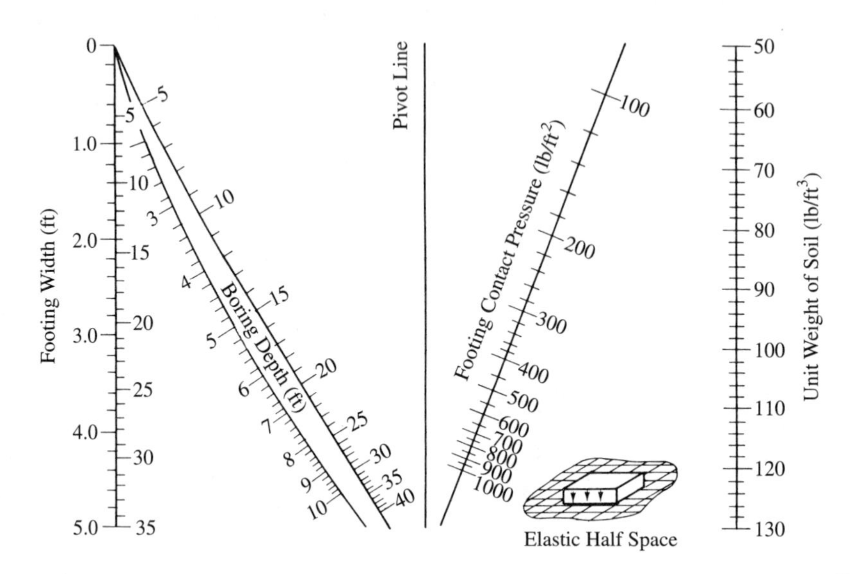

FIGURE 3–9 Square loading (low-pressure)—Boussinesq-type solid (1 ft = 0.3048 m; 1 lb/ft^2 = 47.88 N/m^2; 1 lb/ft^3 = 0.1571 kn/m^3) [3].

made between required depths of boring for shallow and deep groundwater conditions, or the groundwater table can be conservatively assumed to be at the footing's base. It should be noted that on the left sides of Figures 3–7 through 3–9 two scales are given for footing width and minimum test boring depth. In each figure, footing widths given on one side of the width scale correspond with boring depths given on the same side of the boring depth scale [3].

EXAMPLE 3–1 [3]

Given

1. An 8-ft square footing is subjected to a contact pressure of 4000 lb/ft^2.
2. The wet unit weight of the soil supporting the footing is estimated to be 120 lb/ft^3.
3. The water table is estimated to be 30 ft beneath the footing.

Required

The minimum depth of test boring.

Solution

Because the water table is estimated to be 30 ft beneath the footing and the footing's width is 8 ft, the soil's wet unit weight should be used. From Figure 3–8, with a wet unit weight of 120 lb/ft^3, contact pressure between footing and soil equal to 4000 lb/ft^2, and width of footing equal to 8 ft, the minimum depth of test boring is determined to be 22 ft.

Figures 3–7 through 3–9 are quite useful for estimating minimum required test boring depths in cohesive soils. In the final analysis, however, the depth of a specific boring should be determined by the soils engineer based on his or her expertise, experience, judgment, and general knowledge of the specific area. Also, in some cases, the depth (and spacing) of borings may be specified by local codes or company policy.

Sampling

Sampling refers to the taking of soil or rock from bored holes. Samples may be classified as either *disturbed* or *undisturbed*.

As mentioned previously in this section, in both auger borings and wash borings, soil is brought to the ground surface, where samples can be collected. Such samples are obviously disturbed samples, and thus some of their characteristics are changed. (Split-spoon samples, described in Section 3–5, also provide disturbed samples.) Disturbed samples should be placed in an airtight container (plastic bag or airtight jar, for example) and should, of course, be properly labeled as to date, location, borehole number, sampling depth, and so on. Disturbed samples are generally used for soil grain-size analysis, determination of liquid and plastic limits and specific gravity of soil, and other tests, such as the compaction and CBR (California bearing ratio) tests.

For determination of certain other properties of soils, such as strength, compressibility, and permeability, it is necessary that the collected soil sample be exactly the same as it was when it existed in place within the ground. Such a soil sample is referred to as an undisturbed sample. It should be realized, however, that such a sample can never be completely undisturbed (i.e., be exactly the same as it was when it existed in place within the ground).

Undisturbed samples may be collected by several methods. If a test pit is available in clay soil, an undisturbed sample may be obtained by simply carving a sample very carefully out of the side of the test pit. Such a sample should then be coated with paraffin wax and placed in an airtight container. This method is often too tedious, time consuming, and expensive to be done on a large scale, however.

A more common method of obtaining an undisturbed sample is to push a thin tube into the soil, thereby trapping the (undisturbed) sample inside the tube, and then to remove the tube and sample intact. The ends of the tube should be sealed with paraffin wax immediately after the tube containing the sample is brought to the ground surface. The sealed tube should then be sent to the soils laboratory, where subsequent tests can be made on the sample, with the assumption that such test results are indicative of the properties of the soil as it existed in place within the ground. The thin-tube sampler is called a *Shelby tube.* It is a 2- to 3-in. (51- to 76-mm)-diameter 16-gauge seamless steel tube (see Figure 3–10).

When using a thin-tube sampler, the soils engineer should minimize the disturbance of the soil. Pushing the sampler into the soil quickly and with constant speed causes the least disturbance; driving the sampler into the soil by blows of a hammer produces the most.

FIGURE 3–10 Shelby tube.

For samples of a given diameter obtained by a sampler pushed into the soil, the *degree of disturbance* depends on the dimensions of the sampler—in particular, the *area ratio*, which is defined as follows [4]*:

$$A_r = 100 \times (D_e^2 - D_i^2)/D_i^2 \quad \textbf{(3–1)}$$

where A_r = area ratio (in percent)
D_e = external diameter of sampler
D_i = internal diameter of sampler

For disturbance to be nominal, the area ratio should not exceed about 20%.

Normally, samples (both disturbed and undisturbed) are collected at least every 5 ft (1.5 m) in depth of the boring hole. When, however, any change in soil characteristics is noted within 5-ft intervals, additional samples should be taken.

The importance of properly and accurately identifying and labeling each sample cannot be overemphasized.

After a boring has been made and samples taken, an estimate of the groundwater table can be made. It is common practice to cover the hole (for example, with a small piece of plywood) for safety reasons, mark it for identification, leave it overnight, and return the next day to record the groundwater level. The hole should then be filled in to avoid subsequent injury to people or animals (see Section 3–4).

*From Karl Terzaghi, Ralph B. Peck, and Gholamreza Mesri, *Soil Mechanics in Engineering Practice,* 3rd ed., John Wiley & Sons, Inc., New York, 1996. Copyright © 1996, by John Wiley & Sons, Inc. Reprinted by permission of John Wiley & Sons, Inc.

Testing

A large number of tests can be performed to evaluate various soil properties. These include both laboratory and field tests. Some of the most common tests are listed in Table 3–1. As indicated at the beginning of this chapter, the reader is referred to *Soil Properties: Testing, Measurement, and Evaluation,* 4th edition, by Liu and Evett (Prentice Hall, 2000) for specific step-by-step procedures involving these tests. Three tests—the standard penetration test, cone penetration test, and vane test—are described in some detail in Sections 3–5 through 3–7.

3–4 GROUNDWATER TABLE

The term *groundwater table* (or just *water table*) has been mentioned several times earlier in this chapter. Section 3–4 presents more detailed information about this important phenomenon as it relates to the study of soils.

The location of the water table is a matter of importance to soils engineers, particularly when it is near the ground surface. For example, a soil's bearing capacity (see Chapter 9) can be reduced when the water table is at or near a footing. The location of the water table is not fixed at a particular site; it tends to rise and fall during periods of wet and dry weather, respectively. Fluctuations of the water table may result in reduction of foundation stability; in extreme cases, structures may float out of the ground. Accordingly, foundation design and/or methods of construction may be affected by the location of the water table. Knowing the position of the water table is also very important when sites are being chosen for hazardous waste and sanitary landfills, to avoid contaminating groundwater.

The water table can be located by measuring down to the water level in existing wells in an area. It can also be determined from boring holes. The level to which groundwater rises in a boring hole is the groundwater elevation in that area. If adjacent soil is pervious, the water level in a boring hole will stabilize in a short period of time; if the soil is relatively impervious, it may take much longer for this to happen. General practice in soil surveying is to cover the boring hole (e.g., with a small piece of plywood) for safety reasons, leave it for at least 24 hours to allow the water level to rise in the hole and stabilize, and return the next day to locate and record the groundwater table. The hole should then be filled to avoid subsequent injury to people or animals.

3–5 STANDARD PENETRATION TEST

The standard penetration test (SPT) is widely used in the United States. Relatively simple and inexpensive to perform, it is useful in determining certain properties of soils, particularly of cohesionless soils, for which undisturbed samples are not easily obtained.

TABLE 3–1
Common Types of Testing [2]

Property of Soil	Type of Test	ASTM Designation	AASHTO Designation
	(a) *Laboratory testing of soils*		
Grain-size distribution	Mechanical analysis	D421, D422 D1140	T88
Consistency	Liquid limit (*LL*)	D4318	T89
	Plastic limit (*PL*)	D4318	T90
	Plasticity index (*PI*)	D4318	T90
Unit weight	Specific gravity	D854	T100
Moisture	Natural water content		
	Field moisture equivalent	D2216	T93
	Centrifuge moisture equivalent	D425	
Shear strength	Unconfined compression	D2166	T208
	Direct shear	D3080	T236
	Triaxial	D2850	T234
Volume change	Shrinkage factors	D427	T92
Compressibility	Consolidation	D2435	T216
Permeability	Permeability	D2434	T215
Compaction characteristics	Standard Proctor	D698	T99
	Modified Proctor	D1557	T180
California bearing ratio (CBR)		D1883	T193
	(b) *Field testing of soils*		
Compaction control	Moisture–density relations	D698	T99, T180
	In-place density	D1556	T191
		D2167	T205
Shear strength (soft clay)	Vane test	D2573	T223
Relative density (granular soil)	Penetration test	D1586	T206
Permeability	Pumping test		
Bearing capacity			
Pavement	CBR		
Footings	Plate bearing	D1195	T221
Piles (vertical load)		D1196	T222
	Plate bearing	D1194	T235
Batter piles	Load test	D1143	
	Lateral load test		

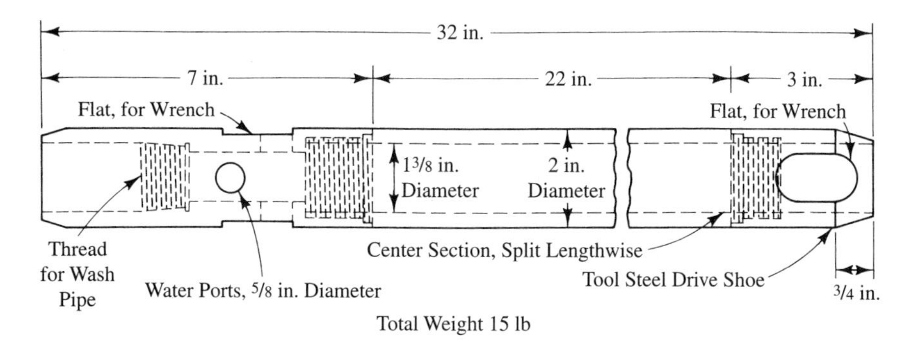

FIGURE 3–11 Split-spoon sampler for the standard penetration test [5].

The SPT utilizes a *split-spoon sampler* (see Figure 3–11). It is a 2-in. (51-mm)-O.D. 1 3/8-in. (35-mm)-I.D. tube, 18 to 24 in. (457 to 610 mm) long, that is split longitudinally down the middle. The split-spoon sampler is attached to the bottom of a drilling rod and driven into the soil with a drop hammer. Specifically, a 140-lb (623-N) hammer falling 30 in. (762 mm) is used to drive the split-spoon sampler 18 in. (457 mm) into the soil. As the sampler is driven the 18 in. (457 mm) into the soil, the number of blows required to penetrate each of the three 6-in. (152-mm) increments is recorded separately. The standard penetration resistance value (or *N*-value) is the number of blows required to penetrate the last 12 in. (305 mm). Thus, the *N*-value represents the number of blows per foot (305 mm). After blow counts have been obtained, the split-spoon sampler can be removed and opened (along the longitudinal split) to obtain a disturbed sample for subsequent examination and testing [6].

SPT results (i.e., *N*-values) are influenced by overburden pressure (effective weight of overlying soil) at locations where blow counts are made. Several methods have been proposed to correct *N*-values to reflect the influence of overburden pressure. Three of these methods are presented here.

One method [7] utilizes the following equations to evaluate C_N, a correction factor to be applied to the *N*-value determined in the field:

$$C_N = 0.77 \log_{10} \frac{20}{p_0} \qquad (p_0 \text{ in tons/ft}^2) \tag{3–2}$$

$$C_N = 0.77 \log_{10} \frac{1915}{p_0} \qquad (p_0 \text{ in kN/m}^2) \tag{3–3}$$

where p_0 is the effective overburden pressure at the elevation of the SPT. These equations are not valid if p_0 is less than 0.25 ton/ft^2 (24 kN/m^2). Figure 3–12 gives a graphic relationship, based in part on Eq. (3–2), for determining a correction factor to be applied to the *N*-value recorded in the field. If p_0 is greater than or equal to 0.25 ton/ft^2, the correction factor may be determined using either Eq. (3–2) or Figure 3–12.

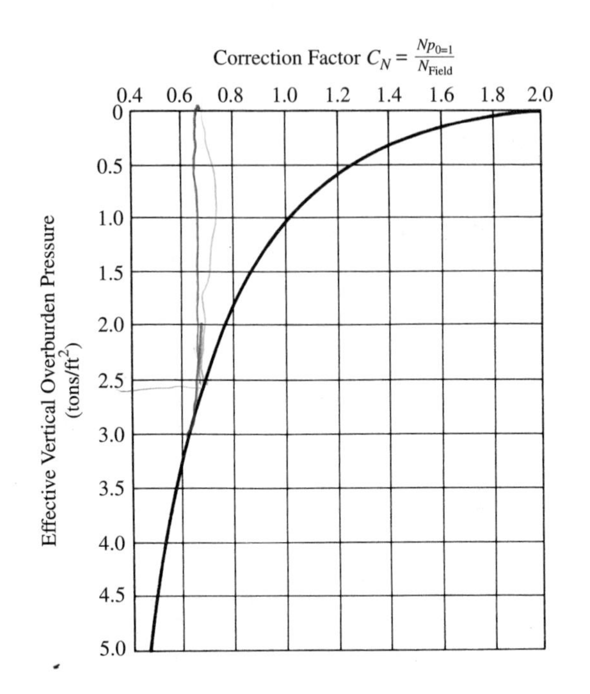

FIGURE 3–12 Chart for correction of N-values in sand for influence of overburden pressure (reference value of effective overburden pressure, 1 ton/ft^2) [7].
Note: 1 ton/ft^2 = 95.76 kn/m^2.

If p_0 is less than 0.25 ton/ft^2, the correction factor should be taken from the figure.

Another method [8] utilizes the following equations:

$$N = \frac{4N'}{1 + 2p_0} \quad [p_0 \text{ in kips per square foot (kips/ft}^2)] \quad \text{if } p_0 \leq 1.5 \text{ kips/ft}^2 \tag{3–4}$$

$$N = \frac{4N'}{3.25 + 0.5p_0} \quad (p_0 \text{ in kips/ft}^2) \text{ if } p_0 \geq 1.5 \text{ kips/ft}^2 \tag{3–5}$$

$$N = \frac{4N'}{1 + 0.0418p_0} \quad (p_0 \text{ in kN/m}^2) \text{ if } p_0 \leq 72 \text{ kN/m}^2 \tag{3–6}$$

$$N = \frac{4N'}{3.25 + 0.0104p_0} \quad (p_0 \text{ in kN/m}^2) \text{ if } p_0 \geq 72 \text{ kN/m}^2 \tag{3–7}$$

where N = corrected N-value
N' = N-value determined in the field
p_0 = effective overburden pressure

A third method for correcting N-values to reflect the influence of overburden pressure [4, 9] utilizes the following equation*:

$$N = N' \times (100/p_0)^{1/2} \tag{3-8}$$

where the terms are the same as in Eqs. (3–4) through (3–7), with p_0 expressed in kN/m^2.

These three methods give comparable results. It should be noted that the first method [Eqs. (3–2) and (3–3)] results in no adjustment of the N-value at a depth where the effective overburden pressure is 1 ton/ft^2 (96 kN/m^2), whereas the second method [Eqs. (3–4) through (3–7)] results in no adjustment at a depth where the effective overburden pressure is 0.75 ton-ft^2, or 1.5 kips/ft^2 (72 kN/m^2). The third method [Eq. (3–8)] yields no adjustment at a depth where the effective overburden pressure is 100 kN/m^2 (1.04 tons/ft^2).

EXAMPLE 3–2

Given

An SPT was performed at a depth of 20 ft in sand of unit weight 135 lb/ft^3. The blow count was 40.

Required

The corrected N-value by each of the three methods presented previously.

Solution

1. By Eq. (3–2),

$$C_N = 0.77 \log_{10} \frac{20}{p_0} \tag{3-2}$$

$$p_0 = \frac{(20\ \text{ft})(135\ \text{lb/ft}^3)}{2000\ \text{lb/ton}} = 1.35\ \text{tons/ft}^2$$

$$C_N = 0.77 \log_{10} \frac{20}{1.35\ \text{tons/ft}^2} = 0.901$$

(This value of 0.901 for C_N can also be obtained using Figure 3–12 by locating 1.35 tons/ft^2 along the ordinate, moving horizontally to the curved line, and then moving upward to obtain the correction factor, C_N.) Therefore,

$$N_{\text{corrected}} = (40)(0.901) = 36$$

2. By Eq. (3–4) or (3–5),

*From Samson S. C. Liao and Robert V. Whitman, "Overburden Correction Factors for SPT in Sand." *J. Geotech. Eng. Div. ASCE,* **112**(3), 373–377 (1986). Reproduced by permission of ASCE.

$$p_0 = \frac{(20 \text{ ft})(135 \text{ lb/ft}^3)}{1000 \text{ lb/kip}} = 2.70 \text{ kips/ft}^2$$

Because $[p_0 = 2.70 \text{ kips/ft}^2] > 1.5 \text{ kips/ft}^2$, use Eq. (3–5).

$$N = \frac{4N'}{3.25 + 0.5p_0} \tag{3–5}$$

$$N = \frac{(4)(40)}{3.25 + (0.5)(2.70 \text{ kips/ft}^2)}$$

$$N_{\text{corrected}} = 35$$

3. By Eq. (3–8),

$$N = N' \times (100/p_0)^{1\backslash 2} \tag{3–8}$$

$$p_0 = (1.35 \text{ tons/ft}^2)\left(\frac{95.76 \text{ kN/m}^2}{1 \text{ ton/ft}^2}\right) = 129.3 \text{ kN/m}^2$$

$$N = (40) \times (100/129.3 \text{ kN/m}^2)^{1/2}$$

$$N_{\text{corrected}} = 35$$

EXAMPLE 3–3

Given

An SPT test was performed at a depth of 8.5 m in sand of unit weight 20.04 kN/m^3. The blow count was 38.

Required

The corrected N-value by each of the three methods presented previously.

Solution

1. By Eq. (3–3),

$$C_N = 0.77 \log_{10} \frac{1915}{p_0} \tag{3–3}$$

$$p_0 = (8.5 \text{ m})(20.04 \text{ kN/m}^3) = 170.3 \text{ kN/m}^2$$

$$C_N = 0.77 \log_{10} \frac{1915}{170.3 \text{kN/m}^2} = 0.809$$

Therefore,

$$N_{\text{corrected}} = (38)(0.809) = 31$$

2. By Eq. (3–7) (because $p_0 > 72 \text{ kN/m}^2$),

$$N = \frac{4N'}{3.25 + 0.0104p_0} \tag{3–7}$$

$$N = \frac{(4)(38)}{3.25 + (0.0104)(170.3 \text{ kN/m}^2}$$

$$N_{\text{corrected}} = 30$$

3. By Eq. (3–8),

$$N = N' \times (100/p_0)^{1/2} \tag{3–8}$$

$$N = (38) \times (100/170.3 \text{ kN/m}^2)^{1/2}$$

$$N_{\text{corrected}} = 29$$

EXAMPLE 3–4

Given

Same data as given in Example 3–2, except that the water table is located 5 ft below the ground surface.

Required

The corrected N-value by the first two methods presented previously.

Solution

1. By Eq. (3–2),

$$C_N = 0.77 \log_{10} \frac{20}{p_0} \tag{3–2}$$

$$p_0 = \frac{(5 \text{ ft})(135 \text{ lb/ft}^3) + (15 \text{ ft})(135 \text{ lb/ft}^3 - 62.4 \text{ lb/ft}^3)}{2000 \text{ lb/ton}}$$

$$= 0.882 \text{ ton/ft}^2$$

$$C_N = 0.77 \log_{10} \frac{20}{0.882 \text{ ton/ft}^2} = 1.04$$

Therefore,

$$N_{\text{corrected}} = (40)(1.04) = 42$$

2. By Eq. (3–4) or (3–5),

$$p_0 = \frac{(5 \text{ ft})(135 \text{ lb/ft}^3) + (15 \text{ ft})(135 \text{ lb/ft}^3 - 62.4 \text{ lb/ft}^3)}{1000 \text{ lb/kip}}$$

$$= 1.76 \text{ kips/ft}^2$$

Because $[p_0 = 1.76 \text{ kips/ft}^2] > 1.5 \text{ kips/ft}^2$, use Eq. (3–5).

$$N = \frac{4N'}{3.25 + 0.5p_0} \qquad \textbf{(3–5)}$$

$$N = \frac{(4)(40)}{3.25 + (0.5)(1.76 \text{ kips/ft}^2)}$$

$$N_{\text{corrected}} = 39$$

In addition to the effect of overburden pressure, SPT results (N-values) are influenced by (1) drill rod lengths, (2) whether or not liners are present in the sampler, and (3) borehole diameters. Table 3–2 gives some corrections that can be applied to measured N-values to adjust for these three influences.

Through empirical testing, correlations between (corrected) SPT N-values and several soil parameters have been established. These are particularly useful for cohesionless soils but are less reliable for cohesive soils. Table 3–3 gives correlations of the relative density of sands with SPT N-values; Table 3–4 gives correlations of the consistency of clays and unconfined compressive strength (q_u). Figure 3–13 gives a graphic relationship between the angle of internal friction of cohesionless soil and SPT N-values. Figure 3–13 also gives graphic relationships between certain bearing

TABLE 3–2
Approximate Corrections to Measured *N*-Values [4][1,2]

Influence	Correction Size	Factor
Rod length	>10 m	1.0
	6–10 m	0.95
	4–6 m	0.85
	3–4 m	0.75
Standard sampler	—	1.0
U.S. sampler without liners	—	1.2
Borehole diameter	65–115 mm	1.0
	150 mm	1.05
	200 mm	1.15

[1]After Skempton [10].

[2]From Karl Terzaghi, Ralph B. Peck, and Gholamreza Mesri, *Soil Mechanics in Engineering Practice,* 3rd ed., John Wiley & Sons, Inc., New York, 1996. Copyright © 1996, by John Wiley & Sons, Inc. Reprinted by permission of John Wiley & Sons, Inc.

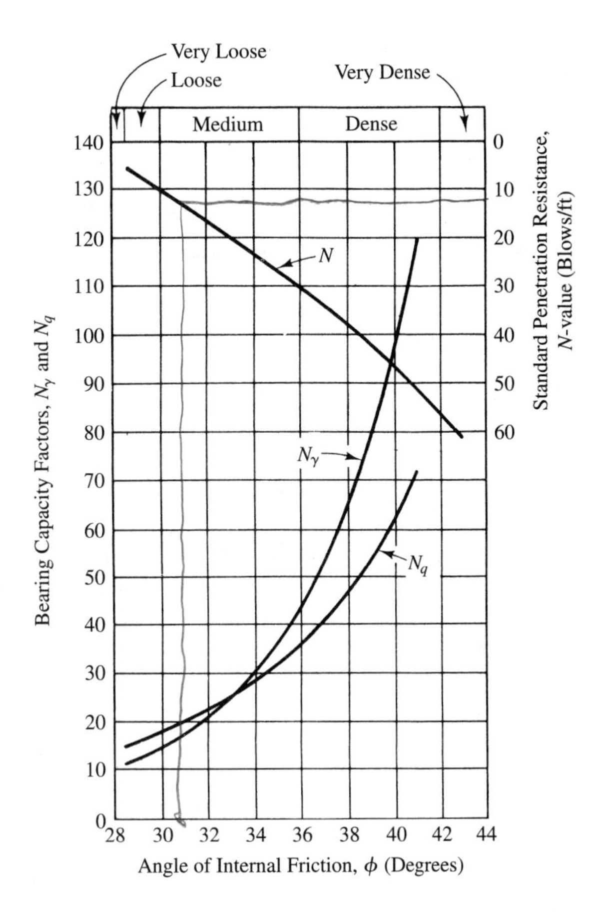

FIGURE 3–13 Curves showing the relationship between bearing capacity factors and φ, as determined by theory, and the rough empirical relationship between bearing capacity factors or φ and values of standard penetration resistance, N [7].

TABLE 3–3
Relative Density of Sands According to Results of Standard Penetration Text [4][1]

SPT *N*-Value	Relative Density
0–4	Very loose
4–10	Loose
10–30	Medium
30–50	Dense
Over 50	Very dense

[1]From Karl Terzaghi, Ralph B. Peck, and Gholamreza Mesri, *Soil Mechanics in Engineering Practice,* 3rd ed., John Wiley & Sons, Inc., New York, 1996. Copyright © 1996, by John Wiley & Sons, Inc. Reprinted by permission of John Wiley & Sons, Inc.

TABLE 3–4
Relation of Consistency of Clay, SPT *N*-Value, and Unconfined Compressive Strength (q_u) [4][1]

	q_u (kN/m^2)					
Consistency:	*Very Soft*	*Soft*	*Medium*	*Stiff*	*Very Stiff*	*Hard*
SPT *N*-value	<2	2–4	4–8	8–15	15–30	>30
q_u	<25	25–50	50–100	100–200	200–400	>400

[1]From Karl Terzaghi, Ralph B. Peck, and Gholamreza Mesri, *Soil Mechanics in Engineering Practice,* 3rd ed., John Wiley & Sons, Inc., New York, 1996. Copyright ©1996, by John Wiley & Sons, Inc. Reprinted by permission of John Wiley & Sons, Inc.

capacity factors for cohesionless soil and SPT *N*-values. These relationships will be utilized in Chapter 9.

The reader is cautioned that, although the standard penetration test is widely used in the United States, results are highly variable and thus difficult to interpret. Nevertheless, it is a useful guide in foundation analysis. Much experience is necessary to properly apply the results obtained. Outside the United States, other techniques are used. For example, in Europe the cone penetration test (Section 3–6) is often preferred.

3–6 CONE PENETRATION TEST

The cone penetration test (CPT) has been widely used in Europe for many years but is now gaining favor in the United States. It has the advantage of accomplishing subsurface exploration rapidly without taking soil samples.

There are two types of mechanical cone penetrometers—the *mechanical cone penetrometer* (see Figure 3–14) and the *mechanical friction-cone penetrometer* (see Figure 3–15). Both types have a conical point with a point angle of 60° and a base diameter of 35.7 mm (1.41 in.), giving a base area of 1000 mm^2 (1.55 in.^{2}). The main difference between the two is that in addition to cone resistance, the friction-cone penetrometer allows for determination of side (sleeve) resistance as the penetrometer is advanced through the soil.

Penetrometers are either pushed (by a hydraulic jack, for example) or driven (such as by blows of a drop hammer) into and through soil. When the penetrometer is pushed, the test is known as a *static cone test* (sometimes referred to as a *Dutch cone test*); when it is driven, the test is called a *dynamic cone test.* In all cases, the penetrometer's resistance to being advanced through the soil is measured and recorded as a function of depth of soil penetrated.

The static test is sensitive to small differences in soil consistency. Because the penetrometer is pushed (rather than driven) in a static test, the procedure probably

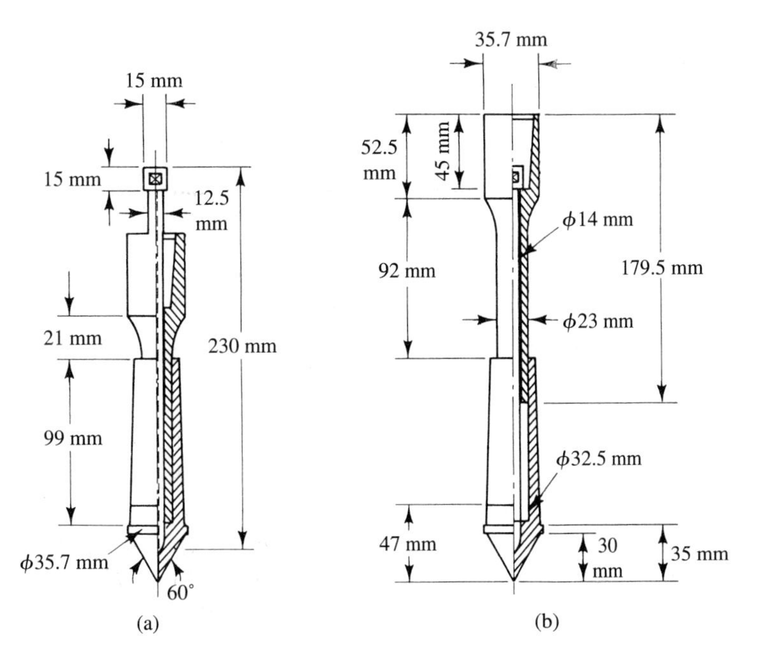

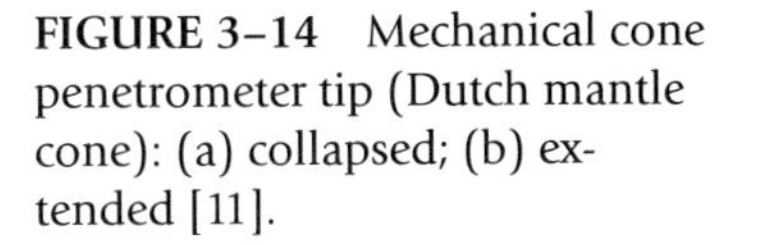
FIGURE 3–14 Mechanical cone penetrometer tip (Dutch mantle cone): (a) collapsed; (b) extended [11].

tends not to alter soil structure significantly for loose sands and sensitive clays. The dynamic test covers a wider range of soil consistencies; and because the penetrometer is driven, penetrations of gravels and soft rock are possible.

When a mechanical cone penetrometer is used, the penetrometer's tip is advanced to the required depth by applying sufficient thrust on the push rods (see Figure 3–14a). Then the tip is extended by applying sufficient thrust on the inner rods (see Figure 3–14b). Cone resistance is obtained at some specific point during the downward movement of the inner rods relative to the stationary push rods. By repeating this two-step procedure again and again, one can obtain cone resistance data as a function of depth. Each increment of depth through which the penetrometer is advanced should not normally exceed 8 in. (203 mm), and the rate of penetration should be approximately 2 to 4 ft/min (10 to 20 mm/s).

The procedure for using a mechanical friction-cone penetrometer is the same as that just described for a mechanical cone penetrometer, except that after cone resistance during the initial phase of the extension is determined, a separate resistance value of cone plus sleeve friction is also measured. Sleeve resistance is obtained by subtracting cone resistance from total resistance.

CPT data are ordinarily presented as plots of cone resistance, q_c; friction resistance, f_s; and friction ratio (ratio of friction resistance to cone resistance), f_s/q_c versus depth (see Figure 3–16), thereby giving a "picture" of the variation in soil types at

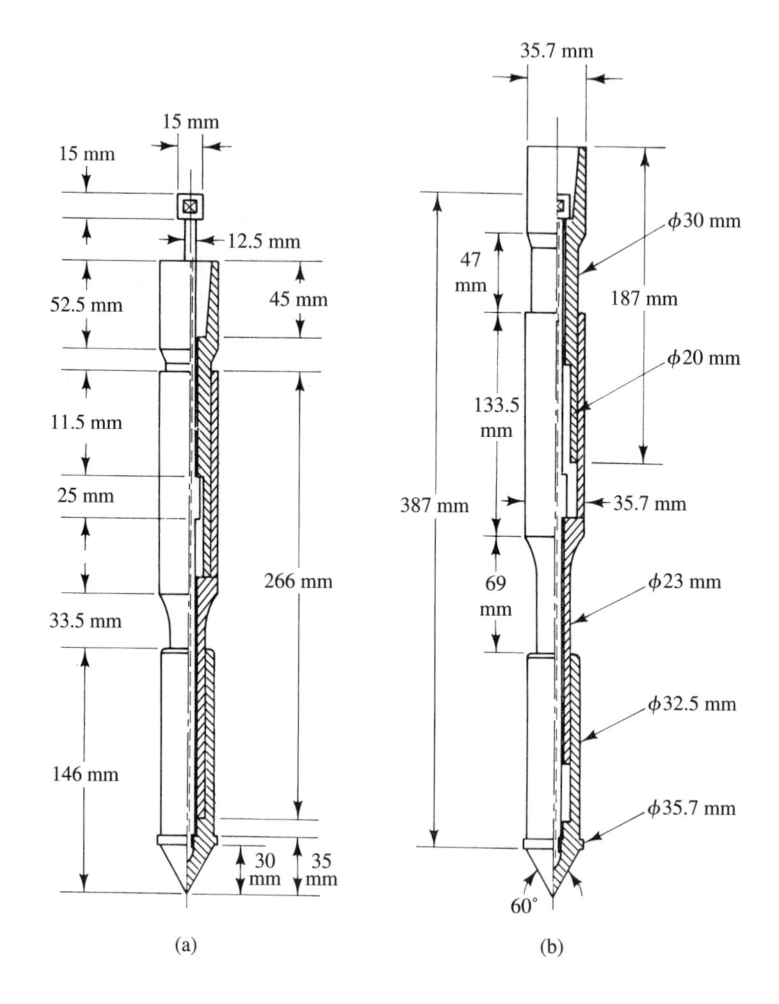

FIGURE 3–15 Mechanical friction-cone penetrometer tip (Begemann friction cone): (a) collapsed; (b) extended [11].

different depths at a test site. In general, the ratio of sleeve resistance to cone resistance is higher in cohesive soils than in cohesionless soils; hence, this ratio can be used to estimate the type of soil being penetrated. For example, Figure 3–17 classifies soils based on cone resistance and friction ratio. This classification is, however, empirical and not necessarily precise; therefore, it should be used cautiously. Certain situations (for example, the presence of thin layers of different materials) can lead to inaccurate classifications. To identify definitively the soil strata that were penetrated by a CPT, supplemental boring is normally required to obtain samples (because no soil samples are taken during a CPT).

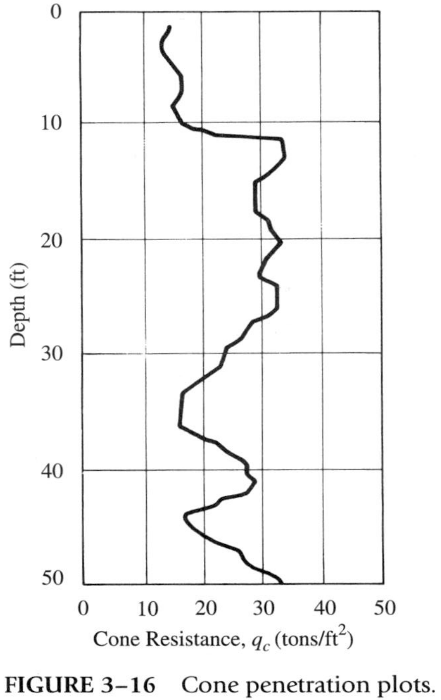

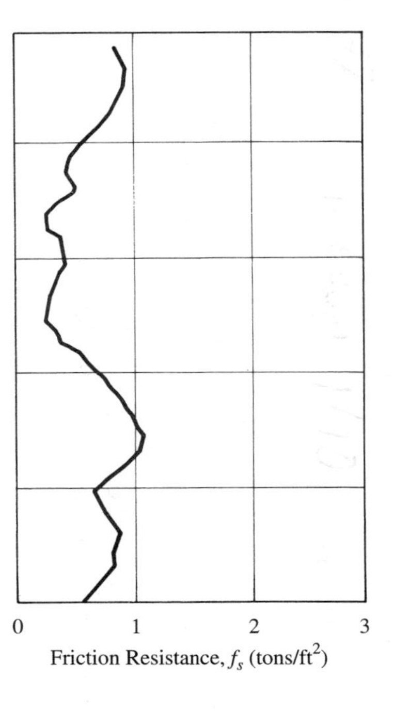

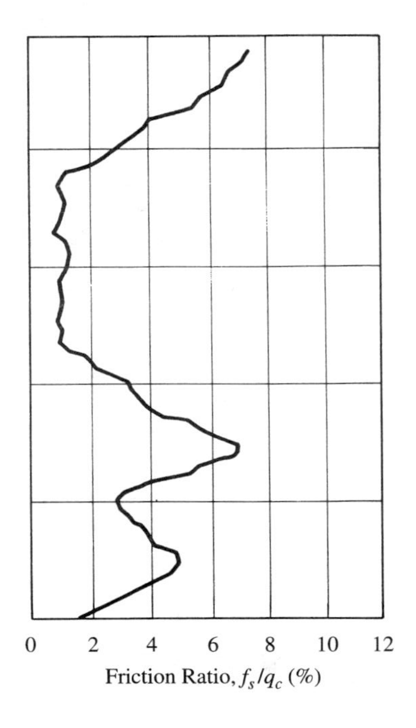

FIGURE 3–16 Cone penetration plots.

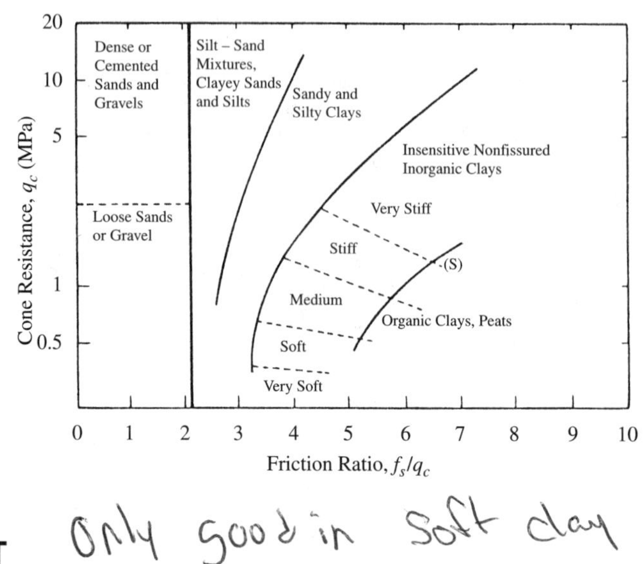

FIGURE 3–17 Soil classification based on Begemann cone penetrometer tests.
Source: From Karl Terzaghi, Ralph B. Peck, and Gholamreza Mesri, *Soil Mechanics in Engineering Practice,* 3rd ed., John Wiley & Sons, Inc., New York, 1996. Copyright © 1996, by John Wiley & Sons, Inc. Reprinted by permission of John Wiley & Sons, Inc.

3–7 VANE TEST

The field vane test is a fairly simple test that can be used to determine in-place shear strength for soft clay soils—particularly those clay soils that lose part of their strength when disturbed (sensitive clays)—without taking an undisturbed sample. A vane tester (see Figure 3–18) is made up of two thin metal blades attached to a vertical shaft. The test is carried out by pushing the vane tester into the soil and then applying a torque to the vertical shaft. The clay's cohesion can be computed by using the following formula [2, 12]:

$$c = \frac{T}{\pi[(d^2h/2) + (d^3/6)]} \tag{3–9}$$

FIGURE 3–18 Vane tester [2].

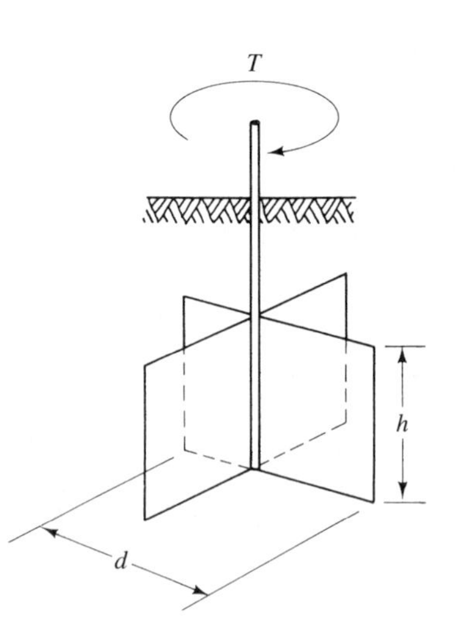

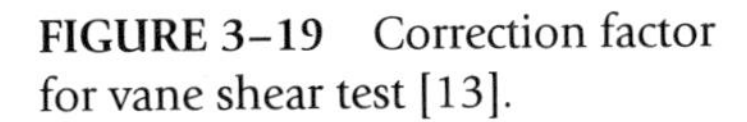

FIGURE 3–19 Correction factor for vane shear test [13].

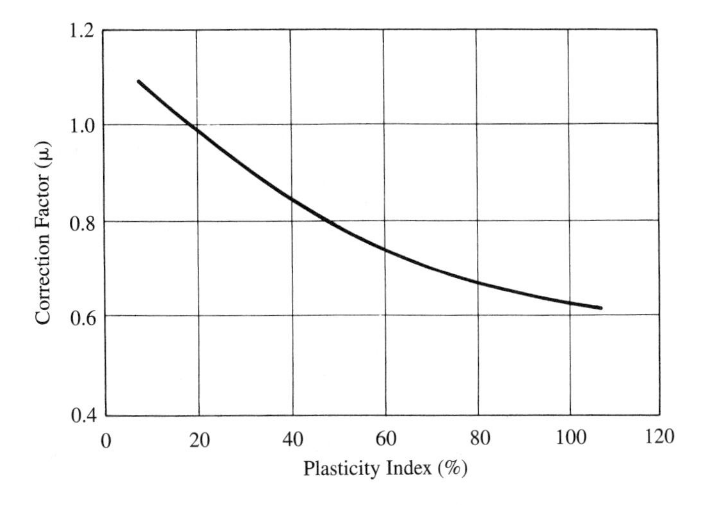

where c = cohesion of the clay, lb/ft^2 or N/m^2
T = torque required to shear the soil, ft-lb or m · N
d = diameter of vane tester, ft or m
h = height of vane tester, ft or m

Bjerrum [13] found a tendency of the vane test to overestimate cohesion in high plasticity clays and developed an empirical relationship for determining a correction factor. This relationship is shown in Figure 3–19, where a correction factor, μ, can be determined if the clay's plasticity index is known.

It should be emphasized that the field vane test is suitable for use only in soft or sensitive clays. Also, no soil sample is obtained for subsequent examination and testing when a field vane test is performed.

EXAMPLE 3–5

Given

A vane tester with diameter and height of 3.625 in. (0.3021 ft) and 7.25 in. (0.6042 ft), respectively, requires a torque of 17.0 ft-lb to shear a clayey soil, the plasticity index of which is 48%.

Required

This soil's cohesion.

Solution

By Eq. (3–9),

$$c = \frac{T}{\pi[(d^2h \div 2) + (d^3 \div 6)]} \tag{3–9}$$

$$c = \frac{17.0 \text{ ft/b}}{\pi\left[\dfrac{(0.3021 \text{ ft})^2(0.6042 \text{ ft})}{2} + \dfrac{(0.3021 \text{ ft})^3}{6}\right]} = 168 \text{ lb/ft}^2$$

From Figure 3–19, with a plasticity index of 48%, a correction factor, μ, of 0.80 is obtained. Hence,

$$c_{\text{corrected}} = (0.80)(168 \text{ lb/ft}^2) = 134 \text{ lb/ft}^2$$

3–8 GEOPHYSICAL METHODS OF SOIL EXPLORATION

Borings and test pits (Section 3–3) afford definitive subsurface exploration. They can, however, be both time consuming and expensive. In addition, they give subsurface conditions only at boring or test pit locations, leaving vast areas in between for which conditions must be interpolated or estimated.

Geophysical methods, which are widely used in highway work and in other applications, can be implemented more quickly and less expensively and can cover greater areas more thoroughly. They tend, however, to yield less definitive results requiring more subjective interpretation by the user. Accordingly, a number of borings are still required to obtain soil samples from which accurate determinations of soil properties can be made in order to verify and complement results determined by geophysical methods.

Two particular geophysical methods—*seismic refraction* and *electrical resistivity*—are discussed in this section. In the former, resistance to flow of a seismic wave through soil is measured; in the latter, resistance of soil to movement of an electrical current is determined. Using values obtained therefrom, a specialist can interpret the depth to and thickness of different soil strata and estimate, with the aid of supplemental borings, some of the engineering properties of the subsurface material.

Seismic Refraction Method

The seismic refraction method is based on the fact that velocities of seismic waves traveling through soil and rock material are related to the material's density and elasticity. In general, the denser the material, the greater will be the velocity of seismic waves moving through it. In carrying out this method, seismic (sound or vibration) waves are created within the soil at a particular location. Ordinarily, these waves are produced either by exploding small charges of dynamite several feet below the ground surface or by striking a heavy hammer against a steel plate. A detector, known as a *geophone,* placed some known (or measurable) distance from the shock source, detects the presence of a wave, and a timing device measures the time required for the wave to travel from the point of impact to the point of detection.

In conducting a seismic refraction field survey, a series of geophone readings is obtained at different distances along a straight line from the point of impact. For detection points relatively close to the impact point, the first shock to reach the geophones travels from the impact point through more direct surface routes to the detection points (see Figure 3–20).

When a harder layer, say rock, underlies the surficial soil layer, a seismic wave traveling downward from the point of impact into the rock layer is refracted to travel longitudinally through the upper part of the rock layer and eventually back to the ground surface (through the surficial layer) to be recorded by the geophones (Figure 3–20). Because seismic wave velocity is much greater through the rock layer than through the surficial soil, for geophones located relatively far from the impact point, the refracted wave will reach the geophone more quickly than the direct wave. The time required for the first shock to reach each geophone is plotted as a function of distance from the shock source, as in Figure 3–21. The wave to the first few geophones closer to the shock source travels directly through the surficial layer; therefore, the slope of the time versus distance graph is inversely equivalent to velocity—that is,

$$v_1 = \frac{L_2 - L_1}{t_2 - t_1} \tag{3–10}$$

where v_1 = wave's velocity through the surficial soil layer (i.e., reciprocal of the slope of line 1 as shown in Figure 3–21)
L_1 and L_2 = distances from shock source to geophones Nos. 1 and 2, respectively (Figure 3–20)
t_1 and t_2 = times required for the first shock wave to reach geophones Nos. 1 and 2, respectively

Similarly, v_2 is the reciprocal of the slope of line 2 as shown in Figure 3–21. The thickness of stratum H_1 is given by

$$H_1 = \frac{L}{2}\sqrt{\frac{v_2 - v_1}{v_2 + v_1}} \tag{3–11}$$

where H_1 = depth of thickness of the upper layer (Figure 3–20)
L = distance taken from the time versus distance graph where the two slopes intersect (Figure 3–21)

As indicated in Table 3–5, wave velocities range from about 800 ft/s (244 m/s) in loose sand above the water table to 20,000 ft/s (6096 m/s) in granite and unweathered gneiss. This wide range makes possible a general assessment of the characteristics of material encountered.

Seismic refraction can be used to estimate depths to successively harder strata, but it will not determine softer strata below harder strata. It can also be used to find the depth to groundwater and to locate sinkholes. However, where boundaries are irregular or poorly defined, interpretation of the results of seismic refraction may be questionable.

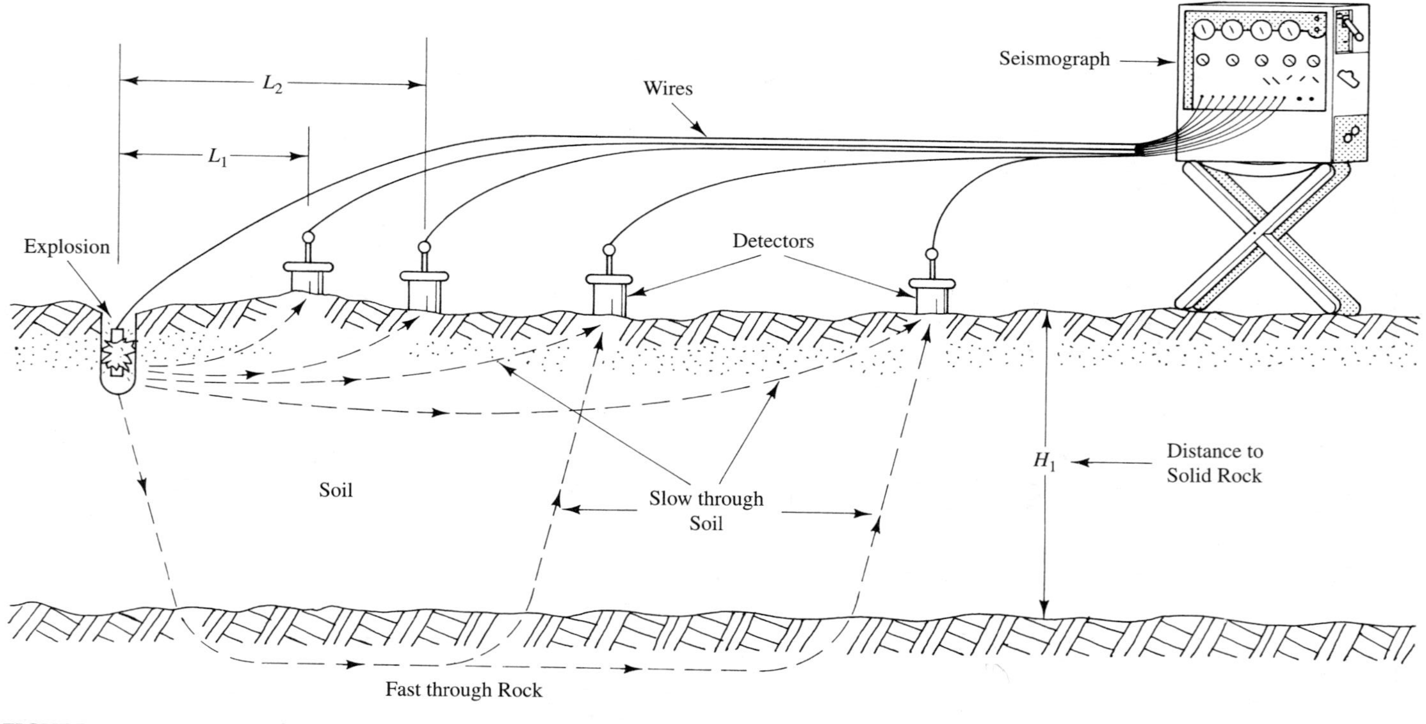

FIGURE 3–20 Seismic refraction test [14].

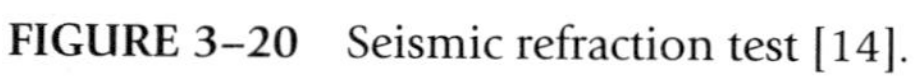

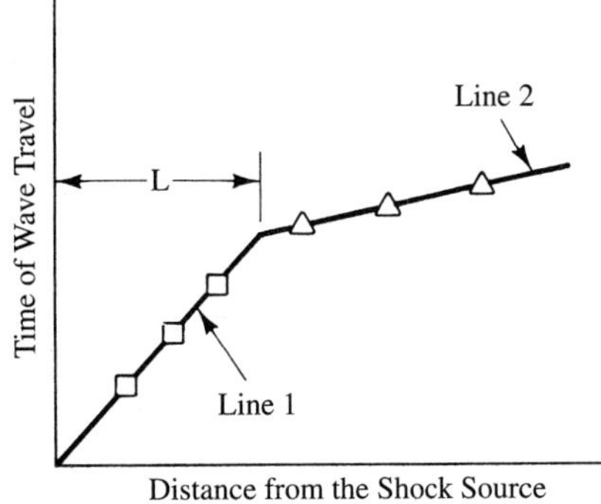

FIGURE 3–21 Time of wave travel as a function of distance from shock source in seismic refraction method.

TABLE 3–5
Representative Velocity Values (ft/s)[1–3]

Unconsolidated materials	
Most unconsolidated materials	Below 3000
Soil	
Normal	800–1500
Hard packed	1500–2000
Water	5000
Loose sand	
Above water table	800–2000
Below water table	1500–4000
Loose mixed sand and gravel, wet	1500–3500
Loose gravel, wet	1500–3000
Consolidated materials	
Most hard rocks	Above 8000
Coal	3000–5000
Clay	3000–6000
Shale	
Soft	4000–7000
Hard	6000–10,000
Sandstone	
Soft	5000–7000
Hard	6000–10,000
Limestone	
Weathered	As low as 4000
Hard	8000–18,000
Basalt	8000–13,000
Granite and unweathered gneiss	10,000–20,000
Compacted glacial tills, hardpan, cemented gravels	4000–7000
Frozen soil	4000–7000
Pure ice	10,000–12,000

[1]Courtesy of Soiltest, Inc.
[2]Occasional formations may yield velocities that lie outside these ranges.
[3]1 ft/s = 0.3048 m/s.

Electrical Resistivity Method

As indicated initially in this section, resistance to movement of an electrical current through soil is determined in the electrical resistivity method. The premise for using this technique in subsurface investigations is that electrical resistance varies significantly enough among different types of soil and rock materials to allow identification of specific types if their resistivities are known.

A soil's resistivity generally varies inversely with its water content and dissolved ion concentration. Because clayey soils exhibit high dissolved ion concentrations, wet clayey soils have the lowest resistivities of all soil materials—as low as 5 ohm-ft (1.5 ohm · m). Coarse, dry sand and gravel deposits and massive bedded and hard bedrocks have the highest resistivities—over 8000 ohm-ft (2438 ohm · m). Table 3–6 gives the resistivity correlation for various types of soil materials.

One specific procedure for conducting an electrical resistivity field survey utilizes four equally spaced electrodes. (This is known as the *Wenner method.*) The four electrodes are placed in a straight line spaced distance D apart, as illustrated in Figure 3–22. An electrical current is supplied (by a battery or small generator) through the outer electrodes (Figure 3–22); its value is measured by an ammeter. The voltage drop in the soil material within the zone created by the electrodes' electric field is measured between the two inner electrodes by a voltmeter (Figure 3–22). The soil material's electrical resistivity can be computed by using the following equation:

$$\rho = 2\pi D \frac{V}{I} = 2\pi D R \tag{3–12}$$

where ρ = resistivity of the soil material, ohm-ft or ohm · m
D = electrode spacing, ft or m
V = voltage drop between the two inner electrodes, volts
I = current supplied through the outer electrodes, amperes
R = resistance, ohms

TABLE 3–6
Resistivity Correlation[1]

Ohm-ft	2π Ohm · cm	Types of Materials
5–10	1000–2000	Wet to moist clayey soils
10–50	3000–15,000	Wet to moist silty clay and silty soils
50–500	15,000–75,000	Moist to dry silty and sandy soils
500–1000	30,000–100,000	Well-fractured to slightly fractured bedrock with moist-soil–filled cracks
1000	100,000	Sand and gravel with silt
1000–8000	100,000–300,000	Slightly fractured bedrock with dry-soil–filled cracks; sand and gravel with layers of silt
8000 (plus)	300,000 (plus)	Massive bedded and hard bedrock; coarse, dry sand and gravel deposits

[1]Courtesy of Soiltest, Inc.

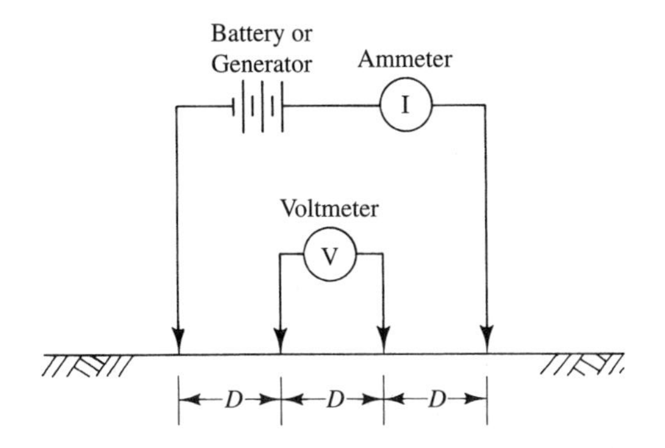

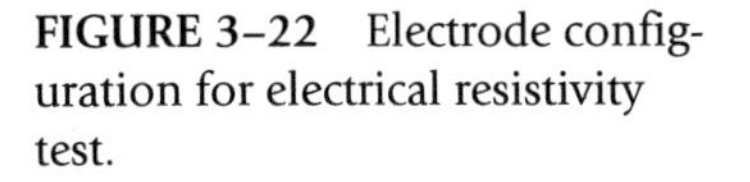

FIGURE 3–22 Electrode configuration for electrical resistivity test.

The zone created by the electrodes' electrical field extends downward to a depth approximately equal to the electrode spacing (i.e., D in Figure 3–22). Consequently, the depth of subsurface material included in a given measurement is approximately equal to the spacing between electrodes. The resistivity determined by this method [computed by Eq. (3–12)] is actually a weighted mean value of all soil material within the zone.

A single application of the procedure just outlined would give an indication of the "average" type of subsurface material within the applicable zone. To determine depths of strata of different resistivities, the procedure is repeated for successively increasing electrode spacings (see Figure 3–23). Because the applicable zone's depth varies directly with electrode spacing, data obtained from successively increasing electrode spacings should indicate changes in resistivity with depth, which in turn serves to locate different soil strata.

Resistivity data can be analyzed by plotting $\Sigma\rho$ (summation of soil resistivity values) versus electrode spacing (D) for increasing electrode spacings. Such a plotting is

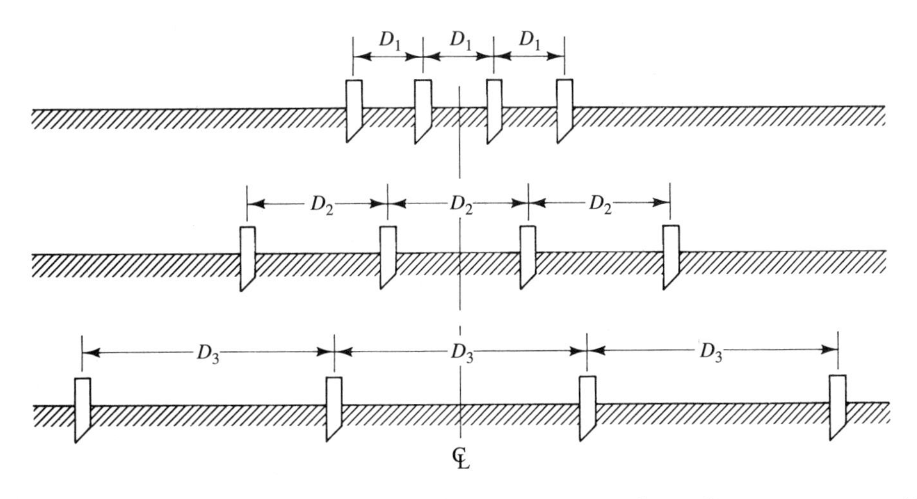

FIGURE 3–23 Representative electrode positions during a sequence of sounding measurements (the position of the center of the spread is fixed).
Source: Courtesy of Soiltest, Inc.

illustrated in Figure 3–24. A straight-line plot indicates a constant soil resistivity (and therefore the same soil type) within the depth range for which the plot is straight. Furthermore, the slope of the straight line is equal to ρ_1/D, and ρ_1 gives the resistivity in the upper layer. Using this value of resistivity, one can estimate the type of soil within this layer. If a different soil type is encountered as additional tests are performed at increasing electrode spacings, a second straight-line plot should result with a slope equal to ρ_2/D, with ρ_2 giving the resistivity of the lower layer, from which the type of soil can be evaluated. Furthermore, the intersection of the two straight lines gives the approximate depth of the boundary between the two layers (Figure 3–24).

The electrical resistivity method can be used to indicate subsurface variations where a hard layer underlies a soft layer; however, unlike the seismic refraction method, it can also be used where a soft layer underlies a hard layer. The electrical re-

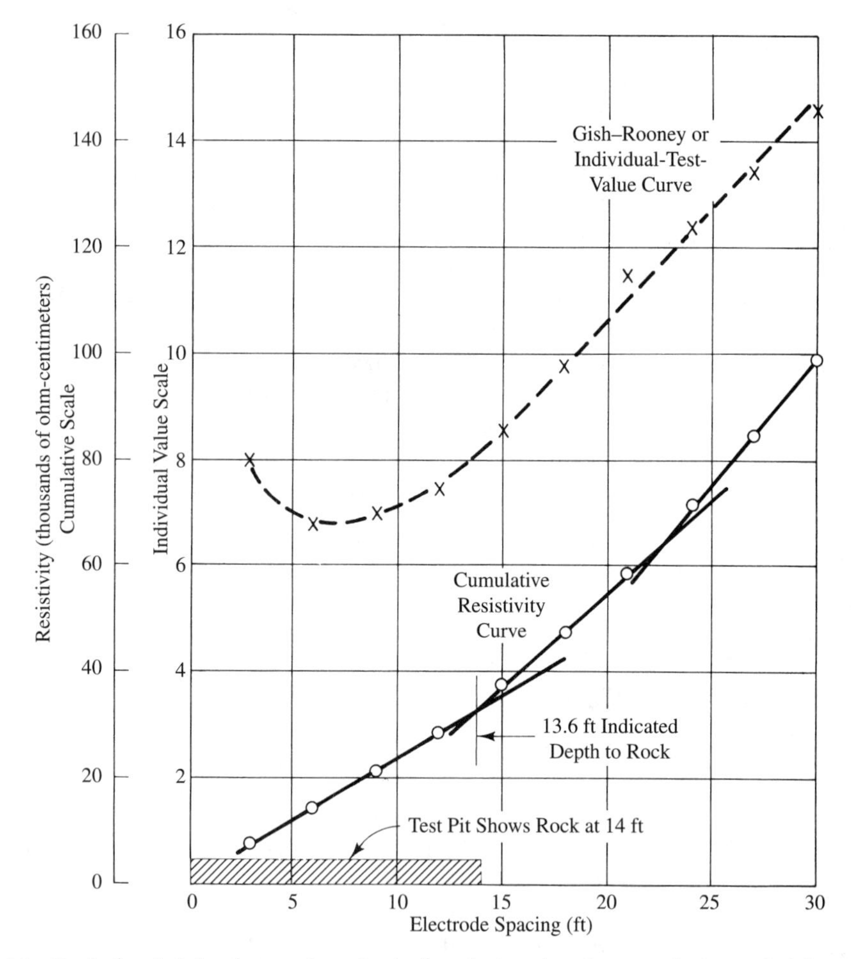

FIGURE 3–24 Typical resistivity data and method of analysis using the cumulative resistivity curve [14].

sistivity method can be used not only to estimate depth to strata of different resistivities, but also to find depth to groundwater and to locate masses of dry sands, gravels, and rock. It should be realized that errors in interpretation can occur because soil resistivity varies with moisture content and identifies soil only indirectly. Hence, the electrical resistivity method should always be used with confirmatory drilling.

As related at the beginning of this section, geophysical methods afford relatively rapid and low-cost subsurface exploration as compared with test borings. However, dependable results from geophysical methods require experienced and skillful interpretation of test data. Geophysical methods have some disadvantages. The greatest is that, because of the subjectivity involved in analyzing, interpreting, and drawing conclusions from collected data, the resulting picture of the area's subsurface features may not be entirely accurate. Accordingly, geophysical methods should always be used in conjunction with test borings—either using sufficient test borings to verify results of geophysical methods or using geophysical methods to provide intermediate subsurface information between adjacent test borings.

3–9 RECORD OF SOIL EXPLORATION

It is of utmost importance that complete and accurate records be kept of all data collected. Boring, sampling, and testing are often costly undertakings, and failure to keep good, accurate records not only is senseless, but also may be dangerous.

To begin with, a good map giving specific locations of all borings should be available. Each boring should be identified (by number, for example), and its location documented by measurement to permanent features. Such a map is illustrated in Figure 3–25.

For each boring, all pertinent data should be recorded in the field on a boring log sheet. Normally, these sheets are preprinted forms containing blanks for filling in appropriate data. An example of a boring log is given in Figure 3–26.

Soil data obtained from a series of test borings can best be presented by preparing a geologic profile, which shows the arrangement of various layers of soil as well

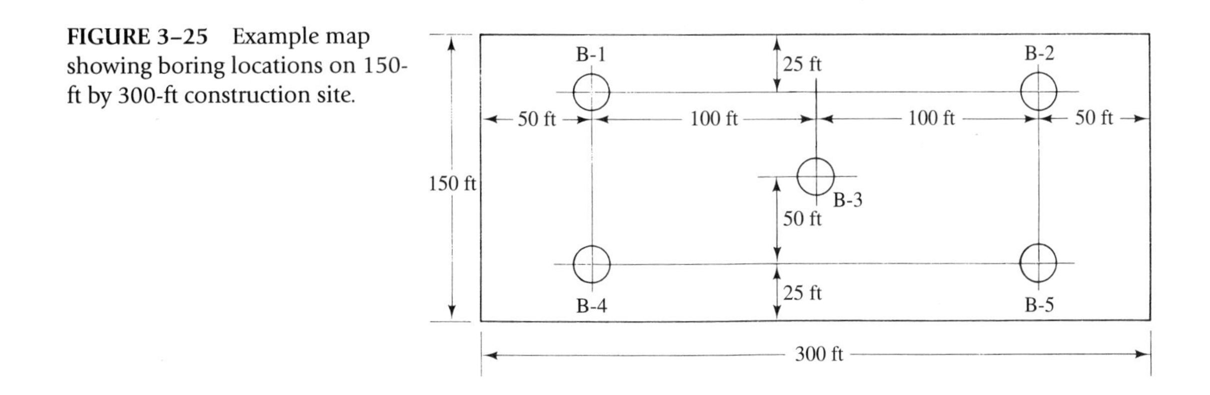

FIGURE 3–25 Example map showing boring locations on 150-ft by 300-ft construction site.

Sheet 1 of 1

ABC DRILLING COMPANY, INC. BORING NO. 5
NEWARK, NEW YORK ORD. ELEV. 372.4

PROJECT: Job No. 459
Name Eureka Warehouse
Address Illion, New York

CASING (Size & Type) 2½" Drive Pipe
SAMPLE SPOON (Size & Type) 2" O.D.S.S.
HAMMER (Csg): Wt. 250 lb, Drop 24 in.
(Spoon): Wt. 140 lb, Drop 30 in.
DATE: Started 7/28/– Completed 7/29/– Driller Henry James

GROUND WATER OBSERVATIONS

Date	Time	Depth	Casing at
7/29/--	3:00 PM	18'3"	15'0"
"	4:00 PM	12'0"	10'0"
"	4:30 PM	8'0"	5'0"
7/30/--	8:30 PM	7'0"	Out

DEPTH FT.	BLOWS CSG.	BLOWS SPOON		Samples	
0–1	2				Black and grey moist FILL: cinders, brick and silt
1–2	16	11 / 8			
2–3	9	4	12	S #1, 1'–2'6"	3'0"
3–4	3	1 / 1			
4–5	3	2	3	S #2, 3'–4'6"	Black PEAT
5–6	3	P / 1			
6–7	5	1	2	S #3, 5'–6'6"	6'0"
7–8	6	3 / 6			
8–9	8	5	11	S #4, 7'–8'6"	Grey moist SILT with embedded fine gravel, trace of fine sand
9–10	9				
10–11	3	4 / 8			
11–12	8	6	14	S #5, 10'–11'6"	12'6"
12–13	15	15			Weathered SHALE
13–14	32	18 / 21			
14–15	78		39	S #6, 12'6"–14'	TOP OF ROCK 15'0"
15–20				Core Boring Series M— double tube core barrel, 2 in. diameter bit	Weathered grey SHALE Run #1, 15'0" – 20'0" Recovered 30" – 50% Lost water @ 16'6" 20'0"
20–25					SHALE and SANDSTONE Run #2, 20'0" – 25'0" Recovered 56" – 93% Steady resistance while drilling 25'0"

FIGURE 3–26 Boring log sheet [15].

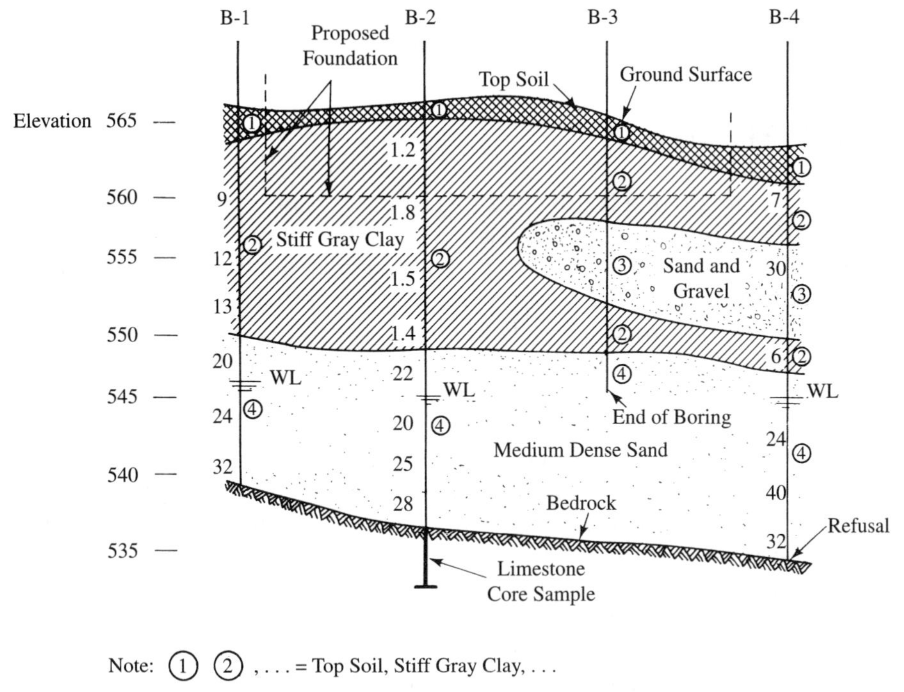

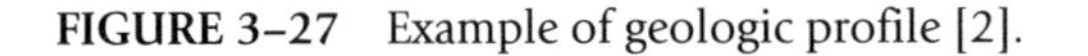
FIGURE 3–27 Example of geologic profile [2].

as the groundwater table, existing and proposed structures, and soil properties data (SPT values, for example). Each borehole is identified and indicated on the geologic profile by a vertical line. An example of a geologic profile is shown in Figure 3–27.

A geologic profile is prepared by indicating on each borehole on the profile (i.e., each vertical line representing a borehole) the data obtained by boring, sampling, and testing. From these data, soil layers can be sketched in. Obviously, the more boreholes and the closer they are spaced, the more accurate the resulting geologic profile.

3–10 CONCLUSION

The subject of this chapter should be considered as one of the most important in this book. Analysis of soil and design of associated structures are of questionable value if the soil exploration data are not accurately determined and reported.

The authors hope this chapter will give the reader an effective introduction to actual soil exploration. However, learning to conduct soil exploration well requires much practice and varied experience under the guidance of experienced practitioners. Not only is it a complex science, it is a difficult art.

3–11 PROBLEMS

3–1. A 4-ft square footing is subjected to a contact pressure of 6000 lb/ft^2. The wet unit weight of the cohesive soil supporting the footing is estimated to be 118 lb/ft^3, and groundwater is known to be at a great depth. Determine the minimum depth of test boring based on the criterion that test borings in cohesive soils should be carried at least to a depth where the increase in stress due to the foundation loading is less than 10% of the effective overburden pressure.

3–2. A standard penetration test (SPT) was performed at a depth of 10 ft in sand of unit weight 120 lb/ft^3. The *N*-value was found to be 26. Determine the corrected *N*-value by all three methods presented in this chapter.

3–3. Rework Problem 3–2 if groundwater is located 8 ft below the ground surface.

3–4. An SPT was performed at a depth of 7 m in sand of unit weight 20.40 kN/m^3. The *N*-value was found to be 22. Compute the corrected *N*-value by all three methods presented in this chapter.

3–5. Rework Problem 3–4 if groundwater is located 2 m below the ground surface.

3–6. A field vane test was performed in a soft, sensitive clay layer. The vane tester's diameter and height are 4 and 8 in., respectively. The torque required to shear the clay was 61 ft-lb. Determine the clayey soil's cohesion if its plasticity index is known to be 40%.

3–7. Soil exploration was conducted at a construction site by seismic refraction, with field readings obtained as listed next:

Distance (ft)	Time (ms)
20	21
40	42
60	62.25
80	83
100	86.75
120	88.25
140	89.25
160	90.75
180	93

Estimate the thickness and type of material of the first soil layer and the type of material in the underlying second layer.

3–8. Soil exploration was conducted at a construction site by the electrical resistivity method, with field data obtained as follows:

Electrode Spacing (ft)	Resistance Readings (ohms)
10	12.73
20	2.79
30	1.46
40	1.15
50	1.05
60	0.84
70	1.21
80	1.00
90	0.97
100	0.95

Estimate the thickness and type of material of the first soil layer and the type of material in the underlying second layer.

References

[1] David F. McCarthy, *Essentials of Soil Mechanics and Foundations,* Reston Publishing Company, Inc., Reston, Va., 1977.

[2] Wayne C. Teng, *Foundation Design,* Prentice-Hall, Inc., Englewood Cliffs, N.J., 1962.

[3] Richard D. Barksdale and Milton O. Schreiber, "Calculating Test-Boring Depths," *Civil Eng., ASCE,* **49**(8), 74–75 (1979).

[4] Karl Terzaghi, Ralph B. Peck, and Gholamreza Mesri, *Soil Mechanics in Engineering Practice,* 3rd ed., John Wiley & Sons, Inc., New York, 1996.

[5] R. H. Karol, *Soils and Soil Engineering,* Prentice-Hall, Inc., Englewood Cliffs, N.J., 1960.

[6] Karl Terzaghi and Ralph B. Peck, *Soil Mechanics in Engineering Practice,* John Wiley & Sons, Inc., New York, 1967. Copyright © 1967, John Wiley & Sons, Inc. Reprinted by permission of John Wiley & Sons, Inc.

[7] Ralph B. Peck, Walter E. Hansen, and Thomas H. Thornburn, *Foundation Engineering,* 2nd ed., John Wiley & Sons, Inc., New York, 1974. Copyright © 1974, by John Wiley & Sons, Inc. Reprinted by permission of John Wiley & Sons, Inc.

[8] Abdel Rahman Sadik Said Bazaraa, "Use of the Standard Penetration Test for Estimating Settlements of Shallow Foundations or Sand," Ph.D. thesis, University of Illinois, 1967.

[9] Samson S. C. Liao and Robert V. Whitman, "Overburden Correction Factors for SPT in Sand," *J. Geotech. Eng. Div. ASCE,* **112**(3), 373–377 (1986).

[10] A. W. Skempton, "Standard Penetration Test Procedures and the Effects in Sands of Overburden Pressure, Relative Density, Particle Size, Ageing, and Overconsolidation," *Geotechnique,* **36**(3), 425–447 (1986).

[11] *1981 Annual Book of ASTM Standards,* ASTM, Philadelphia, 1981. Copyright, American Society for Testing and Materials, 1916 Race Street, Philadelphia, PA 19103. Reprinted with permission.

[12] A. W. Skempton and A. W. Bishop, "The Measurement of the Shear Strength of Soils," *Geotechnique,* **2**(2) (1950).

[13] L. Bjerrum, "Problems of Soil Mechanics and Construction on Soft Clays," *8th Int. Conf. SMFE,* Moscow, 1973. Reprinted in *Norwegian Geotechnical Institute Publ. No. 100,* Oslo, 1974.

[14] R. Woodward Moore, "Geophysics Efficient in Exploring the Subsurface," *J. Soil Mech. Found. Div., Proc. ASCE,* **SM3** (June 1961).

[15] B. K. Hough, *Basic Soils Engineering,* 2nd ed., The Ronald Press Company, New York, 1969. Copyright © 1969, by John Wiley & Sons, Inc. Reprinted by permission of John Wiley & Sons, Inc.

4

Soil Compaction and Stabilization

4–1 DEFINITION AND PURPOSE OF COMPACTION

The general meaning of the verb *compact* is "to press closely together." In soil mechanics, it means to press the soil particles tightly together by expelling air from the void space. Compaction is normally produced deliberately and proceeds rapidly during construction, often by heavy compaction rollers. This is in contrast to *consolidation* (Chapter 7), which also results in a reduction of voids but which is caused by extrusion of water (rather than air) from the void space. Also, consolidation is not rapid.

Compaction of soil increases its density and produces three important effects: (1) an increase in the soil's shear strength, (2) a decrease in future settlement of the soil, and (3) a decrease in its permeability [1]. These three effects are beneficial for various types of earth construction, such as highways, airfields, and earthen dams; and, as a general rule, the greater the compaction, the greater these benefits will be. Compaction is actually a rather cheap and effective way to improve the properties of a given soil.

Compaction is quantified in terms of a soil's dry unit weight, γ_d, which can be computed in terms of wet unit weight, γ, and moisture content, w (expressed as a decimal), by

$$\gamma_d = \frac{\gamma}{1 + w} \tag{4–1}$$

In most cases, dry soils can be best compacted (and thus a greater density achieved) if for each soil a certain amount of water is added to it. In effect, water acts as a lubricant and allows soil particles to be packed together better. If, however, too much water is added, a lesser density results. Thus, for a given compactive effort, there is a particular moisture content at which dry unit weight is greatest and compaction best. This moisture content is called the *optimum moisture content*, and the associated dry unit weight is known as the *maximum dry unit weight.*

Usual practice in a construction project is to perform laboratory compaction tests (covered in Section 4–2) on representative soil samples from the construction site to determine the soil's optimum moisture content and maximum dry unit weight. This maximum dry unit weight is used by the designer in specifying design shear strength, resistance to future settlement, and permeability characteristics. The soil is then compacted by field compaction methods (covered in Section 4–4) until the laboratory maximum dry unit weight (or an acceptable percentage of it) has been achieved. In-place soil unit weight tests (covered in Section 4–6) are used to determine if and when the laboratory maximum dry unit weight (or an acceptable percentage thereof) has been reached. Section 4-7 covers field control of compaction.

4–2 LABORATORY COMPACTION TESTS

As related in the preceding section, laboratory compaction tests are performed to determine a soil's optimum moisture content and maximum dry unit weight. Compaction test equipment, shown in Figure 4–1, consists of a baseplate, collar, and mold, in which soil is placed, and a hammer that is raised and dropped freely onto

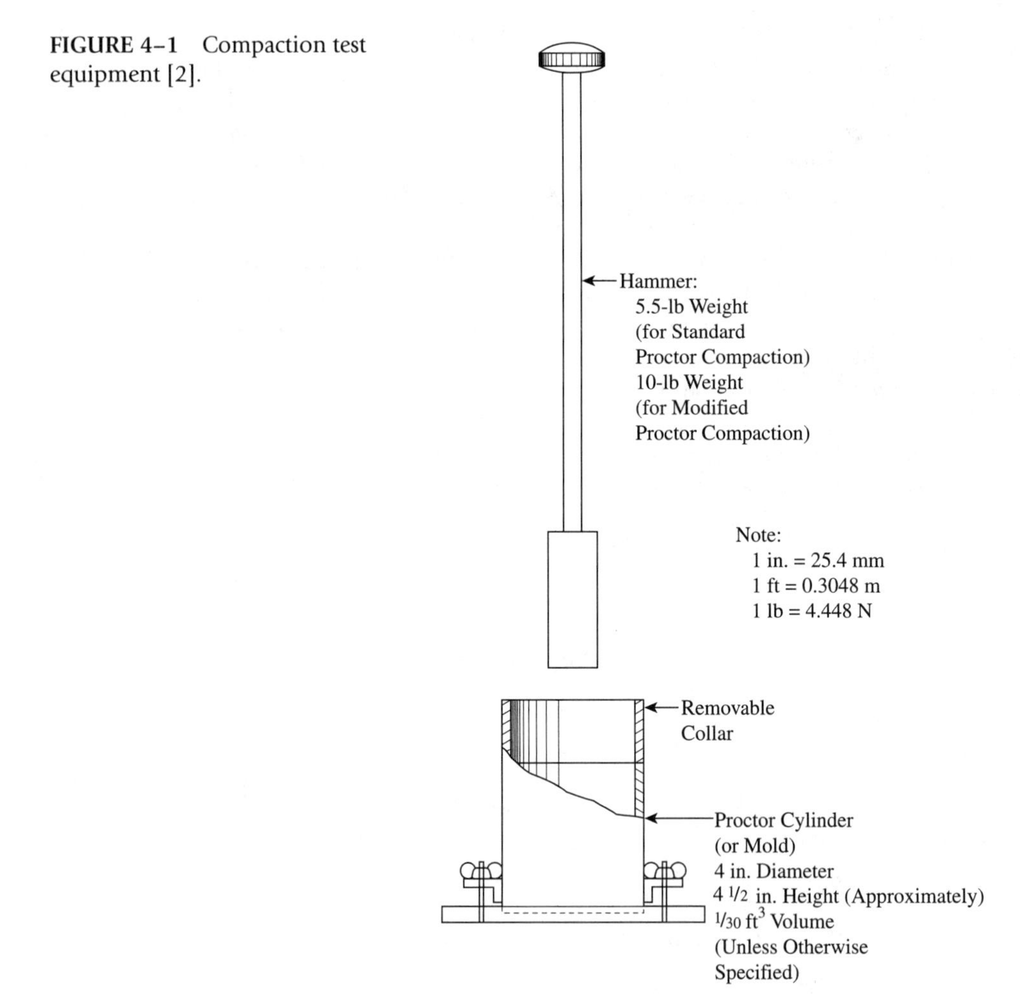

FIGURE 4–1 Compaction test equipment [2].

the soil in the mold. The mold's size and the hammer's weight and drop distance are standardized, with several variations in size and weight available.

Table 4–1 summarizes specifications for compaction testing equipment, compaction effort, and sample fractionation for six test designations. The three on the left side of the table are designated ASTM D 698. Method A under these designations is known as the original *Standard Proctor* compaction test. The three test designations on the right side of the table are designated ASTM D 1557. Method A under these designations is known as the original *Modified Proctor* compaction test and was developed subsequent to the Standard Proctor test to obtain higher values of dry unit weights or densities. It was developed in response to the need for higher unit weights or densities of airfield pavement subgrades, embankments, earthen dams, and so forth, and for compacted soil that is to support large and heavy structures. It can be noted from Table 4–1 that the Standard Proctor test utilizes a 5.5-lb (24.5-N) hammer, which is dropped 12 in. (305 mm), whereas the Modified Proctor test uses a 10-lb (44.5-N) hammer, which is dropped 18 in. (457 mm).

To carry out a laboratory compaction test, the soils engineer allows a soil sample from the field to dry until it becomes friable under a trowel. The soil sample may be dried in air or a drying oven. If an oven is used, its temperature should not exceed 60°C (140°F). After drying, a series of at least four specimens is prepared by adding increasing amounts of water to each sample so that the moisture contents will bracket the optimum moisture content. After a specified curing period, each prepared specimen is placed, in turn, in a compaction mold (with collar attached) and compacted in layers by dropping the hammer onto the specimen in the mold a certain distance and specified number of uniformly distributed blows per layer. This results in a specific energy

TABLE 4–1
Summary of Specifications for Compaction Testing Equipment, Compaction Effort, and Sample Fractionation[1]

	Test Designation					
	ASTM D 698			*ASTM D 1557*		
	Method A[2]	*Method* B	*Method* C	*Method* A[3]	*Method* B	*Method* C
Hammer weight (lb)	5.5	5.5	5.5	10	10	10
Drop (in.)	12	12	12	18	18	18
Size of mold						
Diameter (in.)	4	4	6	4	4	6
Height (in.)	4.584	4.584	4.584	4.584	4.584	4.584
Volume (ft^3)	1/30	1/30	1/13.33	1/30	1/30	1/13.33
Number of layers	3	3	3	5	5	5
Blows per layer	25	25	56	25	25	56
Fraction tested	−No. 4	−3/8 in.	−3/4 in.	−No. 4	−3/8 in.	−3/4 in.

[1] 1 lb = 4.448 N; 1 in. = 25.4 mm; 1 ft^3 = 0.02832 m^3.
[2] This is the original Standard Proctor test.
[3] This is the original Modified Proctor test.

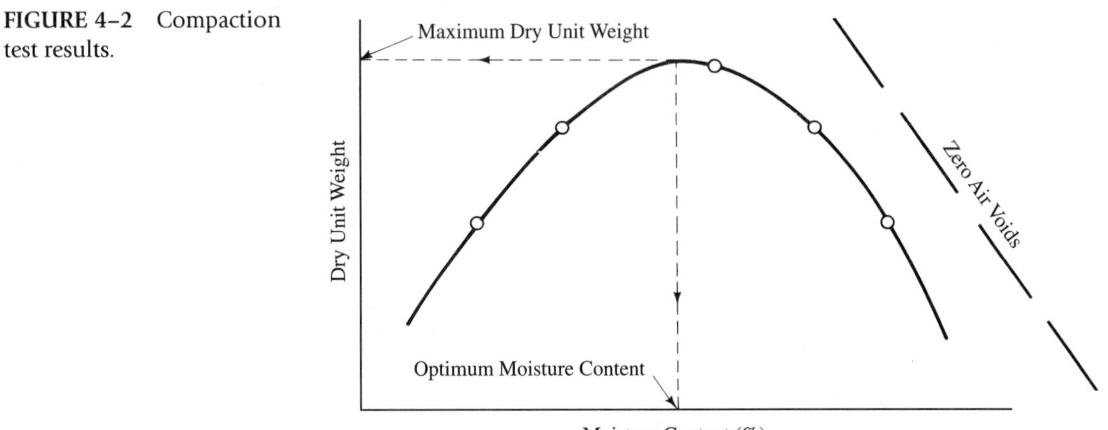

FIGURE 4–2 Compaction test results.

exertion per unit volume of soil. Upon completion of each compaction, the attached collar is removed and the compacted soil trimmed until it is even with the top of the mold. The compacted soil specimen's wet unit weight is then determined by dividing the weight of compacted soil in the mold by the soil specimen's volume, which is the volume of the mold. The compacted soil is subsequently removed from the mold, and its moisture content determined. With the compacted soil's wet unit weight and moisture content known, its dry unit weight is computed using Eq. (4–1).

A plot made of the soil's moisture content versus dry unit weight for the data collected as described in the preceding paragraph will be of a form similar to the curve shown in Figure 4–2. The coordinates of the point at the curve's peak give the soil's maximum dry unit weight and optimum moisture content. Presumably, this gives the maximum expected dry unit weight—the dry unit weight to be used by the designer and to be striven for in the field compaction. To achieve this maximum dry unit weight, field compaction should be done at or near the optimum moisture content.

In Figure 4–2, the right side of the moisture content versus dry unit weight curve roughly parallels the dashed line labeled "Zero air voids." This line represents the dry unit weight when saturation is 100% (i.e., the soil's entire volume is water and solids). This line actually represents, in theory, the upper limit on unit weight at any moisture content. For this reason, the zero-air-voids line is often included on moisture content versus dry unit weight curves. It can be determined from the following equation:

$$\gamma_{ZAV} = \frac{G_s \gamma_w}{1 + wG_s} \tag{4-2}$$

where γ_{ZAV} = dry unit weight at zero air voids
G_s = specific gravity of solids
γ_w = unit weight of water
w = moisture content (expressed as a decimal)

Example 4–1 illustrates computation of the unit weight of a specimen of a laboratory-compacted soil. Example 4–2 illustrates determination of the maximum dry unit weight and optimum moisture content, as the result of a laboratory compaction test.

EXAMPLE 4–1

Given

1. The combined weight of a mold and the specimen of compacted soil it contains is 8.63 lb.
2. The mold's volume is 1/30 ft^3.
3. The mold's weight is 4.35 lb.
4. The specimen's water content is 10%.

Required

1. Wet unit weight of the specimen.
2. Dry unit weight of the specimen.

Solution

1. From Eq. (2–10),

$$\gamma = \frac{W}{V}$$

$$\gamma = \frac{8.63\text{ lb} - 4.35\text{ lb}}{\frac{1}{30}\text{ ft}^3} = 128.4\text{ lb/ft}^3 \tag{2–10}$$

2. From Eq. (4–1),

$$\gamma_d = \frac{\gamma}{1 + w} \tag{4–1}$$

$$\gamma_d = \frac{128.4\text{ lb/ft}^3}{1 + 0.10} = 116.7\text{ lb/ft}^3$$

EXAMPLE 4–2

Given

A set of laboratory compaction test data and results is tabulated as follows. The test was conducted in accordance with the ASTM D 698 Standard Proctor test.

Determination Number	1	2	3	4	5
Dry unit weight (lb/ft^3)	112.2	116.7	118.3	115.2	109.0
Moisture content (%)	7.1	10.0	13.4	16.7	20.1

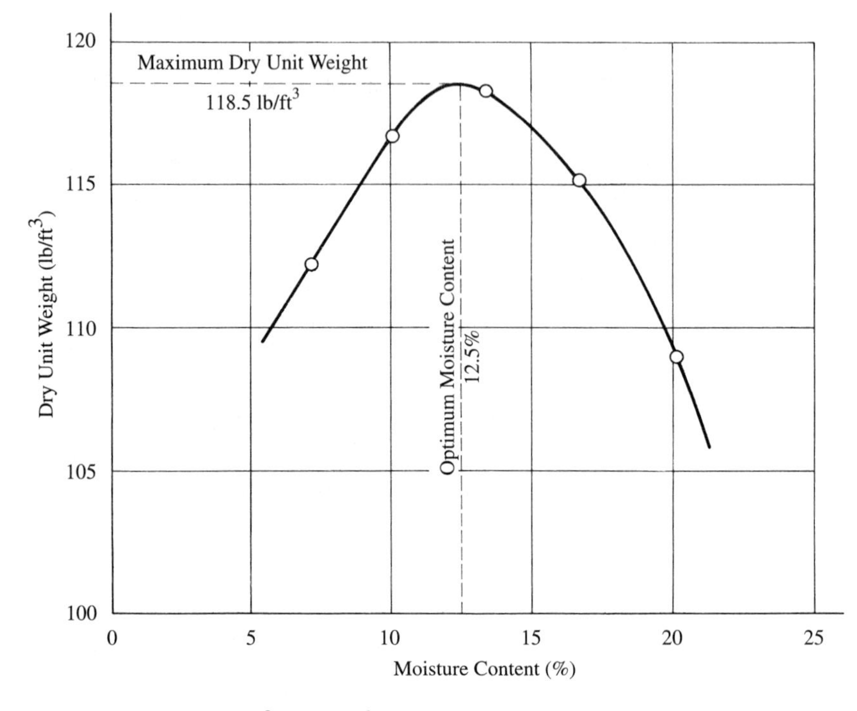

FIGURE 4–3 Proctor curve for Example 4–2.

Required

1. Plot a Proctor curve (i.e., dry unit weight versus moisture content).
2. Determine the soil's maximum dry unit weight and optimum moisture content.

Solution

1. See Figure 4–3.
2. From Figure 4–3,

$$\text{Maximum dry unit weight} = 118.5 \text{ lb/ft}^3$$
$$\text{Optimum moisture content} = 12.5\%$$

4–3 FACTORS AFFECTING COMPACTION OF SOIL

Several factors affect the compaction of soil. These might be categorized as moisture content, compaction effort, and type of soil. Section 4–2 covered the influence of moisture content on the degree of compaction achieved by a given soil sample. This section discusses the effect of compaction effort and soil type on the compaction of soil.

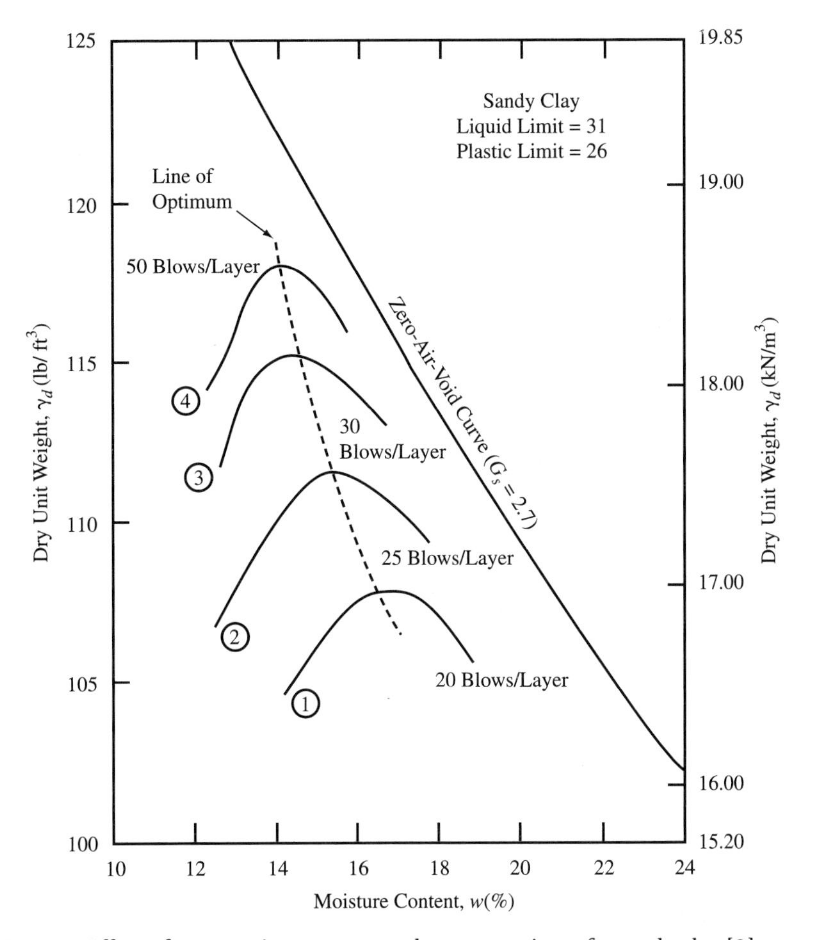

FIGURE 4–4 Effect of compaction energy on the compaction of a sandy clay [3].
Source: From Braja M. Das, *Principles of Geotechnical Engineering,* 3rd ed., PWS Publishing Company, Boston, 1994. *Principles of Geotechnical Engineering,* Third Edition by Braja M. Das is copyright © 1994 by PWS Publishing Company, a division of International Thomson Publishing.

Compaction effort can be quantified in terms of the compaction energy per unit volume. A function of the number of blows per layer, number of layers, weight of the hammer, height of the drop of the hammer, and volume of the mold, compaction energy per unit volume is 12,400 ft-lb/ft^3 (600 kN · m/m^3) for the Standard Proctor test and 56,000 ft-lb/ft^3 (2700 kN · m/m^3) for the Modified Proctor test. Clearly, the greater the compaction energy per unit volume, the greater will be the compaction. In fact, if the compaction energy per unit volume is changed, the Proctor curve (moisture content versus unit weight, see Figure 4–2) will change. Figure 4–4 illustrates the influence of compaction energy on the compaction of a sandy clay; as the number of blows per layer increases (and therefore the compaction energy per unit volume), the maximum dry unit weight increases and the optimum moisture content decreases.

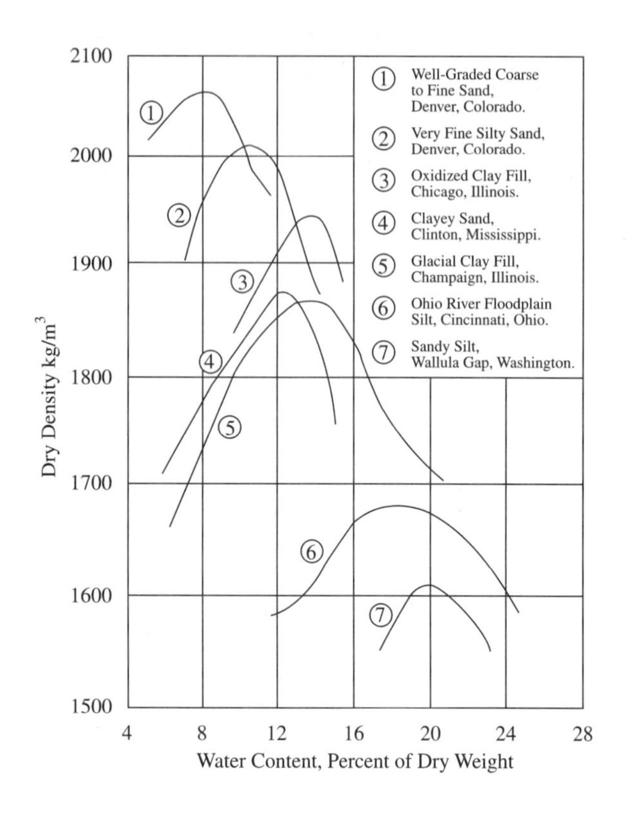

FIGURE 4–5 Moisture–density relations for various types of soils as determined by ASTM Method D 698 [5].
Source: From Karl Terzaghi Ralph B. Peck, and Gholamreza Mesri, *Soil Mechanics in Engineering Practice,* 3rd ed., John Wiley & Sons, Inc., New York, 1996. Copyright © 1996, by John Wiley & Sons, Inc. Reprinted by permission of John Wiley & Sons, Inc.

Clearly, the type of soil will also affect the compaction of soil. The grain-size distribution of soil, and shape, and the specific gravity of solids, as well as the type and amount of clay minerals present, affect maximum dry unit weight and optimum moisture content for a given compactive effort and compaction method. Maximum dry unit weights may range from about 60 lb/ft^3 (9.42 kN/m^3) for organic soils to about 145 lb/ft^3 (22.78 kN/m^3) for well-graded granular material containing just enough fines to fill small voids. Optimum moisture contents may range from around 5% for granular material to about 35% for elastic silts and clays. Higher optimum moisture contents are generally associated with lower dry unit weights. Higher dry unit weights are associated with well-graded granular materials. Uniformly graded sand, clays of high plasticity, and organic silts and clays typically respond poorly to compaction [4].

Moisture versus density curves for various types of soils are given in Figure 4–5. These curves were determined by the Standard Proctor method (ASTM D 698). It should be noted that both the shapes and the positions of the curves change as the texture of the soils varies from coarse to fine.

Table 4–2 presents some general compaction characteristics of various soil types, along with their values as embankment, subgrade, and base material, for soils classified according to the Unified Soil Classification System (USCS). Table 4–3 gives anticipated embankment performance for soils classified according to the American Association of State Highway and Transportation Officials (AASHTO) system.

TABLE 4–2
Compaction Characteristics and Ratings of Unified Soil Classification System Classes for Soil Construction [4, 6]

Class	Compaction Characteristics	Maximum Dry Unit Weight Standard Proctor $(lb/ft^3)^1$	Compressibility and Expansion	Value as Embankment Material	Value as Subgrade Material	Value as Base Course
GW	Good: Tractor, rubber-tired, steel wheel, or vibratory roller	125–135	Almost none	Very stable	Excellent	Good
GP	Good: Tractor, rubber-tired, steel wheel, or vibratory roller	115–125	Almost none	Reasonably stable	Excellent to good	Poor to fair
GM	Good: Rubber-tired or light sheepsfoot roller	120–135	Slight	Reasonably stable	Excellent to good	Fair to poor
GC	Good to fair: Rubber-tired or sheepsfoot roller	115–130	Slight	Reasonably stable	Good	Good to fair
SW	Good: Tractor, rubber-tired, or vibratory roller	110–130	Almost none	Very stable	Good	Fair to poor
SP	Good: Tractor, rubber-tired, or vibratory roller	100–120	Almost none	Reasonably stable when dense	Good to fair	Poor
SM	Good: Rubber-tired or sheepsfoot roller	110–125	Slight	Reasonably stable when dense	Good to fair	Poor
SC	Good to fair: Rubber-tired or sheepsfoot roller	105–125	Slight to medium	Reasonably stable	Good to fair	Fair to poor
ML	Good to poor: Rubber-tired or sheepsfoot roller	95–120	Slight to medium	Poor stability, high density required	Fair to poor	Not suitable
CL	Good to fair: Sheepsfoot or rubber-tired roller	95–120	Medium	Good stability	Fair to poor	Not suitable
OL	Fair to poor: Sheepsfoot or rubber-tired roller	80–100	Medium to high	Unstable, should not be used	Poor	Not suitable
MH	Fair to poor: Sheepsfoot or rubber-tired roller	70–95	High	Poor stability, should not be used	Poor	Not suitable
CH	Fair to poor: Sheepsfoot roller	80–105	Very high	Fair stability, may soften on expansion	Poor to very poor	Not suitable
OH	Fair to poor: Sheepsfoot roller	65–100	High	Unstable, should not be used	Very poor	Not suitable
PT	Not suitable	—	Very high	Should not be used	Not suitable	Not suitable

[1] $1\ lb/ft^3 = 0.1571\ kN/m^3$.

TABLE 4–3
General Guide to Selection of Soils on Basis of Anticipated Embankment Performance [4, 7]

AASHTO Classification	Visual Description	Maximum Dry Unit Weight Range (lb/ft^3)[1]	Optimum Moisture Range (%)	Anticipated Embankment Performance
A-1-a	Granular material	115–142	7–15	Good to excellent
A-1-b				
A-2-4	Granular material	110–135	9–18	Fair to excellent
A-2-5	with soil			
A-2-6				
A-2-7				
A-3	Fine sand and sand	110–115	9–15	Fair to good
A-4	Sandy silts and silts	95–130	10–20	Poor to good
A-5	Elastic silts and clays	85–100	20–35	Unsatisfactory
A-6	Silt–clay	95–120	10–30	Poor to good
A-7-5	Elastic silty clay	85–100	20–35	Unsatisfactory
A-7-6	Clay	90–115	15–30	Poor to fair

[1] 1 lb/ft^3 = 0.1571 kN/m^3.

4–4 FIELD COMPACTION

Normally, soil is compacted in layers. An approximately 8-in. (203-mm) loose horizontal layer of soil is often spread from trucks and then compacted to a thickness of about 6 in. (152 mm). The moisture content can be increased by sprinkling water over the soil if it is too dry and thoroughly mixing the water into the uncompacted soil by disk plowing. If the soil is too wet, its moisture content can be reduced by aeration (i.e., by spreading the soil in the sun and turning it with a disk plow to provide aeration and drying). Actual compaction is done by *tampers* and/or *rollers* and is normally accomplished with a maximum of 6 to 10 complete coverages by the compaction equipment. The surface of each compacted layer should be scarified by disk plowing or other means to provide bonding between layers. Various kinds of field compaction equipment (i.e., tampers and rollers) are discussed briefly in this section.

Tampers are devices that compact soil by delivering a succession of relatively light, vertical blows. Tampers are held in place and operated by hand. They may be powered either pneumatically or by gasoline-driven pistons. Tampers are limited in scope and compacting ability. Therefore, they are most useful in areas not readily accessible to rollers, in which case soil may be placed in loose horizontal layers not exceeding 6 in. (152 mm) and then compacted with tampers.

Rollers come in a variety of forms, such as the smooth wheel roller, sheepsfoot roller, pneumatic roller, and vibratory roller. Some of these are self-propelled, whereas others are towed by tractors. Some are more suited to certain types of soil.

FIGURE 4–6 Smooth wheel roller.
Source: Courtesy of BOMAG (U.S.A.), Inc.

Rollers can easily cover large areas relatively quickly and with great compacting pressures. Following are brief descriptions of the four types of rollers just mentioned.

A *smooth wheel roller* (see Figure 4–6) employs two or three smooth metal rollers. It is useful in compacting base courses and paving mixtures and is also used to provide a smooth finished grade. Generally, smooth wheel rollers are self-propelled and equipped with a reversing gear so that they can be driven back and forth without turning. A smooth wheel roller provides compactive effort primarily through its static weight.

A *sheepsfoot roller* (see Figure 4–7) consists of a drum with metal projecting "feet" attached. Because only the projecting feet come in contact with the soil, the area of contact between roller and soil is smaller (than for a smooth wheel roller), and therefore a greater compacting pressure results (generally more than 200 lb/in.^{2}). A sheepsfoot roller provides kneading action and is effective for compacting fine-grained soils (such as clays and silts).

A *pneumatic roller* (see Figure 4–8) consists of a number of rubber tires, highly inflated. They vary from small rollers to very large and heavy ones. Most large pneumatic rollers are towed, whereas some smaller ones are self-propelled. Some have boxes mounted above their wheels, to which sand or other material can be added for increased compacting pressure. Clayey soils and silty soils may be compacted effectively by pneumatic rollers. These rollers are also effective in compacting granular material containing a small amount of fines.

A *vibratory roller* (see Figure 4–9) contains some kind of vibrating unit that imparts an up-and-down vibration to the roller as it is pulled over the soil. Vibrating

FIGURE 4–7 Sheepsfoot roller.
Source: Courtesy of BOMAG (U.S.A.), Inc.

FIGURE 4–8 Pneumatic roller.
Source: Courtesy of BOMAG (U.S.A.), Inc.

FIGURE 4–9 Vibratory roller.
Source: Courtesy of Hyster Company, Construction Equipment Division.

units can supply frequencies of vibration at 1500 to 2000 cycles per minute, depending on compacting requirements. They are effective in compacting granular materials—particularly clean sands and gravels.

Two means (or possibly a combination of the two) may be used to specify a particular compaction requirement. One is to specify the procedure to be followed by the contractor, such as the type of compactor (i.e., roller) to be used and the number of passes to be made. The other is to simply specify the compacted soil's required final dry unit weight. The first method has the advantage that little testing is required, but it has the disadvantage that the specified procedure may not produce the required result. The second method requires much field testing, but it ensures that the required dry unit weight is achieved. In effect, the second method specifies the required final dry unit weight but leaves it up to the contractor as to how that unit weight is achieved. This (i.e., the second) method is probably more commonly employed.

4–5 DYNAMIC COMPACTION

In cases where existing surface or near-surface soil is poor with regard to foundation support, a field procedure known as *dynamic compaction* may be employed to improve the soil's properties. This method is carried out essentially by repeatedly dropping a very heavy weight onto the soil from a relatively great height. The dropped weight may be an ordinary steel wrecking ball, or it may be a mass especially designed for the dynamic-compaction procedure. Typical weights range from 2 to 20 tons or higher, whereas dropping distances range from 20 to 100 ft. Generally, the heavier the weight and the greater the dropping distance, the greater the compactive

FIGURE 4–10 Drop pattern [9].

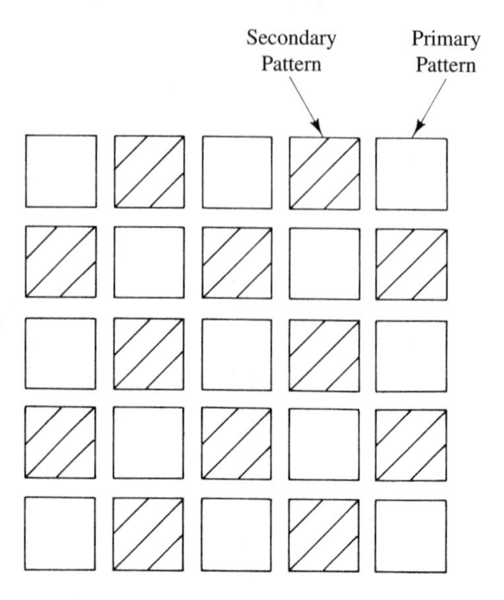

effort will be. For a given situation, however, the weight and dropping distance used may depend on the lifting equipment (such as a crane) available.

Dynamic compaction may be used for both cohesive and cohesionless soils. It can also be utilized to compact buried refuse fill areas. In cohesive soils, the reduction of settlements due to dynamic compaction is more distinct than the increase in bearing capacity. The tamping produces a true presettlement of the soil, well beyond the settlement that would have occurred as a result of construction weight only, without any preliminary consolidation [8]. For cohesionless soils, dynamic compaction densifies loose soil.

Dynamic compaction should not be done by dropping weight randomly. Instead, a closely spaced grid pattern is selected for a given compaction site (see Figure 4–10). Preliminary work is done to determine grid spacing and weight, height, and number of drops. Typically, 5 to 10 drops are made on each grid point. Figure 4–11 shows a photograph of a dynamic-compaction site.

The approximate depth of influence of dynamic compaction (D) may be determined in terms of weight (W) and distance dropped (h). For cohesionless soils [9],

$$D = 0.5\sqrt{Wh} \tag{4–3}$$

For cohesive soils [8],

$$D = \sqrt{Wh} \tag{4–4}$$

These equations give the depth of zone (D) receiving improvement in meters if W is in metric tons (1000 kg) and h in meters. The extent of improvement is greatest near the surface and diminishes with depth. Improvement increases with the number of drops made up to some limit—typically from 5 to 10 drops—beyond which additional drops afford little or no additional improvement.

FIGURE 4–11 Dynamic-compaction site.
Source: Courtesy of Hayward Baker Inc.

With saturated, fine-grained soils, satisfactory results may be obtained by performing a series of drops at intervals of one or several days, the purpose being to provide time for dissipation of pore water pressures created by the previous compaction.

It should be noted that a soil surface may become cratered as a result of dynamic compaction. This is particularly true of "loose" soils. When this happens, the craters must be backfilled and compacted by other means (such as those described in Section 4–4).

4–6 IN-PLACE SOIL UNIT WEIGHT TEST

As related previously, after a fill layer of soil has been compacted by the contractor, it is important that the compacted soil's in-place dry unit weight be determined in order to ascertain whether the maximum laboratory dry unit weight has been attained. If the maximum dry unit weight (or an acceptable percentage thereof) has not been attained, additional compaction effort is required.

There are several methods for determining in-place unit weight. As a general rule, the weight and volume of an in-place soil sample are determined, from which unit weight can be computed. Measurement of the sample's weight is straightforward, but there are several methods for determining its volume. For cohesive soils, a thin-walled cylinder may be driven into the soil to remove a sample. The sample's volume is known from the cylinder's volume. This method is known as *density of soil in-place by the drive cylinder method* and is designated as ASTM D 2937 or AASHTO T 204. The drive cylinder method is not applicable for very hard soil that cannot be easily penetrated. Neither is it applicable for low plasticity or cohesionless soils, which are not readily retained in the cylinder.

For low plasticity or cohesionless soils, a hole can be dug in the ground or compacted fill and the removed soil sample weighed and tested for water content. The volume of soil removed, which is the same as the volume of the hole, can be determined by filling the hole with loose, dry sand of uniform unit weight (such as Ottawa sand). By measuring the weight of sand required to fill the hole and knowing the sand's unit weight, one can find the volume of the hole. With the soil sample's weight, volume, and water content known, its dry unit weight can be easily computed. This method is carried out using a sand-cone apparatus, which consists of a large jar with an attached cone-shaped funnel (see Figures 4–12 and 4–13). With the jar inverted, sand is allowed to pass through the funnel into the hole until the hole is just filled with sand. The weight of sand required to fill the hole can then be determined, from which, with the sand's unit weight known, the soil sample's volume and subsequently its dry unit weight can be computed. Typically, Ottawa sand, a loose, dry sand of uniform unit weight approximating 100 lb/ft^3 (16 kN/m^3), is used in this test. This procedure is called *unit weight of soil in-place by the sand-cone method* and is designated as ASTM D 1556 and AASHTO T 191.

Another method for determining *in situ* dry unit weight is known as *unit weight of soil in-place by the rubber-balloon method* (designated as ASTM D 2167 and AASHTO T 205). In this method, a hole is dug and the removed soil sample weighed and tested

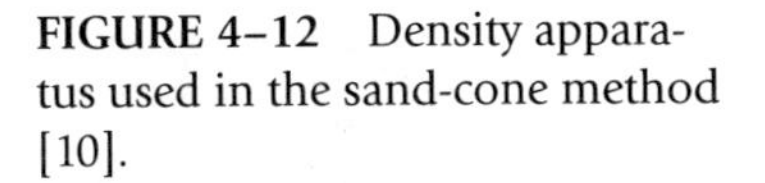

FIGURE 4–12 Density apparatus used in the sand-cone method [10].

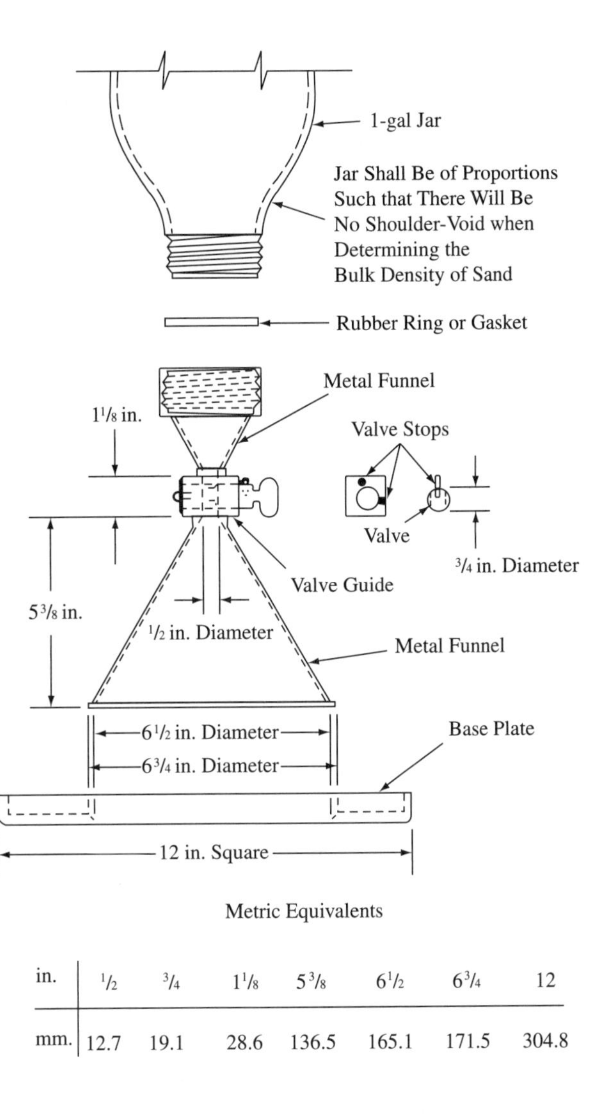

in.	1/2	3/4	1 1/8	5 3/8	6 1/2	6 3/4	12
mm.	12.7	19.1	28.6	136.5	165.1	171.5	304.8

for water content as in the previous method. The volume of soil removed is determined using a balloon apparatus (Figures 4–14 and 4–15), which consists of a vertical cylinder with transparent sides and graduation marks on its side. A rubber membrane or balloon is stretched over the open bottom of the cylinder. In use, the apparatus is placed over the empty hole, and air is pumped into the top of the cylinder above the water level, forcing the balloon and water down into the hole, completely filling it. The volume of water required to fill the hole, which is easily determined by reading the water level in the cylinder before and after forcing the water into the hole, is the same as the volume of the hole and of the soil removed from the hole. As in the previous method, with the soil sample's weight, volume, and water content known, its dry unit weight can be determined.

FIGURE 4–13 Density apparatus used in the sand-cone method.
Source: Courtesy of ELE International, Inc.

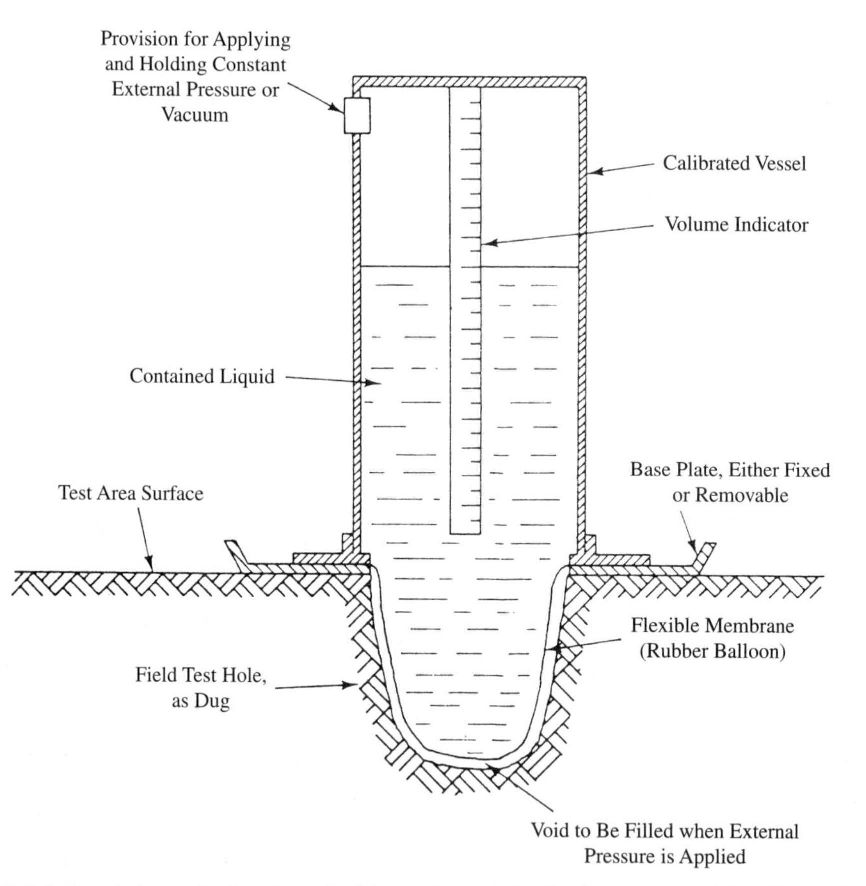

FIGURE 4–14 Schematic drawing of calibrated vessel, indicating the principle of the rubber-balloon method (not to scale) [10].

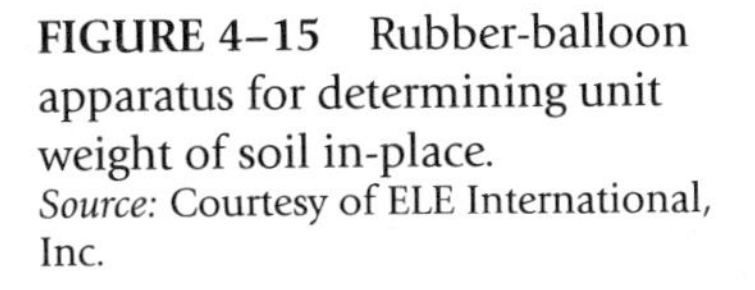

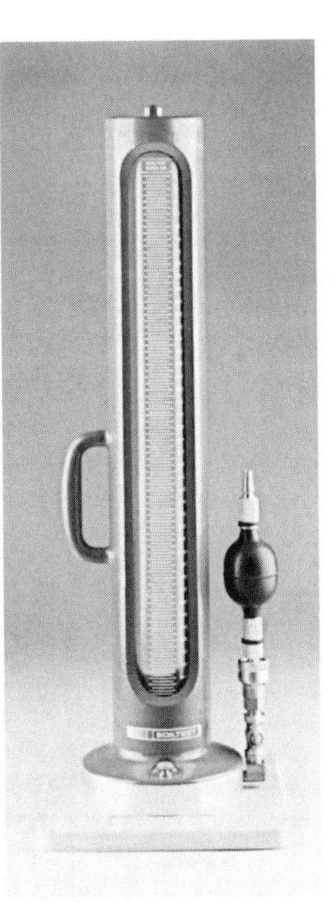

FIGURE 4–15 Rubber-balloon apparatus for determining unit weight of soil in-place.
Source: Courtesy of ELE International, Inc.

Although widely used, the sand-cone and rubber-balloon methods are *destructive* testing methods, in that a sizable hole must be dug in the ground or compacted fill. They are also fairly time consuming, a significant factor when numerous tests must be performed as quickly as possible at a construction site.

A *nondestructive* method for determining *in situ* dry unit weight utilizes a nuclear apparatus (see Figure 4–16). In use, this apparatus is placed on the ground or compacted fill and emits gamma rays through the soil. Some of the gamma rays will be absorbed; others will reach a detector. Soil unit weight is inversely proportional to the amount of radiation that reaches the detector. Through proper calibration, nuclear count rates received at the detector can be translated into values of soil (wet) unit weight. Calibration curves are normally provided by the manufacturer. The nuclear apparatus also determines moisture content by emitting alpha particles that bombard a beryllium target, causing the beryllium to emit fast neutrons. Fast neutrons that strike hydrogen atoms in water molecules lose velocity; the resulting low-velocity neutrons are thermal neutrons. Thermal neutron counts are made, from which—with proper correlation—soil moisture results (as weight of water per unit of volume) can be determined. (*Note:* Moisture determinations by this method can be in error in soils containing iron, boron, or cadmium.) The dry unit weight can then be found by subtracting this moisture result from the wet unit weight previously

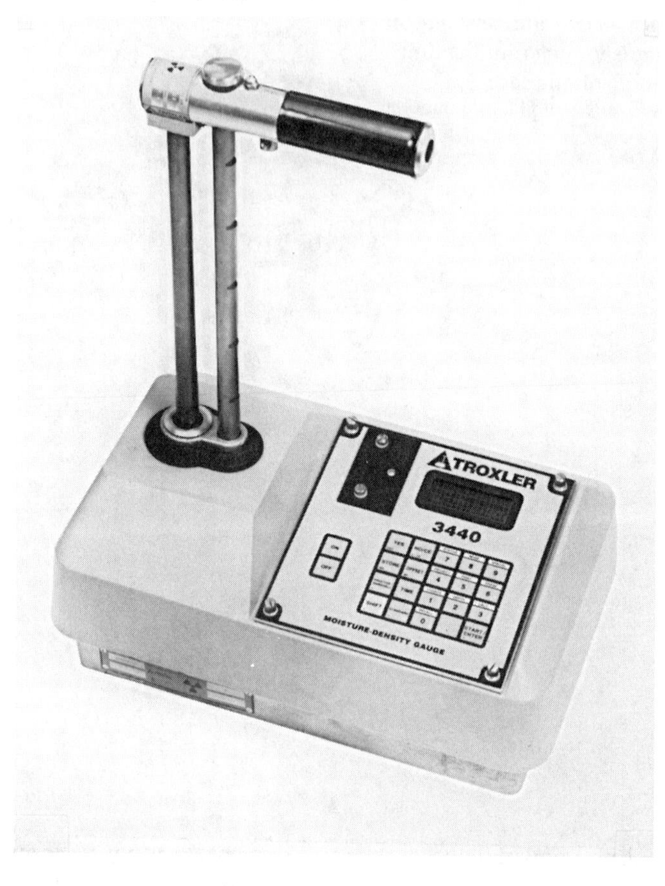

FIGURE 4–16 Nuclear moisture–density apparatus.
Source: Courtesy of Troxler Electronic Laboratories, Inc., North Carolina.

determined. Figure 4–17 illustrates several modes for using a nuclear apparatus. This method for determining *in situ* dry unit weight is known as *unit weight of soil and soil-aggregate in-place by nuclear methods* and is designated as ASTM D 2922.

The nuclear method is considerably faster to perform than the sand-cone and rubber-balloon methods. It has the disadvantage, however, of potential hazards to individuals handling radioactive materials. The nuclear apparatus is also considerably more costly than the apparatuses used in the other two methods.

In addition to determining the in-place wet unit weight of soil (using the sand-cone or balloon method), it is necessary to determine the soil's moisture content in order to compute the compacted soil's dry unit weight. Although the moisture content can be determined by oven drying, this method is often too time consuming because test results are commonly needed quickly. Drying of a soil sample can be accomplished by putting it in a skillet and placing the skillet over the open flame of a camp stove. The Speedy Moisture Tester (see Figure 4–18) can also be used to determine moisture content quickly with fairly good results. Because of the rather small amount of sample utilized in this test, the Speedy Moisture Tester may not be appropriate for use in coarser materials.

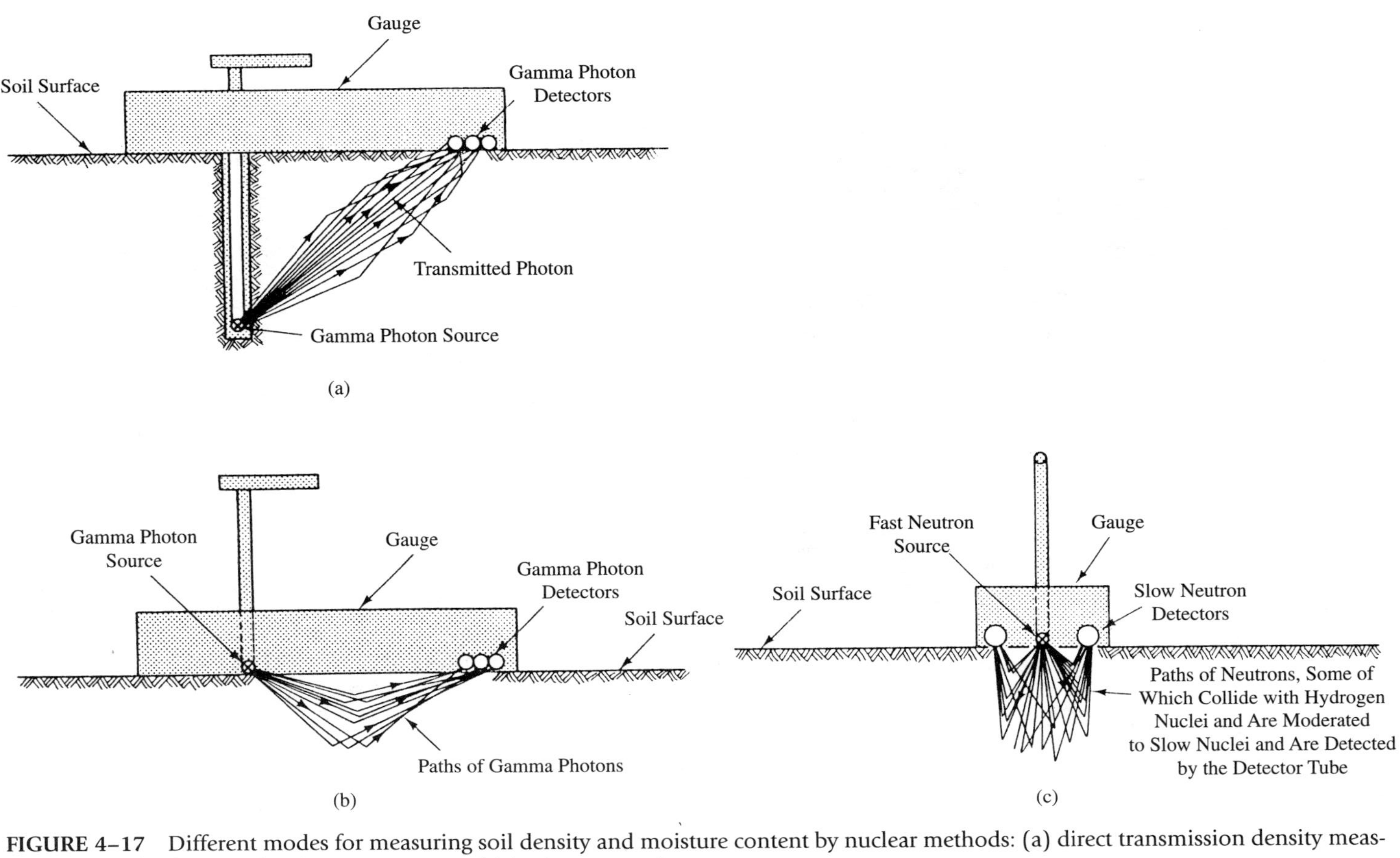

FIGURE 4–17 Different modes for measuring soil density and moisture content by nuclear methods: (a) direct transmission density measurement; (b) backscatter density measurement; (c) backscatter moisture measurement.
Source: Courtesy of Troxler Electronic Laboratories, Inc., North Carolina.

FIGURE 4–18 Speedy Moisture Tester.
Source: Courtesy of Soiltest, Inc.

Step-by-step details of all the aforementioned test procedures are given in *Soil Properties: Testing, Measurement, and Evaluation,* 4th edition, by Liu and Evett (Prentice Hall, 2000).

EXAMPLE 4–3

Given

During construction of a soil embankment, a sand-cone in-place unit weight test was performed in the field. The following data were obtained:

1. Weight of sand used to fill test hole and funnel of sand-cone device = 867 g.
2. Weight of sand to fill funnel = 319 g.
3. Unit weight of sand = 98.0 lb/ft^3.
4. Weight of wet soil from the test hole = 747 g.
5. Moisture content of soil from test hole as determined by Speedy Moisture Tester = 13.7%.

Required

Dry unit weight of the compacted soil.

Solution

Weight of sand used in test hole

= Weight of sand to fill test hole and funnel − Weight of sand to fill funnel
= 867 g − 319 g = 548 g

$$\text{Volume of test hole} = \frac{548\ \text{g}/453.6\ \text{g/lb}}{98.0\ \text{lb/ft}^3} = 0.0123\ \text{ft}^3$$

$$\text{Wet unit weight of soil in-place} = \frac{747\ \text{g}/453.6\ \text{g/lb}}{0.0123\ \text{ft}^3} = 133.9\ \text{lb/ft}^3$$

From Eq. (4–1),

$$\gamma_d = \frac{\gamma}{1 + w} \tag{4–1}$$

$$\gamma_d = \frac{133.9\ \text{lb/ft}^3}{1 + 0.137} = 117.8\ \text{lb/ft}^3$$

4–7 FIELD CONTROL OF COMPACTION

As related previously, after a fill layer of soil has been compacted, an in-place soil unit weight test is usually performed to determine whether the maximum laboratory dry unit weight (or an acceptable percentage thereof) has been attained. It is common to specify a required percent of compaction, which is "the required in-place dry unit weight" divided by "the maximum laboratory dry unit weight" expressed as a percentage, in a contract document. Thus, if the maximum dry unit weight obtained from ASTM or AASHTO compaction in the laboratory is 100 lb/ft^3 and the required percent of compaction is 95% according to a contract, an in-place dry unit weight of 95 lb/ft^3 (or higher) would be acceptable. In theory, this is simple enough to do, but there are some practical considerations that must be taken into account. For example, the type of soil or compaction characteristics of soil taken from borrow pits may vary from one location to another. Also, the degree of compaction may not be uniform throughout.

To deal with the problem of nonuniformity of soil from borrow pits, it is necessary to conduct ASTM or AASHTO compaction tests in the laboratory to establish the maximum laboratory dry unit weight along with the optimum moisture content for each type of soil encountered in a project. Then, as soil is transported from the borrow pit and subsequently placed and compacted in the fill area, it is imperative that the results of each in-place soil unit weight test be checked against the maximum laboratory dry unit weight of the respective type of soil.

To deal with the problem of the variable degree of field compaction of a soil, it is common practice to specify a minimum number of field unit weight tests. For example, for a dam embankment, it might be specified that one test be made for every 2400 yd^3 (loose measure) of fill placed.

To ensure that the required field unit weight is achieved by the field compaction, a specifications contract between the owner and the contractor is prepared. The contract will normally specify the required percent of compaction and minimum number of field unit weight tests required. For compaction adjacent to a structure, where settlement is a serious matter, a higher percent of compaction and a higher minimum number of tests may be specified than for compaction, for example, of the foundation of a parking lot. The specifications contract may also include additional items, such as the maximum thickness of loose lifts (layers) prior to compaction, methods to obtain maximum dry unit weight (e.g., ASTM D 698 or AASHTO T 99), methods to determine in-place unit weight (e.g., ASTM D 1556 or AASHTO T 191), and so on.

As the owner's representative, a soils engineer is responsible for ensuring that contract provisions are carried out precisely and completely. He or she is responsible for the testing and must see that the required compacted dry unit weight is achieved. If a particular test indicates that the required compacted dry unit weight has not been achieved, he or she must require additional compaction effort, possibly including an adjustment in moisture content. In addition, he or she must be knowledgeable and capable of dealing with field situations that arise that may go beyond the "textbook procedure."

EXAMPLE 4–4

Given

1. Soil from a borrow pit to be used for construction of an embankment gave the following laboratory results when subjected to the ASTM D 698 Standard Proctor test (from Example 4–2):

$$\text{Maximum dry unit weight} = 118.5 \text{ lb/ft}^3$$
$$\text{Optimum moisture content} = 12.5\%$$

2. The contractor, during construction of the soil embankment, achieved the following (from Example 4–3):

$$\text{Dry unit weight reached by field compaction} = 117.8 \text{ lb/ft}^3$$
$$\text{Actual water content} = 13.7\%$$

Required

Percent of compaction achieved by the contractor.

Solution

Percent of Standard Proctor compaction achieved

$$= \frac{\text{In-place dry unit weight}}{\text{Maximum laboratory dry unit weight}} \times 100 = \frac{117.8 \text{ lb/ft}^3}{118.5 \text{ lb/ft}^3} \times 100 = 99.4\%$$

EXAMPLE 4–5

Given

1. A borrow pit's soil is being used as earth fill at a construction project.
2. The *in situ* dry unit weight of the borrow pit soil was determined to be 17.18 kN/m^3.

3. The soil at the construction site is to be compacted to a dry unit weight of 18.90 kN/m^3.
4. The construction project requires 15,000 m^3 of compacted soil fill.

Required

Volume of soil required to be excavated from the borrow pit to provide the necessary volume of compacted fill.

Solution

Total dry weight required to furnish the compacted fill

= Total dry weight of soil required to be excavated from the borrow pit

$$= (18.90 \text{ kN/m}^3)(15{,}000 \text{ m}^3) = 283{,}500 \text{ kN}$$

Volume of soil required to be obtained from the borrow pit

$$= \frac{283{,}500 \text{ kN}}{17.18 \text{ kN/m}^3} = 16{,}500 \text{ m}^3$$

EXAMPLE 4–6

Given

1. The *in situ* void ratio (e) of a borrow pit's soil is 0.72.
2. The borrow pit soil is to be excavated and transported to fill a construction site where it will be compacted to a void ratio of 0.42.
3. The construction project requires 10,000 m^3 of compacted soil fill.

Required

Volume of soil that must be excavated from the borrow pit to provide the required volume of fill.

Solution

Let subscript f denote soil in the fill. From Eq. (2–6),

$$e = \frac{V_v}{V_s} \tag{2–6}$$

$$0.42 = \frac{(V_v)_f}{(V_s)_f}$$

$$(0.42)(V_s)_f = (V_v)_f \tag{A}$$

$$(V_s)_f + (V_v)_f = 10{,}000 \text{ m}^3 \tag{B}$$

Substitute Eq. (A) into Eq. (B).

$$(V_s)_f + 0.42(V_s)_f = 10{,}000 \text{ m}^3$$
$$(V_s)_f = 7042 \text{ m}^3$$

Let subscript b denote soil in the borrow pit.

$$(V_s)_b = (V_s)_f = 7042 \text{ m}^3 \tag{C}$$

From Eq. (2–6),

$$0.72 = \frac{(V_v)_b}{(V_s)_b}$$

$$(0.72)(V_s)_b = (V_v)_b \tag{D}$$

From Eq. (C), $(V_s)_b = 7042 \text{ m}^3$; substitute into Eq. (D).

$$(0.72)(7042 \text{ m}^3) = (V_v)_b$$
$$(V_v)_b = 5070 \text{ m}^3$$

Total volume of soil from borrow pit $(V_b) = (V_v)_b + (V_s)_b = 5070 \text{ m}^3 + 7042 \text{ m}^3 = 12{,}112 \text{ m}^3$

4–8 SOIL STABILIZATION

Sections 4–4 and 4–5 described a physical means (field and dynamic compaction) whereby a soil can have its physical properties improved to increase bearing capacity, increase soil shear strength, decrease settlement, and reduce soil permeability. *Soil stabilization* can also be used to improve the properties of a natural soil by preloading the soil or by adding other special soil (mechanical stabilization), chemical material (chemical stabilization), or some kind of fabric materials (geosynthetics) to the soil. These means of achieving soil stabilization are discussed in the remainder of this section.

Preloading

Preloading refers to adding an artificial load to a potential construction site prior to the time the structure is built (and loaded). The soil is improved by causing soil consolidation to occur prior to construction and loading, thereby decreasing subsequent settling of the structure.

Preloading is carried out simply by adding fill or other surcharge to the natural soil *in situ* and allowing the added weight to consolidate the soil naturally over a period of time. In general, the greater the added surcharge and the longer the time it is in place prior to construction, the better the consolidation will be and the better the bearing capacity of the soil will be. In most cases, the amount of material to be used as surcharge and the time available may be limited, however, by practical and/or economic considerations. Transporting soil is expensive, and in some cases suitable surcharge material may not be readily available. The time needed to effect soil improvement may be reduced by including vertical sand and/or gravel drains in the soil during the surcharge period. The amount of time needed varies from several months to several years.

Preloading works best in soft silty and clayey soils. Granular soils, where consolidation is an insignificant phenomenon, are not generally amenable to improvement by preloading.

Mechanical Stabilization

Mechanical stabilization is a relatively simple means of soil stabilization that is carried out by adding soil material to the naturally occurring soil. The added soil material is usually mechanically mixed with the natural soil and worked together, after which the mixture is compacted. Normally, a blending of coarse aggregate and fine-grained soil is achieved in order to get a soil mixture that possesses some internal friction and cohesion and will thereby be workable and subsequently stable when mixing and compaction have been completed.

Chemical Stabilization

Chemical stabilization is achieved by adding a cementing material or some kind of chemical to the soil. The chemical material may be mechanically mixed with the natural soil and the resulting mixture compacted, or the chemical material may be simply applied to the natural soil and allowed to penetrate the soil through the void space. Another process is to inject the stabilizing chemical into or through the soil under pressure; this is known as *grouting*. The procedure of grouting (i.e., injection stabilization) is generally performed where it is necessary to improve soil that cannot be disturbed. Grouting can be effective at relatively large depths of soil formation. Grouting and injection stabilization are generally performed by specialty contractors who have proper and adequate equipment and have developed experience over the years with one or more stabilization procedures.

Many different chemicals have been used for chemical stabilization—sodium chloride, calcium chloride, cement, and lime, to name a few of the more common ones. Sodium chloride and calcium chloride may be added to a soil when it is desired to hold soil water. Sodium chloride spread on the surface of dirt roads can help with dust control on rural highways. Various kinds of cement (Portland cement, asphalt cement) may be added to soil to bond the soil particles together. When Portland cement is added to soil in the presence of water, concrete is formed. In the construction of the soil-cement mixture, the soil needs to be at or near the optimum moisture content for maximum compaction as determined by a compaction test (i.e., ASTM D 698). In soil stabilization using Portland cement, the amount of cement added is quite small (on the order of 7% to 14% by weight for sandy to clayey soils, respectively) and the result is a stabilized soil that is stronger than the natural soil but not nearly as strong as concrete. Cement-stabilized soils may be used as road bases when traffic is relatively light and not of heavy weight. Lime and calcium chloride may be used as additives for high-plasticity, clayey soils where they serve to reduce plasticity. This technique can be effective in reducing volume changes associated with certain expansive clays. The construction of lime stabilization requires mixing lime with natural soil, curing for a few days, then remixing followed by compaction.

There are available additional chemical stabilizers that can be used for soil stabilization; some are marketed under their trade names. The chemical stabilizers described here are among the more common ones. Soils engineers' practical experience may be invaluable in deciding what specific type of chemical stabilization to use in any give situation.

Geosynthetics

Geosynthetics refers to a family of manufactured materials (sheet or netlike products) made of plastics or fiberglass. Geosynthetics may be used to stabilize and reinforce soil masses, such as erosion control of earth slope surfaces, reinforcing backfill of retaining walls, reinforcing slopes or embankments, slope protection of open channels, and drainage control, to name a few. Figure 4–19 illustrates a number of examples of the use of geosynthetics. Geosynthetics may come in the form of *geotextiles, geogrids, geonets,* and *geomembranes.*

Geotextiles are similar to woven fabric, or textiles. Common usages for geotextiles are for strata separation, soil reinforcement, filtration, etc. In strata separation, the geotextile is simply placed between two different soil strata where it serves to retain the strata separation and to preserve each stratum's individual properties and function. A typical example is to place a geotextile between a fine-grained subgrade soil and aggregate base course to prevent the fine-grained soil from intruding into the aggregate base course. In reinforcement, the geotextile may be placed over a weak soil with a layer of "good" fill placed on the geotextile. In filtration, small openings in the geotextile allow water, but not soil particles, to move through the geotextile. For example, a geotextile may be used in this capacity to protect a drain from soil infiltration.

Geogrids have larger openings (1 to 4 in.) than geotextiles and therefore resemble nets. They are used (in conjunction with geotextiles) to reinforce relatively poor soils over which paved surfaces, such as roads and parking lots, are to be constructed. Geogrids can be used for improved slope stability (to prevent potential slip failure) and also as a reinforcement to construct an earth wall, similar to a reinforced wall.

Geonets are similar to geogrids but have intersecting ribs. They may be used for drainage purposes under roadways and landfills and behind retaining walls. They too are often used in conjunction with geotextiles. Geonets are usually installed on a slope toward a perforated drain pipe or a ditch.

Geomembranes are impervious, thin plastic sheets. They are used to prohibit, or greatly restrict, the movement of water within soil masses. A common example of their use is as landfill liners, to prevent the movement of wastewater (leachate) from within the landfill into surrounding soil strata.

The preceding covers only a few of numerous applications of geosynthetics in soil stabilization and other soil usages. Reference [11] at the end of the chapter gives extensive information about geosynthetics.

(a) (b)

(c) (d)

(e)

FIGURE 4–19 Examples of use of Geosynthetics (Courtesy of C.F.P. Inc.). (a) Geotextile stabilization, (b) Geocomposite drainage panel, (c) High strength geotextile (slide repair) (d) Geogrid reinforced steepened slope and (e) Geosynthetic clay liner

4–8 PROBLEMS

4–1. A compaction test was conducted in a soils laboratory and the Standard Proctor compaction procedure (ASTM D 698) was used. The weight of a compacted soil specimen plus mold was determined to be 3815 g. The volume and weight of the mold were 1/30 ft^3 and 2050 g, respectively. The water content of the specimen was 9.1%. Compute both the wet and dry unit weights of the compacted specimen.

4–2. A soil sample was taken from the site of a proposed borrow pit and sent to the laboratory for a Standard Proctor test (ASTM D 698). Results of the test are as follows:

Determination Number	1	2	3	4	5
Dry unit weight (lb/ft^3)	107.0	109.8	112.0	111.6	107.3
Moisture content (%)	9.1	11.8	14.0	16.5	18.9

Plot a moisture content versus dry unit weight curve and determine the soil's maximum dry unit weight and optimum moisture content.

4–3. Using the results of the Standard Proctor test as given in Problem 4–2, determine the range of water content most likely to attain 95% or more of the maximum dry unit weight.

4–4. A laboratory compaction test was performed on a soil sample taken from a proposed cut area. The maximum dry unit weight and optimum moisture content were determined to be 104.8 lb/ft^3 and 20.7%, respectively. Estimate the possible type (or classification) of soil for this sample.

4–5. During construction of a highway project, a soil sample was taken from compacted earth fill for a sand-cone in-place density test. The following data were obtained during the test:

1. Weight of sand used to fill test hole and funnel of sand-cone device = 845 g.
2. Weight of sand to fill funnel = 323 g.
3. Unit weight of sand = 100 lb/ft^3.
4. Weight of wet soil from test hole = 648 g.
5. Moisture content of soil from test hole = 16%.

Calculate the dry unit weight of the compacted earth fill.

4–6. A soil sample was taken from a proposed cut area in a highway construction project and sent to a soils laboratory for a compaction test, using the Standard Proctor compaction procedure. Results of the test are as follows:

Maximum dry unit weight = 112.6 lb/ft^3
Optimum moisture content = 15.5%

The contractor, during construction of the soil embankment, achieved the following:

Dry unit weight reached by field compaction = 107.1 lb/ft^3
Actual water content = 16.0%

Determine the percent compaction achieved by the contractor.

4–7. Soil having a void ratio of 0.68 as it exists in a borrow pit is to be excavated and transported to a fill site where it will be compacted to a void ratio of 0.45. The volume of fill required is 2500 m^3. Find the volume of soil that must be excavated from the borrow pit to furnish the required volume of fill.

References

[1] T. William Lambe, *Soil Testing for Engineers,* John Wiley & Sons, Inc., New York, 1951. Copyright © 1951, by John Wiley & Sons, Inc. Reprinted by permission of John Wiley & Sons, Inc.

[2] B. K. Hough, *Basic Soils Engineering,* 2nd ed., The Ronald Press Company, New York, 1969. Copyright © 1969, by John Wiley & Sons, Inc. Reprinted by permission of John Wiley & Sons, Inc.

[3] Braja M. Das, *Principles of Geotechnical Engineering,* 3rd ed., PWS Publishing Company, Boston, 1994.

[4] Robert D. Krebs and Richard D. Walker, *Highway Materials,* McGraw-Hill Book Company, New York, 1971.

[5] Karl Terzaghi, Ralph B. Peck, and Gholamreza Mesri, *Soil Mechanics in Engineering Practice,* 3rd ed., John Wiley & Sons, Inc., New York, 1996.

[6] *The Unified Soil Classification System,* Waterways Exp. Sta. Tech. Mem. 3-357 (including Appendix A, 1953, and Appendix B, 1957), U.S. Army Corps of Engineers, Vicksburg, Miss., 1953.

[7] L. E. Gregg, "Earthwork," in K. B. Woods, ed., *Highway Engineering Handbook,* McGraw-Hill Book Company, New York, 1960.

[8] L. Menard and Y. Broise, "Theoretical and Practical Aspects of Dynamic Consolidation," *Geotechnique,* **25**(1), 3–18 (1975).

[9] Gerald A. Leonards, William A. Cutter, and Robert D. Holtz, "Dynamic Compaction of Granular Soils," *J. Geotech. Eng. Div. ASCE,* **106**(GT1), 35–44 (1980).

[10] *1995 Annual Book of ASTM Standards,* ASTM, Philadelphia, 1995. Copyright, American Society for Testing and Materials, 1916 Race Street, Philadelphia, PA 19103. Reprinted with permission.

[11] R. M. Koerner, *Design with Geosynthetics,* 2nd ed., Prentice-Hall, Englewood Cliffs, N.J., 1990.

5 WATER IN SOIL

5–1 INTRODUCTION

As indicated in Chapter 2, water is a component of soil, and its presence in a given soil may range from virtually none to *saturation,* the latter case occurring when the soil's void space is completely filled with water. When the voids are only partially filled with water, a soil is said to be *partially saturated.* Any soil's characteristics and engineering behavior are greatly influenced by its water content. This is especially true for fine-grained soils. A clayey soil may be "hard as a rock" when dry but become soft and plastic when wet. In contrast, a very sandy soil, such as is found on a beach, may be relatively loose when dry but rather hard and more stable when wet. It may be somewhat ironic that one can generally walk and drive rather easily on dry clay and wet sand but more difficultly on saturated clay and very dry, loose sand.

The effects of water in soil are very important in the study of soil mechanics. Cohesive soils in particular tend to shrink when dry and swell when wet—some types of clay expanding greatly when saturated. In addition, fine-grained soils are significantly weakened at high water contents. Such factors must be considered in most soil engineering problems and foundation design.

The effects of water movement within soil are also very important in many geotechnical engineering applications. Factors such as highway sub-drainage, wells as a source of water supply, capillary and frost action, seepage flow analysis, and pumping water for underground construction all require the consideration of in-soil water movement.

5–2 FLOW OF WATER IN SOILS

As indicated in the preceding section, water movement within soil is an important consideration in geotechnical engineering. The facility with which water flows through soil is an engineering property known as *permeability.* Because water movement within soil is through interconnected voids, in general, the larger a soil's void spaces, the greater will be its permeability. Conversely, the smaller the void spaces,

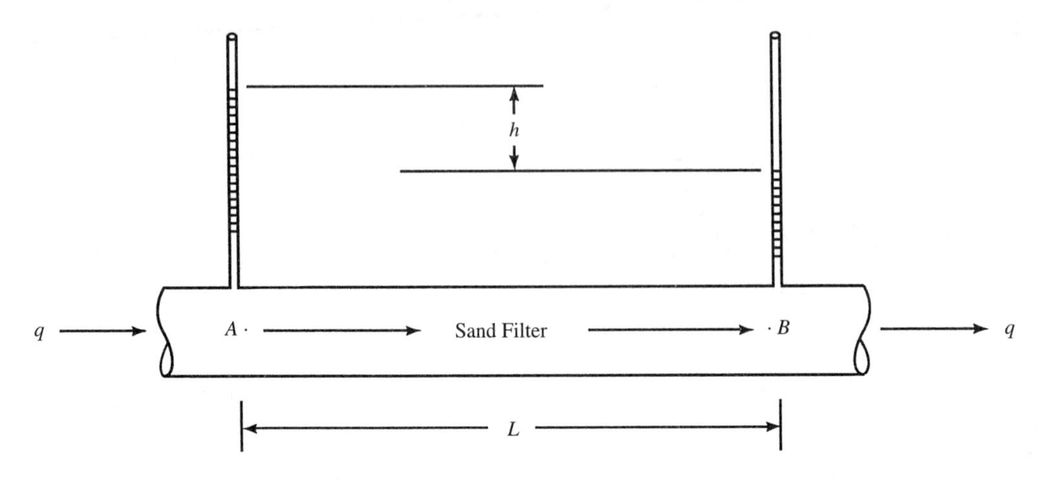

FIGURE 5–1 Illustration of Darcy's experiment.

the lesser will be its permeability. Thus, coarse-grained soils such as sand commonly exhibit high permeabilities, whereas fine-grained soils like clay ordinarily have lower permeabilities.

Flow of water in soil between two points occurs as a result of a pressure (or *hydraulic head*) difference between two points, with the direction of flow being from the higher to the lower pressure. Furthermore, the velocity of flow varies directly with the magnitude of the difference between hydraulic heads as well as with soil permeability.

Flow of water in soil can be analyzed quantitatively using *Darcy's law,* which was developed by Darcy in the eighteenth century based on experiments involving the flow of water through sand filters. Figure 5–1 illustrates Darcy's experiment in which water moves through a soil sample contained in a cylindrical conduit. His tests indicated that the flow rate through the soil in the conduit varied directly with both the hydraulic head difference (h in Figure 5–1) and the cross-sectional area of the soil, and inversely with the length over which the hydraulic head difference occurred (L in Figure 5–1). Accordingly,

$$q \propto \frac{hA}{L}$$

where
q = flow rate (volume per unit time)
h = hydraulic head difference (between points A and B in Figure 5–1)
A = soil sample's cross-sectional area
L = length of soil sample (between points A and B)

If a constant of proportionality, k, is supplied, the preceding proportionality becomes

$$q = k\frac{h}{L}A \tag{5–1}$$

The constant of proportionality (k) in Eq. (5–1) is known as the *coefficient of permeability* and has the same units as velocity. The hydraulic head difference divided by the length of the soil sample (h/L) is known as the *hydraulic gradient* and is denoted by i. With this substitution, Eq. (5–1) can be rewritten as follows:

$$q = kiA \tag{5–2}$$

If the velocity of flow, v, is desired, because $q = Av$,

$$v = ki \tag{5–3}$$

This velocity is an average velocity because it represents flow rate divided by gross cross-sectional area of the soil. This area, however, includes both solid soil material and voids. Because water moves only through the voids, the actual (interstitial) velocity is

$$v_{\text{actual}} = \frac{v}{n} \tag{5–4}$$

where n is porosity. Because $n = e/(1 + e)$, where e is the soil's void ratio,

$$v_{\text{actual}} = \frac{v(1 + e)}{e} \tag{5–5}$$

EXAMPLE 5–1

Given

1. Water flows through the sand filter shown in Figure 5–1.
2. The cross-sectional area and length of the soil mass are 0.250 m^2 and 2.00 m, respectively.
3. The hydraulic head difference is 0.160 m.
4. The coefficient of permeability is 6.90×10^{-4} m/s.

Required

Flow rate of water through the soil.

Solution

From Eq. (5–2),

$$q = kiA \tag{5–2}$$

$$i = \frac{h}{L} = \frac{0.160\ m}{2.00\ m} = 0.0800$$

$$q = (6.90 \times 10^{-4}\ \text{m/s})(0.0800)(0.250\ \text{m}^2) = 1.38 \times 10^{-5}\ \text{m}^3/\text{s}$$

EXAMPLE 5–2

Given

In a soil test, it took 16.0 min for 1508 cm^3 of water to flow through a sand sample, the cross-sectional area of which was 50.3 cm^2. The void ratio of the soil sample was 0.68.

Required

1. Velocity of water through the soil.
2. Actual (interstitial) velocity.

Solution

1. $v = \text{Volume/Time/Area}$

   $v = 1508 \text{ cm}^3/16.0 \text{ min}/50.3 \text{ cm}^2 = 1.874 \text{ cm/min, or } 0.0312 \text{ cm/s}$

2. $$v_{\text{actual}} = \frac{v(1+e)}{e} \quad (5\text{–}5)$$

   $$v_{\text{actual}} = \frac{(0.0312 \text{ cm/s})(1 + 0.68)}{0.68} = 0.0771 \text{ cm/s}$$

In predicting the flow of water in soils, it becomes necessary to evaluate the coefficient of permeability for given soils. Both laboratory and field tests are available for doing this.

Laboratory Tests for Coefficient of Permeability

Laboratory tests are relatively simple and inexpensive to carry out and are ordinarily performed following either the *constant-head method* or the *falling-head method.* Brief descriptions of each of these methods follow.

The constant-head method for determining the coefficient of permeability can be used for granular soils. It utilizes a device known as a *constant-head permeameter,* as depicted in Figure 5–2. The general test procedure is to allow water to move through the soil specimen under a stable-head condition while the soils engineer determines and records the time required for a certain quantity of water to pass through the soil specimen. By measuring and recording the quantity (volume) of water discharged during a test (Q), length of the specimen (distance between manometer outlets) (L), cross-sectional area of the specimen (A), time required for the quantity of water Q to be discharged (t), and head (difference in manometer levels) (h), the soils engineer can derive the coefficient of permeability (k) as follows:

$$Q = Avt \quad (5\text{–}6)$$

Because $v = ki$ [from Eq. (5–3)] and $i = h/L$,

$$Q = A\frac{kh}{L}t \quad (5\text{–}7)$$

Solving for k gives

$$k = \frac{QL}{Ath} \quad (5\text{–}8)$$

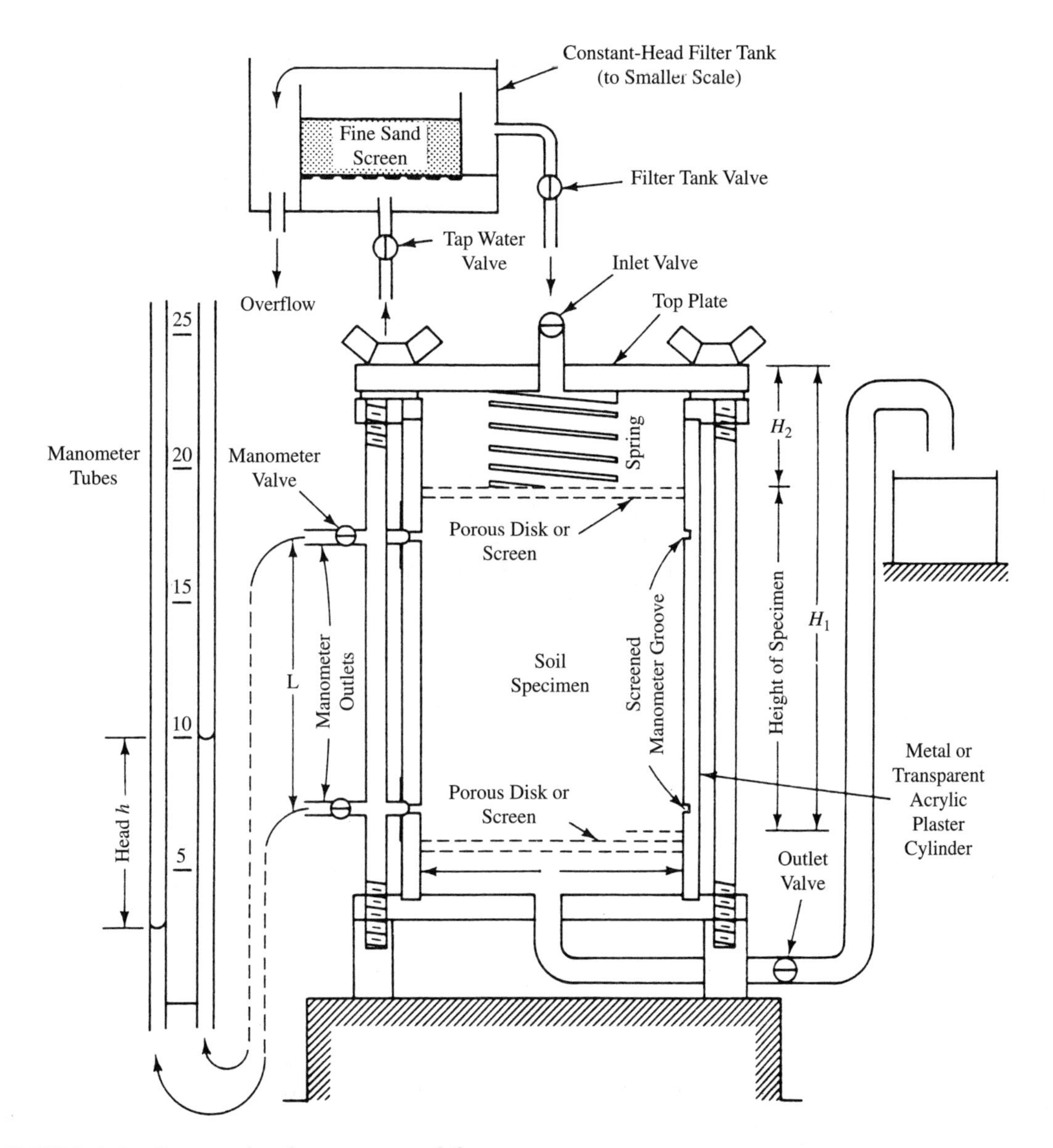

FIGURE 5–2 Constant-head permeameter [1].

The falling-head method can be used to find the coefficient of permeability for both fine-grained soils and coarse-grained, or granular, soils. It utilizes a permeameter like that depicted in Figure 5–3. The general test procedure does not vary a great deal from that of the constant-head method. The specimen is first saturated with water. Water is then allowed to move through the soil specimen under a falling-head

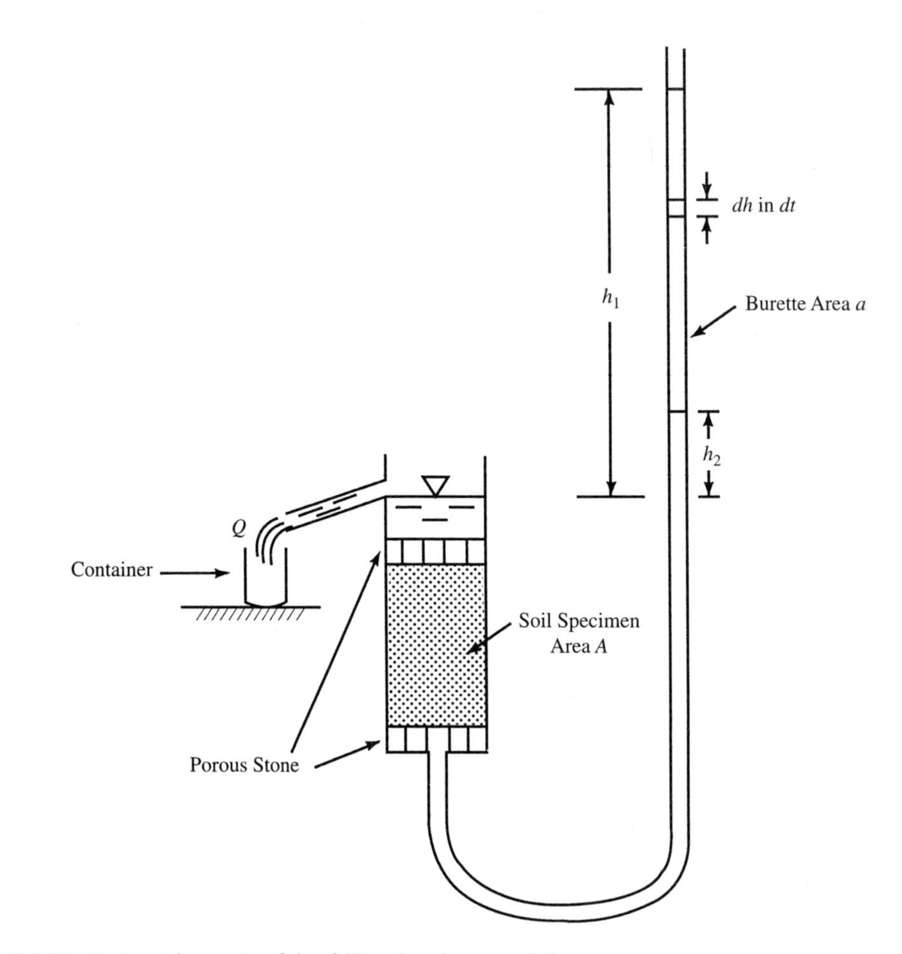

FIGURE 5–3 Schematic of the falling-head permeability setup.

condition (rather than a stable-head condition) while the time required for a certain quantity of water to pass through the soil specimen is determined and recorded. If a is the cross-sectional area of the burette, and h_1 and h_2 are the hydraulic heads at the beginning and end of the test, respectively (Figure 5–3), the coefficient of permeability can be derived as follows:

As shown in Figure 5–3, the velocity of fall in the burette is given by $v = -dh/dt$, with the minus sign used to indicate a falling (and therefore decreasing) head. The flow of water into the specimen is therefore $q_{in} = -a(dh/dt)$, and the flow through and out of the specimen is, from Eq. (5–1), $q_{out} = k(h/L)A$. Equating q_{in} and q_{out} gives

$$-a\frac{dh}{dt} = k\frac{h}{L}A \tag{5-9}$$

$$-a\frac{dh}{h} = k\frac{A}{L}dt \tag{5-10}$$

$$-a\int_{h_1}^{h_2}\frac{dh}{h} = k\frac{A}{L}\int_{t_1}^{t_2}dt \tag{5-11}$$

$$-a[\ln h]_{h_1}^{h_2} = k\frac{A}{L}[t]_0^t \tag{5-12}$$

$$a\ln\frac{h_1}{h_2} = k\frac{A}{L}t \tag{5-13}$$

Therefore,

$$k = \frac{aL}{At}\ln\frac{h_1}{h_2} \tag{5-14}$$

or

$$k = \frac{2.3aL}{At}\log\frac{h_1}{h_2} \tag{5-15}$$

The coefficient of permeability as determined by both methods is the value for the particular water temperature at which the test was conducted. This value is ordinarily corrected to that for 20 °C by multiplying the computed value by the ratio of the viscosity of water at the test temperature to the viscosity of water at 20 °C.

Permeability determined in a laboratory may not be truly indicative of the *in situ* permeability. There are several reasons for this in addition to the fact that the soil in the permeameter does not exactly duplicate the structure of the soil *in situ*, particularly that of nonhomogeneous soils and granular materials. For one thing, the flow of water in the permeameter is downward, whereas flow in the soil *in situ* may be more nearly horizontal or in a direction between horizontal and vertical. Indeed, the permeability of a natural soil in the horizontal direction can be considerably greater than that in its vertical direction. For another thing, naturally occurring strata in the *in situ* soil will not be duplicated in the permeameter. Also, the relatively smooth walls of the permeameter afford different boundary conditions from those of the *in situ* soil. Finally, the hydraulic head in the permeameter may differ from the field gradient.

Another concern with the permeability test is any effect from entrapped air in the water and test specimen. To avoid this, the water to be used in the test should be de-aired by boiling distilled water and keeping it covered and nonagitated until used.

EXAMPLE 5–3

Given

In a laboratory, a constant-head permeability test was conducted on a brown sand with a trace of mica. For the constant-head permeameter (Figure 5–2), the following data were obtained:

1. Quantity of water discharged during the test = 250 cm^3.
2. Length of specimen between manometer outlets = 11.43 cm.
3. Time required for given quantity of water to be discharged = 65.0 s.
4. Head (difference between manometer levels) = 5.5 cm.
5. Temperature of water = 20°C.
6. Diameter of specimen = 10.16 cm.

Required

Coefficient of permeability.

Solution

From Eq. (5–8),

$$k = \frac{QL}{Ath} \tag{5–8}$$

$$A = \frac{(\pi)(10.16\text{ cm})^2}{4} = 81.07\text{ cm}^2$$

$$k = \frac{(250\text{ cm}^3)(11.43\text{ cm})}{(81.07\text{ cm}^2)(65.0\text{ s})(5.5\text{ cm})} = 0.0986\text{ cm/s}$$

EXAMPLE 5–4

Given

In a laboratory, a falling-head permeability test was conducted on a silty soil. For the falling-head apparatus (Figure 5–3), the following data were obtained:

1. Length of specimen = 15.80 cm.
2. Diameter of specimen = 10.16 cm.
3. Cross-sectional area of burette = 1.83 cm^2.
4. Hydraulic head at beginning of test (h_1) = 120.0 cm.
5. Hydraulic head at end of test (h_2) = 110.0 cm.
6. Time required for water in the burette to drop from h_1 to h_2 = 20.0 min (1200 s).
7. Temperature of water = 20°C.

Required

Coefficient of permeability.

Solution

From Eq. (5–15),

$$k = \frac{2.3aL}{At}\log\frac{h_1}{h_2} \tag{5–15}$$

$$A = \frac{(\pi)(10.16\text{ cm})^2}{4} = 81.07\ cm^2$$

$$k = \frac{(2.3)(1.83\text{ cm}^2)(15.80\text{ cm})}{(81.07\text{ cm}^2)(1200\text{ s})}\log\frac{120.0\text{ cm}}{110.0\text{ cm}} = 2.58 \times 10^{-5}\text{cm/s}$$

Field Tests for Coefficient of Permeability

As noted previously, permeability determined in a laboratory may not be truly indicative of the *in situ* permeability. Thus, field tests are generally more reliable than laboratory tests for determining soil permeability, the main reason being that field tests are performed on the undisturbed soil exactly as it occurs *in situ* at the test location. Other reasons are that soil stratification, overburden stress, location of the groundwater table, and certain other factors that might influence permeability test results are virtually unchanged with field tests, which is not the case for laboratory tests.

There are several field methods for evaluating permeability, such as pumping, borehole, and tracer tests. The latter use dye, salt, or radioactive tracers to find the time it takes a given tracer to travel between two wells or borings; by finding the differential head between the two, the soils engineer can determine the coefficient of permeability. The pumping method is detailed next.

Figure 5–4 illustrates a well extending downward through an impermeable layer and then a permeable layer (an aquifer) to another impermeable layer. If water is pumped from the well at a constant discharge (q), flow will enter the well only from the aquifer, and the piezometric surface will be drawn down toward the well as shown in Figure 5–4. At some time after pumping begins, an equilibrium condition will be reached. The piezometric surface can be located by auxiliary observation wells located at distances r_1 and r_2 from the pumping well (Figure 5–4). The piezometric surface is located at distance h_1 above the top of the aquifer at point r_1 from the pumping well and at distance h_2 at point r_2. All parameters noted in this discussion and on Figure 5–4 can be measured during a pumping test, and from these data the coefficient of permeability can be computed, as follows. It should be noted that the permeability so determined is that of the soil in the aquifer in the direction of flow (i.e., in horizontal radial directions).

Equation (5–2) can be applied to the equilibrium pumping condition in Figure

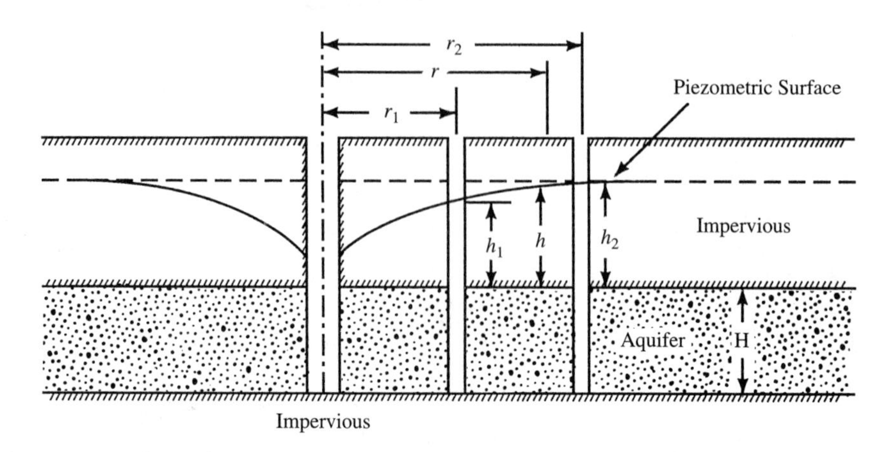

FIGURE 5–4 Flow of water toward pumping well (confined aquifer).

5–4. Hydraulic gradient i in the equation is given for any point on the piezometric surface by dh/dr. The soil's cross-sectional area at any point on the piezometric surface through which water flows [A in Eq. (5–2)] is that of a cylinder with radius r and height H (Figure 5–4). Substituting these into Eq. (5–2) gives

$$q = kiA = k\frac{dh}{dr}2\pi rH \tag{5–16}$$

$$\int_{r_1}^{r_2} q\frac{dr}{r} = \int_{h_1}^{h_2} 2\pi kH\,dh \tag{5–17}$$

Integrating gives

$$q\,[\ln r]_{r_1}^{r_2} = 2\pi kH[h]_{h_1}^{h_2} \tag{5–18}$$

$$q\ln\frac{r_2}{r_1} = 2\pi kH(h_2 - h_1) \tag{5–19}$$

Solving for k yields

$$k = \frac{q\ln(r_2/r_1)}{2\pi H(h_2 - h_1)} \tag{5–20}$$

Figure 5–5 illustrates a pumping well located in an unconfined, homogeneous aquifer. In this case, the piezometric surface lies within the aquifer. The analysis of this type of well is the same as that for the confined aquifer (i.e., Figure 5–4), except that the A term in Eq. (5–2) becomes $2\pi rh$. Hence,

$$q = k\frac{dh}{dr}2\pi rh \tag{5–21}$$

$$\int_{r_1}^{r_2} q\frac{dr}{r} = \int_{h_1}^{h_2} 2\pi kh\,dh \tag{5–22}$$

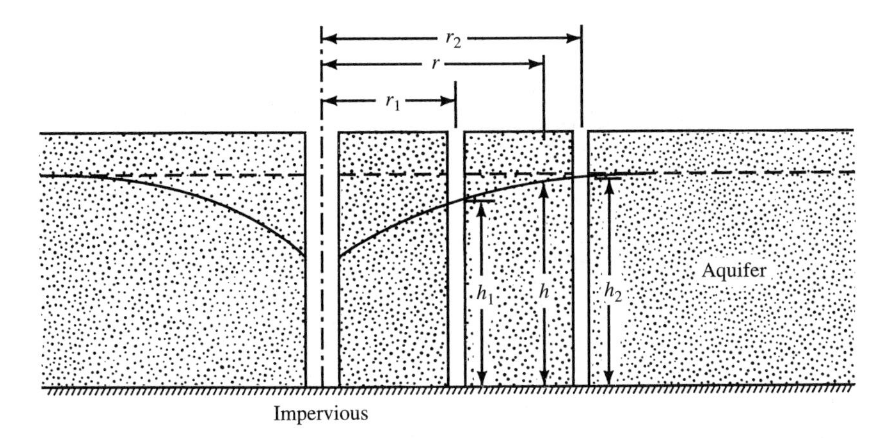

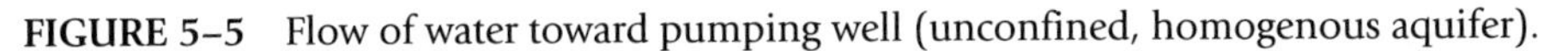
FIGURE 5–5 Flow of water toward pumping well (unconfined, homogenous aquifer).

$$q[\ln r]_{r_1}^{r_2} = 2\pi k \left[\frac{h^2}{2}\right]_{h_1}^{h_2} \tag{5–23}$$

$$q \ln \frac{r_2}{r_1} = \pi k \,(h_2^2 - h_1^2) \tag{5–24}$$

$$k = \frac{q \ln (r_2/r_1)}{\pi(h_2^2 - h_1^2)} \tag{5–25}$$

EXAMPLE 5–5

Given

A pumping test was performed in a well penetrating a confined aquifer (Figure 5–4) to evaluate the coefficient of permeability of the soil in the aquifer. When equilibrium flow was reached, the following data were obtained:

1. Equilibrium discharge of water from the well = 200 gal/min.
2. Water levels (h_1 and h_2) = 15 and 18 ft at distances from the well (r_1 and r_2) of 60 and 180 ft, respectively.
3. Thickness of aquifer = 20 ft.

Required

Coefficient of permeability of the soil in the aquifer.

Solution

From Eq. (5–20),

$$k = \frac{q \ln (r_2/r_1)}{2\pi H(h_2 - h_1)} \tag{5–20}$$

$$q = (200 \text{ gal/min})(1 \text{ ft}^3/7.48 \text{ gal})(1 \text{ min}/60 \text{ s}) = 0.4456 \text{ ft}^3/\text{s}$$

$$k = \frac{(0.4456 \text{ ft}^3/\text{s}) \ln (180 \text{ ft}/60 \text{ ft})}{(2)(\pi)(20 \text{ ft})(18 \text{ ft} - 15 \text{ ft})} = 0.00130 \text{ ft/s}$$

EXAMPLE 5–6

Given

Same conditions as in Example 5–5, except that the well is located in an unconfined aquifer (Figure 5–5).

Required

Coefficient of permeability of the soil in the aquifer.

Solution

From Eq. (5–25),

$$k = \frac{q \ln (r_2/r_1)}{\pi(h_2^2 - h_1^2)} \tag{5–25}$$

$$k = \frac{(0.4456 \text{ ft}^3/\text{s}) \ln (180 \text{ ft}/60 \text{ ft})}{(\pi)[(18 \text{ ft})^2 - (15 \text{ ft})^2]} = 0.00157 \text{ ft/s}$$

Empirical Relationships for Coefficient of Permeability

Through the years, investigators have studied the flow of water through soil in tubes and conduits in an attempt to relate permeability to a soil's grain size. Because permeability is related to pore area, and pore area is related to grain size, it follows that the coefficient of permeability might be quantified in terms of grain size. Some relationships have been found that are somewhat valid for granular soils. Two such permeability-grain-size relationships are presented next.

The coefficient of permeability for uniform sands in a loose state can be estimated by using an empirical formula proposed by Hazen as follows [2]*

$$k = C_1 D_{10}^2 \tag{5–26}$$

where k = coefficient of permeability (cm/s)
C_1 = 100 to 150 (1/cm · s)
D_{10} = effective grain size (soil particle diameter corresponding to 10% passing on the grain-size distribution curve; see Section 2–2) (cm)

For dense or compacted sands, the coefficient of permeability can be approximated by using the following equation [3]**

$$k = 0.35 D_{15}^2 \tag{5–27}$$

* From Karl Terzaghi and Ralph B. Peck, *Soil Mechanics in Engineering Practice*, 2nd ed., Copyright © 1967 by John Wiley & Sons, Inc., New York.

** From James L. Sherard, Lorn P. Dunnigan, and James R. Talbot, "Basic Properties of Sand and Gravel Filters," *J. Geotech. Eng. Div. ASCE*, **110**(6), 684–700 (June 1984). Reproduced by permission of ASCE.

TABLE 5–1
Classification of Soils According to Their Coefficients of Permeability [4][1]

Degree of Permeability	Value of k (m/s)
High	Over 10^{-3}
Medium	10^{-3} to 10^{-5}
Low	10^{-5} to 10^{-7}
Very low	10^{-7} to 10^{-9}
Practically impermeable	Less than 10^{-9}

[1] From Karl Terzaghi, Ralph B. Peck, and Gholamreza Mesri, *Soil Mechanics in Engineering Practice,* 3rd ed., John Wiley & Sons, Inc., New York, 1996. Copyright © 1996, by John Wiley & Sons, Inc. Reprinted by permission of John Wiley & Sons, Inc.

where k = coefficient of permeability (cm/s)
D_{15} = soil particle diameter corresponding to 15% passing on the grain-size distribution curve (mm)

If silts and/or clays are present in a sandy soil, even in small amounts, the coefficient of permeability may change significantly, because the fine silt–clay particles clog the sand's pore area.

Permeability varies greatly among the types of soils encountered in practice. Table 5–1 gives a broad classification of soils according to their coefficients of permeability. Table 5–2 gives ranges of coefficients of permeability to be expected for common natural soil formations. Table 5–3 gives additional information with regard to the range of the coefficient of permeability, drainage characteristics, and the most suitable methods for determining coefficients of permeability for various soils.

Permeability in Stratified Soils

In the preceding discussion in this section, soil was assumed to be homogeneous, with the same value of permeability k throughout. In reality, natural soil deposits are often nonhomogeneous, and the value of k varies, sometimes greatly, within a given soil mass. When one tries to analyze permeability in a nonhomogeneous soil, a simplification can be made to consider an aquifer consisting of layers of soils with differing permeabilities. Figure 5–6 depicts such a case, with layers of soils having permeabilities $k_1, k_2, k_3, \ldots, k_n$ and thicknesses $H_1, H_2, H_3, \ldots, H_n$. The general procedure is to find and use an average value of k. Because flow can occur in either the horizontal or vertical (x or y) direction, each of these cases is considered separately. (Of course, the flow could be in some oblique direction as well, but that case is not considered here.)

Consider first the case where flow is in the y direction (Figure 5–6). Because the water must travel successively through layers 1, 2, 3, . . . , n, the flow rate and velocity through each layer must be equal. If i denotes the overall hydraulic gradient, i_1,

TABLE 5–2
Coefficient of Permeability of Common Natural Soil Formations [4][1]

Formation	Value of *k* (m/s)
River deposits	
Rhône at Genissiat	Up to 4×10^{-3}
Small streams, eastern Alps	2×10^{-4} to 2×10^{-3}
Missouri	2×10^{-4} to 2×10^{-3}
Mississippi	2×10^{-4} to 10^{-3}
Glacial deposits	
Outwash plains	5×10^{-4} to 2×10^{-2}
Esker, Westfield, Mass.	10^{-4} to 10^{-3}
Delta, Chicopee, Mass.	10^{-6} to 1.5×10^{-4}
Till	Less than 10^{-6}
Wind deposits	
Dune sand	10^{-3} to 3×10^{-3}
Loess	$10^{-5}\pm$
Loess loam	$10^{-6}\pm$
Lacustrine and marine offshore deposits	
Very fine uniform sand, $C_U = 5$ to 2	10^{-6} to 6×10^{-5}
Bull's liver, Sixth Ave., N.Y., $C_U = 5$ to 2	10^{-6} to 5×10^{-5}
Bull's liver, Brooklyn, $C_U = 5$	10^{-7} to 10^{-6}
Clay	Less than 10^{-9}

[1] From Karl Terzaghi, Ralph B. Peck, and Gholamreza Mesri, *Soil Mechanics in Engineering Practice*, 3rd ed., John Wiley & Sons, Inc., New York, 1996. Copyright © 1996, by John Wiley & Sons, Inc. Reprinted by permission of John Wiley & Sons, Inc.

$i_2, i_3, \ldots, i_n$ represent gradients for each respective layer, and k_y is the average permeability of the entire stratified soil system in the y direction, application of Eq. (5–3) gives

$$v_y = k_y i = k_1 i_1 = k_2 i_2 = k_3 i_3 = \cdots = k_n i_n \tag{5–28}$$

Because total head loss is the sum of head losses in all layers,

$$iH = i_1 H_1 + i_2 H_2 + i_3 H_3 + \cdots + i_n H_n \tag{5–29}$$

or

$$i = \frac{i_1 H_1 + i_2 H_2 + i_3 H_3 + \cdots + i_n H_n}{H} \tag{5–30}$$

From Eq. (5–28),

$$i_1 = k_y i/k_1; \qquad i_2 = k_y i/k_2; \qquad i_3 = k_y i/k_3; \qquad \cdots \qquad i_n = k_y i/k_n \tag{5–31}$$

Substitute these values of $i_1, i_2, i_3, \cdots i_n$ into Eq. (5–30).

$$i = \frac{(k_y i/k_1)H_1 + (k_y i/k_2)H_2 + (k_y i/k_3)H_3 + \cdots + (k_y i/k_n)H_n}{H} \tag{5–32}$$

TABLE 5–3
Permeability and Drainage Characteristics of Soils[1] [2][2]

<table>
<tr><th></th><th colspan="12">Coefficient of Permeability k (cm/s)(Log Scale)</th></tr>
<tr><th></th><th>10^2</th><th>10^1</th><th>1.0</th><th>10^{-1}</th><th>10^{-2}</th><th>10^{-3}</th><th>10^{-4}</th><th>10^{-5}</th><th>10^{-6}</th><th>10^{-7}</th><th>10^{-8}</th><th>10^{-9}</th></tr>
<tr><td>Drainage</td><td colspan="6">Good</td><td colspan="2">Poor</td><td colspan="4">Practically impervious</td></tr>
<tr><td rowspan="2">Soil types</td><td colspan="2" rowspan="2">Clean gravel</td><td colspan="3">Clean sands, clean sand and gravel mixtures</td><td colspan="4">Very fine sands, organic and inorganic silts, mixtures of sand silt and clay, glacial till, stratified clay deposits, etc.</td><td colspan="3">"Impervious" soils (e.g., homogeneous clays below zone of weathering)</td></tr>
<tr><td colspan="2"></td><td colspan="5">"Impervious" soils modified by effects of vegetation and weathering</td><td colspan="3"></td></tr>
<tr><td rowspan="2">Direct determination of k</td><td colspan="6">Direct testing of soil in its original position—pumping tests; reliable if properly conducted; considerable experience required</td><td colspan="6"></td></tr>
<tr><td colspan="5">Constant-head permeameter; little experience required</td><td colspan="7"></td></tr>
<tr><td rowspan="2">Indirect determination of k</td><td colspan="2"></td><td colspan="3">Falling-head permeameter; reliable; little experience required</td><td colspan="3">Falling-head permeameter; unreliable; much experience required</td><td colspan="4">Falling-head permeameter; fairly reliable; considerable experience necessary</td></tr>
<tr><td colspan="5">Computation from grain-size distribution; applicable only to clean cohesionless sands and gravels</td><td colspan="4"></td><td colspan="3">Computation based on results of consolidation tests; reliable; considerable experience required</td></tr>
</table>

1 After Casagrande and Fadum (1940).

2 From Karl Terzaghi and Ralph B. Peck, *Soil Mechanics in Engineering Practice,* 2nd ed., Copyright © 1967 by John Wiley & Sons, Inc., New York.

$$i = \frac{(k_y i)(H_1/k_1 + H_2/k_2 + H_3/k_3 + \cdots + H_n/k_n)}{H} \tag{5–33}$$

Therefore,

$$k_y = \frac{H}{(H_1/k_1) + (H_2/k_2) + (H_3/k_3) + \cdots + (H_n/k_n)} \tag{5–34}$$

For flow in the x direction, let k_x denote the average permeability of the entire stratified soil system in that direction. In this case, total flow is the sum of the flows in all layers. Applying Eq. (5–2) and using H for the A term yields the following:

$$q = k_x iH = (k_1H_1 + k_2H_2 + k_3H_3 + \cdots + k_nH_n)i \tag{5–35}$$

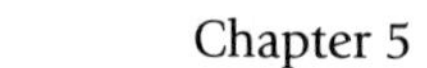

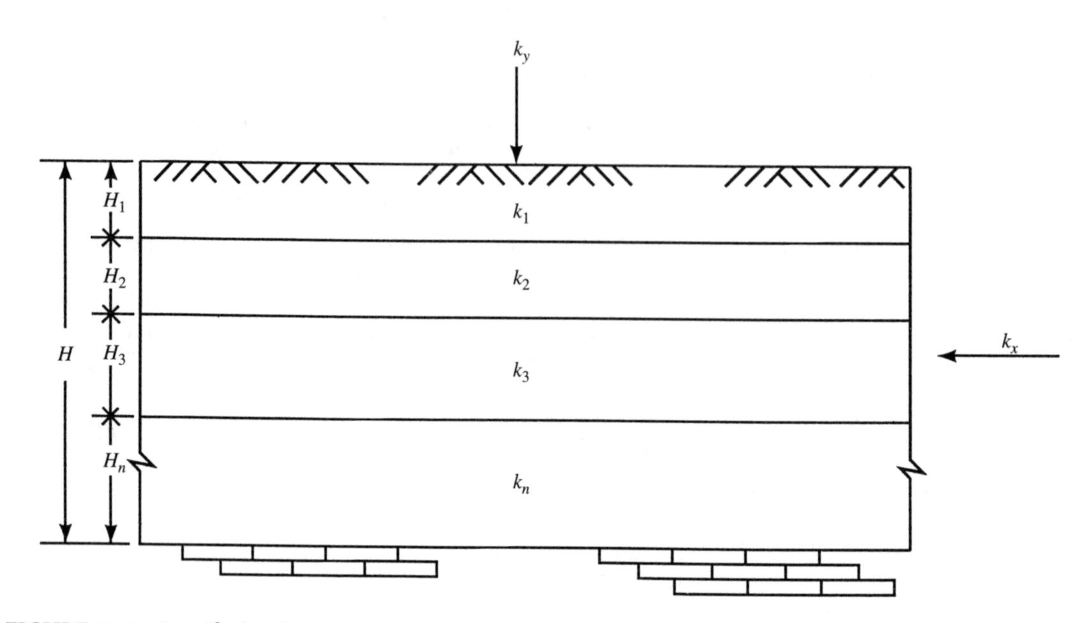

FIGURE 5–6 Stratified soil consisting of layers with various permeabilities.

or

$$k_x = \frac{k_1H_1 + k_2H_2 + k_3H_3 + \cdots + k_nH_n}{H} \tag{5–36}$$

In stratified soils, average horizontal permeability (k_x) is greater than average vertical permeability (k_y).

EXAMPLE 5–7

Given

A nonhomogeneous soil consisting of layers of soil with different permeabilities as shown in Figure 5–7.

FIGURE 5–7

1.5 m — $k_x = 1.2 \times 10^{-3}$ cm/s, $k_y = 2.4 \times 10^{-4}$ cm/s

2.0 m — $k_x = 2.8 \times 10^{-4}$ cm/s, $k_y = 3.1 \times 10^{-5}$ cm/s

2.5 m — $k_x = 5.5 \times 10^{-5}$ cm/s, $k_y = 4.7 \times 10^{-6}$ cm/s

Required

1. Estimate the average coefficient of permeability in the horizontal direction (k_x).
2. Estimate the average coefficient of permeability in the vertical direction (k_y).

Solution

1. From Eq. (5–36),

$$k_x = \frac{k_1H_1 + k_2H_2 + k_3H_3 + \ldots + k_nH_n}{H} \tag{5-36}$$

$$k_x = \frac{(1.2 \times 10^{-3}\text{cm/s})(1.5\text{ m}) + (2.8 \times 10^{-4}\text{cm/s})(2.0\text{ m}) + (5.5 \times 10^{-5}\text{cm/s})(2.5\text{ m})}{1.5\text{ m} + 2.0\text{ m} + 2.5\text{ m}}$$

$$k_x = 4.16 \times 10^{-4}\text{ cm/s}$$

2. From Eq. (5–34),

$$k_y = \frac{H}{(H_1/k_1) + (H_2/k_2) + (H_3/k_3) + \cdots (H_n/k_n)} \tag{5-34}$$

$$k_y = \frac{1.5\text{ m} + 2.0\text{ m} + 2.5\text{ m}}{(1.5\text{ m})/(2.4 \times 10^{-4}\text{cm/s}) + (2.0\text{ m})/(3.1 \times 10^{-5}\text{cm/s}) + (2.5\text{ m})/(4.7 \times 10^{-6}\text{cm/s})}$$

$$k_y = 9.96 \times 10^{-6}\text{ cm/s}$$

5–3 CAPILLARY RISE IN SOILS

As introduced in Chapter 2, *capillarity* refers to the rise of water (or another liquid) in a small-diameter tube inserted into the water, the rise being caused by both cohesion of the water's molecules and adhesion of the water to the tube's walls. Figure 5-8 illustrates the capillary rise of water in a tube. In equilibrium, the weight of water in the capillary tube (a downward force) must be exactly offset by the ability of the surface film to adhere to the tube's walls and hold the water in the tube (the upward force).

The weight of water in the tube is simply the volume of water multiplied by the unit weight of water (γ), or $\pi r^2 h\gamma$, where r is the tube's radius and h is the height of rise. The upward force (adhesion) is equal to the surface tension force developed around the circumference of the tube; it is computed by multiplying the value of surface tension T (a property of water defined as a force per unit length of free surface) by the tube's circumference, by the cosine of the angle formed between a tangent to the meniscus and the capillary wall. For water and a glass tube, the meniscus at the capillary wall is tangent to the wall surface; hence, the angle is zero and its cosine is one. The upward force is therefore $2\pi rT$. Equating the downward and upward forces gives the following:

$$\pi r^2 h\gamma = 2\pi rT \tag{5-37}$$

Solving for h yields

$$h = \frac{2T}{r\gamma} \tag{5-38}$$

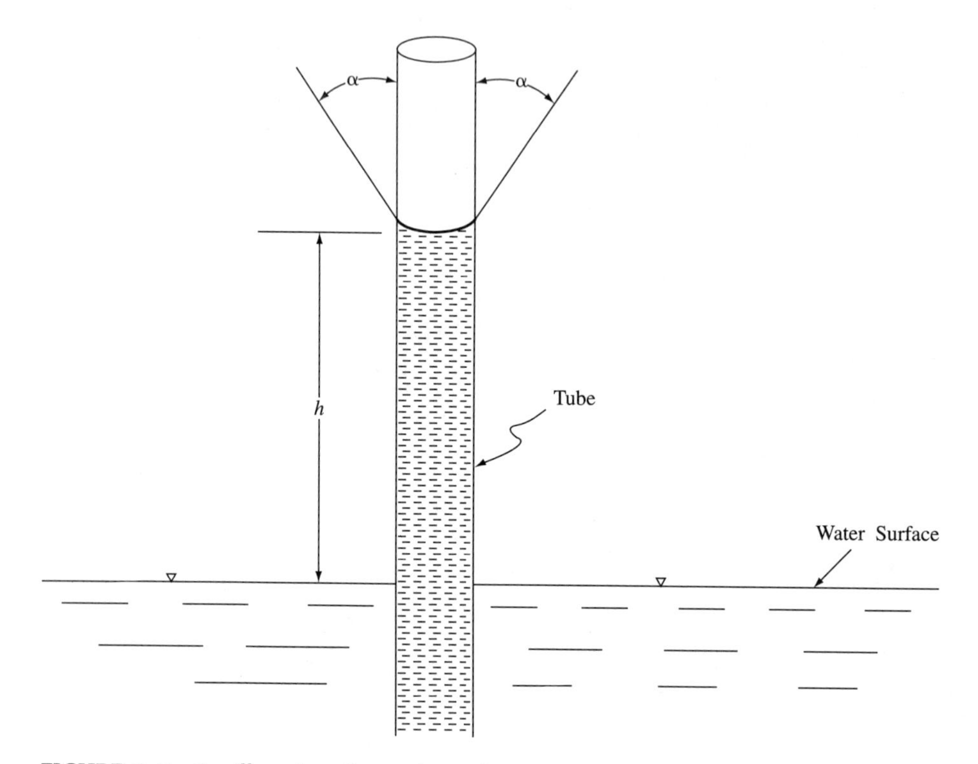

FIGURE 5–8 Capillary rise of water in a tube.

or

$$h = \frac{4T}{d\gamma} \tag{5-39}$$

where
h = height of rise
T = surface tension
r = tube radius
d = tube diameter
γ = unit weight of water

Equation (5–39) is applicable only to the rise of pure water in clean glass tubes. At 20°C (68°F), the values of surface tension and unit weight of water are approximately 0.0728 N/m (0.00501 lb/ft) and 9790 N/m^3 (62.4 lb/ft^3), respectively. If these values are substituted into Eq. (5–39), the resulting equation is

$$h = \frac{0.030}{d} \tag{5-40}$$

where h is in meters and d in millimeters. Equation (5–40) is, of course, valid only for water at 20°C, but that is roughly room temperature, and the equation gives generally adequate results for temperatures between 0 and 30°C.

EXAMPLE 5–8

Given

A clean glass capillary tube with a diameter of 0.5 mm is inserted into water with a surface tension of 0.073 N/m.

Required

The height of capillary rise in the tube.

Solution

From Eq. (5–39),

$$h = \frac{4T}{d\gamma} \tag{5–39}$$

$$h = \frac{(4)(0.073\ \text{N/m})}{[(0.5\ \text{mm})(1\ \text{m}/1000\ \text{mm})](9790\ \text{N/m}^3)} = 0.060\ \text{m}$$

or,

$$h = \frac{0.030}{d} \tag{5–40}$$

$$h = \frac{0.030}{0.5\ \text{mm}} = 0.060\ \text{m}$$

With soils, capillarity occurs at the groundwater table when water rises from saturated soil below into dry or partially saturated soil above the water table. The "capillary tubes" through which water rises in soils are actually the void spaces among soil particles. Because the voids interconnect in varying directions (not just vertically) and are irregular in size and shape, accurate calculation of the height of capillary rise is virtually impossible. It is known, however, that the height of capillary rise is associated with the mean diameter of a soil's voids, which is in turn related to average grain size. In general, the smaller the grain size, the smaller the void space, and consequently the greater will be the capillary rise. Thus, clayey soils, with the smallest grain size, should theoretically experience the greatest capillary rise, although the rate of rise may be extremely slow because of the characteristically low permeability of such soils. In fact, the largest capillary rise for any particular length of time generally occurs in soils of medium grain sizes (such as silts and very fine sands).

A crude approximation of the maximum height of capillary rise of water in a particular soil can be determined from the following equation [5]*

* From Ralph B. Peck, Walter E. Hansen, and Thomas H. Thornburn, *Foundation Engineering,* 2nd ed., Copyright © 1974 by John Wiley & Sons, Inc., New York.

$$h = \frac{C}{eD_{10}} \tag{5-41}$$

where h = maximum height of capillary rise
C = empirical coefficient
e = soil's void ratio
D_{10} = effective grain size (see Section 2–2)

With D_{10} expressed in centimeters and C, which depends on surface impurities and the shape of grains, ranging from 0.1 to 0.5 cm^2, the computed value of h will be in centimeters. Equation (5–41) gives the maximum height of capillary rise for smaller voids. Larger voids overlying smaller voids may interfere with the capillary process and thereby cause values of h from Eq. (5–41) to be invalid.

5–4 FROST ACTION IN SOILS

It is well known from physics that water expands when it is cooled and freezes. When the temperature in a soil mass drops below water's freezing point, water in the voids freezes and therefore expands, causing the soil mass to move upward. This vertical expansion of soil caused by freezing water within is known as *frost heave.* Serious damage may result from frost heave when structures such as pavements and building foundations supported by soil are lifted. Because the amount of frost heave (i.e., upward soil movement) is not necessarily uniform in a horizontal direction, cracking of pavements, building walls, and floors may occur. When the temperature rises above the freezing point, frozen soil thaws from the top downward. Because resulting melted water near the surface cannot drain through underlying frozen soil, an increase in water content of the upper soil, a decrease in its strength, and subsequent settlement of the structure may occur. Clearly, such alternate lifting and settling of pavements and structures as a result of freezing and thawing of soil pore water are undesirable, may cause serious structural damage, and should be avoided or at least minimized.

The actual amount of frost heave in any particular soil is difficult to compute or even estimate or predict accurately. Although pore water freezes in soil when the temperature is low enough, the frozen water is not necessarily uniform; and ice layers, or lenses, may occur. Capillary water rising from the water table can add to an ice lens, thereby increasing its volume and causing large heaves to occur. Frost heaves of a few inches are common in the northern half of the United States and may, in extreme cases, be much greater. Figure 5–9 gives maximum depths of frost penetration (in inches) for the conterminous United States.

Because frost heave is a natural phenomenon and is virtually unpreventable, the best defense against structural damage therefrom is to construct foundations deep enough to escape the effects of frost heave. A rule of thumb is to place foundations to a depth equal to or greater than the depth of frost penetration (Figure 5–9) in a given area. In making such a judgment, one must remember that the location of the water table is not fixed. Of course, if a given soil is not susceptible to frost action or if no water is pres-

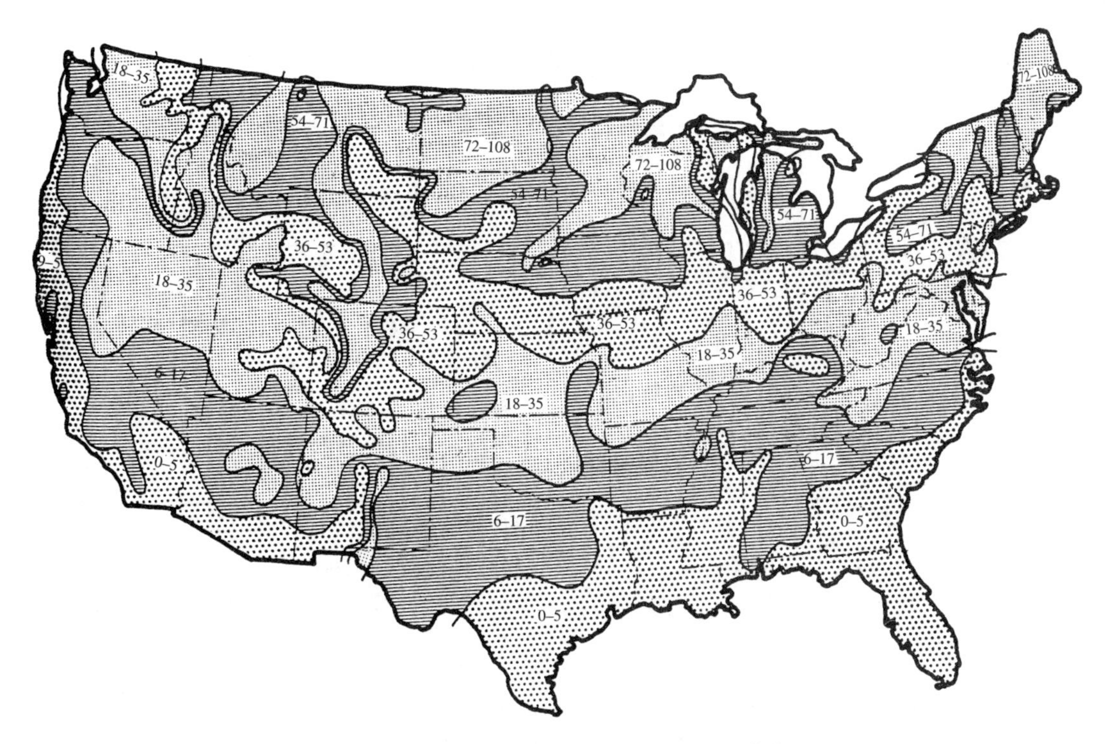

FIGURE 5–9 Maximum depth of frost penetration in the United States [6].

ent (and is never expected to be present), severe frost heave problems may not occur. However, it is still good practice to construct foundations below the depth of frost penetration rather than risk structural failure resulting from possible future frost heave.

5–5 FLOW NETS AND SEEPAGE

When water flows underground through well-defined aquifers over long distances, the flow rate can be computed by using Darcy's law [Eq. (5–2)] if the individual terms in the equation can be evaluated. In cases where the path of flow is irregular or if the water entering and leaving the permeable soil is over a short distance, flow boundary conditions may not be so well defined; and analytic solutions, such as the use of Eq. (5–2), become difficult. In such cases, flow may be evaluated by using *flow nets.*

Figure 5–10 illustrates a flow net. In the figure, water seeps through the permeable stratum beneath the wall from the upstream side (left) to the downstream side (right). The solid lines below the wall are known as *flow lines.* Each flow line represents the path along which a given water particle travels in moving from the upstream

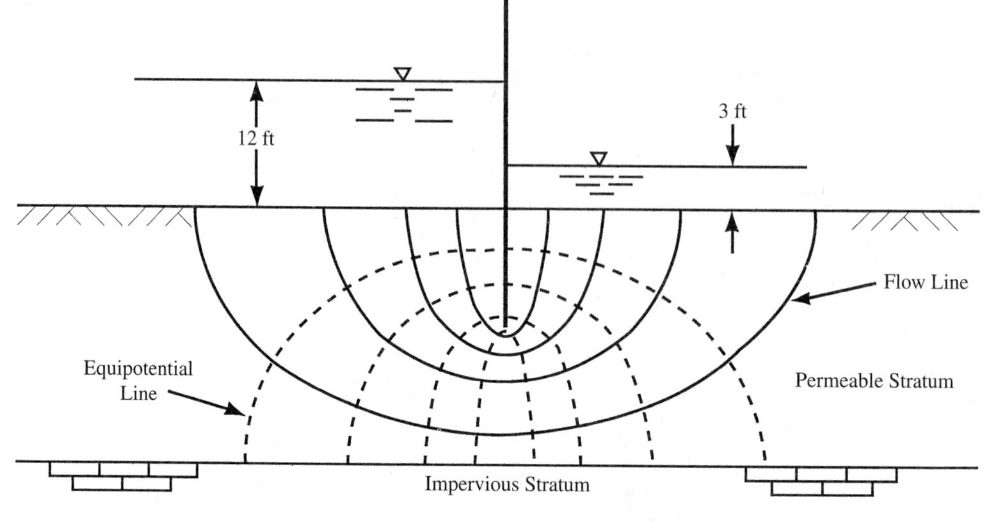

FIGURE 5–10 Flow net.

side through the permeable stratum to the downstream side. The dashed lines in Figure 5–10 represent *equipotential lines.* They connect points on different flow lines having equal total energy heads. A collection of flow lines intersecting equipotential lines, as shown in Figure 5–10, constitutes a flow net; as demonstrated subsequently, it is a useful tool in evaluating seepage through permeable soil.

Construction of Flow Nets

Construction of a flow net requires, as a first step, a scale drawing of a cross section of the flow path, as shown in Figure 5–11a. In addition to the pervious soil mass, the drawing shows the impervious boundaries that restrict flow and the pervious boundaries through which water enters and exits the soil.

The second step is to sketch several (generally, two to four) flow lines. As indicated previously, they represent paths along which given water particles travel in moving through the permeable stratum. As shown in Figure 5–11b, they should be drawn approximately parallel to the impervious boundaries and perpendicular to the pervious boundaries.

The next step is to sketch equipotential lines. Because they connect points on different flow lines having equal total energy heads, they should be drawn approximately perpendicular to the flow lines, as illustrated in Figure 5–11c. Furthermore, they should be drawn to form quasi-squares where equipotential lines and flow lines intersect. In other words, intersecting equipotential lines and flow lines should form figures that each have approximately equal lengths and widths.

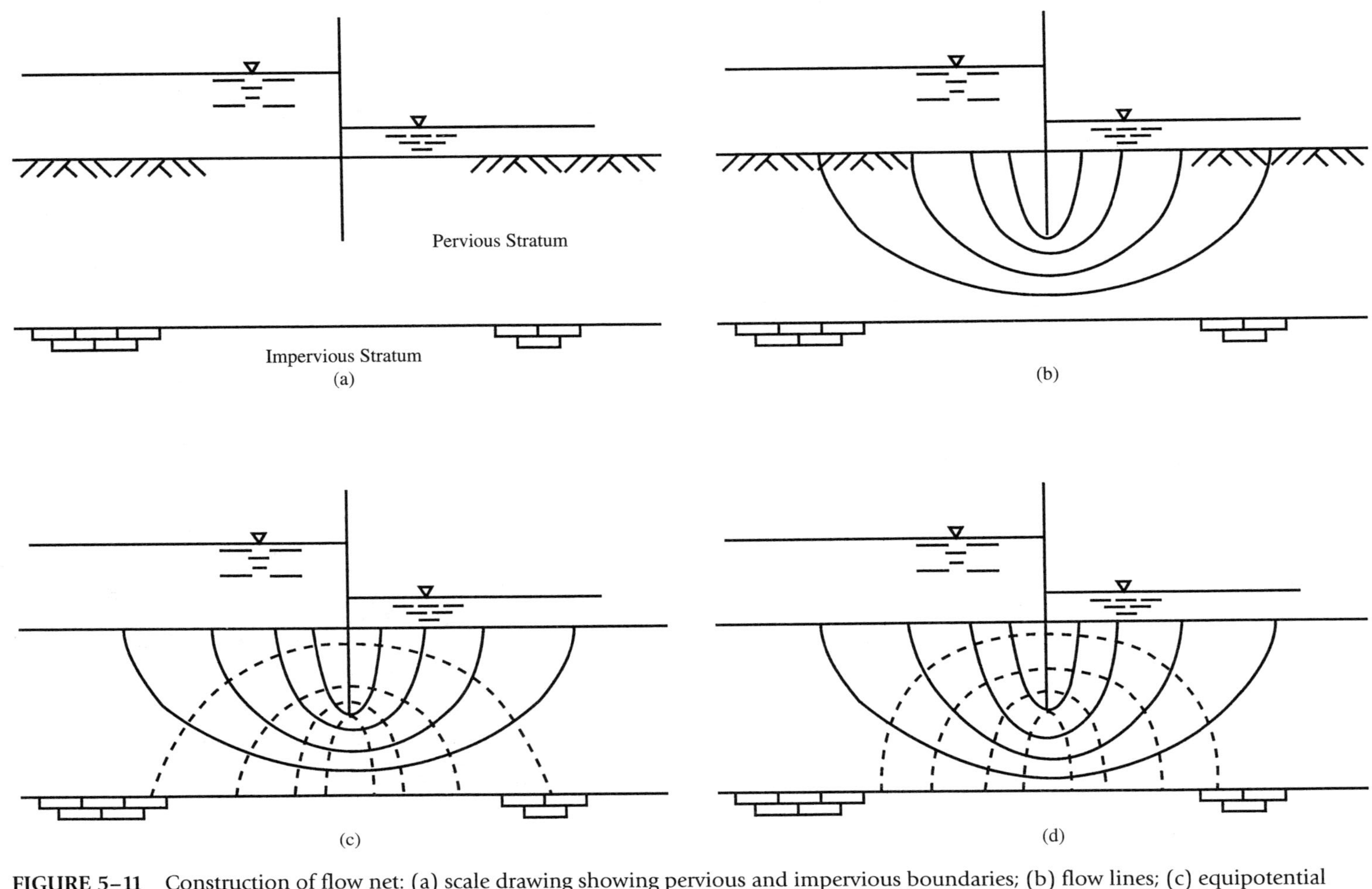

FIGURE 5–11 Construction of flow net: (a) scale drawing showing pervious and impervious boundaries; (b) flow lines; (c) equipotential lines; (d) final flow net.

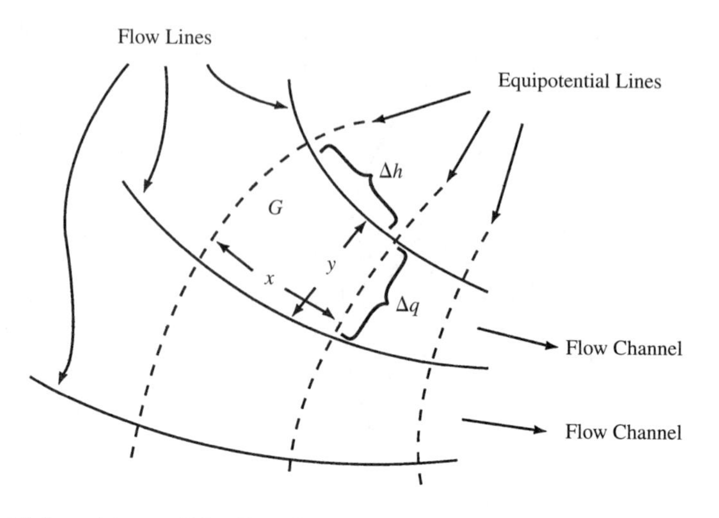

FIGURE 5–12 Flow channel and equipotential drops.

Because the initial positions of the flow lines represent guesses, the first attempt at constructing a flow net will usually not be totally accurate (i.e., will not result in the necessary quasi-squares). Hence, the fourth and final step is to use the first attempted flow net as a guide to adjust the equipotential lines and the flow lines so that all figures have equal widths and lengths and all intersections are at right angles as nearly as possible. Figure 5–11d shows the final flow net achieved by adjusting the initial flow net attempt (Figure 5–11c). It should be noted from Figure 5–11d that the figures formed are generally not all perfect squares because their lengths and widths are not all equal, their sides are seldom straight lines, and the lines forming them do not always intersect at precise right angles. Nevertheless, they should be drawn to approximate square figures.

Calculation of Seepage Flow

Once a suitable flow net has been prepared as described in the preceding paragraphs, seepage flow can be determined by modifying Darcy's law, as follows.

$$q = kiA \tag{5–2}$$

Consider one square in a flow net—for example, the one labeled G in Figure 5–12. Let Δq and Δh denote the flow rate and drop in head (energy), respectively, for this square. Because each square is x units wide and y units long and has a unit width perpendicular to the figure, term i in Eq. (5–2) is given by $\Delta h/x$, and term A is equal to y. Hence,

$$\Delta q = k\frac{\Delta h}{x}y \tag{5–42}$$

However, because the figure is square, y/x is unity and

$$\Delta q = k\Delta h \tag{5–43}$$

If N_d represents the number of equipotential increments (spaces between equipotential lines), then Δh equals h/N_d and

$$\Delta q = \frac{kh}{N_d} \tag{5-44}$$

If N_f denotes the number of flow paths (spaces between flow lines), then Δq equals q/N_f (where q is the total flow rate of the flow net per unit width) and

$$\frac{q}{N_f} = \frac{kh}{N_d} \tag{5-45}$$

or

$$q = \frac{khN_f}{N_d} \tag{5-46}$$

Example 5–9 illustrates the computation of seepage through a flow net using Eq. (5–46).

EXAMPLE 5–9

Given

For the flow net depicted in Figure 5–10, the coefficient of permeability of the permeable soil stratum is 4.80×10^{-3} cm/s.

Required

The total rate of seepage per unit width of sheet pile through the permeable stratum.

Solution

From Eq. (5–46),

$$q = \frac{khN_f}{N_d} \tag{5-46}$$

$$k = (4.80 \times 10^{-3}\ \text{cm/s})(1\ \text{in.}/2.54\ \text{cm})(1\ \text{ft}/12\ \text{in.})$$
$$= 1.57 \times 10^{-4}\ \text{ft/s}$$
$$h = 12\ \text{ft} - 3\ \text{ft} = 9\ \text{ft}$$
$$N_f = 5$$
$$N_d = 9$$
$$q = \frac{(1.57 \times 10^{-4}\ \text{ft/s})(9\ \text{ft})(5)}{9} = 7.85 \times 10^{-4}\ \text{ft}^3/\text{s per foot of sheet pile}$$

In the foregoing discussion of flow nets, it was assumed that soil was isotropic—that is, equal soil permeability in all directions. In actuality, natural soils are not isotropic, but often soil permeabilities in vertical and horizontal directions are similar enough that the assumption of isotropic soil is acceptable for finding flow without appreciable error. In stratified soil deposits, however, where horizontal permeability is much greater than vertical permeability, the flow net must be modified and Eq. (5–46) altered to compute flow. For the situation where k_y and k_x (representing average vertical and horizontal coefficients of permeability, respectively) differ appreciably, the method for constructing the flow net can be modified by use of a *transformed section* to account for the different permeabilities. The modification is done when the scale drawing of the cross section of the flow path is prepared. Vertical lengths are plotted in the usual manner to fit the scale selected for the sketch, but horizontal dimensions are first altered by multiplying all horizontal lengths by the factor $\sqrt{k_y/k_x}$ and plotting the results to scale. The resulting drawing will appear somewhat distorted, with apparently shortened horizontal dimensions. The conventional flow net is then sketched on the transformed section in the manner described previously. In analyzing the resulting flow net to compute seepage flow, one must replace the k term in Eq. (5–46) with the factor $\sqrt{k_y/k_x}$, which was used in plotting the drawing. Thus, for flow through stratified, nonisotropic soil, the seepage equation becomes

$$q = \sqrt{\frac{k_y}{k_x}}\frac{hN_f}{N_d} \tag{5–47}$$

5–6 PROBLEMS

5–1. Water flows through a sand filter as shown in Figure 5–13. The soil mass's cross-sectional area and length are 400 in.^{2} and 5.0 ft, respectively. If the coefficient of permeability of the sand filter is 3.6×10^{-2} cm/s, find the flow rate of water through the soil.

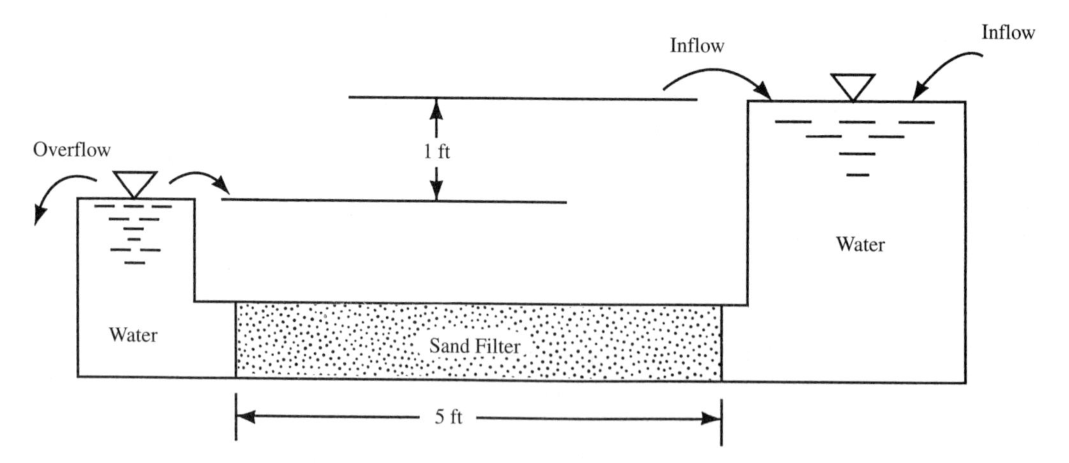

FIGURE 5–13

5–2. A quantity of 2000 ml of water required 20 min to flow through a sand sample, the cross-sectional area of which was 60.0 cm^2. The void ratio of the sand was 0.71. Compute the velocity of water moving through the soil and the actual (interstitial) velocity.

5–3. A constant-head permeability test was conducted on a clean sand sample (Figure 5–2). The diameter and length of the test specimen were 10.0 and 12.0 cm, respectively. The head difference between manometer levels was 4.9 cm during the test, and the water temperature was 20°C. If it took 152 s for 500 ml of water to discharge, determine the soil's coefficient of permeability.

5–4. A falling-head permeability test was conducted on a silty clay sample (Figure 5–3). The diameter and length of the test specimen were 10.20 and 16.20 cm, respectively. The cross-sectional area of the standpipe was 1.95 cm^2, and the water temperature was 20°C. If it took 35 min for the water in the standpipe to drop from a height of 100.0 cm at the beginning of the test to 92.0 cm at the end, determine the soil's coefficient of permeability.

5–5. A pump test was conducted on a test well in an unconfined aquifer, with the results as shown in Figure 5–14. If water was pumped at a steady flow of 185 gal/min, determine the coefficient of permeability of the permeable soil.

5–6. A pump test was conducted on a test well drilled into a confined aquifer, with the results as shown in Figure 5–15. If water was pumped at a steady flow of 205 gal/min, determine the coefficient of permeability of the permeable soil in the aquifer.

5–7. A grain-size analysis for a uniform sand in a loose state indicated that the soil particle diameter corresponding to 10% passing on the grain-size distribution curve is 0.18 mm. Estimate the coefficient of permeability.

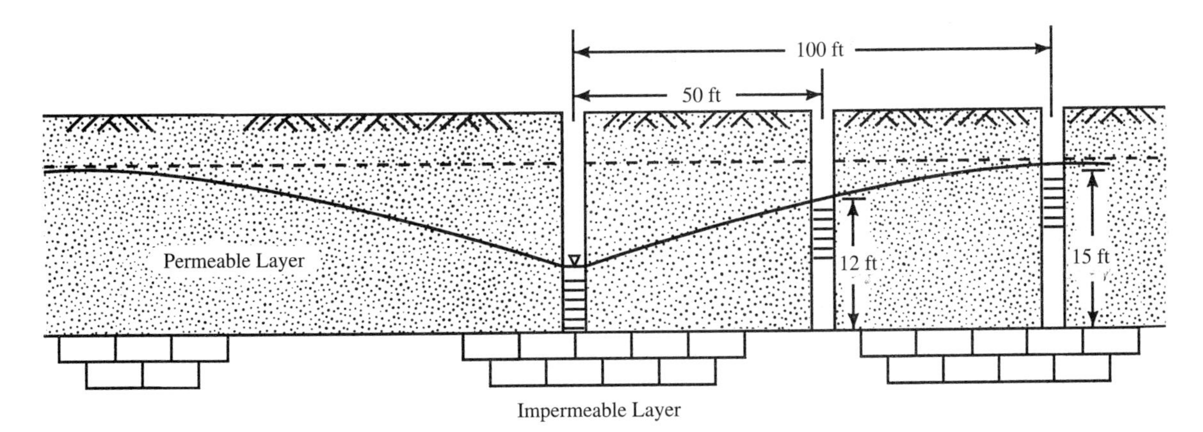

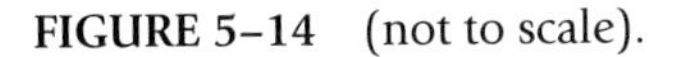
FIGURE 5–14 (not to scale).

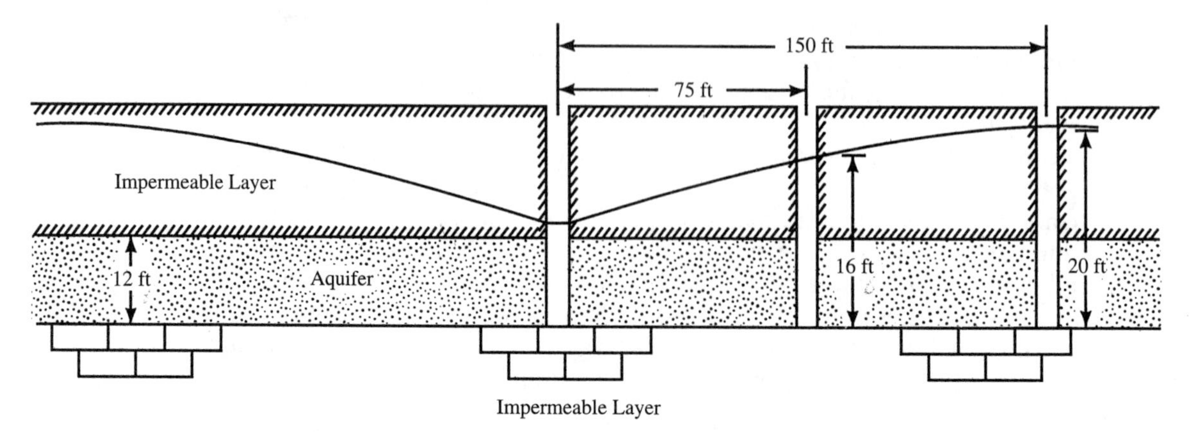

FIGURE 5–15 (not to scale).

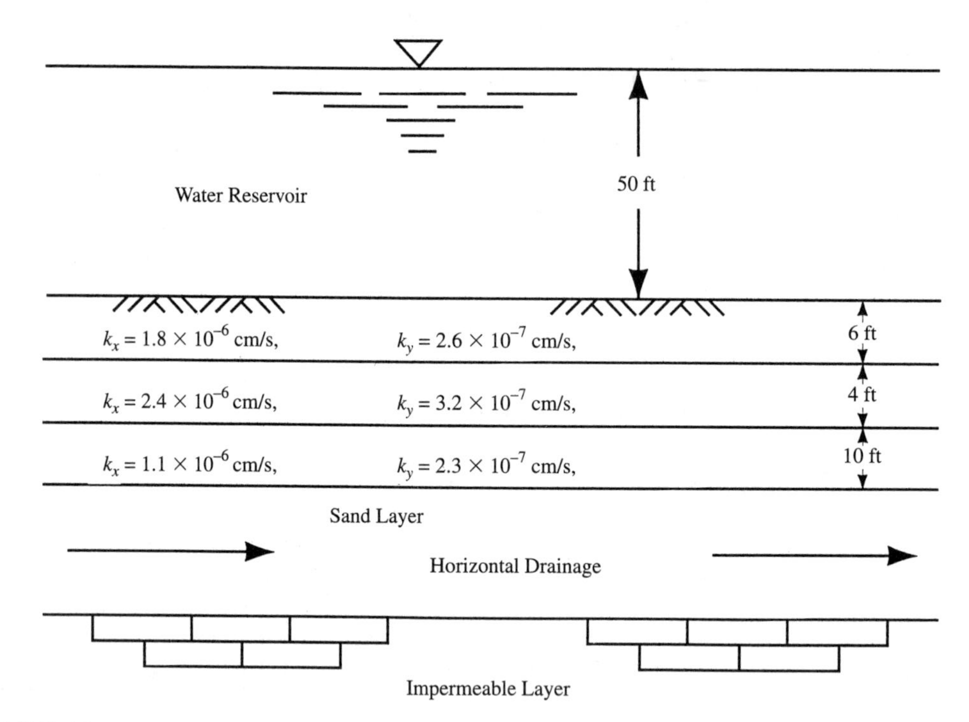

FIGURE 5–16

5–8. A grain-size analysis for a dense filter sand indicated that the soil particle diameter corresponding to 15% passing on the grain-size distribution curve is 0.25 mm. Estimate the coefficient of permeability.

5–9. A clean glass capillary tube having a diameter of 0.008 in. was inserted into water with a surface tension of 0.00504 lb/ft. Calculate the height of capillary rise in the tube.

5–10. A reservoir with a 35,000-ft^2 area is underlain by layers of stratified soils as depicted in Figure 5–16. Compute the water loss from the reservoir in 1 year. Assume that the pore pressure at the bottom sand layer is zero.

5–11. For the reservoir described in Problem 5–10, estimate the average coefficient of permeability in the horizontal direction.

5–12. Construct a flow net for the sheet pile shown in Figure 5–17. Estimate the seepage per foot of width of the sheet pile.

5–13. Construct a flow net for the concrete dam shown in Figure 5–18. Estimate the seepage per foot of width of the dam.

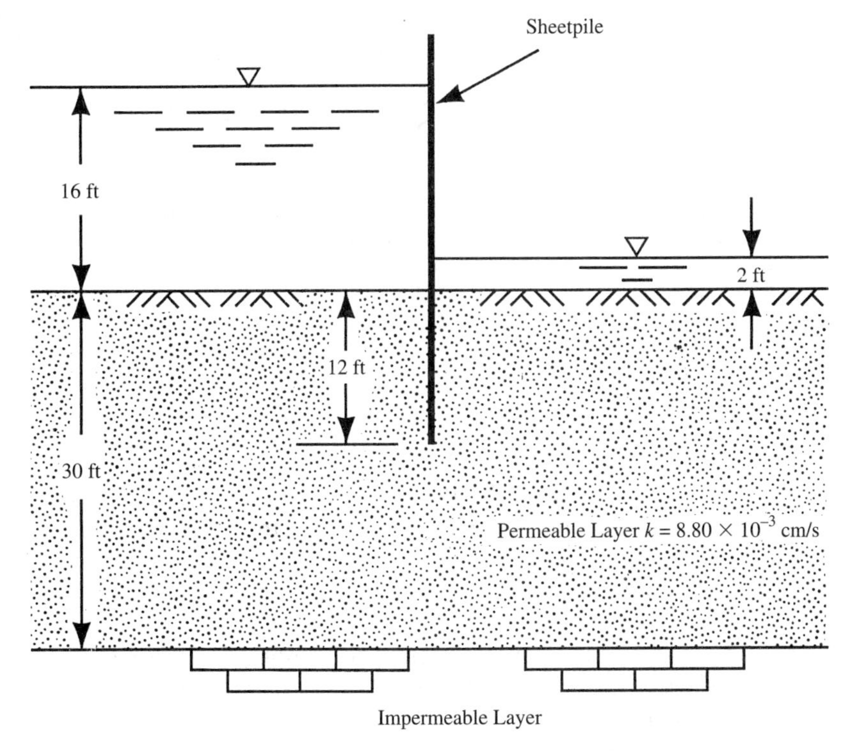

FIGURE 5–17

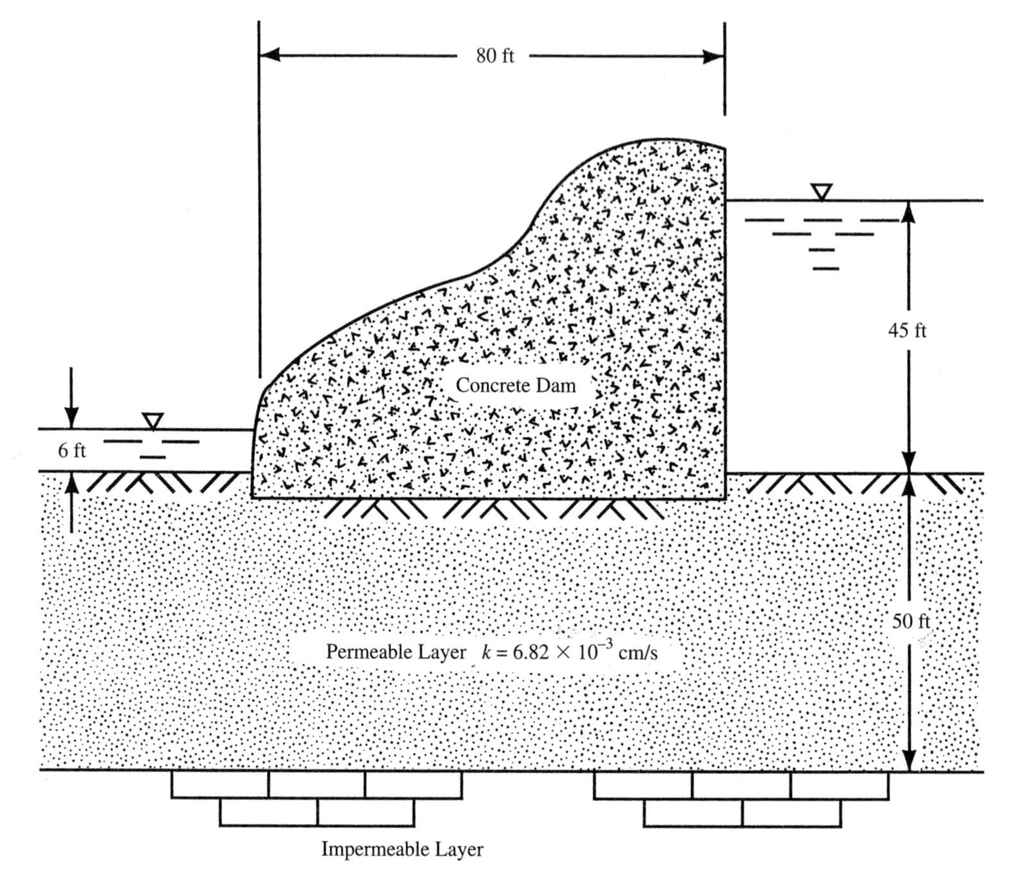

FIGURE 5–18

References

[1] *1989 Annual Book of ASTM Standards,* ASTM, Philadelphia, 1989. Copyright, American Society for Testing and Materials, 1916 Race Street, Philadelphia, PA 19103. Reprinted with permission.

[2] Karl Terzaghi and Ralph B. Peck, *Soil Mechanics in Engineering Practice,* 2nd ed., John Wiley & Sons, Inc., New York, 1967.

[3] James L. Sherard, Lorn P. Dunnigan, and James R. Talbot, "Basic Properties of Sand and Gravel Filters," *J. Geotech. Eng. Div. ASCE,* **110**(6), 684–700 (June 1984).

[4] Karl Terzaghi, Ralph B. Peck, and Gholamreza Mesri, *Soil Mechanics in Engineering Practice,* 3rd ed., John Wiley & Sons, Inc., New York, 1996.

[5] Ralph B. Peck, Walter E. Hansen, and Thomas H. Thornburn, *Foundation Engineering,* 2nd ed., John Wiley & Sons, Inc., New York, 1974.

[6] *Frost Action in Roads and Airfields,* Highway Research Board, Special Report No. 1, Publ. 211, National Academy of Sciences–National Research Council, Washington, D.C., 1952.

6

STRESS DISTRIBUTION IN SOIL

6–1 INTRODUCTION

If a vertical load of 1 ton is applied to a column of 1-ft^2 cross-sectional area, and the column rests directly on a soil surface, the vertical pressure exerted by the column onto the soil would be, on average, 1 ton/ft^2 (neglecting the column's weight). In addition to this pressure at the area of contact between column and soil, stress influence extends both downward and outward within the soil in the general area where the load is applied. The increase in pressure in the soil at any horizontal plane below the load is greatest directly under the load and diminishes outwardly (see Figure 6–1). The pressure's magnitude decreases with increasing depth. This is illustrated in Figure 6–1, where pressure p_2 at depth d_2 is less than pressure p_1 at depth d_1. Figure 6–1 also illustrates the increase in the area of stress influence outward with increase in depth.

Stress distribution in soil is quite important to soils engineers—particularly with regard to stability analysis and the settlement analysis of foundations. The remainder of this chapter deals with quantitative analyses of stress distribution in soil.

6–2 VERTICAL PRESSURE BELOW A CONCENTRATED LOAD

Two methods for calculating pressure below a concentrated load are presented here—the Westergaard equation and the Boussinesq equation. Both of these result from the theory of elasticity, which assumes that stress is proportional to strain. Implicit in this assumption is a homogeneous material, although soil is seldom homogeneous. The Westergaard equation is based on alternating thin layers of an elastic material between layers of an inelastic material. The Boussinesq equation assumes a homogeneous soil throughout [1].

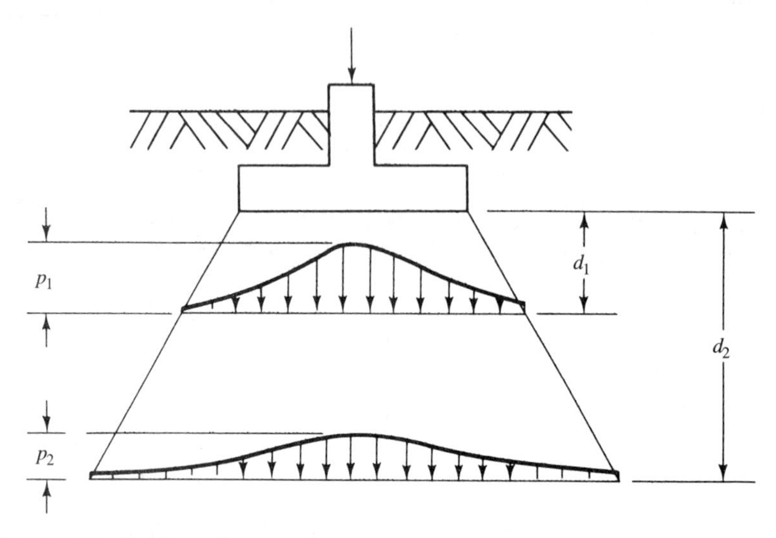

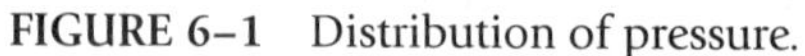
FIGURE 6–1 Distribution of pressure.

Westergaard Equation [1, 2]

The Westergaard equation is as follows:

$$q = \frac{Q\sqrt{(1 - 2\mu)/(2 - 2\mu)}}{2\pi z^2[(1 - 2\mu)/(2 - 2\mu) + (r/z)^2]^{3/2}} \qquad \textbf{(6–1)}$$

where q = vertical stress at depth z
Q = concentrated load
μ = Poisson's ratio (ratio of the strain in a material in a direction normal to an applied stress to the strain parallel to the applied stress)
z = depth
r = horizontal distance from point of application of Q to point at which q is desired

Note: q, the vertical stress at depth z resulting from load Q, is sometimes referred to as the *vertical stress increment* because it represents stress added by the load to the stress existing prior to application of the load. (The stress existing prior to application of the load is the *overburden pressure.*) This equation gives stress q as a function of both the vertical distance z and horizontal distance r between the point of application of Q and the point at which q is desired (see Figure 6–2). If Poisson's ratio is taken to be zero, Eq. (6–1) reduces to

$$q = \frac{Q}{\pi z^2[(1 + 2(r/z)^2]^{3/2}} \qquad \textbf{(6–2)}$$

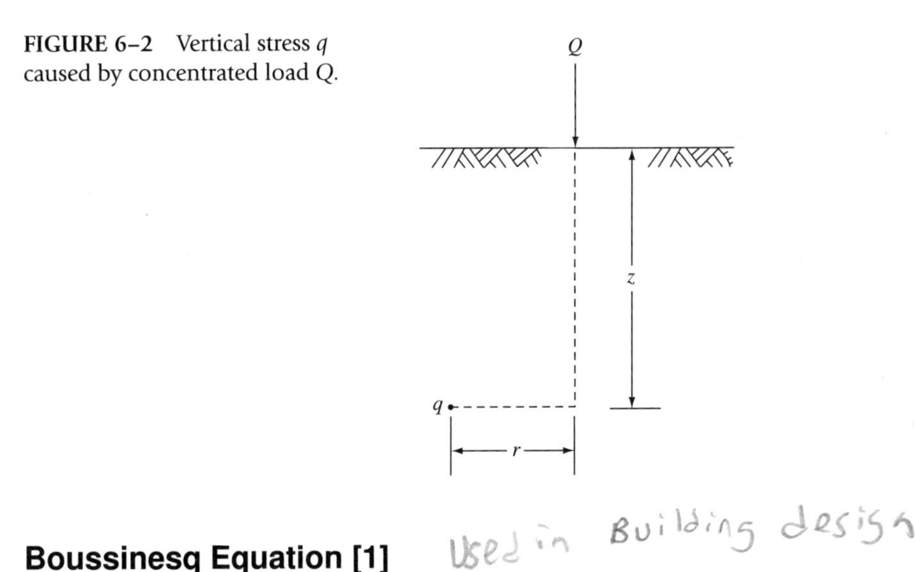

FIGURE 6–2 Vertical stress q caused by concentrated load Q.

Boussinesq Equation [1]

The Boussinesq equation is as follows:

$$q = \frac{3Q}{2\pi z^2[1 + (r/z)^2]^{5/2}} \tag{6–3}$$

where the terms are the same as those in Eq. (6–1). This equation also gives stress q as a function of both the vertical distance z and horizontal distance r. For low r/z ratios, the Boussinesq equation gives higher values of q than those resulting from the Westergaard equation. The Boussinesq equation is more widely used.

Although the Westergaard and Boussinesq equations are not excessively difficult to solve mathematically, computations of vertical stress (q) can be simplified by using *stress influence factors,* which are related to r/z. For example, the Westergaard equation can be written as follows:

$$q = \frac{Q}{\pi z^2[1 + 2(r/z)^2]^{3/2}} = \frac{Q}{z^2} I_W \tag{6–4}$$

where I_W is the stress influence factor for the Westergaard equation and the other terms are as in Eq. (6–2). Values of I_W for different values of r/z can be determined from Figure 6–3. Similarly, the Boussinesq equation can be written as follows:

$$q = \frac{3Q}{2\pi z^2[1 + (r/z)^2]^{5/2}} = \frac{Q}{z^2} I_B \tag{6–5}$$

where I_B is the stress influence factor for the Boussinesq equation. Values of I_B for different values of r/z can also be determined from Figure 6–3.

Examples 6–1 and 6–2 illustrate the use of the Boussinesq equation to calculate vertical stress below a concentrated load.

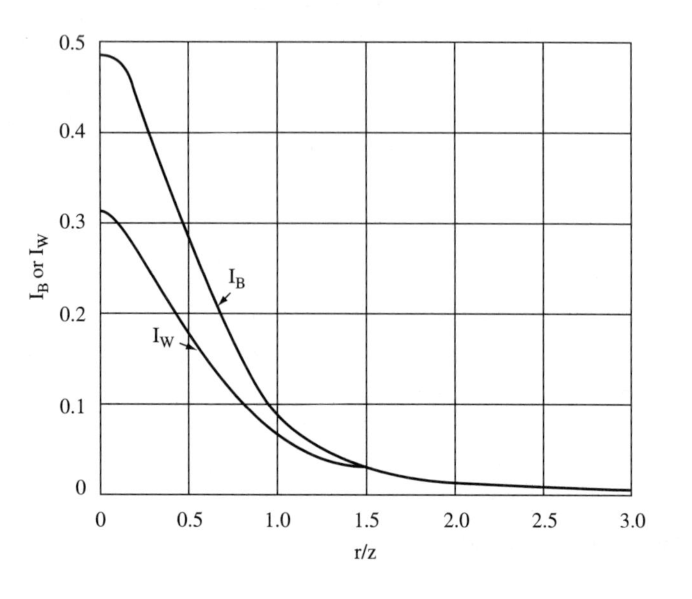

FIGURE 6–3 Chart for calculating vertical stresses caused by surface load Q.

EXAMPLE 6–1

Given

A concentrated load of 250 tons is applied to the ground surface.

Required

The vertical stress increment due to this load at a depth of 20 ft directly below the load.

Solution

From Eq. (6–3),

$$q = \frac{3Q}{2\pi z^2[1 + (r/z)^2]^{5/2}} \tag{6–3}$$

From given, $z = 20$ ft
$r = 0$
$Q = 250$ tons $= 500{,}000$ lb

Thus,

$$q = \frac{(3)(500{,}000 \text{ lb})}{(2)(\pi)(20 \text{ ft})^2[1 + (0/20 \text{ ft})^2]^{5/2}} = 597 \text{ lb/ft}^2$$

Alternatively, with $r/z = 0/20$ ft $= 0$, from Figure 6–3 $I_B = 0.48$. From Eq. (6–5),

$$q = \frac{Q}{z^2} I_B \qquad \text{(6–5)}$$

$$q = \frac{500{,}000 \text{ lb}}{(20 \text{ ft})^2} \times 0.48 = 600 \, lb/ft^2$$

EXAMPLE 6–2

Given

A concentrated load of 250 tons is applied to the ground surface.

Required

The vertical stress increment due to this load at a point 20 ft below the ground surface and 16 ft from the line of the concentrated load (i.e., $r = 16$ ft, $z = 20$ ft, as illustrated in Figure 6–4).

Solution

From Eq. (6–3),

$$q = \frac{3Q}{2\pi z^2[1 + (r/z)^2]^{5/2}} \tag{6–3}$$

From given, $z = 20$ ft
$r = 16$ ft
$Q = 250 \text{ tons} = 500{,}000$ lb

Thus,

$$q = \frac{(3)(500{,}000 \text{ lb})}{(2)(\pi)(20 \text{ ft})^2[1 + (16 \text{ ft}/20 \text{ ft})^2]^{5/2}} = 173 \text{ lb/ft}^2$$

Alternatively, with $r/z = (16 \text{ ft})/(20 \text{ ft}) = 0.80$, from Figure 6–3 $I_B = 0.14$. From Eq. (6–5),

$$q = \frac{Q}{z^2} I_B \tag{6–5}$$

$$q = \frac{500{,}000 \text{ lb}}{(20 \text{ ft})^2} \times 0.14 = 175 \text{ lb/ft}^2$$

FIGURE 6–4.

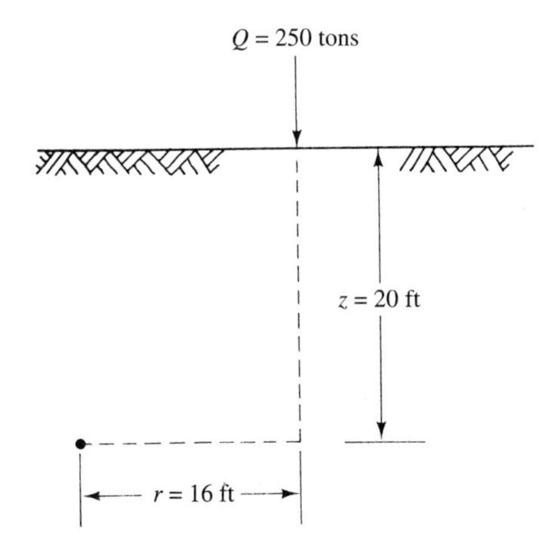

6–3 VERTICAL PRESSURE BELOW A LOADED SURFACE AREA (UNIFORM LOAD)

The methods presented in Section 6–2 deal with determination of vertical pressure below a concentrated load. Usually, however, concentrated loads are not applied directly onto soil. Instead, concentrated loads rest on footings, piers, and the like, and the load is applied to soil through footings or piers in the form of a loaded surface area (uniform load). Analysis of stress distributions resulting from loaded surface areas is generally more complicated than those resulting from concentrated loads.

Two methods for computing vertical pressure below a loaded surface area are discussed in this section. One is called the *approximate method;* the other is based on elastic theory.

Approximate Method [1]

The approximate method is based on the assumption that the area (in a horizontal plane) of stress below a concentrated load increases with depth, as shown in Figure 6–5. With the 2:1 slope shown, it is apparent that at any depth z, both L and B are increased by the amount z. Accordingly, stress at depth z is given by

$$q = \frac{Q}{(B + z)(L + z)} \tag{6–6}$$

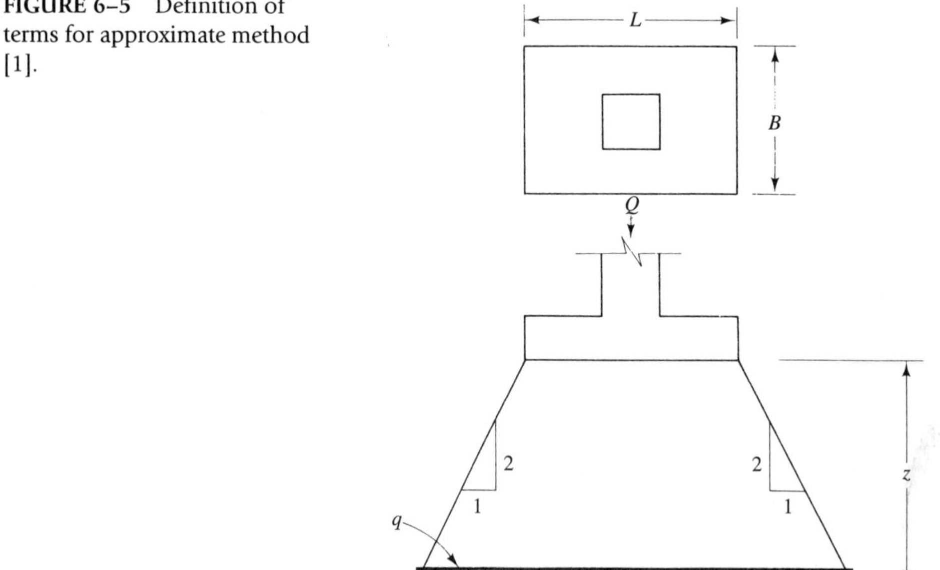

FIGURE 6–5 Definition of terms for approximate method [1].

where q = approximate vertical stress at depth z
Q = total load
B = width
L = length
z = depth

Because Q, L, and B are constants for a given application, it is obvious that the stress at depth z (q) decreases as depth increases. This method should be considered crude at best. It may be useful for preliminary stability analysis of footings; however, for settlement analysis the approximate method may likely not be accurate enough, and a more accurate approach based on elastic theory (discussed later in this section) may be required.

Examples 6–3 and 6–4 illustrate the use of the approximate method to calculate vertical pressure below a uniform load.

EXAMPLE 6–3

Given

A 10-ft by 15-ft rectangular area carrying a uniform load of 5000 lb/ft^2 is applied to the ground surface.

Required

The vertical stress increment due to this load at a depth of 20 ft below the ground surface by the approximate method.

Solution

From Eq. (6–6),

$$q = \frac{Q}{(B + z)(L + z)} \tag{6–6}$$

From given,
Q = (5000 lb/ft^2)(10 ft)(15 ft) = 750,000 lb
B = 10 ft
L = 15 ft
z = 20 ft

Thus,

$$q = \frac{750{,}000 \text{ lb}}{(10 \text{ ft} + 20 \text{ ft})(15 \text{ ft} + 20 \text{ ft})} = 714 \text{ lb/ft}^2$$

EXAMPLE 6–4

Given

A 3-m by 4-m rectangular area carrying a uniform load of 200 kN/m^2 is applied to the ground surface.

Required

The vertical stress increment due to this load at a depth of 6 m below the ground surface by the approximate method.

Solution

From Eq. (6–6),

$$q = \frac{Q}{(B + z)(L + z)} \tag{6-6}$$

From given, $Q = (200 \text{ kN/m}^2)(3 \text{ m})(4 \text{ m}) = 2400 \text{ kN}$
$B = 3 \text{ m}$
$L = 4 \text{ m}$
$z = 6 \text{ m}$

Thus,

$$q = \frac{2400 \text{ kN}}{(3 \text{ m} + 6 \text{ m})(4 \text{ m} + 6 \text{ m})} = 26.7 \text{ kN/m}^2$$

Method Based on Elastic Theory

Uniform Load on a Circular Area. Vertical pressure below a uniform load on a circular area can be determined utilizing Table 6–1 or Figure 6–6. Here, z and r

TABLE 6–1
Influence Coefficients for Points under Uniformly Loaded Circular Area [3]

	r/a									
z/a (1)	0 (2)	0.25 (3)	0.50 (4)	1.0 (5)	1.5 (6)	2.0 (7)	2.5 (8)	3.0 (9)	3.5 (10)	4.0 (11)
0.25	0.986	0.983	0.964	0.460	0.015	0.002	0.000	0.000	0.000	0.000
0.50	0.911	0.895	0.840	0.418	0.060	0.010	0.003	0.000	0.000	0.000
0.75	0.784	0.762	0.691	0.374	0.105	0.025	0.010	0.002	0.000	0.000
1.00	0.646	0.625	0.560	0.335	0.125	0.043	0.016	0.007	0.003	0.000
1.25	0.524	0.508	0.455	0.295	0.135	0.057	0.023	0.010	0.005	0.001
1.50	0.424	0.413	0.374	0.256	0.137	0.064	0.029	0.013	0.007	0.002
1.75	0.346	0.336	0.309	0.223	0.135	0.071	0.037	0.018	0.009	0.004
2.00	0.284	0.277	0.258	0.194	0.127	0.073	0.041	0.022	0.012	0.006
2.5	0.200	0.196	0.186	0.150	0.109	0.073	0.044	0.028	0.017	0.011
3.0	0.146	0.143	0.137	0.117	0.091	0.066	0.045	0.031	0.022	0.015
4.0	0.087	0.086	0.083	0.076	0.061	0.052	0.041	0.031	0.024	0.018
5.0	0.057	0.057	0.056	0.052	0.045	0.039	0.033	0.027	0.022	0.018
7.0	0.030	0.030	0.029	0.028	0.026	0.024	0.021	0.019	0.016	0.015
10.00	0.015	0.015	0.014	0.014	0.013	0.013	0.013	0.012	0.012	0.011

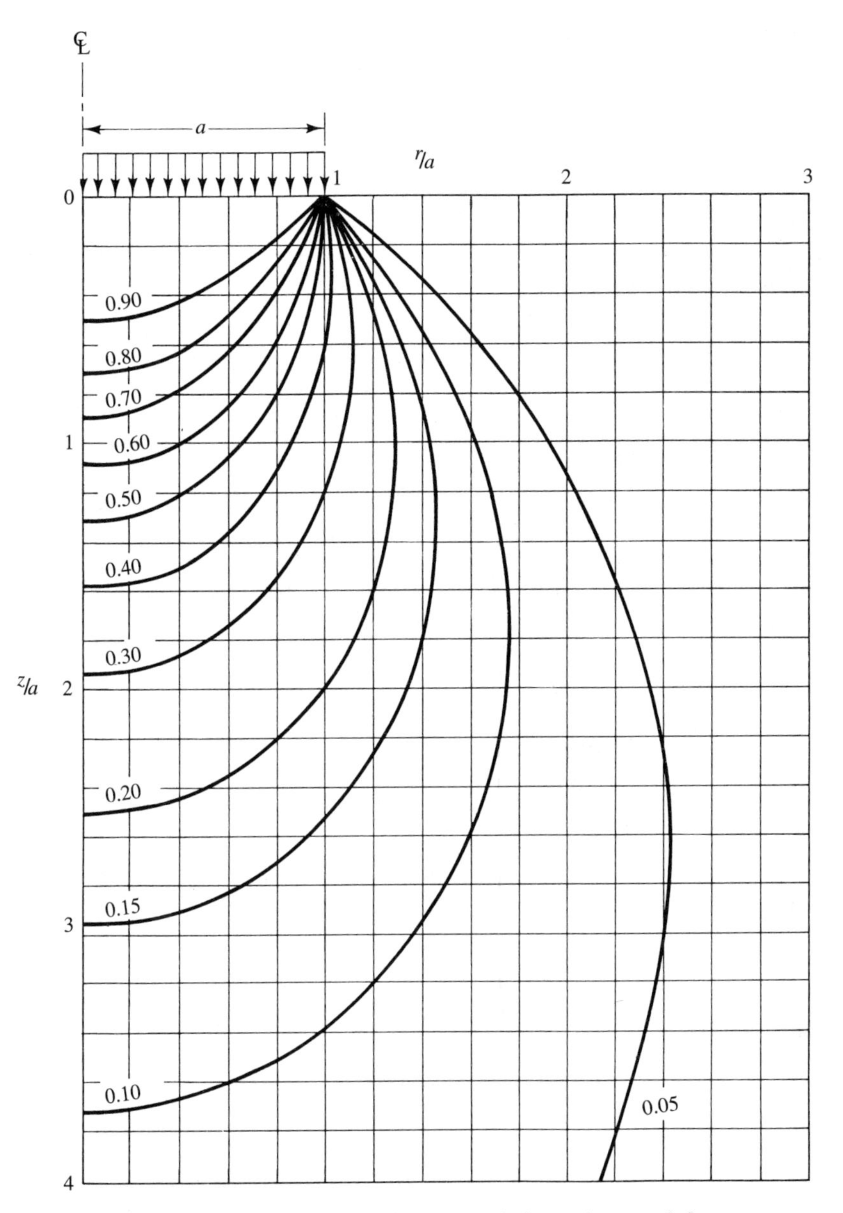

FIGURE 6–6 Influence coefficients for uniformly loaded circular area [4].

represent, respectively, the depth and radial horizontal distance from the center of the circle to the point at which pressure is desired (these are similar to the z and r shown in Figure 6–2); and a represents the radius of the circle on which the uniform load acts. To calculate vertical pressure below a uniform load on a circular area, one computes the ratios z/a and r/a, then an *influence coefficient* is determined from Table 6–1 or Figure 6–6. This influence coefficient is simply multiplied by the uniform load applied to the circular area to determine the pressure at the desired point. Examples 6–5 and 6–6 illustrate this method.

EXAMPLE 6–5

Given

1. A circular area carrying a uniformly distributed load of 2000 lb/ft^2 is applied to the ground surface.
2. The radius of the circular area is 10 ft.

Required

The vertical stress increment due to this uniform load:

1. At a point 20 ft below the center of the circular area.
2. At a point 20 ft below the ground surface at a horizontal distance of 5 ft from the center of the circular area (i.e., $r = 5$ ft, $z = 20$ ft).
3. At a point 20 ft below the edge of the circular area.
4. At a point 20 ft below the ground surface at a horizontal distance of 18 ft from the center of the circular area (i.e., $r = 18$ ft, $z = 20$ ft).

Solution

1. q = Influence coefficient × Uniform load

With $a = 10$ ft (radius of circle)
$r = 0$ ft
$z = 20$ ft

$$\frac{z}{a} = \frac{20 \text{ ft}}{10 \text{ ft}} = 2.00$$
$$\frac{r}{a} = \frac{0 \text{ ft}}{10 \text{ ft}} = 0$$

The influence coefficient from Table 6–1 or Figure 6–6 = 0.284, thus

$$q = (0.284)(2000 \text{ lb/ft}^2) = 568 \text{ lb/ft}^2$$

2. With $a = 10$ ft
$r = 5$ ft
$z = 20$ ft

$$\frac{z}{a} = \frac{20 \text{ ft}}{10 \text{ ft}} = 2.00$$
$$\frac{r}{a} = \frac{5 \text{ ft}}{10 \text{ ft}} = 0.5$$

The influence coefficient from Table 6–1 or Figure 6–6 = 0.258, thus

$$q = (0.258)(2000 \text{ lb/ft}^2) = 516 \text{ lb/ft}^2$$

3. With $a = 10$ ft
$r = 10$ ft
$z = 20$ ft

$$\frac{z}{a} = \frac{20\text{ ft}}{10\text{ ft}} = 2.00$$
$$\frac{r}{a} = \frac{10\text{ ft}}{10\text{ ft}} = 1.00$$

The influence coefficient from Table 6–1 or Figure 6–6 = 0.194, thus

$$q = (0.194)(2000\text{ lb/ft}^2) = 388\text{ lb/ft}^2$$

4. With $a = 10$ ft
$r = 18$ ft
$z = 20$ ft

$$\frac{z}{a} = \frac{20\text{ ft}}{10\text{ ft}} = 2.00$$
$$\frac{r}{a} = \frac{18\text{ ft}}{10\text{ ft}} = 1.8$$

From Table 6–1,

when $\frac{z}{a} = 2.00$ and $\frac{r}{a} = 1.5$, influence coefficient = 0.127

when $\frac{z}{a} = 2.00$ and $\frac{r}{a} = 2.00$, influence coefficient = 0.073

By interpolation between 0.127 and 0.073, the desired influence coefficient for $z/a = 2.00$ and $r/a = 1.8$ is

$$0.073 + \left(\frac{0.127 - 0.073}{5}\right)(2) = 0.095$$

or

$$0.127 - \left(\frac{0.127 - 0.073}{5}\right)(3) = 0.095$$

Or, from Figure 6–6, the influence coefficient is determined to be 0.095. Thus,

$$q = (0.095)(2000\text{ lb/ft}^2) = 190\text{ lb/ft}^2$$

EXAMPLE 6–6

Given

Soil with a unit weight of 16.97 kN/m^3 is loaded on the ground surface by a uniformly distributed load of 300 kN/m^2 over a circular area 4 m in diameter (see Figure 6–7).

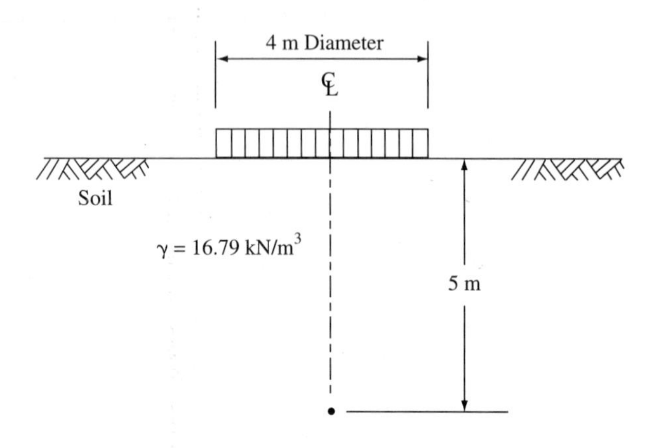

FIGURE 6–7.

Required

1. The vertical stress increment due to this uniform load at a depth of 5 m below the center of the circular area.
2. The total vertical pressure at the same location.

Solution

1. With $a = 2$ m
   $r = 0$ m
   $z = 5$ m

$$\frac{r}{a} = \frac{0 \text{ m}}{2 \text{ m}} = 0$$
$$\frac{z}{a} = \frac{5 \text{ m}}{2 \text{ m}} = 2.50$$

The influence coefficient (from Table 6–1 or Figure 6–6) = 0.200, thus

$$q = (0.200)(300 \text{ kN/m}^2) = 60.0 \text{ kN/m}^2$$

2. Total vertical pressure = Overburden pressure (p_0)
   + Vertical stress increment (q)

$$\text{Overburden pressure } (p_0) = \gamma z = (16.97 \text{ kN/m}^3)(5 \text{ m}) = 84.8 \text{ kN/m}^2$$

$$\text{Total vertical pressure} = 84.8 \text{ kN/m}^2 + 60.0 \text{ kN/m}^2 = 144.8 \text{ kN/m}^2$$

Uniform Load on a Rectangular Area. Vertical pressure below a uniform load on a rectangular area can be determined utilizing Table 6–2. In the table, z, A, and B represent, respectively, depth below the loaded surface and width and length of the rectangle on which the uniform load acts. To calculate vertical pressure below a uniform load on a rectangular area, one computes the ratios $n = B/z$ and $m = A/z$, then an influence coefficient is determined from Table 6–2. Either m or n can be read along the first column, and the other (n or m) is read across the top. The influence coefficient can also be determined utilizing Figure 6–8. The influence

TABLE 6–2
Influence Coefficients for Points under Uniformly Loaded Rectangular Areas [3, 5]

$m = A/z$ or $n = B/z$	$n = B/z$ or $m = A/z$																	
	0.1	0.2	0.3	0.4	0.5	0.6	0.7	0.8	0.9	1.0	1.2	1.5	2.0	2.5	3.0	5.0	10.0	∞
0.1	0.005	0.009	0.013	0.017	0.020	0.022	0.024	0.026	0.027	0.028	0.029	0.030	0.031	0.031	0.032	0.032	0.032	0.032
0.2	0.009	0.018	0.026	0.033	0.039	0.043	0.047	0.050	0.053	0.055	0.057	0.059	0.061	0.062	0.062	0.062	0.062	0.062
0.3	0.013	0.026	0.037	0.047	0.056	0.063	0.069	0.073	0.077	0.079	0.083	0.086	0.089	0.090	0.090	0.090	0.090	0.090
0.4	0.017	0.033	0.047	0.060	0.071	0.080	0.087	0.093	0.098	0.101	0.106	0.110	0.113	0.115	0.115	0.115	0.115	0.115
0.5	0.020	0.039	0.056	0.071	0.084	0.095	0.103	0.110	0.116	0.120	0.126	0.131	0.135	0.137	0.137	0.137	0.137	0.137
0.6	0.022	0.043	0.063	0.080	0.095	0.107	0.117	0.125	0.131	0.136	0.143	0.149	0.153	0.155	0.156	0.156	0.156	0.156
0.7	0.024	0.047	0.069	0.087	0.103	0.117	0.128	0.137	0.144	0.149	0.157	0.164	0.169	0.170	0.171	0.172	0.172	0.172
0.8	0.026	0.050	0.073	0.093	0.110	0.125	0.137	0.146	0.154	0.160	0.168	0.176	0.181	0.183	0.184	0.185	0.185	0.185
0.9	0.027	0.053	0.077	0.098	0.116	0.131	0.144	0.154	0.162	0.168	0.178	0.186	0.192	0.194	0.195	0.196	0.196	0.196
1.0	0.028	0.055	0.079	0.101	0.120	0.136	0.149	0.160	0.168	0.175	0.185	0.193	0.200	0.202	0.203	0.204	0.205	0.205
1.2	0.029	0.057	0.083	0.106	0.126	0.143	0.157	0.168	0.178	0.185	0.196	0.205	0.212	0.215	0.216	0.217	0.218	0.218
1.5	0.030	0.059	0.086	0.110	0.131	0.149	0.164	0.176	0.186	0.193	0.205	0.215	0.223	0.226	0.228	0.229	0.230	0.230
2.0	0.031	0.061	0.089	0.113	0.135	0.153	0.169	0.181	0.192	0.200	0.212	0.223	0.232	0.236	0.238	0.239	0.240	0.240
2.5	0.031	0.062	0.090	0.115	0.137	0.155	0.170	0.183	0.194	0.202	0.215	0.226	0.236	0.240	0.242	0.244	0.244	0.244
3.0	0.032	0.062	0.090	0.115	0.137	0.156	0.171	0.184	0.195	0.203	0.216	0.228	0.238	0.242	0.244	0.246	0.247	0.247
5.0	0.032	0.062	0.090	0.115	0.137	0.156	0.172	0.185	0.196	0.204	0.217	0.229	0.239	0.244	0.246	0.249	0.249	0.249
10.0	0.032	0.062	0.090	0.115	0.137	0.156	0.172	0.185	0.196	0.205	0.218	0.230	0.240	0.244	0.247	0.249	0.250	0.250
∞	0.032	0.062	0.090	0.115	0.137	0.156	0.172	0.185	0.196	0.205	0.218	0.230	0.240	0.244	0.247	0.249	0.250	0.250

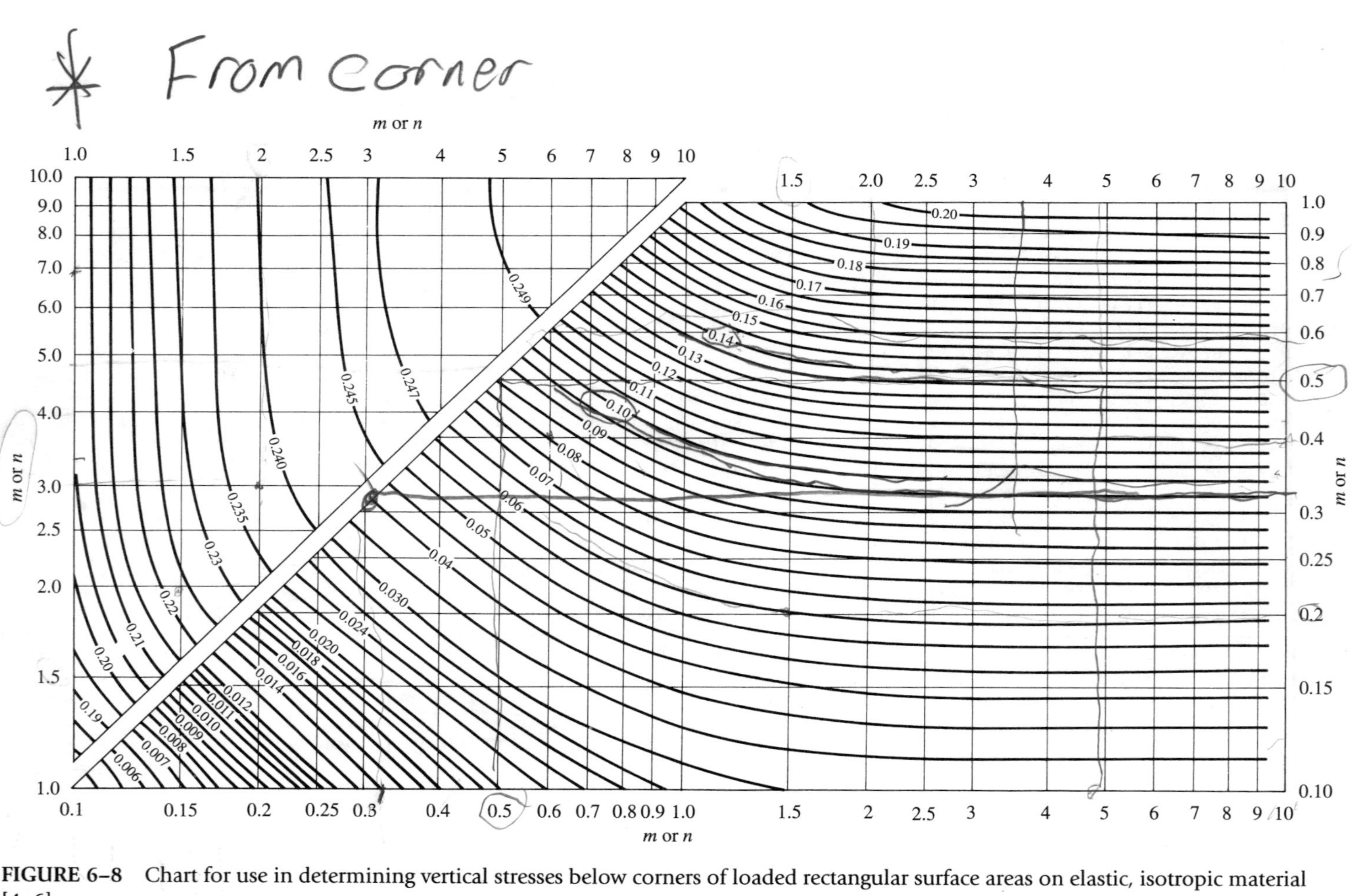

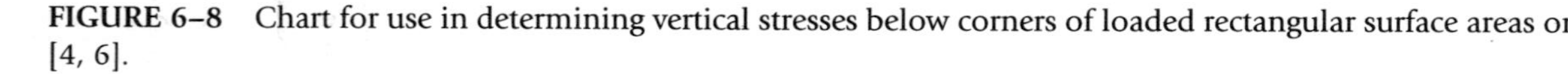

FIGURE 6–8 Chart for use in determining vertical stresses below corners of loaded rectangular surface areas on elastic, isotropic material [4, 6].

FIGURE 6–8 (*continued*)

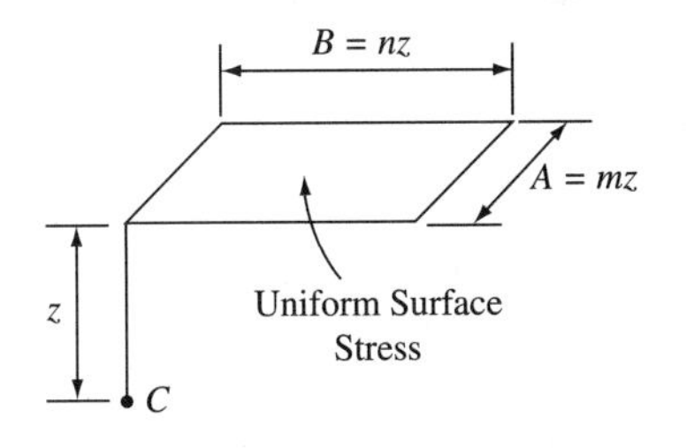

coefficient is multiplied by the uniform load applied to the rectangular area to determine the pressure at depth z below each *corner* of the rectangle. Example 6–7 illustrates this method.

EXAMPLE 6–7

Given

A 15-ft by 20-ft rectangular foundation carrying a uniform load of 4000 lb/ft^2 is applied to the ground surface.

Required

The vertical stress increment due to this uniform load at a point 10 ft below the corner of the rectangular loaded area.

Solution

From Table 6–2 or Figure 6–8, with

$$A = mz \text{ or } m = \frac{A}{z} \qquad A = 15 \text{ ft} \qquad z = 10 \text{ ft} \qquad m = \frac{15 \text{ ft}}{10 \text{ ft}} = 1.5$$

$$B = nz \text{ or } n = \frac{B}{z} \qquad B = 20 \text{ ft} \qquad z = 10 \text{ ft} \qquad n = \frac{20 \text{ ft}}{10 \text{ ft}} = 2.0$$

$$\text{Influence coefficient} = 0.223$$

$$q = (0.223)(4000 \text{ lb/ft}^2) = 892 \text{ lb/ft}^2$$

It should be emphasized that the pressure determined by using the influence coefficients in Table 6–2 or Figure 6–8 (as in Example 6–7) is acting at depth z *directly below a corner of the rectangular area.* This is shown in Figure 6–9, where such a computed stress acts at point C. It is sometimes necessary to determine the pressure below a rectangular loaded area at points other than directly below a corner of the rectangular area. For example, it may be necessary to determine the pressure at some depth directly below the center of a rectangular area, or at some point outside the downward projection of the rectangular area. This can be accomplished by dividing the area into rectangles, each of which has one corner directly above the point at which the pressure is desired at depth z. The pressure is computed for each rectangle in the usual manner and the results added or subtracted to get the total pressure. Figure 6–10 should facilitate

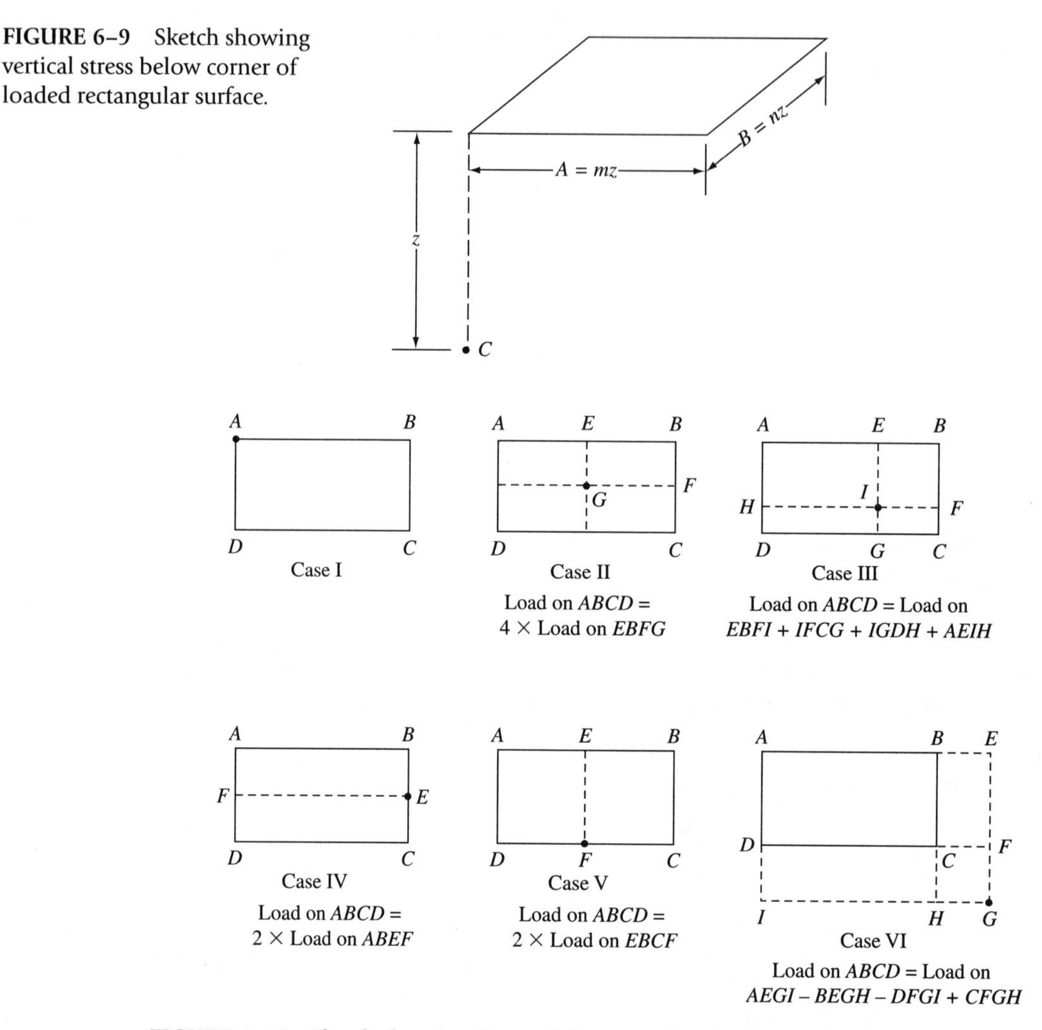

FIGURE 6–9 Sketch showing vertical stress below corner of loaded rectangular surface.

FIGURE 6–10 Sketch showing the combination of rectangles used to obtain the stress below a specific point caused by a uniform surface pressure over area *ABCD*.

understanding of this procedure. In each case of Figure 6–10, the heavy dot indicates the point at which the pressure at depth z is required. Examples 6–8 through 6–11 illustrate this procedure.

EXAMPLE 6–8

Given

A 20-ft by 30-ft rectangular foundation carrying a uniform load of 6000 lb/ft^2 is applied to the ground surface.

FIGURE 6–11.

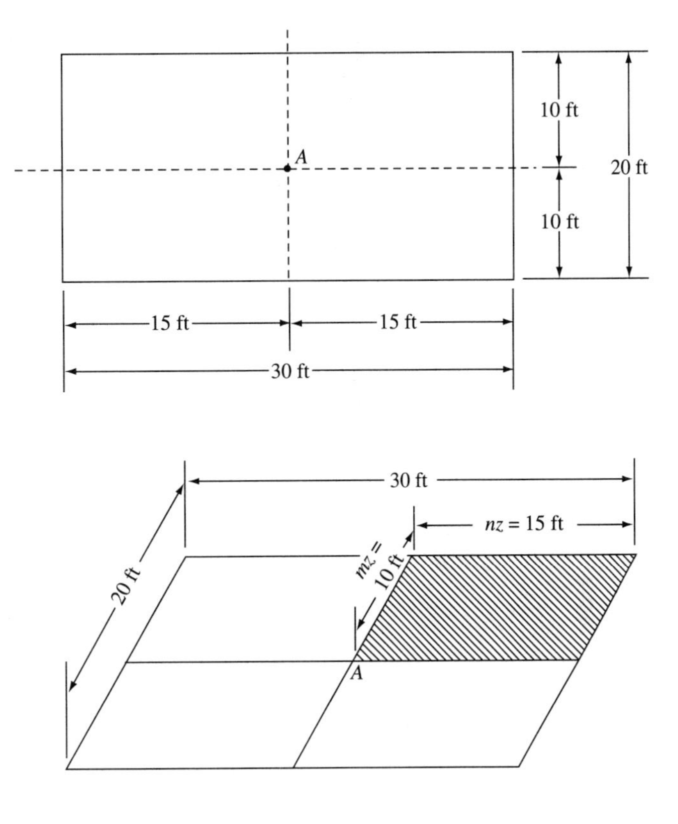

Required

The vertical stress increment due to this uniform load at a depth of 20 ft below the center of the loaded area. (See point A in Figure 6–11.)

Solution

This corresponds to case II of Figure 6–10, so the area is divided into four equal parts.

$$A = mz \quad \text{or} \quad m = \frac{A}{z} \qquad A = 10\text{ ft} \qquad z = 20\text{ ft} \qquad m = \frac{10\text{ ft}}{20\text{ ft}} = 0.5$$

$$B = nz \quad \text{or} \quad n = \frac{B}{z} \qquad B = 15\text{ ft} \qquad z = 20\text{ ft} \qquad n = \frac{15\text{ ft}}{20\text{ ft}} = 0.75$$

From Table 6–2 or Figure 6–8, the influence coefficient = 0.107 for a 10-ft by 15-ft loaded area. Because the original area of 20 ft by 30 ft consists of four smaller equal areas of 10 ft by 15 ft and each of these four areas shares a corner at point A,

$$q = (4)(0.107)(6000\text{ lb/ft}^2) = 2570\text{ lb/ft}^2$$

FIGURE 6–12.

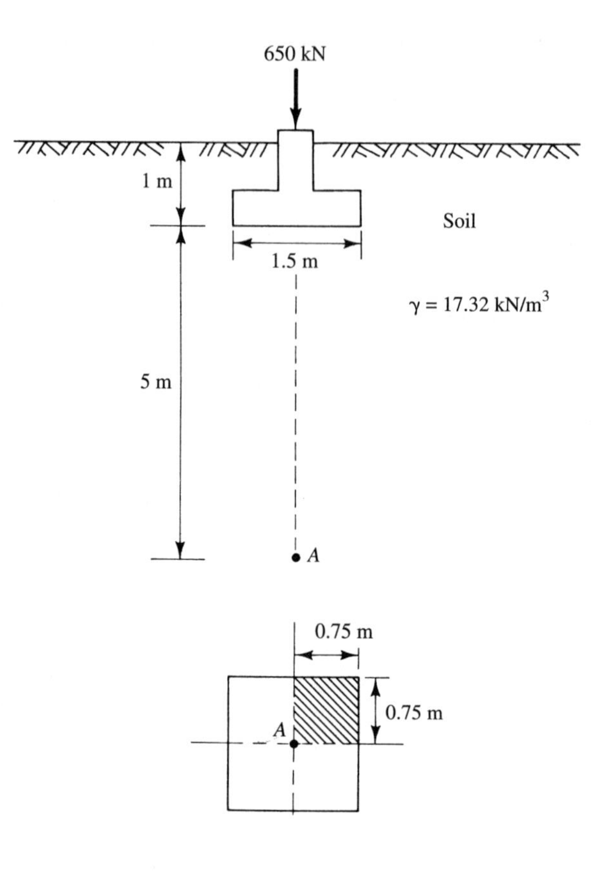

EXAMPLE 6–9

Given

A 1.5-m by 1.5-m footing located 1 m below the ground surface as shown in Figure 6–12 carries a load of 650 kN (including column load and weight of footing and soil surcharge).

Required

The net vertical stress increment due to this load at a depth of 5 m below the center of the footing (i.e., at point A in Figure 6–12).

Solution

As in Example 6–8, the total area is divided into four equal areas, 0.75 m by 0.75 m, as shown in Figure 6–12.

$$A = mz \quad \text{or} \quad m = \frac{A}{z} \qquad A = 0.75 \text{ m} \qquad z = 5 \text{ m} \qquad m = \frac{0.75 \text{ m}}{5 \text{ m}} = 0.150$$

$$B = nz \quad \text{or} \quad n = \frac{B}{z} \qquad B = 0.75 \text{ m} \qquad z = 5 \text{ m} \qquad n = \frac{0.75 \text{ m}}{5 \text{ m}} = 0.150$$

From Table 6–2 or Figure 6–8, the influence coefficient = 0.0103 for a 0.75-m by 0.75-m loaded area. Because the 1.5-m by 1.5-m footing consists of four smaller equal areas of 0.75 m by 0.75 m and each of these four areas shares a corner at point *A*,

$$q = (4)(0.0103)(\text{Net vertical stress increment at the footing's base})$$

Net vertical stress increment at footing's base

$$= \frac{650 \text{ kN}}{(1.5 \text{ m})(1.5 \text{ m})} - (17.32 \text{ kN/m}^3)(1 \text{ m}) = 271.6 \text{ kN/m}^2$$

Thus,

$$q = (4)(0.0103)(271.6 \text{ kN/m}^2) = 11.2 \text{ kN/m}^2$$

EXAMPLE 6–10

Given

1. An L-shaped building (in plan) shown in Figure 6–13.
2. The load exerted by the structure is 1400 lb/ft^2.

Required

Determine the vertical stress increment due to the structure load at a depth of 15 ft below interior corner *A* of the L-shaped building. Assume that the foundation is under the entire building.

Solution

Divide the L-shaped building into three smaller areas, *ABCD, ADEF,* and *AFGH.* Note that these three areas share a common corner at point *A* (or corner *A*).

Area ABCD

From Table 6–2 or Figure 6–8, with

$$\text{A} = mz \quad \text{or} \quad m = \frac{\text{A}}{z} \qquad \text{A} = 60 \text{ ft} \qquad z = 15 \text{ ft} \qquad m = \frac{60 \text{ ft}}{15 \text{ ft}} = 4$$

$$\text{B} = nz \quad \text{or} \quad n = \frac{\text{B}}{z} \qquad \text{B} = 45 \text{ ft} \qquad z = 15 \text{ ft} \qquad n = \frac{45 \text{ ft}}{15 \text{ ft}} = 3$$

Influence coefficient = 0.245

Area ADEF

From Table 6–2 or Figure 6–8, with

$$\text{A} = 30 \text{ ft} \qquad z = 15 \text{ ft} \qquad m = \frac{30 \text{ ft}}{15 \text{ ft}} = 2$$

$$\text{B} = 45 \text{ ft} \qquad z = 15 \text{ ft} \qquad n = \frac{45 \text{ ft}}{15 \text{ ft}} = 3$$

FIGURE 6–13

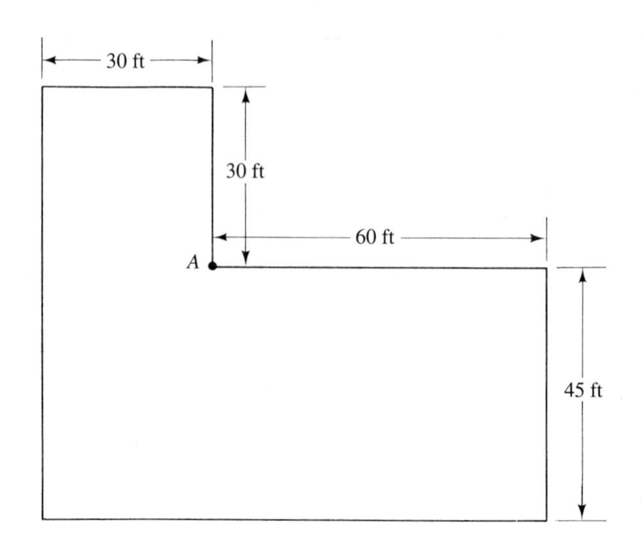

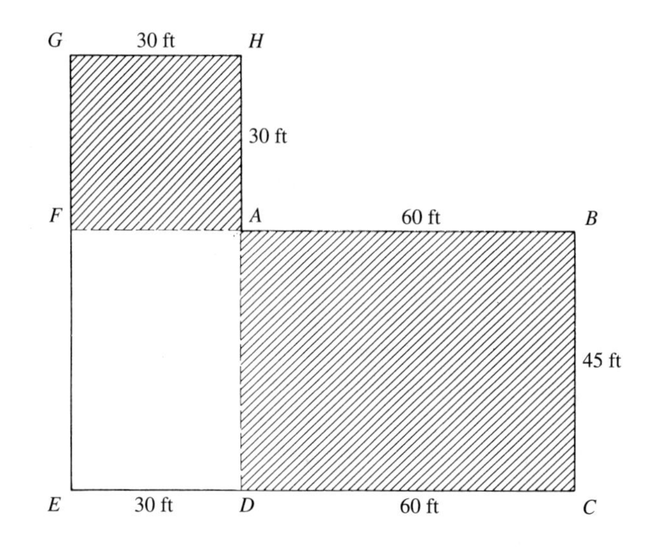

Influence coefficient = 0.238

Area AFGH

From Table 6–2 or Figure 6–8, with

$$A = 30 \text{ ft} \qquad z = 15 \text{ ft} \qquad m = \frac{30 \text{ ft}}{15 \text{ ft}} = 2$$

$$B = 30 \text{ ft} \qquad z = 15 \text{ ft} \qquad n = \frac{30 \text{ ft}}{15 \text{ ft}} = 2$$

Influence coefficient = 0.232

$$q = [\Sigma \text{ Influence coefficients}] \times \text{Uniform load}$$

$$q = (0.245 + 0.238 + 0.232)(1400 \text{ lb/ft}^2) = 1001 \text{ lb/ft}^2$$

EXAMPLE 6–11

Given

1. A rectangular loaded area *ABCD* shown in plan in Figure 6–14.
2. The load exerted on the area is 80 kN/m^2.

FIGURE 6–14

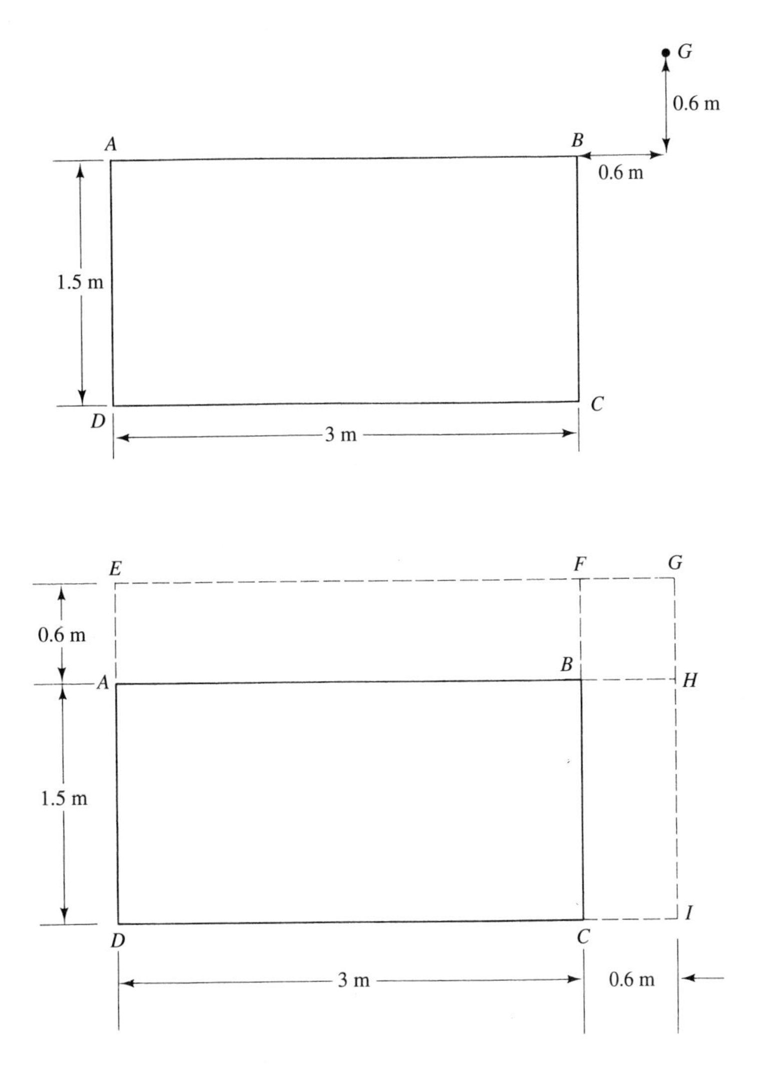

Required

Vertical stress increment due to the exerted load at a depth of 3 m below point G (Figure 6–14).

Solution

This corresponds to case VI of Figure 6–10. The influence coefficient for the vertical stress increment under point G due to the uniform load on area *ABCD* may be obtained from the coefficients for various rectangles as follows:

$$\text{Load on } ABCD = \text{Load on } DEGI - AEGH - CFGI + BFGH$$

(*Note:* In the preceding equation, the last term, *BFGH*, is added because when *AEGH* is subtracted, area *BFGH* is included in it and when *CFGI* is subtracted, area *BFGH* is also included in it. Thus, the effect of area *BFGH* has been subtracted twice. Hence, it must be added in order that its effect be subtracted only one time.)

Area DEGI

From Table 6–2 or Figure 6–8, with

$$A = mz \quad \text{or} \quad m = \frac{A}{z} \qquad A = 2.1 \text{ m} \qquad z = 3 \text{ m} \qquad m = \frac{2.1 \text{ m}}{3 \text{ m}} = 0.7$$

$$B = nz \quad \text{or} \quad n = \frac{B}{z} \qquad B = 3.6 \text{ m} \qquad z = 3 \text{ m} \qquad n = \frac{3.6 \text{ m}}{3 \text{ m}} = 1.2$$

Influence coefficient for area $DEGI = 0.157$

Area AEGH

From Table 6–2 or Figure 6–8, with

$$A = 0.6 \text{ m} \qquad z = 3 \text{ m} \qquad m = \frac{0.6 \text{ m}}{3 \text{ m}} = 0.2$$

$$B = 3.6 \text{ m} \qquad z = 3 \text{ m} \qquad n = \frac{3.6 \text{ m}}{3 \text{ m}} = 1.2$$

Influence coefficient for area $AEGH = 0.057$

Area CFGI

From Table 6–2 or Figure 6–8, with

$$A = 2.1 \text{ m} \qquad z = 3 \text{ m} \qquad m = \frac{2.1 \text{ m}}{3 \text{ m}} = 0.7$$

$$B = 0.6 \text{ m} \qquad z = 3 \text{ m} \qquad n = \frac{0.6 \text{ m}}{3 \text{ m}} = 0.2$$

Influence coefficient for area *CFGI* = 0.047

Area BFGH

From Table 6–2 or Figure 6–8, with

$$A = 0.6\text{ m} \qquad z = 3\text{ m} \qquad m = \frac{0.6\text{ m}}{3\text{ m}} = 0.2$$

$$B = 0.6\text{ m} \qquad z = 3\text{ m} \qquad n = \frac{0.6\text{ m}}{3\text{ m}} = 0.2$$

Influence coefficient for area *BFGH* = 0.018

$$q = (0.157 - 0.057 - 0.047 + 0.018)(80\text{ kN/m}^2) = 5.68\text{ kN/m}^2$$

Uniform Load on a Strip Area. Vertical pressure below a uniform load on a strip area can be determined utilizing Figure 6–15. Use of Figure 6–15 is similar to that of Figure 6–6 for a loaded circular area, except that B and r represent strip width and radial horizontal distance from the strip footing's center line, respectively (z denotes depth in both cases).

EXAMPLE 6–12

Given

1. Soil with a unit weight of 17.92 kN/m^3 is loaded on the ground surface by a wall footing 1 m wide.
2. The load of the wall footing is 295 kN/m of wall length.

Required

1. The vertical stress increment due to the wall footing at a point 3 m below the edge of the strip (see Figure 6–16).
2. The total vertical load at the same location.

Solution

1. From Figure 6–15, with

$$\frac{r}{B} = \frac{0.5\text{ m}}{1\text{ m}} = 0.5$$

$$\frac{z}{B} = \frac{3\text{ m}}{1\text{ m}} = 3.0$$

Influence coefficient = 0.20

$$q = (0.20)(295\text{ kN/m}) = 59.0\text{ kN/m of wall length}$$

2. Total vertical load = Overburden pressure (p_0) + Vertical stress increment (q)

$$\text{Overburden pressure } (p_0) = \gamma z = (17.92\text{ kN/m}^3)(3\text{ m}) = 53.8\text{ kN/m}^2\text{, or } 53.8\text{ kN/m of wall length}$$

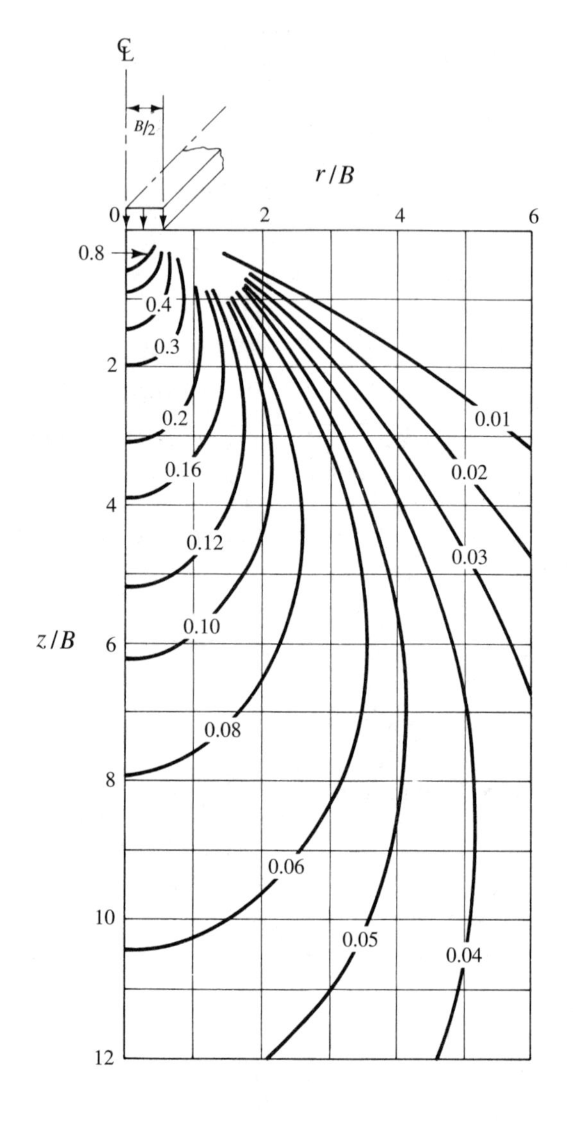

FIGURE 6–15 Influence coefficients for uniformly loaded strip area [7].

$$\text{Total vertical load} = 53.8\ \text{kN/m} + 59.0\ \text{kN/m}$$
$$= 112.8\ \text{kN/m of wall length}$$

Uniform Load on Any Area. Vertical pressure below a uniform load on any area can be determined using an influence chart (see Figure 6–17) developed by Newmark [6] based on Boussinesq's equation. To utilize this method, one must make a sketch (plan view) of the loaded area on tracing paper and draw it to such a scale that distance *AB* on Figure 6–17 equals the depth at which the pressure is desired. This sketch is placed on the chart (Figure 6–17) so that the point below

FIGURE 6–16.

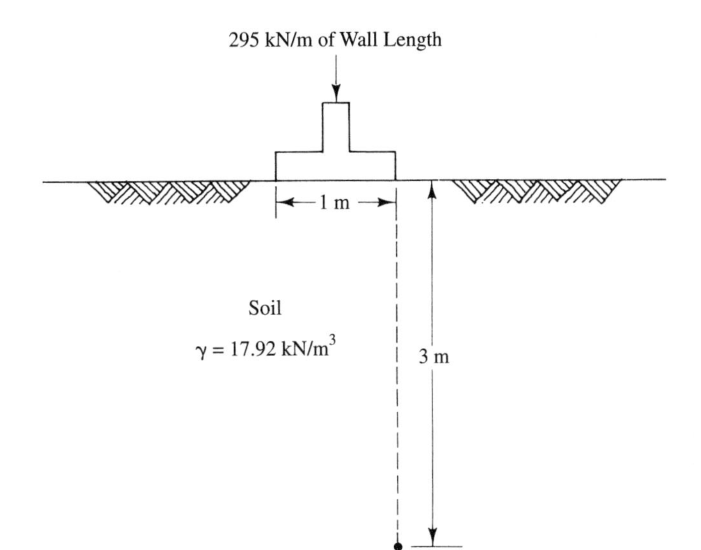

which pressure is desired coincides with the chart's center. The next step is to count the quasi-rectangles enclosed by the loaded area. The pressure at the indicated point at the desired depth is determined by multiplying the number of quasi-rectangles by the applied uniform load by 0.001. As indicated on Figure 6–17, the number 0.001 is the *influence value* for this particular chart. The same sketch may be used to determine pressure at other points at the same depth by shifting the sketch until a desired point coincides with the chart's center and counting the quasi-rectangles. If, however, pressure at some other depth is required, a new sketch must be drawn to such a scale that distance *AB* on Figure 6–17 equals the depth at which the pressure is desired.

6–4 PROBLEMS

6–1. A concentrated load of 200 kips is applied to the ground surface. What is the vertical stress increment due to the load at a depth of 15 ft directly below the load?

6–2. A concentrated load of 200 kips is applied to the ground surface. What is the vertical stress increment due to the load at a point 15 ft below the ground surface at a horizontal distance of 10 ft from the line of the concentrated load?

6–3. A 10-ft by 7.5-ft rectangular area carrying a uniform load of 5000 lb/ft^2 is applied to the ground surface. Determine the vertical stress increment due to this uniform load at a depth of 12 ft below the ground surface by the approximate method (i.e., 2:1 slope method).

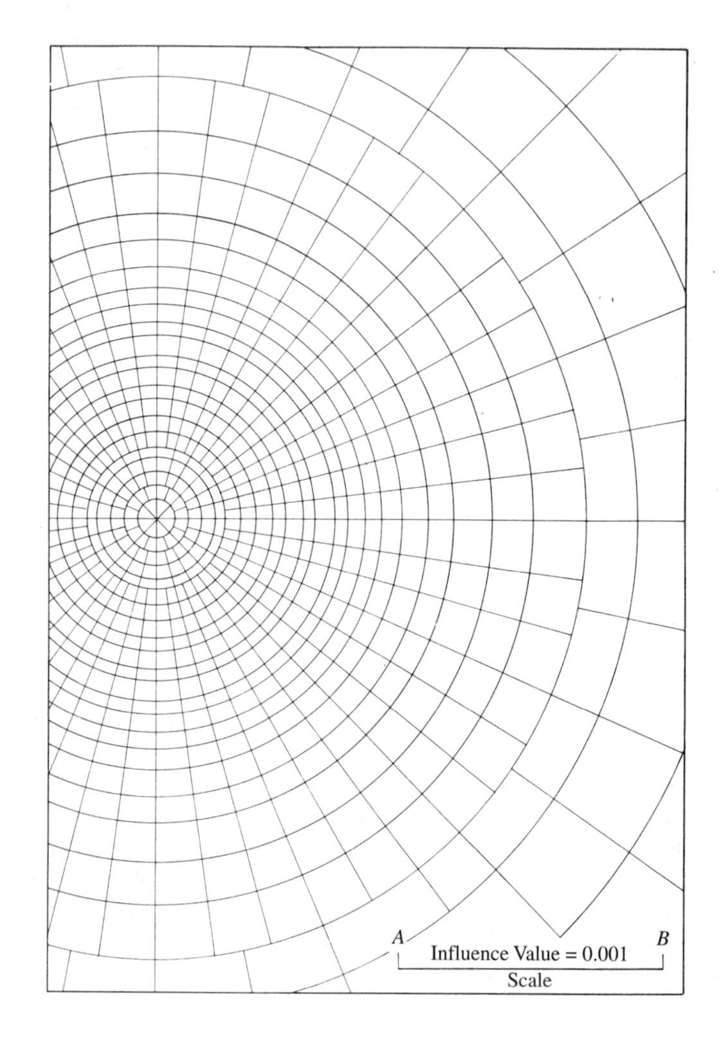

FIGURE 6–17 Newmark influence chart for computing vertical pressure [8].

Source: After Corps of Engineers.

6–4. A rectangular area 2 m by 3 m carrying a uniform load of 195 kN/m^2 is applied to the ground surface. Determine the vertical stress increment due to the uniform load at (a) 1, (b) 3, and (c) 5 m below the area by the approximate method.

6–5. A circular area carrying a uniform load of 4500 lb/ft^2 is applied to the ground surface. The area's radius is 12 ft. What is the vertical stress increment due to this uniform load (a) at a point 18 ft below the area's center, and (b) at a point 18 ft below the ground surface at a horizontal distance of 6 ft from the area's center?

FIGURE 6–18.

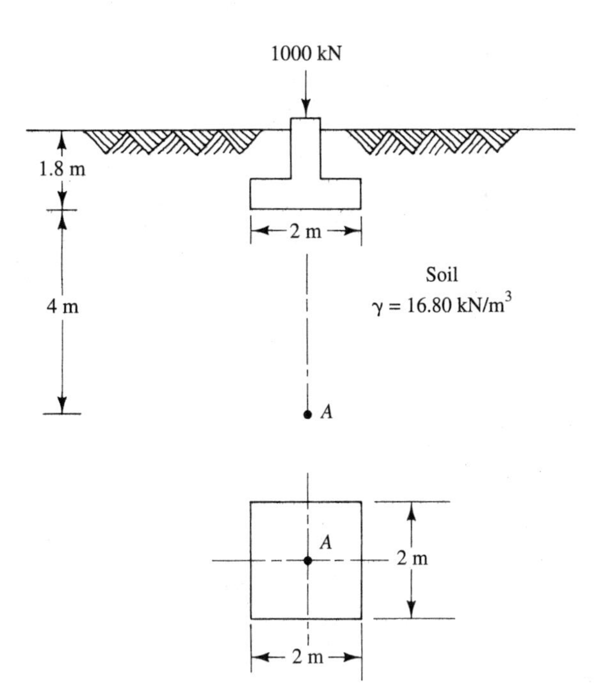

6–6. Soil with a unit weight of 16.38 kN/m^3 is loaded on the ground surface by a uniformly distributed load of 250 kN/m^2 over a circular area 3 m in diameter. Determine (a) the vertical stress increment due to the uniform load, and (b) the total vertical pressure at a depth of 3 m under the edge of the circular area.

6–7. An 8-ft by 12-ft rectangular area carrying a uniform load of 6000 lb/ft^2 is applied to the ground surface. What is the vertical stress increment due to the uniform load at a depth of 15 ft below the corner of the rectangular loaded area?

6–8. A 12-ft by 12-ft square area carrying a uniform load of 5000 lb/ft^2 is applied to the ground surface. Find the vertical stress increment due to the load at a depth of 25 ft below the center of the loaded area.

6–9. A 2-m by 2-m square footing is located 1.8 m below the ground surface and carries a load of 1000 kN. Determine the net vertical stress increment due to the uniform load at a depth of 4 m below the center of the footing (see Figure 6–18).

6–10. The L-shaped area shown in Figure 6–19 carries a 2000-lb/ft^2 uniform load. Find the vertical stress increment due to the structure load at a depth of 24 ft (a) below corner *A*, and (b) below corner *E*.

FIGURE 6–19

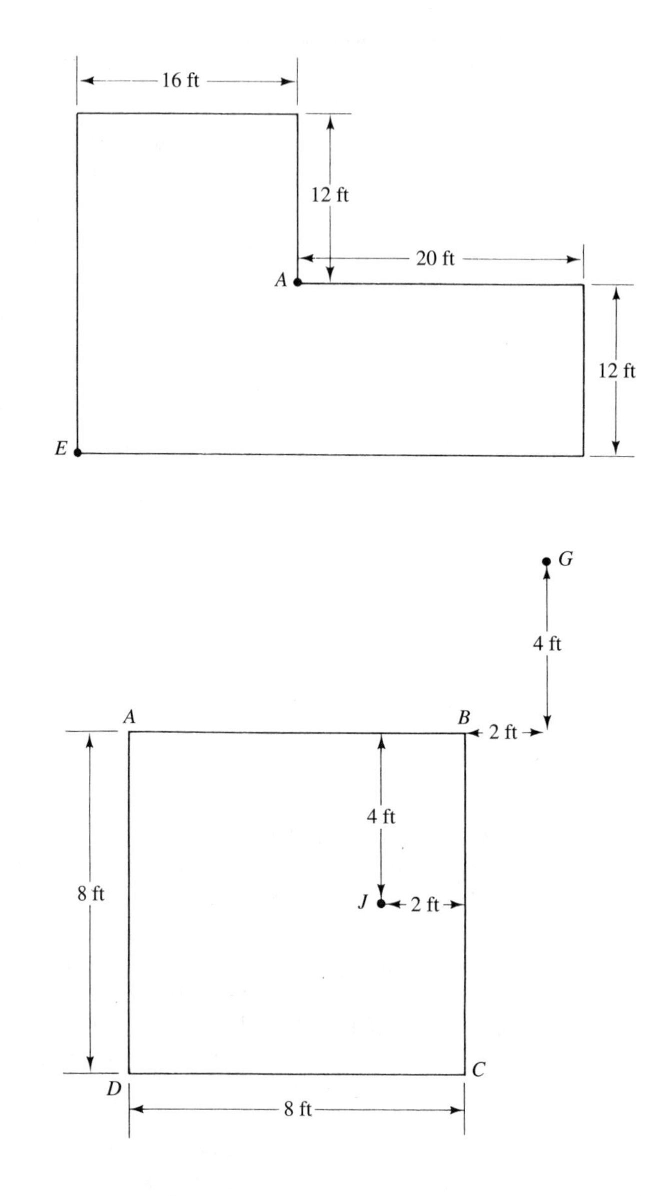

FIGURE 6–20

6–11. The square area *ABCD* shown in Figure 6–20 carries a 2500-lb/ft^2 uniform load. Find the vertical stress increment due to the exerted load at a depth of 12 ft (a) below point *G*, and (b) below point *J*.

6–12. Soil with a unit weight of 19.65 kN/m^3 is loaded on the ground surface by a strip load 1.5 m wide. The strip load is 365 kN/m of wall length. Determine (a) the vertical stress increment due to the strip load, and (b) the total pressure—both at a point 3 m below the center of the strip load.

6–13. Moist sand having a unit weight of 18.60 kN/m^3 is to be excavated to a depth of 5 m to accommodate a rectangular building 58 m long by 38 m wide. Find the reduction in vertical pressure, due to removal of the sand from the excavated area, at one corner of the building at a depth 15 m below the original ground surface.

6–14. Soil to be compacted 10 ft deep to cover a shopping-center construction site has a unit weight of 118 lb/ft^3. A 20-ft by 15-ft foundation is to be built at floor level (i.e., at the top of the fill) and is to support a total structure load of 1500 kips. Find the net stress increase that will result 6 ft below the original ground surface directly beneath the foundation's center.

References

[1] Joseph E. Bowles, *Foundation Analysis and Design,* McGraw-Hill Book Company, New York, 1968.

[2] H. M. Westergaard, "A Problem of Elasticity Suggested by a Problem in Soil Mechanics: Soft Material Reinforced by Numerous Strong Horizontal Sheets," in *Contributions to the Mechanics of Solids,* Stephen Timoshenko 60th Anniversary Volume, Macmillan Publishing Co., Inc., New York, 1938.

[3] Merlin G. Spangler and Richard L. Handy, *Soil Engineering,* 3rd ed., Intext Educational Publishers, New York, 1973. Copyright © 1951, 1960, 1973 by Harper & Row, Publishers, Inc. Reprinted by permission of the publisher.

[4] T. William Lamb and Robert V. Whitman, *Soil Mechanics, SI Version,* John Wiley & Sons, Inc., New York, 1979. Copyright © 1979, by John Wiley & Sons, Inc. Reprinted by permission of John Wiley & Sons, Inc.

[5] Nathan M. Newmark, *Simplified Computation of Vertical Pressures in Elastic Foundations,* Circ. No. 24, Eng. Exp. Sta., Univ. Ill., 1935.

[6] Nathan M. Newmark, *Influence Charts for Computation of Stresses in Elastic Foundations,* Univ. Ill. Bull. 338, 1942.

[7] Irving S. Dunn, Loren R. Anderson, and Fred W. Kiefer, *Fundamentals of Geotechnical Analysis,* John Wiley & Sons, Inc., New York, 1980. Copyright © 1980, by John Wiley & Sons, Inc. Reprinted by permission of John Wiley & Sons, Inc.

[8] Wayne C. Teng, *Foundation Design,* Prentice-Hall, Inc., Englewood Cliffs, N.J., 1962.

7

Consolidation of Soil and Settlement of Structures

7–1 INTRODUCTION

Structures built on soil are subject to settlement. Some settlement is often inevitable, and, depending on the circumstances, some settlement is tolerable. For example, small uniform settlement of a building throughout the floor area might be tolerable, whereas nonuniform settlement of the same building might not be. Or, settlement of a garage or warehouse building might be tolerable, whereas the same settlement (especially differential settlement) of a luxury hotel building would not be because of damage to walls, ceilings, and so on. In any event, a knowledge of the causes of settlement and a means of computing (or predicting) settlement quantitatively are important to the soils engineer.

Although there are several possible causes of settlement (e.g., dynamic forces, changes in the groundwater table, adjacent excavation, etc.), probably the major cause is compressive deformation of soil beneath a structure. Compressive deformation generally results from reduction in void volume, accompanied by rearrangement of soil grains and compression of the material in the voids. If soil is dry, its voids are filled with air; and because air is compressible, rearrangement of soil grains can occur rapidly. If soil is saturated, its voids are filled with incompressible water, which must be extruded from the soil mass before soil grains can rearrange themselves. In soils of high permeability (i.e., coarse-grained soils), this process requires a short time interval for completion, and almost all settlement occurs by the time construction is complete. However, in soils of low permeability (i.e., fine-grained soils), the process requires a long time interval for completion. The result is that the strain occurs very slowly; thus, settlement takes place slowly and continues over a long period of time. The latter case (fine-grained soil) is of more concern because of long-term uncertainty.

As indicated, the process of compression due to extrusion of water from the voids in a fine-grained soil as a result of increased loading (such as the weight of a structure above) is very slow and continues over a long period of time. This phenomenon is called *primary consolidation* [1]. Associated settlement is referred to as *primary consolidation settlement.* (These are commonly referred to simply as *consolidation* and *consolidation settlement*, respectively.) After primary consolidation has ended, soil compression and additional associated settlement continue at a very slow rate, the result of plastic readjustment of soil grains due to new, changed stresses in the soil and progressive breaking of clayey particles and their interparticle bonds. This phenomenon is known as *secondary compression*, and associated settlement is called *secondary compression settlement.*

In analyzing clayey soil, one must differentiate between two types of clay—*normally consolidated clay* and *overconsolidated clay.* In the case of normally consolidated clay, the clay formation has never been subjected to any loading larger than the present effective overburden pressure. This is the case whenever the height of soil above the clay formation (and therefore the weight of the soil above, which causes the pressure) has been more or less constant through time. With overconsolidated clay, the clay formation has been subjected at some time to a loading greater than the present effective overburden pressure. This occurs whenever the present height of soil above the clay formation is less than it was at some time in the past. Such a situation could exist if significant erosion has occurred at the ground surface. (Because of the erosion, the present height of soil above the clay formation is less than it was prior to the erosion.) It might be noted that overconsolidated clay is generally less compressible. As is related in this chapter, the analysis of clays for settlement differs somewhat depending on whether the clay is normally consolidated or overconsolidated.

This chapter deals primarily with the determination of settlement of structures. Sections 7–2 through 7–7 deal with settlement on clay. Section 7–2 covers the laboratory testing required for analyzing settlement. Section 7–3 shows how laboratory data are analyzed to determine if the clay is normally consolidated, and Section 7–4 shows how they are analyzed to determine if the clay is overconsolidated. Section 7–5 demonstrates the development of a *field consolidation line,* which in turn is used to calculate consolidation settlement on clay (Section 7–6). Section 7–7 covers secondary compression and associated secondary compression settlement. Section 7–8 deals with settlement on sand.

7–2 CONSOLIDATION TEST

As a means of estimating both the amount and time of consolidation and resulting settlement, consolidation tests are run in a laboratory. For complete and detailed instructions for conducting a consolidation test, the reader is referred to *Soil Properties: Testing, Measurement, and Evaluation,* 4th edition, by Liu and Evett (Prentice Hall, 2000). A generalized discussion is given here.

To begin with, an undisturbed soil sample is placed in a metal ring. One porous disk is placed above the sample and another is placed beneath the sample. The purpose of the disks is to allow water to flow vertically into and out of the soil sample. This assembly is immersed in water. As a load is applied to the upper disk, the sample is compressed and deformation is measured by a dial gauge (see Figure 7–1).

To begin a particular test, a specific pressure (e.g., 500 lb/ft^2) is applied to the soil sample, and dial readings (reflecting deformation) and corresponding time observations are made and recorded until deformation has nearly ceased. Normally,

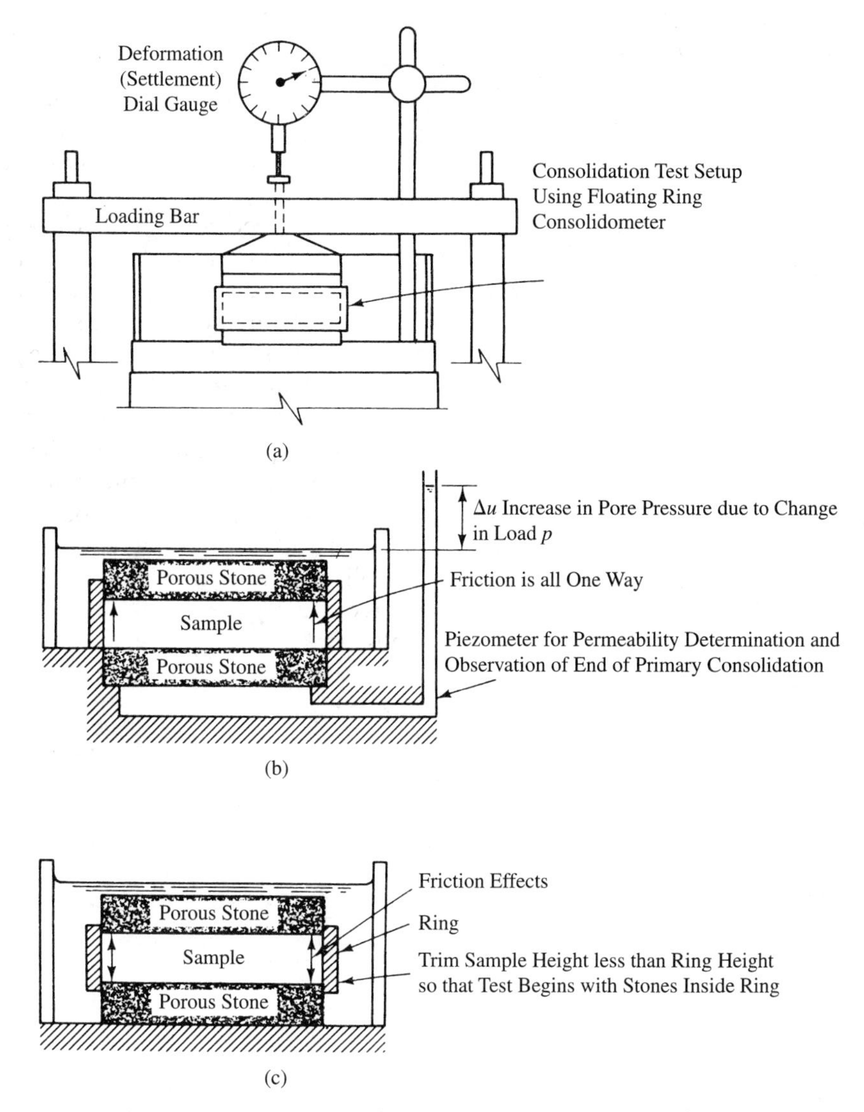

FIGURE 7–1 (a) Consolidometer; (b) fixed-ring consolidometer, which may be used to obtain information during a consolidation test if a piezometer is installed; (c) floating-ring consolidometer [2].

this is done over a 24-hour period. Then, a graph is prepared using these data, with time along the abscissa on a logarithmic scale and dial readings along the ordinate on an arithmetic scale. An example of such a graph is given in Figure 7–2.

The procedure is repeated after the applied pressure is doubled, giving another graph of time versus dial readings corresponding to the new pressure. The procedure is repeated for additional doublings of applied pressure until the applied pressure is in excess of the total pressure to which the clay formation is expected to be subjected

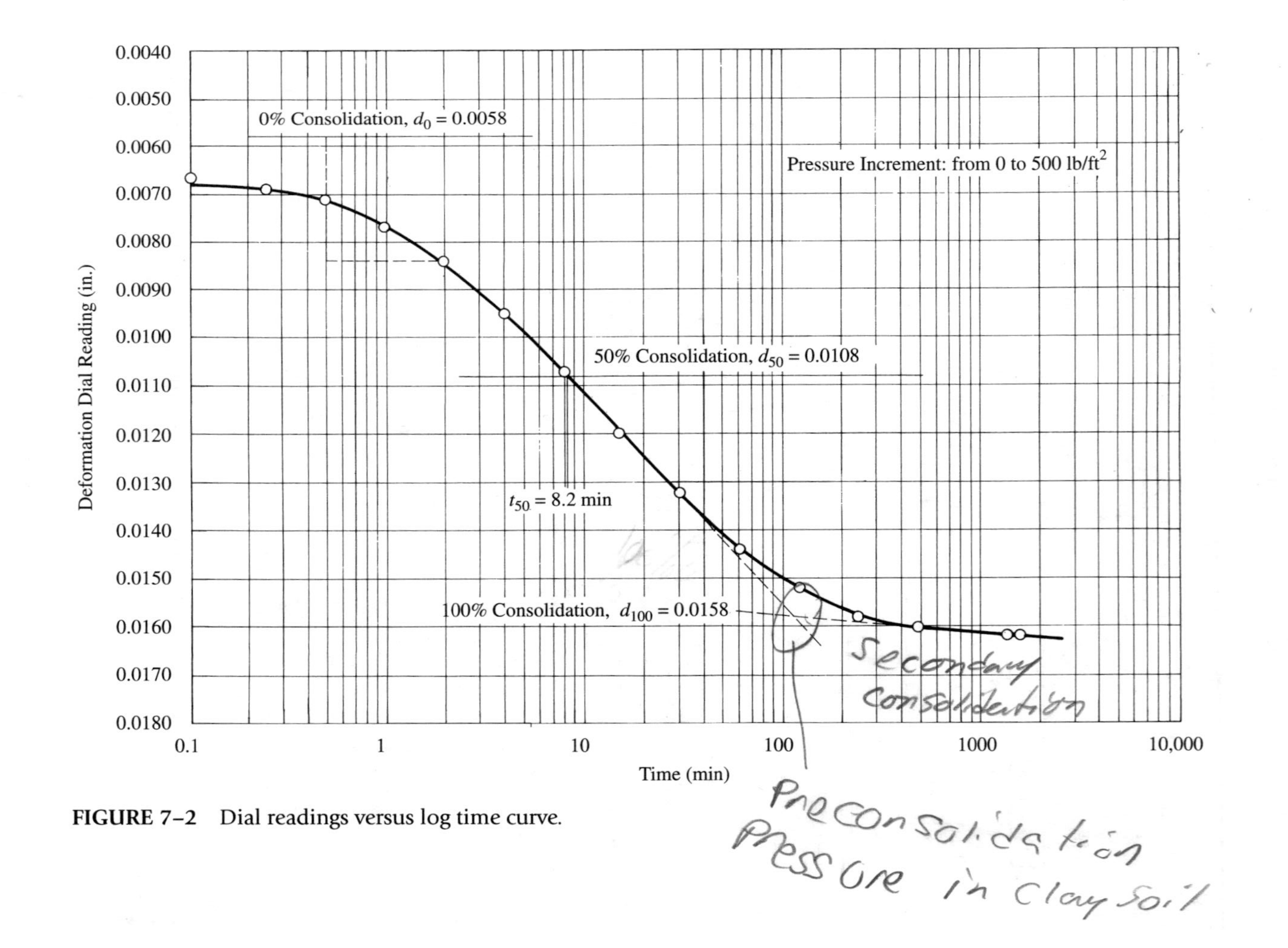

FIGURE 7–2 Dial readings versus log time curve.

when the proposed structure is built. [The total pressure includes effective overburden pressure and net additional pressure (or consolidation pressure) due to the structure.]

From each graph of time versus dial readings, the void ratio (e) and coefficient of consolidation (c_v) that correspond to the specific applied pressure (p) for that graph are determined using the following steps.

1. Find the deformation representing 100% consolidation for each load increment. First, draw a straight line through the points representing the final readings that exhibit a straight-line trend and a flat slope. Draw a second straight line tangent to the steepest part of the deformation versus log time curve. The intersection of the two lines represents the deformation corresponding to 100% consolidation. Compression that occurs subsequent to 100% consolidation is defined as secondary compression (Figure 7–2).
2. Find the deformation representing 0% consolidation by selecting the deformations at any two times that have a ratio of 1:4. The deformation corresponding to the larger of the two times should be greater than one-fourth but less than one-half of the total change in deformation for the load increment. The deformation corresponding to 0% consolidation is equal to the deformation corresponding to the smaller time interval less the difference in the deformations for the two selected times (Figure 7–2).
3. The deformation corresponding to 50% consolidation for each load increment is equal to the average of the deformations corresponding to the 0 and 100% deformations. The time required for 50% consolidation under any load increment may be found graphically from the deformation versus log time curve for that load increment by observing the time that corresponds to 50% consolidation [3] (Figure 7–2).
4. To obtain the change in thickness of the specimen, subtract the initial dial reading at the beginning of the first loading from the dial reading corresponding to 100% consolidation for the given loading. Find the change in void ratio (Δe) for the given loading by dividing the change in thickness of the specimen by the height of solid in the specimen. Determine the void ratio (e) for this loading by subtracting the change in void ratio (Δe) from the initial void ratio (e_0).
5. Compute the coefficient of consolidation (c_v) for this loading using the following equation:

$$c_v = \frac{0.196H^2}{t_{50}} \tag{7–1}$$

where H = thickness of test specimen at 50% consolidation (i.e., initial height of specimen at beginning of test minus deformation dial reading at 50% consolidation). Use half the thickness if the specimen is drained on both top and bottom during the test.

t_{50} = time to 50% consolidation

By using values of e and c_v determined from the various graphs of time versus dial readings corresponding to the different test loadings, one can prepare two graphs—one of void ratio versus pressure (e–log p curve), with pressure along the abscissa on a logarithmic scale and void ratio along the ordinate on an arithmetic scale, and another of consolidation coefficient versus pressure (c_v–log p curve), with pressure along the abscissa on a logarithmic scale and coefficient of consolidation along the ordinate on an arithmetic scale. An example of an e–log p curve is given in Figure 7–3, and an example of a c_v–log p curve is given in Figure 7–4. As is related subsequently, the e–log p curve is used to determine the amount of consolidation settlement, and the c_v–log p curve is used to determine the timing of the consolidation settlement.

In Figure 7–3, the upper curve exhibits the relationship between void ratio and pressure as the pressure is increased. As is shown in Section 7–5, in the case of overconsolidated clay, it is necessary to have a "rebound curve." Exhibited by the lower curve in Figure 7–3, the rebound curve is obtained by unloading the soil sample during the consolidation test after the maximum pressure has been reached. As the sample is unloaded, the soil tends to swell, causing movement and associated dial readings to reverse direction.

The primary results of a laboratory consolidation test are (1) the e–log p curve, (2) the c_v–log p curve, and (3) the initial void ratio of the soil *in situ* (e_0).

FIGURE 7–3 Void ratio versus logarithm of pressure.

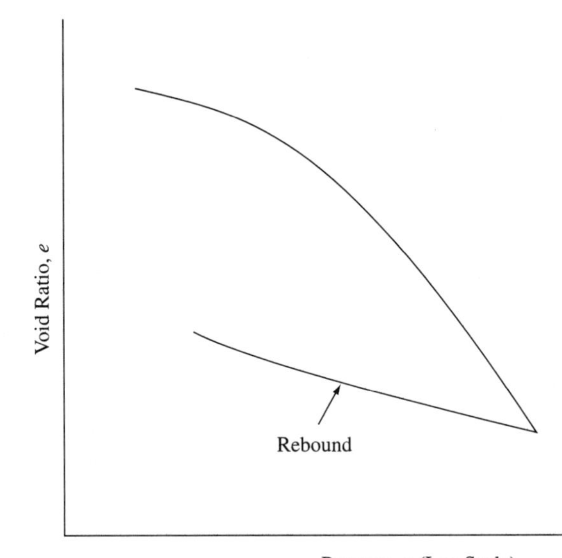

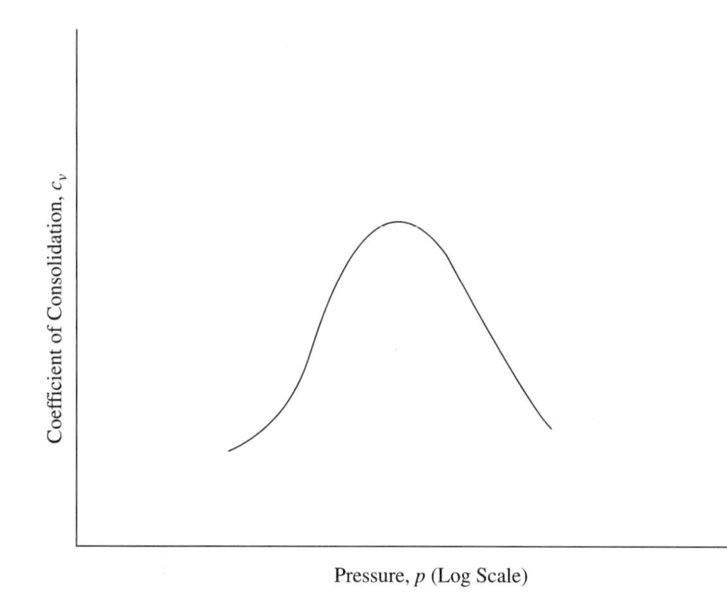

FIGURE 7–4 Coefficient of consolidation versus logarithm of pressure.

EXAMPLE 7–1

Given

A clayey soil obtained from the field was subjected to a laboratory consolidation test. The test results are as follows:

1. Diameter of test specimen = 2.50 in.
2. Initial height of specimen = 0.780 in.
3. Specific gravity of solids = 2.72.
4. Dry mass of specimen = 75.91 g.
5. Pressure versus deformation dial readings are as given in Table 7–1.

Required

1. Initial void ratio.
2. The e–log p curve.

Solution

1. Volume of solid in specimen $(V_s) = \dfrac{\text{Dry mass of solid}}{\text{Unit mass of solid}}$

$$= \frac{\text{Dry mass of solid}}{\text{(Specific gravity of solids)(Unit mass of water)}}$$

$$= \frac{75.91\ \text{g}}{(2.72)(1.0\ \text{g/cm}^3)} = 27.91\ \text{cm}^3$$

TABLE 7–1
Pressure Versus Deformation Dial Readings for Example 7–1

Pressure, p (lb/ft^2)	Initial Deformation Dial Reading at Beginning of First Loading (in.)	Deformation Dial Reading Representing 100% Primary Consolidation (in.)
0	0	0
500	0	0.0158
1000	0	0.0284
2000	0	0.0490
4000	0	0.0761
8000	0	0.1145
16,000	0	0.1580

$$\text{Initial volume of specimen } (V) = \frac{(0.780 \text{ in.})(\pi)(2.50 \text{ in.})^2}{4}$$

$$= 3.829 \text{ in.}^3, \text{ or } 62.74 \text{ cm}^3$$

$$\text{Initial volume of void in specimen } (V_v) = 62.74 \text{ cm}^3 - 27.91 \text{ cm}^3$$

$$= 34.83 \text{ cm}^3$$

$$\text{Initial void ratio } (e_0) = \frac{V_v}{V_s} = \frac{34.83 \text{ cm}^3}{27.91 \text{ cm}^3} = 1.248$$

2. To develop the e–log p curve, one must determine the height of solid in the specimen.

$$\text{Height of solid in specimen } (H_s) = \frac{V_s}{\text{Area of specimen}}$$

$$= \frac{27.91 \text{ cm}^3}{(\pi)[(2.50 \text{ in.})(2.54 \text{ cm/in.})]^2/4}$$

$$= 0.881 \text{ cm}$$

The change in thickness of the specimen (ΔH) can be found by subtracting the initial deformation dial reading from the deformation dial reading representing 100% primary consolidation. For the 500 lb/ft^2 pressure,

$$\Delta H = (0.0158 \text{ in.} - 0 \text{ in.})(2.54 \text{ cm/in.}) = 0.0401 \text{ cm}$$

The change in void ratio (Δe) can be determined by dividing ΔH by H_s. For the 500 lb/ft^2 pressure,

$$\Delta e = \frac{0.0401 \text{ cm}}{0.881 \text{ cm}} = 0.046$$

TABLE 7–2
Computed Void Ratio–Pressure Relation for Example 7–1

Pressure, p (lb/ft^2)	Initial Deformation Dial Reading at Beginning of First Loading (in.)	Deformation Dial Reading Representing 100% Primary Consolidation (in.)	Change in Thickness of Specimen, ΔH (cm)	Change in Void Ratio, Δe $\left[\Delta e = \frac{\Delta H}{H_s}\right]$	Void Ratio, e $[e = e_0 - \Delta e]$
(1)	*(2)*	*(3)*	*(4) = [(3) − (2)] × 2.54*	$(5) = \frac{(4)}{0.881}$	*(6) = 1.248 − (5)*
0	0	0	0	0	1.248
500	0	0.0158	0.0401	0.046	1.202
1000	0	0.0284	0.0721	0.082	1.166
2000	0	0.0490	0.1245	0.141	1.107
4000	0	0.0761	0.1933	0.219	1.029
8000	0	0.1145	0.2908	0.330	0.918
16,000	0	0.1580	0.4013	0.456	0.792

Finally, the void ratio (e) can be computed by subtracting Δe from e_0. For the 500 lb/ft^2 pressure,

$$e = 1.248 - 0.046 = 1.202$$

Similar computations can be made for the remaining pressures and are presented in Table 7–2.

The e–log p curve is prepared by plotting void ratio versus pressure, with the latter on a logarithmic scale. This curve is given in Figure 7–5.

EXAMPLE 7–2

Given

Additional test results from the consolidation test on the clayey soil presented in Example 7–1 are as given in Table 7–3. In addition, the specimen was drained on both top and bottom during the test.

Required

The c_v–log p curve.

Solution

The coefficient of consolidation is computed using Eq. (7–1).

$$c_v = \frac{0.196H^2}{t_{50}} \tag{7–1}$$

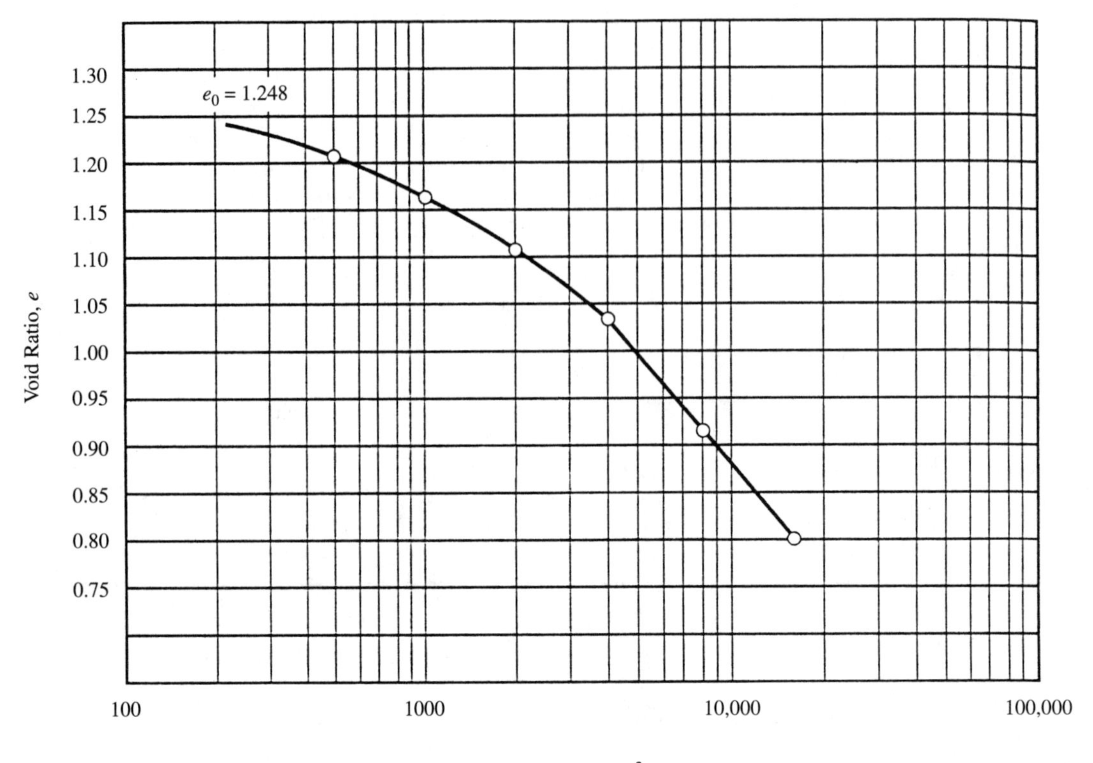

FIGURE 7–5 Void ratio versus logarithm of pressure for Example 7–1.

TABLE 7–3
Pressure Versus Deformation Dial Readings and Time for 50% Consolidation for Example 7–2

Pressure, p (lb/ft^2)	Initial Height of Specimen at Beginning of Test, H_0 (in.)	Deformation Dial Reading at 50% Consolidation (in.)	Time for 50% Consolidation (min)
0	0.780	—	—
500	0.780	0.0108	8.2
1000	0.780	0.0233	6.4
2000	0.780	0.0398	4.0
4000	0.780	0.0644	3.4
8000	0.780	0.0982	3.5
16,000	0.780	0.1387	4.0

The thickness of the specimen at 50% consolidation can be determined by subtracting the deformation dial reading at 50% consolidation from the initial height of the specimen (0.780 in., from Example 7–1). For the 500 lb/ft^2 pressure,

$$\text{Thickness of specimen} = 0.780 \text{ in.} - 0.0108 \text{ in.} = 0.769 \text{ in.}$$

Because the specimen was drained on both top and bottom, half the thickness of the specimen (0.385 in.) must be used for H in Eq. (7–1). For the 500 lb/ft^2 pressure, the value of t_{50} is given to be 8.2 min. Substituting these values into Eq. (7–1) gives the following:

$$c_v = \frac{(0.196)(0.385 \text{ in.})^2}{8.2 \text{ min}} = 3.54 \times 10^{-3} \text{ in.}^2/\text{min}$$

Similar computations can be made for the remaining pressures and are presented in Table 7–4.

The c_v–log p curve is prepared by plotting the coefficient of consolidation versus pressure, with the latter on a logarithmic scale. This curve is given in Figure 7–6.

7–3 NORMALLY CONSOLIDATED CLAY

As indicated in Section 7–1, in the case of a clayey soil it is necessary to determine whether the clay is normally consolidated or overconsolidated. This section shows how to determine if a given clayey soil is normally consolidated.

It is first necessary, however, to determine the present effective overburden pressure (p_0). This pressure is the result of the (effective) weight of soil above midheight of the consolidating clay layer. Although the reader probably knows how to calculate the effective overburden pressure, the procedure is illustrated in Example 7–3.

EXAMPLE 7–3

Given

The soil profile shown in Figure 7–7.

Required

Present effective overburden pressure (p_0) at midheight of the compressible clay layer.

Solution

The elevation of the midheight of the clay layer = (732 ft + 710 ft)/2 = 721 ft.

$$p_0 = (132.0 \text{ lb/ft}^3)(760 \text{ ft} - 752 \text{ ft}) + (132.0 \text{ lb/ft}^3 - 62.4 \text{ lb/ft}^3)(752 \text{ ft} - 732 \text{ ft}) + (125.4 \text{ lb/ft}^3 - 62.4 \text{ lb/ft}^3)(732 \text{ ft} - 721 \text{ ft})$$

$$p_0 = 3141 \text{ lb/ft}^2 = 1.57 \text{ tons/ft}^2$$

TABLE 7–4
Computed Coefficient of Consolidation–Pressure Relation for Example 7–2

Pressure, p (lb/ft^2)	Initial Height of Specimen at Beginning of Test, H_0 (in.)	Deformation Dial Reading at 50% Consolidation (in.)	Thickness of Specimen at 50% Consolidation (in.)	Half-Thickness of Specimen at 50% Consolidation (in.)	Time for 50% Consolidation (min)	Coefficient of Consolidation (in.^{2}/min)
(1)	*(2) [from Example 7–1]*	*(3) [from dial readings versus log of time curves)*	$(4) = (2) - (3)$	$(5) = \frac{(4)}{2}$	*(6) [from dial readings versus log of time curves]*	$(7) = \frac{0.196 \times (5)^2}{(6)}$
0	0.780	—	—	—	—	—
500	0.780	0.0108	0.769	0.385	8.2	3.54×10^{-3}
1000	0.780	0.0233	0.757	0.378	6.4	4.38×10^{-3}
2000	0.780	0.0398	0.740	0.370	4.0	6.71×10^{-3}
4000	0.780	0.0644	0.716	0.358	3.4	7.39×10^{-3}
8000	0.780	0.0982	0.682	0.341	3.5	6.51×10^{-3}
16,000	0.780	0.1387	0.641	0.320	4.0	5.02×10^{-3}

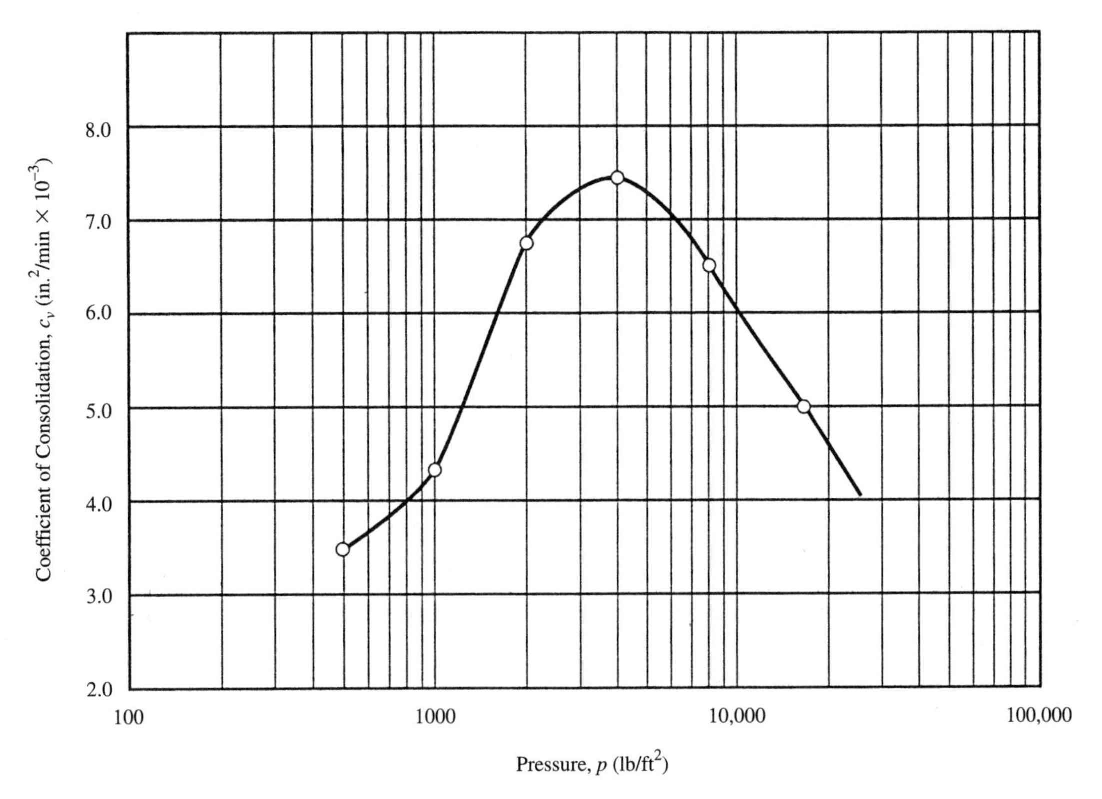

FIGURE 7–6 Coefficient of consolidation versus logarithm of pressure for Example 7–2.

Elevation 760 ft

Sand and Gravel
Unit Weight = 132.0 lb/ft^3

Water Table

Elevation 752 ft

Sand and Gravel
Unit Weight = 132.0 lb/ft^3

Elevation 732 ft

Clay
Unit Weight = 125.4 lb/ft^3

Elevation 710 ft

FIGURE 7–7.

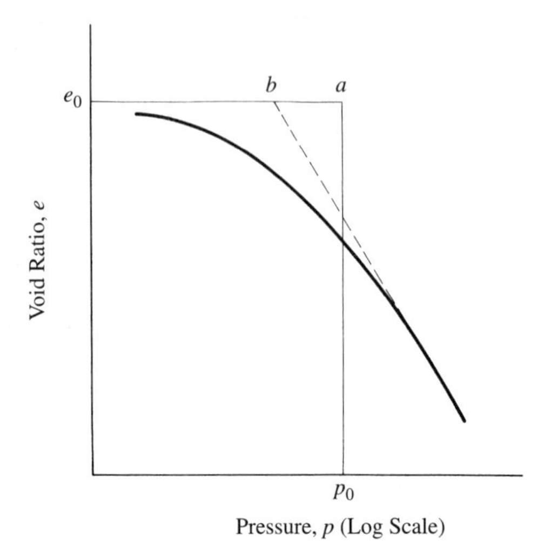

FIGURE 7–8 Typical e–log p curve for normally consolidated clay [4].

The first step in determining if a given clayey soil is normally consolidated is to locate the point designated by a pressure of p_0 (distance along the abscissa) and void ratio of e_0 (distance along the ordinate). (p_0 is the present effective overburden pressure at midheight of the compressible clay layer, and e_0 is the initial void ratio of the soil *in situ*.) This point is labeled a in Figure 7–8. The next step is to project the lower right straight-line portion of the e–log p curve in a straight line upward and to the left. This is the dashed line in Figure 7–8; it will intersect a horizontal line drawn at $e = e_0$. The point of intersection of these two lines is labeled b in Figure 7–8. If point b is to the left of point a (as in Figure 7–8), the soil is normally consolidated clay [4].

7–4 OVERCONSOLIDATED CLAY

The procedure for determining if a given clay is overconsolidated is essentially the same as that for determining if it is normally consolidated. The point designated by a pressure of p_0 and void ratio of e_0 is located and labeled a. The lower right portion of the e–log p curve is projected in a straight line upward and to the left until it intersects a horizontal line drawn at $e = e_0$, with the point of intersection labeled b. If point b is to the right of point a (as in Figure 7–9), the soil is overconsolidated clay [4].

If the given clay is found to be overconsolidated, it is necessary to determine (for subsequent analysis of consolidation settlement) the maximum overburden pressure at the consolidated clay layer (p_0'). The following procedure, developed by Casagrande [5], can be used to determine p_0'. The first step is to locate the point on the e–log p curve where the curvature is greatest (where the radius of curvature is smallest). This is indicated by point g in Figure 7–10. From this point, two straight lines are drawn—one horizontal line (line gh in Figure 7–10) and one tangent to the e–log p curve (line gj in Figure 7–10). The next step is to draw a line that bisects the angle between lines gh and gj (line gi in Figure 7–10). The final step is to project the lower right straight-line portion of the e–log p curve in a straight line upward and to the left.

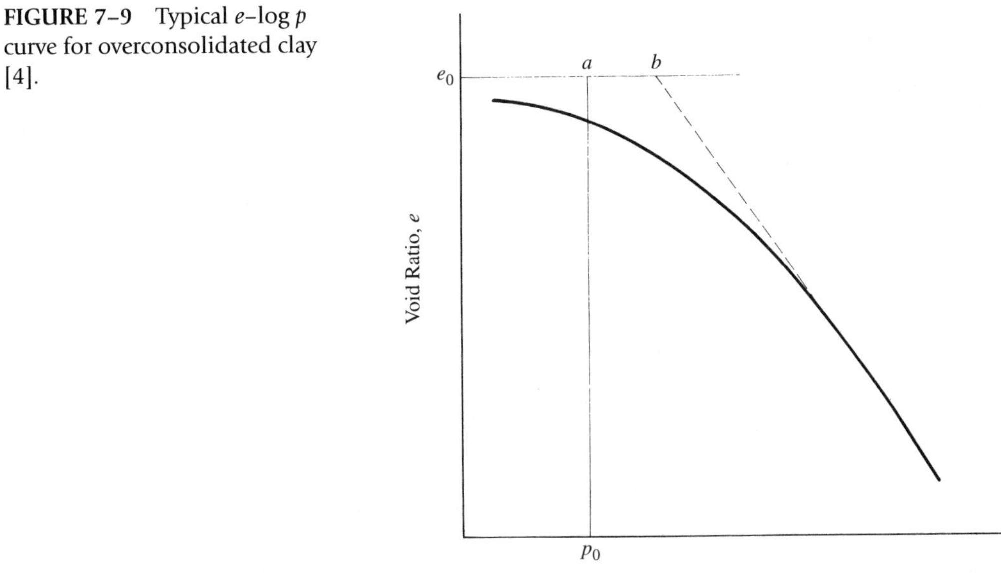

FIGURE 7–9 Typical e–log p curve for overconsolidated clay [4].

This projected line will intersect line *gi* at a point such as *k* in Figure 7–10. The value of p corresponding to point k (p coordinate of point k along the abscissa) is taken as p_0' [4].

The overconsolidation ratio (OCR) can now be defined as the ratio of overconsolidation pressure (p_0') to present overburden pressure (p_0). Hence,

$$\text{OCR} = \frac{p_0'}{p_0} \tag{7-2}$$

The OCR is used to indicate the degree of overconsolidation.

7–5 FIELD CONSOLIDATION LINE

The e–log p curves considered in previous sections give, of course, the relationship between void ratio and pressure for a given soil. Such a relationship is used in calculating consolidation settlement. The e–log p curves of Figure 7–3 and Figures 7–8 through 7–10 reflect, however, the relationship between void ratio and pressure for the soil sample in the laboratory. Although an "undisturbed sample" is used in the laboratory test, it is not generally possible to duplicate soil in the laboratory exactly as it exists in the field. Thus, the e–log p curves developed from laboratory consolidation tests are modified to give an e–log p curve that is presumed to reflect actual field conditions. This modified e–log p curve is called the *field consolidation line.* Two methods for determining the field consolidation line follow—one for normally consolidated clay, the other for overconsolidated clay.

FIGURE 7–10 Graphic construction for estimating maximum overburden pressure, p_0' from e–log p curve [4, 5].

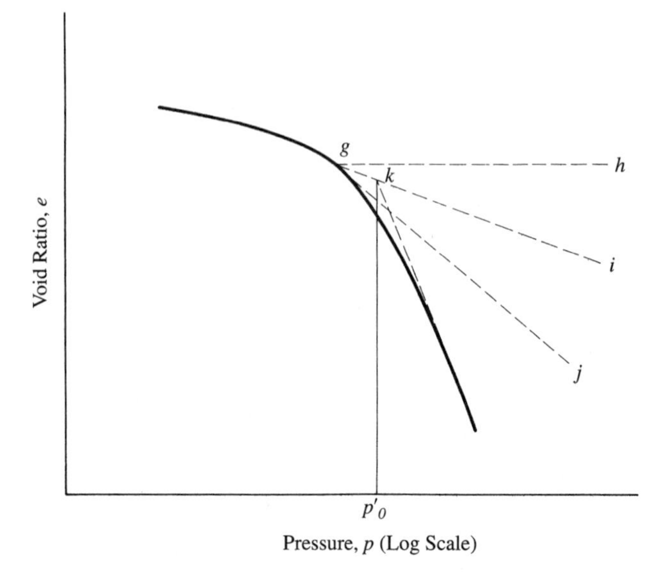

In the case of normally consolidated clay, determination of the field consolidation line is fairly simple. With the given e–log p curve developed from the laboratory test (Figure 7–11), the point on the e–log p curve corresponding to $0.4e_0$ is determined (point f in Figure 7–11). A straight line connecting points a and f gives the field consolidation line for the normally consolidated clay [4, 6]. (The reader will recall that, as related in Section 7–3 and Figure 7–8, point a is the point designated by a pressure of p_0 and void ratio of e_0.)

For overconsolidated clay, finding the field consolidation line is somewhat more difficult. With the given e–log p curve developed from the laboratory test (Figure 7–12), the point on the e–log p curve corresponding to $0.4e_0$ is determined (point f in Figure 7–12). Point a (the point designated by a pressure of p_0 and void ratio of e_0) is located, and a line is drawn through point a parallel to the rebound line. This line through point a parallel to the rebound line is shown as a dashed line in Figure 7–12; it will intersect a vertical line drawn at $p = p_0'$. (The procedure for evaluating p_0' was given in Section 7–4.) This point of intersection is designated by m in Figure 7–12. Points m and f are connected by a straight line, and points a and m are connected by a curved line that follows the same general shape of the e–log p curve. This curved line from a to m and the straight line from m to f give the field consolidation line for the overconsolidated clay [7, 8].

It is the field consolidation line—the dark line in Figure 7–11 (normally consolidated clay) and Figure 7–12 (overconsolidated clay)—that is used in calculating consolidation settlement. The other curves (dial readings versus time and e–log p curve) are required only as a means of determining the field consolidation line. Once the field consolidation line is established, these other curves are no longer used in determining the amount of consolidation settlement.

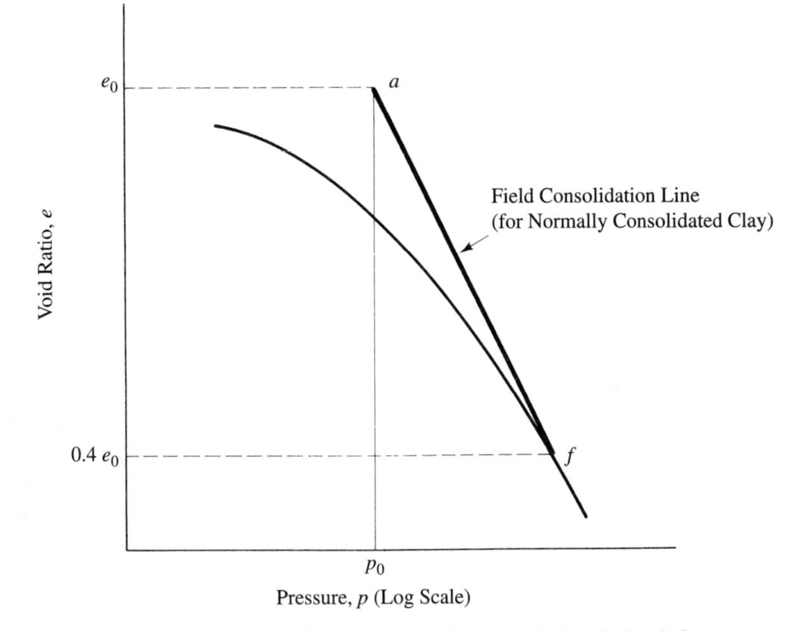

FIGURE 7–11 Field consolidation line for normally consolidated clay [4].

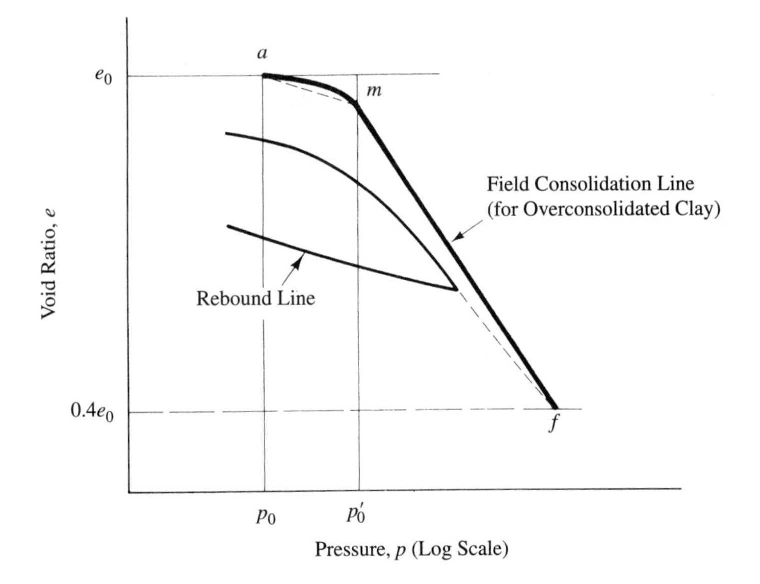

FIGURE 7–12 Field consolidation line for overconsolidated clay [7, 8].

Of special significance is the *slope* of the field consolidation line. This slope is called the *compression index* (C_c) and may be evaluated by finding coordinates of any two points on the field consolidation line [(p_1, e_1) and (p_2, e_2)] and substituting these values into the following equation [4]:

$$C_c = \frac{e_1 - e_2}{\log p_2 - \log p_1} = \frac{e_1 - e_2}{\log (p_2/p_1)} \tag{7–3}$$

Skempton has shown that the compression index can be approximated in terms of the liquid limit (*LL*, in percent) by the following equation [4, 9]:

$$C_c = 0.009(LL - 10) \tag{7–4}$$

for normally consolidated clays.

It should be emphasized that the value of C_c computed from Eq. (7–3) is obtained from the field consolidation line, which is based on the results of a consolidation test, whereas that computed from Eq. (7–4) is based solely on the liquid limit. The consolidation test is much more lengthy, difficult, and expensive to perform than the test to determine the liquid limit. Also, calculation of C_c using results of a consolidation test is much more involved than calculation using the liquid limit. However, calculation of C_c using the liquid limit [Eq. (7–4)] is merely an approximation and should be used only when very rough values of settlement are acceptable (such as in a preliminary design).

EXAMPLE 7–4

Given

When the total pressure acting at midheight of a consolidating clay layer is 200 kN/m^2, the corresponding void ratio of the clay is 0.98. When the total pressure acting at the same location is 500 kN/m^2, the corresponding void ratio decreases to 0.81.

Required

The void ratio of the clay if the total pressure acting at midheight of the consolidating clay layer is 1000 kN/m^2.

Solution

From Eq. (7–3),

$$C_c = \frac{e_1 - e_2}{\log (p_2/p_1)} \tag{7–3}$$

$$e_1 = 0.98$$

$$e_2 = 0.81$$

$$p_2 = 500\ \text{kN/m}^2$$

$$p_1 = 200\ \text{kN/m}^2$$

$$C_c = \frac{0.98 - 0.81}{\log \left(\dfrac{500\ \text{kN/m}^2}{200\ \text{kN/m}^2} \right)} = 0.427$$

Substituting the computed value of C_c and the same values of e_1 and p_1 into Eq. (7–3) gives

$$0.427 = \frac{0.98 - e_2}{\log\left(\frac{1000\ \text{kN/m}^2}{200\ \text{kN/m}^2}\right)}$$

$$e_2 = 0.68$$

EXAMPLE 7–5

Given

A normally consolidated clay has a liquid limit of 51.2%.

Required

Estimate the compression index (C_c).

Solution

From Eq. (7–4),

$$C_c = 0.009(LL - 10) \tag{7–4}$$

$$C_c = 0.009(51.2 - 10) = 0.371$$

7–6 SETTLEMENT OF LOADS ON CLAY DUE TO PRIMARY CONSOLIDATION

Once the field consolidation line has been defined for a given clayey soil, the total expected primary consolidation settlement of the load on the clay can be determined. Consider Figure 7–13, where a mass of soil is depicted before (Figure 7–13a) and after (Figure 7–13b) consolidation settlement has occurred in a clay layer of initial thickness H. From Figure 7–13,

$$\frac{\Delta H}{H} = \frac{\Delta H}{H_s + (H_v)_0} \tag{7–5}$$

where ΔH represents the amount of clay settlement and H_s and $(H_v)_0$ denote the height of solids and initial height of voids, respectively. By definition, the initial void ratio, e_0 in Figure 7–13b, is given by the following:

$$e_0 = \frac{(V_v)_0}{V_s} \tag{7–6}$$

where $(V_v)_0$ and V_s represent the original volume of voids and the volume of solids, respectively. Because each volume can be replaced by the soil's cross-sectional area (A) times the height of the soil, this equation can be modified as follows:

$$e_0 = \frac{(A)(H_v)_0}{(A)(H_s)} = \frac{(H_v)_0}{H_s} \tag{7–7}$$

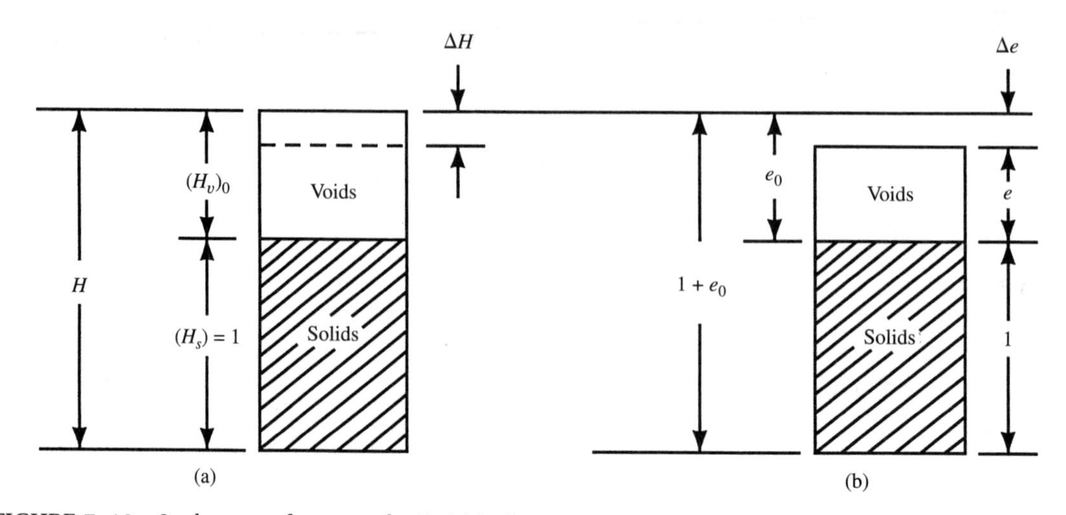

FIGURE 7–13 Settlement of a mass of soil: (a) before consolidation settlement; (b) after consolidation settlement.

Also,

$$\Delta e = \frac{\Delta H}{H_s} \tag{7–8}$$

where Δe represents the change in void ratio as a result of consolidation settlement. If we let the height of solids (H_s) equal unity, then Eqs. (7–5), (7–7), and (7–8) become

$$\frac{\Delta H}{H} = \frac{\Delta H}{1 + (H_v)_0} \tag{7–9}$$

$$e_0 = (H_v)_0 \tag{7–10}$$

$$\Delta e = \Delta H \tag{7–11}$$

Substituting Eqs. (7–10) and (7–11) into Eq. (7–9) yields

$$\frac{\Delta H}{H} = \frac{\Delta e}{1 + e_0} \tag{7–12}$$

or,

$$\Delta H = \frac{\Delta e}{1 + e_0}(H) \tag{7–13}$$

Because $\Delta e = e_0 - e$ and $\Delta H =$ settlement S,

$$S = \frac{e_0 - e}{1 + e_0}(H) \tag{7–14}$$

where S = total settlement due to primary consolidation
e_0 = initial void ratio of the soil *in situ*
e = void ratio of the soil corresponding to the total pressure (p) acting at midheight of the consolidating clay layer
H = thickness of the consolidating clay layer

In practice, the value of e_0 is obtained from the laboratory consolidation test, and the value of e is obtained from the field consolidation line based on total pressure (i.e., effective overburden pressure plus net additional pressure due to the structure—both at midheight of the consolidating clay layer). The value of H is obtained from soil exploration (Chapter 3).

An alternative equation for computing total expected primary consolidation settlement using the compression index (i.e., slope of the field consolidation line) can be derived by recalling Eq. (7–3):

$$C_c = \frac{e_1 - e_2}{\log (p_2/p_1)} \tag{7–3}$$

Because (p_1, e_1) and (p_2, e_2) can be the coordinates of any two points on the field consolidation line, let

p_1 = present effective overburden pressure at midheight of the consolidating clay layer (i.e., p_0)
e_1 = initial void ratio of the soil *in situ* [i.e., e_0 in Eq. (7–14)]
p_2 = total pressure acting at midheight of the consolidating clay layer [$p_0 + \Delta p$ (i.e., p)]
e_2 = void ratio of the soil corresponding to the total pressure (p) acting at midheight of the consolidating clay layer [i.e., e in Eq. (7–14)]

Making these substitutions in Eq. (7–3) gives the following:

$$C_c = \frac{e_0 - e}{\log (p/p_0)} \tag{7–15}$$

Rearranging this equation results in

$$e_0 - e = C_c[\log (p/p_0)] \tag{7–16}$$

Substituting Eq. (7–16) into Eq. (7–14) yields

$$S = \frac{C_c[\log (p/p_0)]}{1 + e_0} (H) \tag{7–17}$$

or

$$S = C_c\left(\frac{H}{1 + e_0}\right) \log \frac{p}{p_0} \tag{7–18}$$

where C_c = slope of the field consolidation line (compression index)
p = total pressure acting at midheight of the consolidating clay layer ($= p_0 + \Delta p$)
p_0 = present effective overburden pressure at midheight of the consolidating clay layer
Δp = net additional pressure at midheight of the consolidating clay layer due to the structure

The value of C_c can be determined by evaluating the slope of the field consolidation line [Eq. (7–3)] or approximated based on the liquid limit [Eq. (7–4)]. If the latter method is used, computed settlement should be considered as a rough approximation.

The time rate of settlement due to primary consolidation can be computed from the following equation [1, 7]:

$$t = \frac{T_v}{c_v} H^2 \tag{7–19}$$

where t = time to reach a particular percent of consolidation; percent of consolidation is defined as the ratio of the amount of settlement at a certain time during the process of consolidation to the total settlement due to consolidation
T_v = time factor, a coefficient depending on the particular percent of consolidation
c_v = coefficient of consolidation corresponding to the total pressure ($p = p_0 + \Delta p$) acting at midheight of the clay layer
H = thickness of the consolidating clay layer [however, if the clay layer *in situ* is drained on both top and bottom, half the thickness of the layer should be substituted for H in Eq. (7–19)]

In practice, the value of T_v is determined from Figure 7–14, based on the desired percent of consolidation (U), and the value of c_v is determined from the c_v–log p curve (e.g., Figure 7–4) based on the total pressure acting at midheight of the clay layer. It will be recalled that the c_v–log p curve is a product of the laboratory consolidation test.

As may be noted from Eq. (7–19), the coefficient of consolidation (c_v) indicates how rapidly (or slowly) the process of consolidation takes place. This consolidation property can also be expressed as follows [10]*

$$c_v = \frac{k}{\gamma_w m_v} \tag{7–20}$$

* From Karl Terzaghi, Ralph B. Peck, and Gholamreza Mesri, *Soil Mechanics in Engineering Practice,* 3rd ed., John Wiley & Sons, Inc., New York, 1996. Copyright © 1996, by John Wiley & Sons, Inc. Reprinted by permission of John Wiley & Sons, Inc.

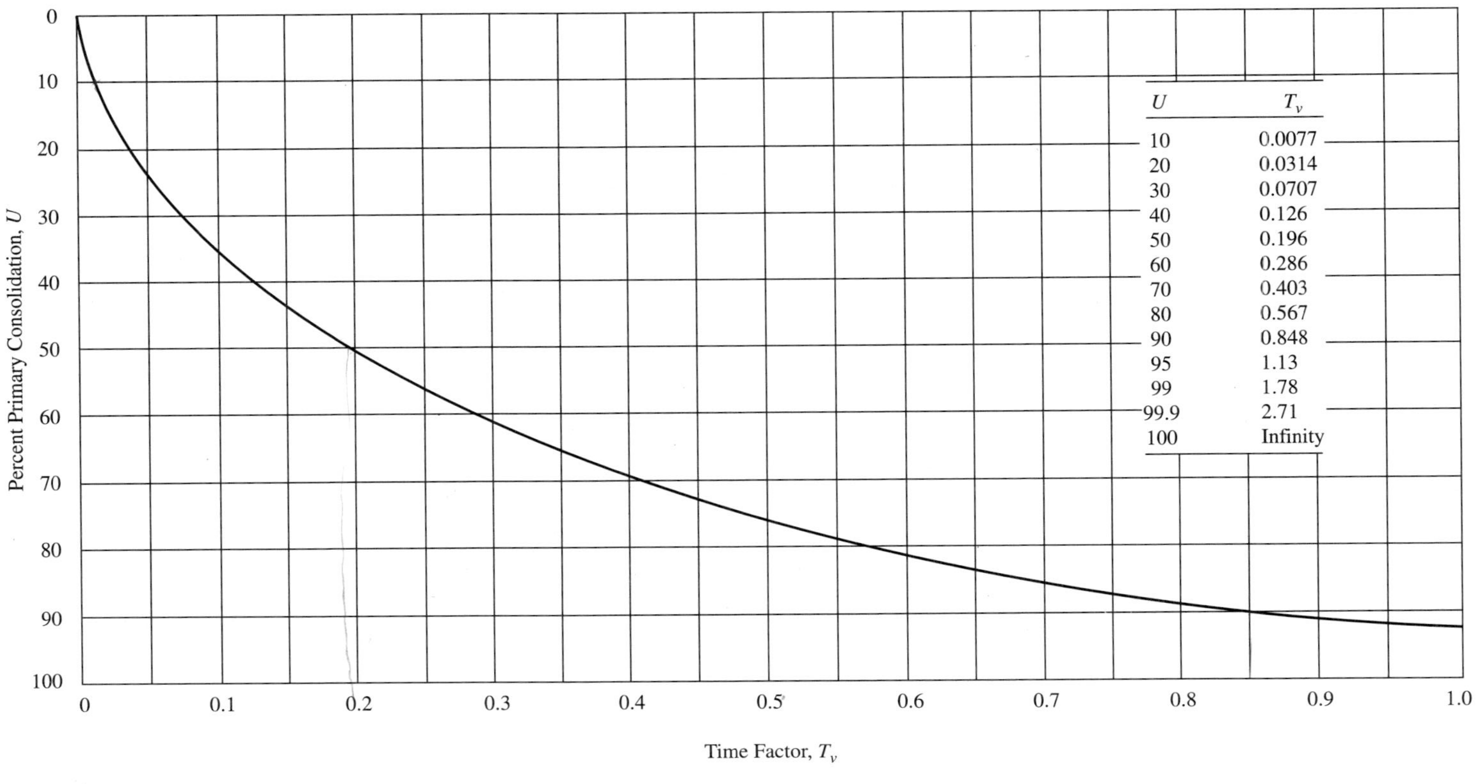

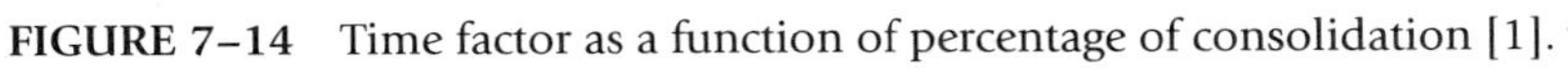

FIGURE 7–14 Time factor as a function of percentage of consolidation [1].

where k = coefficient of permeability
γ_w = unit weight of water
m_v = coefficient of volume compressibility

The last term can be determined from the following:

$$m_v = \frac{a_v}{1 + e} \tag{7–21}$$

where a_v = coefficient of compressibility
e = void ratio

Substituting Eq. (7–21) into Eq. (7–20) gives the following:

$$c_v = \frac{k(1 + e)}{a_v \gamma_w} \tag{7–22}$$

For small strains, change in the void ratio (Δe) may be taken as directly proportional to change in the effective pressure (Δp). Therefore, a_v in Eq. (7–22) can be expressed in equation form as follows:

$$a_v = \frac{\Delta e}{\Delta p} \tag{7–23}$$

Values of k, e, and a_v can be evaluated separately and substituted into Eq. (7–22) to find c_v. However, it is common practice to evaluate c_v directly from the results of a laboratory consolidation test (see Section 7–2 and Example 7–2).

To summarize the means of finding settlement of loads on clay due to primary consolidation, one can use either Eq. (7–14) or Eq. (7–18) to compute total settlement; then Eq. (7–19) can be used to find the time required to reach a particular percentage of that consolidation settlement. For example, if total settlement due to consolidation is computed to be 3.0 in., the time required for the structure to settle 1.5 in. could be determined from Eq. (7–19) by substituting a value of T_v of 0.196 (along with applicable values of c_v and H). The value of 0.196 is obtained from Figure 7–14 for a value of U of 50%. U is 50% because the particular settlement being considered (1.5 in.) is 50% of total settlement (3.0 in.).

EXAMPLE 7–6

Given

A compressible normally consolidated clay layer is 7.40 m thick and has an initial void ratio *in situ* of 0.988. Consolidation tests and subsequent computations indicate that the void ratio of the clay layer corresponding to the total pressure acting at midheight of the consolidating clay layer after construction of a building is 0.942.

Required

Total expected primary consolidation settlement.

Solution

From Eq. (7–14),

$$S = \frac{e_0 - e}{1 + e_0}(H) \tag{7-14}$$

$$e_0 = 0.988$$

$$e = 0.942$$

$$H = 7.40 \text{ m}$$

$$S = \frac{0.988 - 0.942}{1 + 0.988}(7.40 \text{ m}) = 0.171 \text{ m}$$

EXAMPLE 7–7

Given

1. A sample of normally consolidated clay was obtained by a Shelby tube sampler from the midheight of a compressible clay layer (see Figure 7–15).
2. A consolidation test was conducted on a portion of this sample. Results of the consolidation test are as follows:

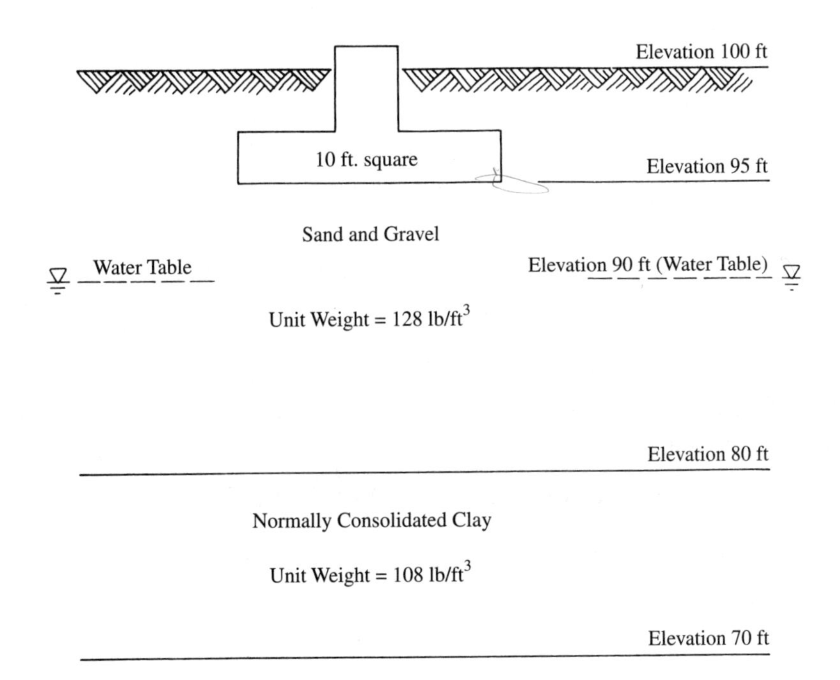

FIGURE 7–15.

a. Natural (or initial) void ratio of the clay existing in the field (e_0) is 1.65.
b. Pressure–void ratio relations are as follows:

p (tons/ft^2)	e
0.8	1.50
1.6	1.42
3.2	1.30
6.4	1.12
12.8	0.94

3. A footing is to be located 5 ft below ground level, as shown in Figure 7–15. The base of the square footing is 10 ft by 10 ft and it exerts a total load of 250 tons, which includes column load, weight of footing, and weight of soil surcharge on the footing.

Required

1. From the given results of the consolidation test, prepare an e–log p curve and construct a field consolidation line, assuming that point f is located at $0.4e_0$.
2. Compute the total expected primary consolidation settlement for the clay layer.

Solution

1. Present effective overburden pressure (p_0) at midheight of clay layer $= (128 \text{ lb/ft}^3)(100 \text{ ft} - 90 \text{ ft}) + (128 \text{ lb/ft}^3 - 62.4 \text{ lb/ft}^3) \times (90 \text{ ft} - 80 \text{ ft}) + (108 \text{ lb/ft}^3 - 62.4 \text{ lb/ft}^3)\left(\dfrac{80 \text{ ft} - 70 \text{ ft}}{2}\right)$

$$p_0 = 2164 \text{ lb/ft}^2, \text{ or } 1.08 \text{ tons/ft}^2$$

$$e_0 = 1.65 \text{ (given)}$$

$$0.4e_0 = (0.4)(1.65) = 0.66$$

The e–log p curve is shown in Figure 7–16 together with the field consolidation line.

2. Effective weight of excavation $= (128 \text{ lb/ft}^3)(5 \text{ ft}) = 640 \text{ lb/ft}^2$, or 0.32 ton/ft^2

$$\text{Net consolidation pressure at base of footing} = \frac{250 \text{ tons}}{(10 \text{ ft})(10 \text{ ft})} - 0.32 \text{ ton/ft}^2 = 2.18 \text{ tons/ft}^2$$

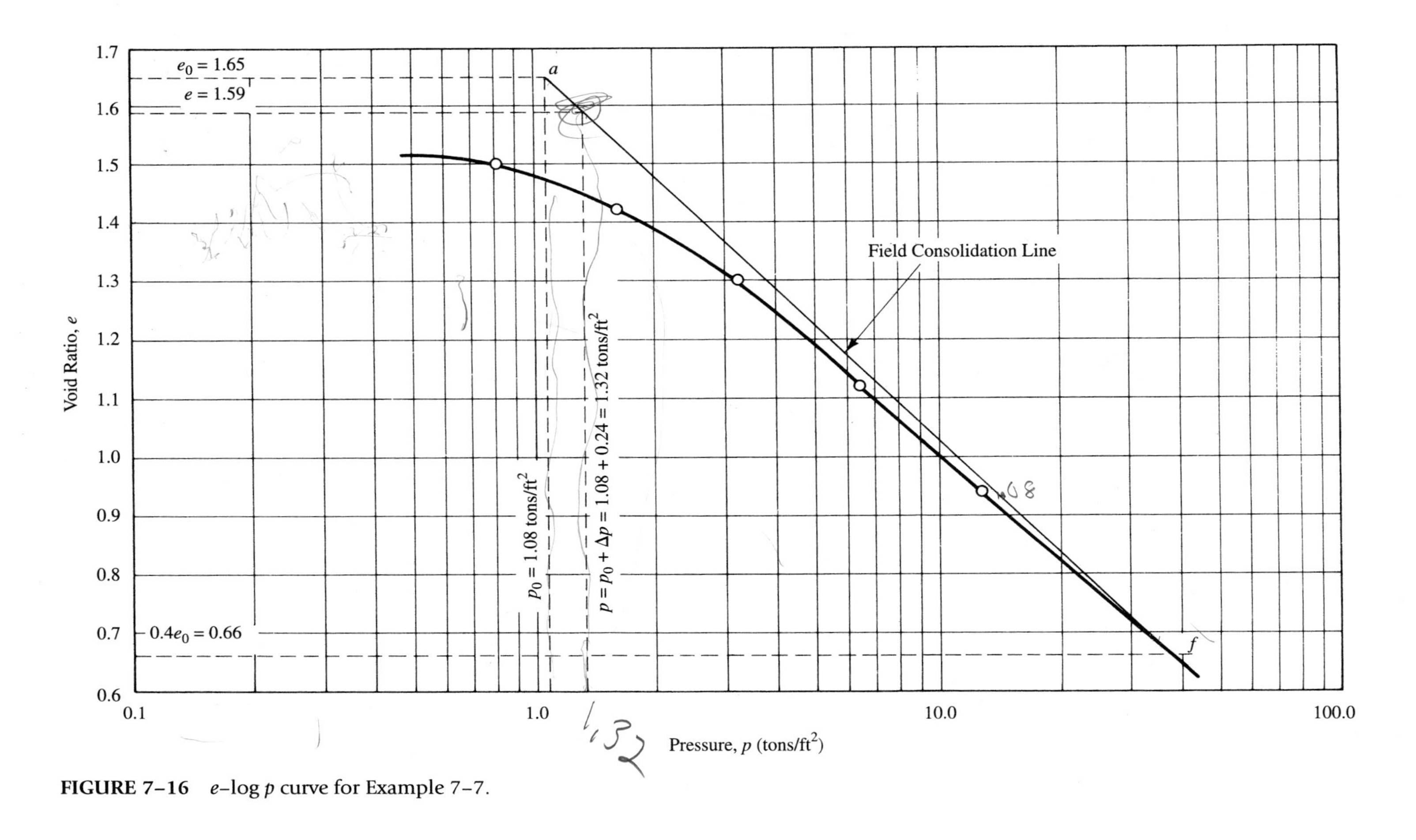

FIGURE 7–16 e–log p curve for Example 7–7.

To determine net consolidation pressure at midheight of the clay layer under the center of the footing, one must divide the base of the footing into four equal 5 ft by 5 ft square areas. Because each of these square areas has a common corner at the footing's center, the desired net consolidation pressure at midheight of the clay layer can be calculated upon determining an influence coefficient by using either Table 6–2 or Figure 6–8. Referring to Figure 6–8, we see that

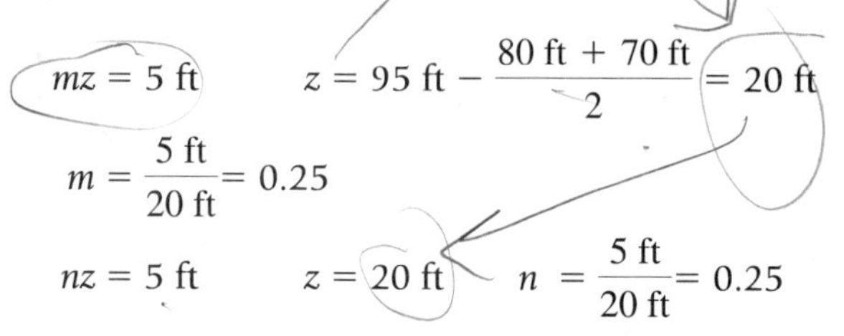

$$mz = 5 \text{ ft} \qquad z = 95 \text{ ft} - \frac{80 \text{ ft} + 70 \text{ ft}}{2} = 20 \text{ ft}$$

$$m = \frac{5 \text{ ft}}{20 \text{ ft}} = 0.25$$

$$nz = 5 \text{ ft} \qquad z = 20 \text{ ft} \qquad n = \frac{5 \text{ ft}}{20 \text{ ft}} = 0.25$$

From Figure 6–8, the influence coefficient = 0.027. Net consolidation pressure at midheight of the clay layer under the center of the footing is as follows:

$$\Delta p = (4)(0.027)(2.18 \text{ tons/ft}^2) = 0.24 \text{ ton/ft}^2$$

Thus, the final pressure at midheight of the clay layer is as follows:

$$p = p_0 + \Delta p = 1.08 \text{ tons/ft}^2 + 0.24 \text{ ton/ft}^2 = 1.32 \text{ tons/ft}^2$$

Locate $p = 1.32 \text{ tons/ft}^2$ along the abscissa of the e–log p curve (Figure 7–16) and move upward vertically until the field consolidation line is intersected. Then turn left and move horizontally to read a void ratio e of 1.59 on the ordinate of the e–log p curve. With

$$e_0 = 1.65$$
$$e = 1.59$$
$$H = 10 \text{ ft} = 120 \text{ in.}$$

substitute into Eq. (7–14):

$$S = \frac{e_0 - e}{1 + e_0}(H) \tag{7–14}$$

$$S = \frac{1.65 - 1.59}{1 + 1.65}(120 \text{ in.}) = 2.72 \text{ in.}$$

The total expected primary consolidation settlement is 2.72 in.

EXAMPLE 7–8

Given

1. Same as Example 7–7; total consolidation settlement = 2.72 in.
2. Results of the laboratory consolidation test also indicated that the coefficient of consolidation (c_v) for the clay sample is 3.28×10^{-3} in.^{2}/min for the pressure increment from 0.8 to 1.6 tons/ft^2.

Required

Time of primary consolidation settlement if the clay layer is underlain by

1. Permeable sand and gravel (double drainage).
2. Impermeable bedrock (single drainage).

Take U at 10% increments and plot these values on a settlement–log time curve.

Solution

1. *Clay layer is underlain by permeable sand and gravel (double drainage).* Use Eq. (7–19):

$$t = \frac{T_v}{c_v} H^2 \qquad (7\text{–}19)$$

where $c_v = 3.28 \times 10^{-3}$ in.^{2}/min
$H = 10$ ft/2 $= 5$ ft $= 60$ in. (double drainage)

a. When $U = 10\%$ (i.e., 10% of total settlement, $S_{10} = 2.72$ in. $\times$ 0.10 $=$ 0.27 in.),

$$T_v = 0.0077 \quad \text{(from Figure 7–14)}$$

$$t_{10} = \frac{(0.0077)(60\text{ in.})^2}{3.28 \times 10^{-3}\text{ in.}^2/\text{min}} = 8451\text{ min} = 0.016\text{ yr}$$

This indicates that the footing will settle approximately 0.27 in. in 0.016 yr.

b. When $U = 20\%$ (i.e., 20% of total settlement, $S_{20} = 0.54$ in.),

$$T_v = 0.0314 \quad \text{(from Figure 7–14)}$$

$$t_{20} = \frac{(0.0314)(60\text{ in.})^2}{3.28 \times 10^{-3}\text{ in.}^2/\text{min}} = 34{,}463\text{ min} = 0.066\text{ yr}$$

This indicates that the footing will settle approximately 0.54 in. in 0.066 yr.

c. When $U = 30\%$ (i.e., 30% of total settlement),

$$T_v = 0.0707 \quad \text{(from Figure 7–14)}$$

$$t_{30} = \frac{(0.0707)(60\text{ in.})^2}{3.28 \times 10^{-3}\text{ in.}^2/\text{min}} = 77{,}598\text{ min} = 0.15\text{ yr}$$

d. When $U = 40\%$ (i.e., 40% of total settlement),

$$T_v = 0.126 \quad \text{(from Figure 7–14)}$$

$$t_{40} = \frac{(0.126)(60\text{ in.})^2}{3.28 \times 10^{-3}\text{ in.}^2/\text{min}} = 1.383 \times 10^5\text{ min} = 0.26\text{ yr}$$

e. When $U = 50\%$ (i.e., 50% of total settlement),

$$T_v = 0.196 \quad \text{(from Figure 7–14)}$$

$$t_{50} = \frac{(0.196)(60\text{ in.})^2}{3.28 \times 10^{-3}\text{ in.}^2/\text{min}} = 2.151 \times 10^5\text{ min} = 0.41\text{ yr}$$

f. When $U = 60\%$ (i.e., 60% of total settlement),

$$T_v = 0.286 \quad \text{(from Figure 7–14)}$$

$$t_{60} = \frac{(0.286)(60 \text{ in.})^2}{3.28 \times 10^{-3} \text{ in.}^2/\text{min}} = 3.139 \times 10^5 \text{ min} = 0.60 \text{ yr}$$

g. When $U = 70\%$ (i.e., 70% of total settlement),

$$T_v = 0.403 \quad \text{(from Figure 7–14)}$$

$$t_{70} = \frac{(0.403)(60 \text{ in.})^2}{3.28 \times 10^{-3} \text{ in.}^2/\text{min}} = 4.423 \times 10^5 \text{ min} = 0.84 \text{ yr}$$

h. When $U = 80\%$ (i.e., 80% of total settlement),

$$T_v = 0.567 \quad \text{(from Figure 7–14)}$$

$$t_{80} = \frac{(0.567)(60 \text{ in.})^2}{3.28 \times 10^{-3} \text{ in.}^2/\text{min}} = 6.223 \times 10^5 \text{ min} = 1.18 \text{ yr}$$

i. When $U = 90\%$ (i.e., 90% of total settlement),

$$T_v = 0.848 \quad \text{(from Figure 7–14)}$$

$$t_{90} = \frac{(0.848)(60 \text{ in.})^2}{3.28 \times 10^{-3} \text{ in.}^2/\text{min}} = 9.307 \times 10^5 \text{ min} = 1.77 \text{ yr}$$

2. *Clay layer is underlain by impermeable bedrock (single drainage).* Eq. (7–19) is still applicable.

$$t = \frac{T_v}{c_v} H^2 \tag{7–19}$$

where $c_v = 3.28 \times 10^{-3} \text{ in.}^2/\text{min}$

$H = 10 \text{ ft} = 120 \text{ in.}$ (single drainage)

a. When $U = 10\%$,

$$T_v = 0.0077$$

$$t_{10} = \frac{(0.0077)(120 \text{ in.})^2}{3.28 \times 10^{-3} \text{ in.}^2/\text{min}} = 33{,}805 \text{ min} = 0.064 \text{ yr}$$

b. When $U = 20\%$,

$$T_v = 0.0314$$

$$t_{20} = \frac{(0.0314)(120 \text{ in.})^2}{3.28 \times 10^{-3} \text{ in.}^2/\text{min}} = 1.379 \times 10^5 \text{ min} = 0.26 \text{ yr}$$

c. When $U = 30\%$,

$$T_v = 0.0707$$

$$t_{30} = \frac{(0.0707)\,(120\ \text{in.})^2}{3.28 \times 10^{-3}\text{in.}^2/\text{min.}} = 3.104 \times 10^5\text{min} = 0.59\ \text{yr}$$

d. When $U = 40\%$,

$$T_v = 0.126$$

$$t_{40} = \frac{(0.126)(120\ \text{in.})^2}{3.28 \times 10^{-3}\ \text{in.}^2/\text{min}} = 5.532 \times 10^5\ \text{min} = 1.05\ \text{yr}$$

e. When $U = 50\%$,

$$T_v = 0.196$$

$$t_{50} = \frac{(0.196)\,(120\ \text{in.})^2}{3.28 \times 10^{-3}\text{in.}^2/\text{min.}} = 8.605 \times 10^5\text{min} = 1.64\ \text{yr}$$

f. When $U = 60\%$,

$$T_v = 0.286$$

$$t_{60} = \frac{(0.286)(120\ \text{in.})^2}{3.28 \times 10^{-3}\ \text{in.}^2/\text{min}} = 1.256 \times 10^6\ \text{min} = 2.39\ \text{yr}$$

g. When $U = 70\%$,

$$T_v = 0.403$$

$$t_{70} = \frac{(0.403)(120\ \text{in.})^2}{3.28 \times 10^{-3}\ \text{in.}^2/\text{min}} = 1.769 \times 10^6\ \text{min} = 3.37\ \text{yr}$$

h. When $U = 80\%$,

$$T_v = 0.567$$

$$t_{80} = \frac{(0.567)(120\ \text{in.})^2}{3.28 \times 10^{-3}\ \text{in.}^2/\text{min}} = 2.489 \times 10^6\ \text{min} = 4.74\ \text{yr}$$

i. When $U = 90\%$,

$$T_v = 0.848$$

$$t_{90} = \frac{(0.848)(120\ \text{in.})^2}{3.28 \times 10^{-3}\ \text{in.}^2/\text{min}} = 3.723 \times 10^6\ \text{min} = 7.08\ \text{yr}$$

The results of these computations are tabulated in Table 7–5 and are shown graphically by a settlement–log time curve in Figure 7–17.

EXAMPLE 7–9

Given

1. An 8-ft clay layer beneath a building is overlain by a stratum of permeable sand and gravel and is underlain by impermeable bedrock.

TABLE 7–5
Computed Time–Settlement Relation for Example 7–8

Fraction of Total Consolidation Settlement, U (%)	Consolidation Settlement (in.)	Time (yr)	
		Double Drainage	Single Drainage
10	0.27	0.016	0.064
20	0.54	0.066	0.26
30	0.82	0.15	0.59
40	1.09	0.26	1.05
50	1.36	0.41	1.64
60	1.63	0.60	2.39
70	1.90	0.84	3.37
80	2.18	1.18	4.74
90	2.45	1.77	7.08
100	2.72	∞	∞

2. The total expected primary consolidation settlement for the clay layer due to the footing load is 2.50 in.
3. The coefficient of consolidation (c_v) is 2.68×10^{-3} in.^{2}/min.

Required

1. How many years will it take for 90% of the total expected primary consolidation settlement to take place?
2. Compute the amount of primary consolidation settlement that will occur in 1 yr.
3. How many years will it take for primary consolidation settlement of 1 in. to take place?

Solution

1. From Eq. (7–19),

$$t = \frac{T_v}{c_v} H^2 \tag{7–19}$$

$T_v = 0.848$ (for $U = 90\%$; see Figure 7–14)

$c_v = 2.68 \times 10^{-3}$ in.^{2}/min (given)

$H = 8$ ft $= 96$ in. (single drainage)

$$t_{90} = \frac{(0.848)(96 \text{ in.})^2}{2.68 \times 10^{-3} \text{ in.}^2/\text{min}} = 2.916 \times 10^6 \text{ min} = 5.55 \text{ yr}$$

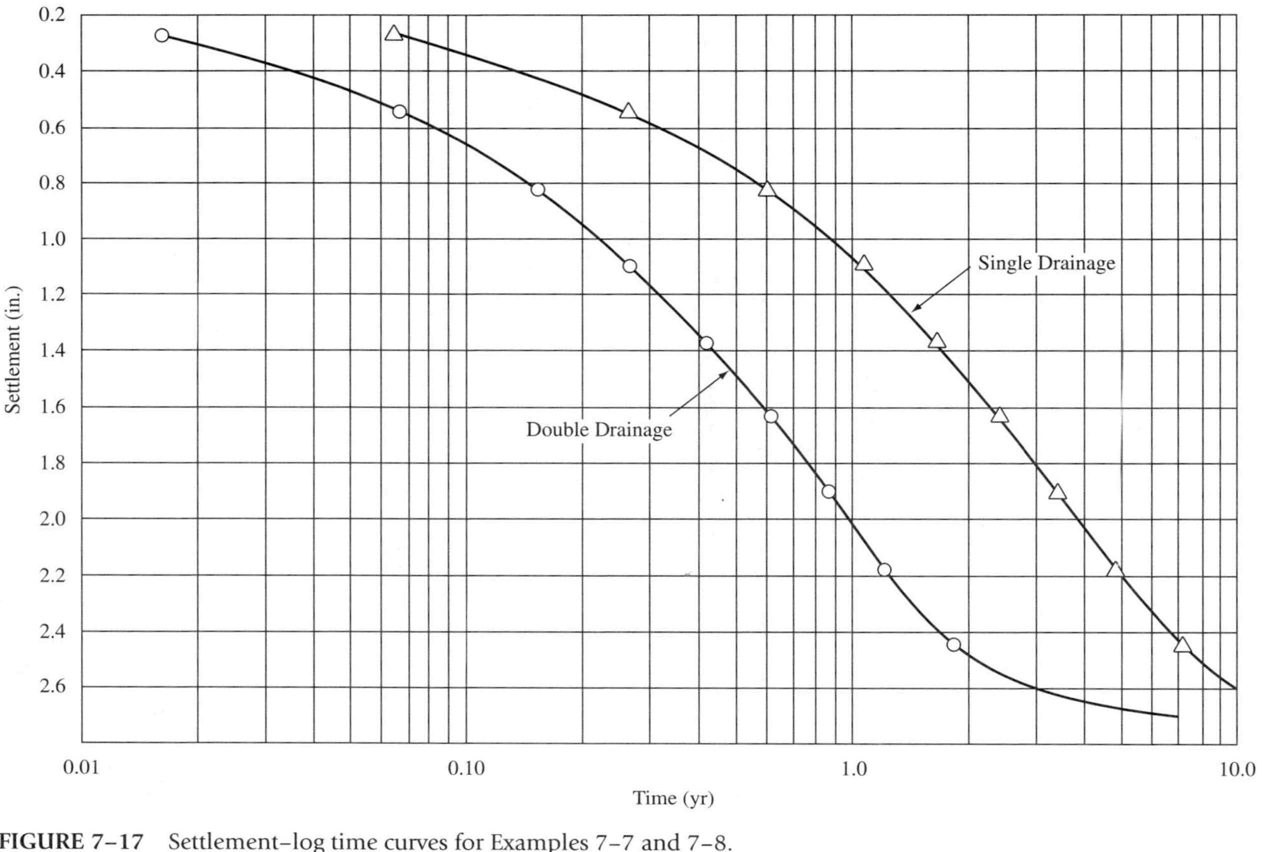

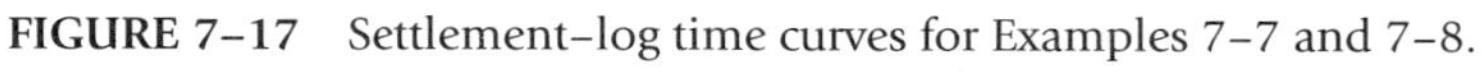

FIGURE 7–17 Settlement–log time curves for Examples 7–7 and 7–8.

2. From Eq. (7–19),

$$t = \frac{T_v}{c_v} H^2 \tag{7–19}$$

$$t = 1 \text{ yr}$$

$$c_v = 2.68 \times 10^{-3} \text{ in.}^2/\text{min}$$

$$H = 8 \text{ ft} = 96 \text{ in.}$$

$$1 \text{ yr} = \frac{T_v}{2.68 \times 10^{-3} \text{ in.}^2/\text{min}} (96 \text{ in.})^2 \times \frac{1}{(60 \text{ min/hr})(24 \text{ hr/day})(365 \text{ days/yr})}$$

$$T_v = 0.15$$

From Figure 7–14, with $T_v = 0.15$, $U = 43\%$.

Amount of primary consolidation settlement that will occur in 1 yr

$$= \text{Total primary consolidation settlement} \times U\%$$

$$= (2.50 \text{ in.})(0.43) = 1.08 \text{ in.}$$

3. $U\%$ = Fraction of total primary consolidation settlement

$$U = \frac{1 \text{ in.}}{2.50 \text{ in.}} \times 100 = 40\%$$

From Figure 7–14, with $U = 40\%$, $T_v = 0.126$. From Eq. (7–19),

$$t = \frac{T_v}{c_v} H^2 \tag{7–19}$$

$$t = \frac{(0.126)(96 \text{ in.})^2}{2.68 \times 10^{-3} \text{ in.}^2/\text{min}} = 4.333 \times 10^5 \text{ min} = 0.82 \text{ yr}$$

EXAMPLE 7–10

Given

1. A foundation is to be constructed at a site where the soil profile is as shown in Figure 7–18.
2. The base of the foundation is 3 m by 6 m, and it exerts a total load of 5400 kN, which includes the weight of the structure, foundation, and soil surcharge on the foundation.
3. The initial void ratio *in situ* (e_0) of the compressible clay layer is 1.38.
4. The compression index (C_c) of the clay layer is 0.68.

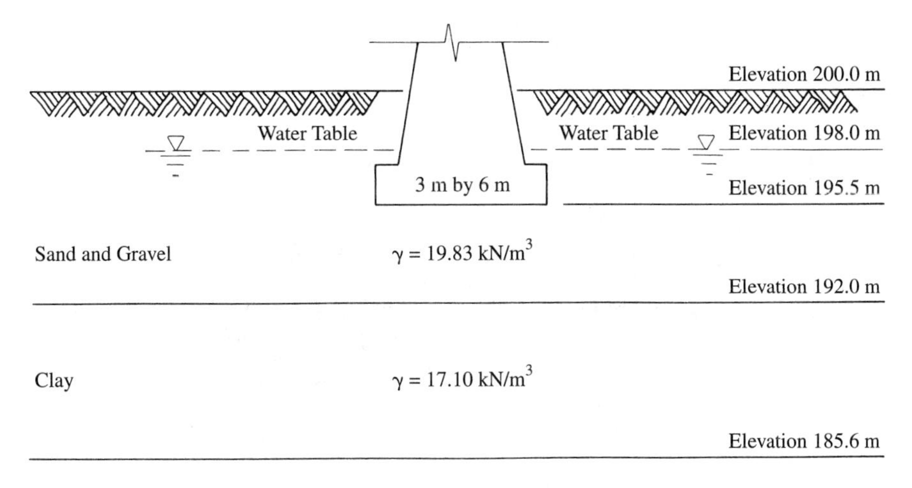

FIGURE 7–18.

Required

Expected primary consolidation settlement of the clay layer.

Solution

Present effective overburden pressure (p_0) at midheight of clay layer

$$= (19.83 \text{ kN/m}^3)(200.0 \text{ m} - 198.0 \text{ m}) + (19.83 \text{ kN/m}^3 - 9.81 \text{ kN/m}^3) \times (198.0 \text{ m} - 192.0 \text{ m}) + (17.10 \text{ kN/m}^3 - 9.81 \text{ kN/m}^3) \times \left(\frac{192.0 \text{ m} - 185.6 \text{ m}}{2}\right) = 123.1 \text{ kN/m}^2$$

$$\text{Effective weight of excavation} = (19.83 \text{ kN/m}^3)(200.0 \text{ m} - 198.0 \text{ m}) + (19.83 \text{ kN/m}^3 - 9.81 \text{ kN/m}^3)(198.0 \text{ m} - 195.5 \text{ m}) = 64.7 \text{ kN/m}^2$$

Net consolidation pressure at the foundation's base

$$= \frac{5400 \text{ kN}}{(3 \text{ m})(6 \text{ m})} - 64.7 \text{ kN/m}^2 = 235.3 \text{ kN/m}^2$$

To determine the net consolidation pressure at midheight of the clay layer under the center of the foundation, one must divide the foundation's base into four equal 1.5-m by 3.0-m rectangular areas. Because each of these areas has a common corner at the foundation's center, the desired net consolidation pressure at midheight of the clay layer can be calculated by determining an influence coefficient using either Table 6–2 or Figure 6–8.

$$mz = 1.5 \text{ m} \qquad nz = 3.0 \text{ m}$$

$$z = 195.5 \text{ m} - \frac{192.0 \text{ m} + 185.6 \text{ m}}{2} = 6.7 \text{ m}$$

$$m = \frac{1.5 \text{ m}}{6.7 \text{ m}} = 0.224 \qquad n = \frac{3.0 \text{ m}}{6.7 \text{ m}} = 0.448$$

From Figure 6–8, the influence coefficient is 0.04. Therefore,

Net consolidation pressure at midheight of clay layer under center of foundation $(\Delta p) = (4)(0.04)(235.3 \text{ kN/m}^2) = 37.6 \text{ kN/m}^2$

Final pressure at midheight of clay layer $(p) = p_0 + \Delta p$
$= 123.1 \text{ kN/m}^2 + 37.6 \text{ kN/m}^2 = 160.7 \text{ kN/m}^2$

From Eq. (7–18),

$$S = C_c \left(\frac{H}{1 + e_0} \right) \log \frac{p}{p_0} \tag{7–18}$$

$C_c = 0.68$ (given)
$H = 192.0 \text{ m} - 185.6 \text{ m} = 6.4 \text{ m}$
$e_0 = 1.38$ (given)
$p = 160.7 \text{ kN/m}^2$
$p_0 = 123.1 \text{ kN/m}^2$

$$S = (0.68)\left(\frac{6.4 \text{ m}}{1 + 1.38} \right) \log \left(\frac{160.7 \text{ kN/m}^2}{123.1 \text{ kN/m}^2} \right) = 0.212 \text{ m}$$

EXAMPLE 7–11

Given

1. Same data as for Example 7–10, including the computed primary consolidation settlement of 0.212 m.
2. Coefficient of consolidation (c_v) is $4.96 \times 10^{-6} \text{ m}^2/\text{min}$.

Required

How long will it take for half the expected consolidation settlement to take place if the clay layer is underlain by

1. Permeable sand and gravel?
2. Impermeable bedrock?

Solution

1. *Clay layer underlain by permeable sand and gravel.* From Eq. (7–19),

$$t = \frac{T_v}{c_v} H^2 \tag{7–19}$$

From Figure 7–14, for $U = 50\%$, $T_v = 0.196$.

$$H = \frac{192.0 \text{ m} - 185.6 \text{ m}}{2} = 3.2 \text{ m}$$

$$t_{50} = \left(\frac{0.196}{4.96 \times 10^{-6} \text{ m}^2/\text{min}}\right)(3.2 \text{ m})^2 = 404{,}645 \text{ min, or } 0.77 \text{ yr}$$

2. *Clay layer underlain by impermeable bedrock.* Equation (7–19) is still applicable with $T_v = 0.196$ and $c_v = 4.96 \times 10^{-6}$ m²/min, but with $H = 192.0$ m $- 185.6$ m $= 6.4$ m.

$$t_{50} = \left(\frac{0.196}{4.96 \times 10^{-6} \text{ m}^2/\text{min}}\right)(6.4 \text{ m})^2 = 1{,}618{,}581 \text{ min, or } 3.08 \text{ yr}$$

7–7 SETTLEMENT OF LOADS ON CLAY DUE TO SECONDARY COMPRESSION

After primary consolidation has ended (i.e., all water has been extruded from the voids in a fine-grained soil) and all primary consolidation settlement has occurred, soil compression (and additional associated settlement) continues very slowly at a decreasing rate. This phenomenon is known as *secondary compression* and perhaps results from plastic readjustment of soil grains due to new stresses in the soil and progressive breaking of clayey particles and their interparticle bonds.

Figure 7–19 gives a plot of void ratio as a function of the logarithm of time. Clearly, as the void ratio decreases, settlement increases. Secondary compression begins immediately after primary consolidation ends; it appears in Figure 7–19 as a straight line with a relatively flat slope. The void ratio corresponding to the end of primary consolidation (or the beginning of secondary compression) can be determined graphically as the point of intersection of the secondary compression line extended backward and a line tangent to the primary consolidation curve (i.e., point A in Figure 7–19).

Secondary compression settlement can be computed from the following equation [11]:

$$S_s = C_\alpha H \log \frac{t_s}{t_p} \qquad (7\text{–}24)$$

where S_s = secondary compression settlement
C_α = coefficient of secondary compression
H = (initial) thickness of the clay layer
t_s = life of the structure (or time for which settlement is required)
t_p = time to completion of primary consolidation

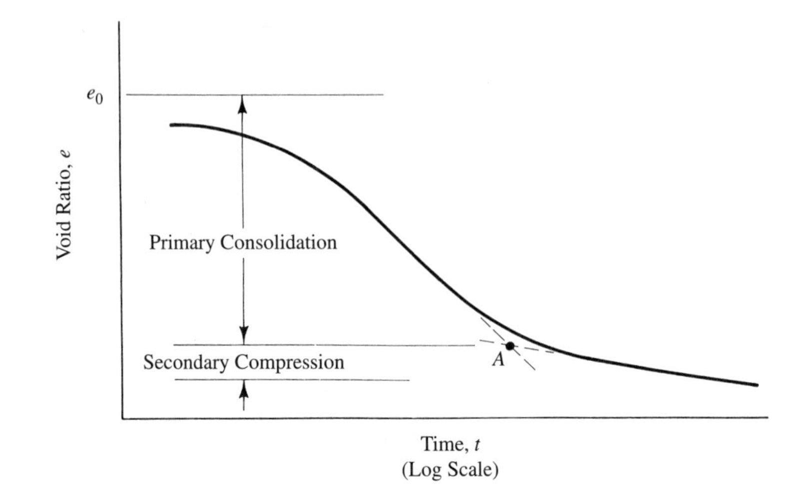

FIGURE 7–19 Sketch showing primary consolidation and secondary compression.

The coefficient of secondary compression (C_α) varies with the clay layer's natural water content and can be determined from Figure 7–20.

The amount of secondary compression settlement may be quite significant for highly compressible clays, highly micaceous soils, and organic materials. On the other hand, it is largely insignificant for inorganic clay with moderate compressibility.

EXAMPLE 7–12

Given

1. A foundation is to be built on a sand deposit underlain by a highly compressible clay layer 5.0 m thick.
2. The clay layer's natural water content is 80%.
3. Primary consolidation is estimated to be complete in 10 yr.

Required

Secondary compression settlement expected to occur from 10 to 50 yr after construction of the foundation.

Solution

From Eq. (7–24),

$$S_s = C_\alpha H \log \frac{t_s}{t_p} \qquad \textbf{(7–24)}$$

$$C_\alpha = 0.015$$

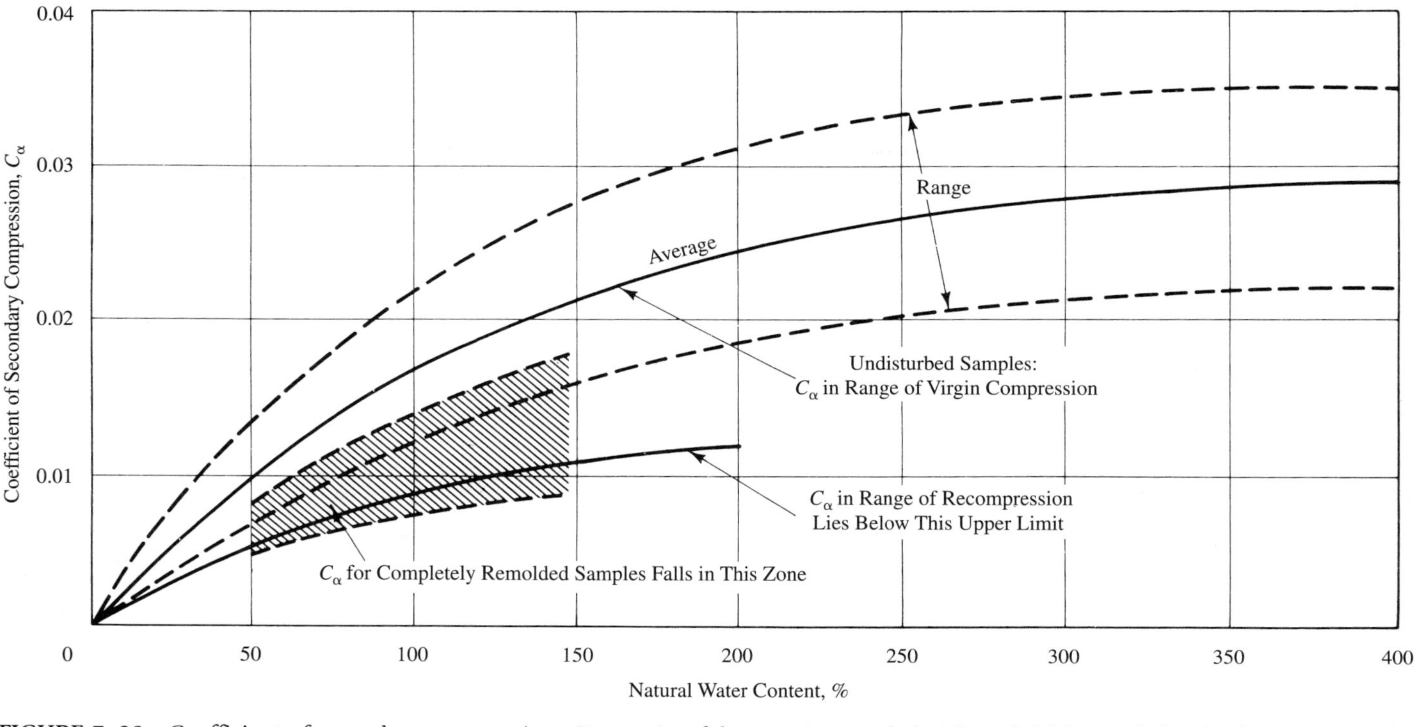

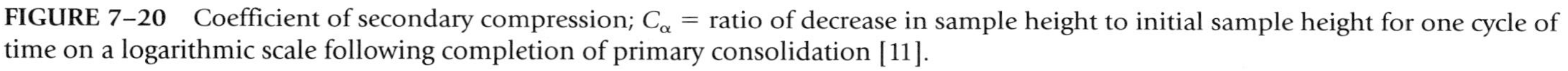

FIGURE 7–20 Coefficient of secondary compression; C_α = ratio of decrease in sample height to initial sample height for one cycle of time on a logarithmic scale following completion of primary consolidation [11].

(from Figure 7–20, with a natural water content of 80%)

$$H = 5.0 \text{ m}$$

$$t_s = 50 \text{ yr}$$

$$t_p = 10 \text{ yr}$$

$$S_s = (0.015)(5.0 \text{ m}) \log\left(\frac{50 \text{ yr}}{10 \text{ yr}}\right) = 0.052 \text{ m}$$

EXAMPLE 7–13

Given

1. Same data as for Example 7–10.
2. Assume that primary consolidation will be complete in 15 yr.
3. Natural water content of the clay layer is 50%.

Required

Estimated total settlement 50 yr after construction.

Solution

From Eq. (7–24),

$$S_s = C_\alpha H \log \frac{t_s}{t_p} \tag{7–24}$$

From Figure 7–20, with a 50% natural water content of the clay layer,

$$C_\alpha = 0.010$$

$$H = 192.0 \text{ m} - 185.6 \text{ m} = 6.4 \text{ m}$$
$$t_s = 50 \text{ yr}$$
$$t_p = 15 \text{ yr}$$
$$S_s = (0.010)(6.4 \text{ m}) \log\left(\frac{50 \text{ yr}}{15 \text{ yr}}\right) = 0.033 \text{ m}$$

The estimated total settlement is the sum of the primary consolidation settlement (determined in Example 7–10) and the secondary compression settlement (S_s). Hence,

$$\text{Estimated total settlement} = 0.212 \text{ m} + 0.033 \text{ m} = 0.245 \text{ m}$$

7–8 SETTLEMENT OF LOADS ON SAND

Most of the settlement of loads on sand has occurred by the time construction is complete. Thus, the time rate of settlement is not a factor as it is with clay. Settlement criteria rather than ultimate bearing capacity (see Chapter 9) commonly govern allowable bearing capacity for footings on sand; furthermore, settlement on sand is not amenable to solution based on laboratory consolidation tests. Indeed, settlement on sand is generally calculated by empirical means.

One empirical method is based on the standard penetration test (SPT), which was discussed in Section 3–5. To determine settlement on sand, one makes SPT determinations at various depths at the test site, normally at depth intervals of 2½ ft (0.76 m), beginning at a depth corresponding to the proposed footing's base. The SPT N-values must be corrected for overburden pressure (see Chapter 3). The next step is to compute the average corrected N-value for each boring for the sand between the footing's base and a depth B below the base, where B is the footing's width. The lowest of the average corrected N-values for all borings at the site is noted and designated N_{lowest}. Maximum settlement can then be computed from the following equation [12]:

$$s_{\max} = \frac{2q}{N_{\text{lowest}}}\left[\frac{2B}{1+B}\right]^2 \tag{7–25}$$

where $s_{\max}$ = maximum settlement on dry sand, inches
q = applied pressure, tons/ft^2
B = width of footing, ft

Equation (7–25) is applicable to settlement on dry sand. If the groundwater table is located at a depth below the base of the footing less than half the footing's width, the settlement computed from Eq. (7–25) should be corrected by multiplying it by x_B, where [12]

$$x_B = \frac{p_d}{p_w} \tag{7–26}$$

where p_d = effective overburden pressure at depth $B/2$ below the footing's base, assuming that the groundwater table is not present

p_w = effective overburden pressure at the same depth with the groundwater table present

Examples 7–14 through 7–17 demonstrate the calculation of settlement on sand.

EXAMPLE 7–14

Given

1. A 10 ft by 10 ft footing carrying a total load of 280 tons is to be constructed on sand as shown in Figure 7–21.
2. Standard penetration tests were conducted on the site. Test results were corrected for overburden pressure (see Chapter 3), and the corrected N-values are listed next.

FIGURE 7–21

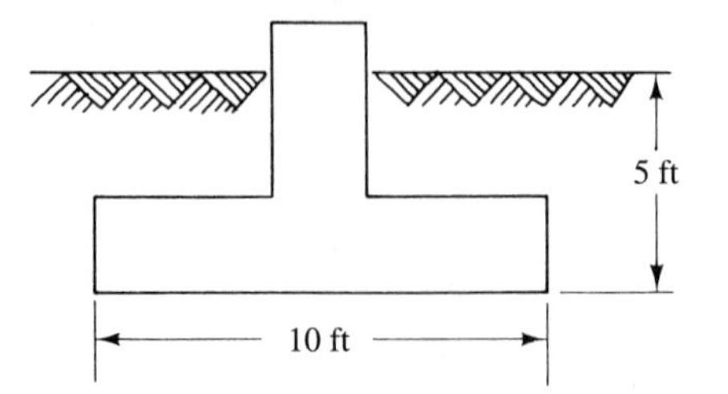

Depth (ft)	Corrected N-Values
5.0	31
7.5	36
10.0	30
12.5	28
15.0	35
17.5	33
20.0	31

Required

Maximum expected settlement of this footing.

Solution

Average Corrected N-Values

The average corrected N-value is determined for each boring for the soil located between the level of the footing's base and a depth B below this level, where B is the footing's width. In this example, appropriate depths for calculating average corrected N-values are 5 to 15 ft. The average corrected N-value is a cumulative average down to the depth indicated.

For a depth of 5 ft,

$$\text{Average corrected } N\text{-value} = 31$$

For a depth of 7.5 ft,

$$\text{Average corrected } N\text{-value} = \frac{31 + 36}{2} = 33$$

For a depth of 10.0 ft,

$$\text{Average corrected } N\text{-value} = \frac{31 + 36 + 30}{3} = 32$$

For a depth of 12.5 ft,

$$\text{Average corrected } N\text{-value} = \frac{31 + 36 + 30 + 28}{4} = 31$$

For a depth of 15.0 ft,

$$\text{Average corrected } N\text{-value} = \frac{31 + 36 + 30 + 28 + 35}{5} = 32$$

Lowest Average Corrected N-Value for Design

Subsurface soil conditions generally vary somewhat at most construction sites. The N-value selected for design is usually the lowest average corrected N-value, which in this example is 31 (at depth 12.5 ft). From Eq. (7–25),

$$s_{\max} = \frac{2q}{N_{\text{lowest}}}\left[\frac{2B}{1 + B}\right]^2 \qquad \text{(7–25)}$$

$$q = \frac{280 \text{ tons}}{(10 \text{ ft})(10 \text{ ft})} = 2.8 \text{ tons/ft}^2$$

$$N_{\text{lowest}} = 31$$

$$B = 10 \text{ ft}$$

$$s_{\max} = \frac{(2)(2.8 \text{ tons/ft}^2)}{31}\left[\frac{2 \times 10 \text{ ft}}{1 + 10 \text{ ft}}\right]^2 = 0.60 \text{ in. on dry sand}$$

EXAMPLE 7–15

Given

Same conditions as in Example 7–14, except that the groundwater table is located 7 ft below ground level (see Figure 7–22).

Required

Maximum expected settlement of the footing.

Solution

From Example 7–14,

$$s_{\max} = 0.60 \text{ in. on dry sand}$$

From Eq. (7–26),

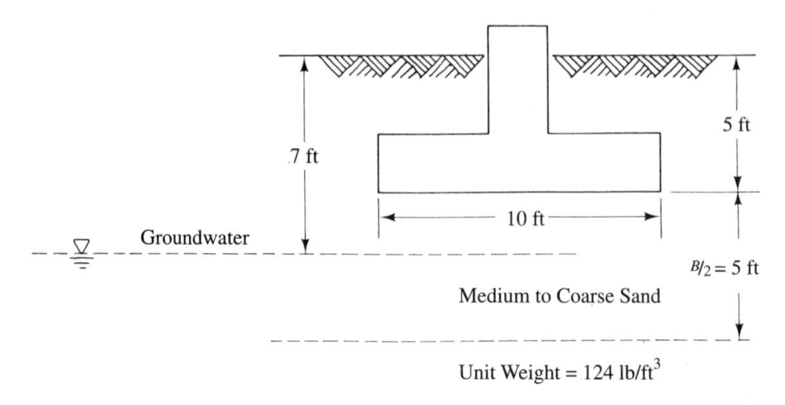

FIGURE 7–22

$$x_B = \frac{p_d}{p_w} \tag{7–26}$$

$$p_d = (124 \text{ lb/ft}^3)\left(5 \text{ ft} + \frac{10 \text{ ft}}{2}\right) = 1240 \text{ lb/ft}^2$$

$$p_w = (124 \text{ lb/ft}^3)(7 \text{ ft}) + (124 \text{ lb/ft}^3 - 62.4 \text{ lb/ft}^3)\left(5 \text{ ft} + \frac{10 \text{ ft}}{2} - 7 \text{ ft}\right)$$

$$= 1053 \text{ lb/ft}^2$$

$$x_B = \frac{1240 \text{ lb/ft}^2}{1053 \text{ lb/ft}^2} = 1.178$$

$$s_{\max} = (0.60 \text{ in.})(1.178) = 0.71 \text{ in. on wet sand}$$

EXAMPLE 7–16

Given

1. A square footing 8 ft by 8 ft located 5 ft below ground level is to be constructed on sand.
2. Standard penetration tests were conducted on the site. Test results were corrected for overburden pressures, and the lowest average corrected N-value was determined to be 41.
3. Groundwater was not encountered.

Required

Allowable soil pressure for a maximum settlement of 1 in.

Solution

From Eq. (7–25),

$$s_{\text{max}} = \frac{2q}{N_{\text{lowest}}}\left[\frac{2B}{1+B}\right]^2 \tag{7–25}$$

$$s_{\text{max}} = 1 \text{ in.}$$

$$N_{\text{lowest}} = 41$$

$$B = 8 \text{ ft}$$

$$1 \text{ in.} = \frac{2q}{41}\left[\frac{2 \times 8 \text{ ft}}{1 + 8 \text{ ft}}\right]^2$$

$$q = 6.49 \text{ tons/ft}^2$$

EXAMPLE 7–17

Given

Same conditions as in Example 7–16, except that the groundwater table is located 6 ft below ground level and the sand's unit weight is 128 lb/ft^3 (see Figure 7–23).

Required

Allowable soil pressure for a maximum settlement of 1 in.

Solution

From Eq. (7–26),

$$x_B = \frac{p_d}{p_w} \tag{7–26}$$

$$p_d = (128 \text{ lb/ft}^3)\left(5 \text{ ft} + \frac{8 \text{ ft}}{2}\right) = 1152 \text{ lb/ft}^2$$

$$p_w = (128 \text{ lb/ft}^3)(6 \text{ ft}) + (128 \text{ lb/ft}^3 - 62.4 \text{ lb/ft}^3)(3 \text{ ft}) = 964.8 \text{ lb/ft}^2$$

$$x_B = \frac{1152 \text{ lb/ft}^2}{964.8 \text{ lb/ft}^2} = 1.194$$

FIGURE 7–23

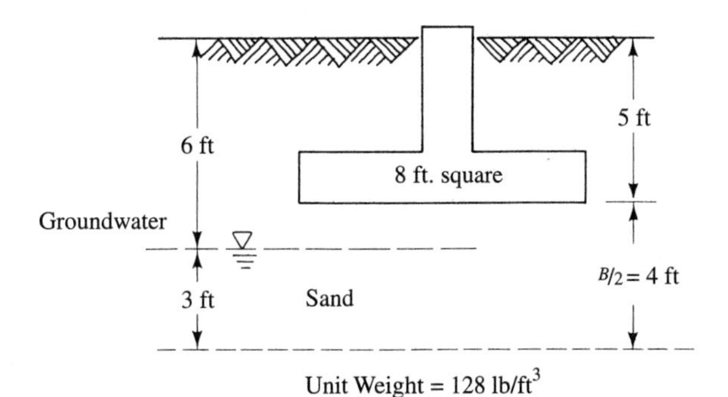

From Example 7–16, allowable soil pressure (q) is 6.49 tons/ft^2 for a settlement of 1 in. when no groundwater is encountered. When the groundwater table is at a depth below the base of the footing less than $B/2$, s_{max} computed from Eq. (7–25) should be multiplied by x_B. Therefore, in this example an allowable soil pressure of 6.49 tons/ft^2 will produce a settlement of 1.194 × 1 in., or 1.194 in. Because settlement varies directly with bearing pressure,

$$\frac{6.49 \text{ tons/ft}^2}{1.194 \text{ in.}} = \frac{\text{Allowable soil pressure for a settlement of 1 in.}}{1 \text{ in.}}$$

Allowable soil pressure for a settlement of 1 in. = 5.44 tons/ft^2.

Another empirical method for estimating foundation settlement on sand, which also uses SPT N-values, was developed by Burland and Burbidge [10, 13]. A footing to be placed in a sandy soil at some depth below ground surface requires removal of sand above the level of the base of the foundation. As a result of such sand removal, sand located below the level of the base of the foundation becomes precompressed. Recompression, however, is assumed for bearing pressure up to the preconstruction effective vertical pressure at the foundation's base.

For sands normally compressed with respect to the original ground surface and for values of foundation contact pressure (q) greater than the preconstruction effective overburden pressure (p_0) at the foundation's base [10]*,

$$S_c = B^{0.75} \frac{1.7}{\overline{N}^{1.4}} (q - 2p_0/3) \tag{7–27}$$

where S_c = settlement at end of construction and application of permanent live load (mm)

B = width of footing (m)

$\overline{N}$ = arithmetic mean of SPT N-values measured within the zone of influence (i.e., $B^{0.75}$ m below the foundation's base)

q = bearing (contact) pressure over the foundation's base (kN/m^2)

p_0 = *in situ* effective overburden pressure at base of foundation (kN/m^2)

For values of foundation contact pressure less than the preconstruction effective overburden pressure at the foundation's base [10]*,

$$S_c = \frac{1}{3} B^{0.75} \frac{1.7}{\overline{N}^{1.4}} q \tag{7–28}$$

where the terms are the same as those in Eq. (7–27).

---

* From Karl Terzaghi, Ralph B. Peck, and Gholamreza Mesri, *Soil Mechanics in Engineering Practice,* 3rd ed., John Wiley & Sons, Inc., New York, 1996.

Equations (7–27) and (7–28) apply to foundations having a length-to-breadth ratio of unity ($L/B = 1$). Burland and Burbidge presented the following empirical relationship between settlement of foundations with $L/B > 1$ and $L/B = 1$ [10]

$$S_c(L/B>1) = S_c(L/B = 1)\left[\frac{1.25(L/B)}{(L/B) + 0.25}\right]^2 \quad \textbf{(7–29)}$$

[Equation (7–29) gives the value of S_c for $L/B > 1$ by multiplying the value of S_c for $L/B = 1$ by the square of the value in brackets.] In the case of strip loading, L/B becomes very large, and the square of the value in brackets in Eq. (7–29) approaches 1.56. Equations (7–27) and (7–28) also apply only to foundations on sands with a factor of safety against bearing capacity failure of at least 3; otherwise, excessive settlements associated with an approaching bearing capacity failure may develop.

In addition, Eqs. (7–27) and (7–28) are applicable only when the groundwater table is below the zone of influence (i.e., $B^{0.75}$ m below the foundation's base). If the groundwater table lies within the zone of influence, settlement will be increased because the effective confining pressure is reduced. On the other hand, however, the reduced confining pressure results in a decrease in SPT N-values. Burland and Burbidge found that these two opposite effects more or less cancel each other; hence, the location of the groundwater table within the zone of influence can generally be neglected when one applies this method [i.e., Eqs. (7–27) and (7–28)] and no corrections are applied. If, however, the groundwater table rose into the zone of influence after the SPTs were performed, actual settlement could be considerably greater than (as great as twice as much) that computed from Eqs. (7–27) and (7–28) neglecting the groundwater table.

For saturated very dense fine or silty sand, measured SPT N-values should be reduced according to the following [10,* 13]:

$$N' = 15 + \frac{(N - 15)}{2} \quad \textbf{(7–30)}$$

Ordinarily, foundations are designed to limit maximum settlement of any footing supporting a building to some acceptable value, such as 1 in. (25 mm). Because of the variability of sandy-soil deposits, settlements of equally loaded footings of any given size can vary from the mean by a factor of 1.6, or perhaps as large as 2.0. Therefore, to be reasonably sure that the largest footing will not settle more than about 1 in. (25 mm), one should strive for a value of S_c of 25/1.6, or 16 mm, when applying Eqs. (7–27), (7–28), and (7–29).

Figure 7–24 may be used to facilitate computations involving Eqs. (7–27) and (7–28). Let

$$Q = \frac{\overline{N}^{1.4}}{1.7B^{0.75}} \quad \textbf{(7–31)}$$

* From Karl Terzaghi, Ralph B. Peck, and Gholamreza Mesri, *Soil Mechanics in Engineering Practice,* 3rd ed., John Wiley & Sons, Inc., New York, 1996. Copyright © 1996, by John Wiley & Sons, Inc. Reprinted by permission of John Wiley & Sons, Inc.

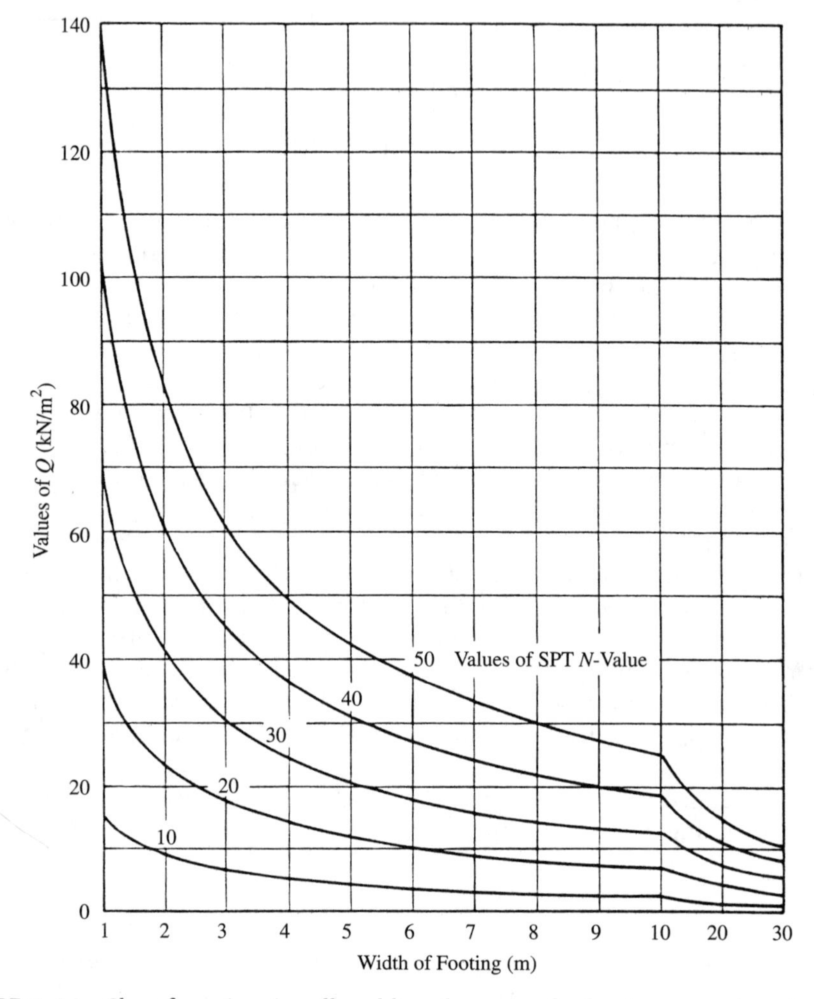

FIGURE 7–24 Chart for estimating allowable soil pressure for footing on sand on the basis of results of the standard penetration test [10].

Then, from Eqs. (7–27) and (7–28), with a value of 16 mm for S_c, for $q > p_0$,

$$q = 16Q + 2p_0/3 \tag{7-32}$$

for $q < p_0$,

$$q = 3 \times 16Q \tag{7-33}$$

To use Figure 7–24, locate a given width of footing and mean SPT N-value and find the corresponding value of Q. Using this value of Q and, if needed, the given value of p_0, compute q from Eq. (7–32) or (7–33). This value of q is the bearing pressure corresponding to a maximum settlement of approximately 1 in. (25 mm) at the end of construction.

The relationship of Figure 7–24 is for square footings of side B. For rectangular footings, the value of q should be reduced in accord with Eq. (7–29).

EXAMPLE 7–18

Given

A square footing 3 m by 3 m located 1.5 m below ground level is to be constructed on sand having a unit weight of 18.30 kN/m^3. The arithmetic mean of the SPT N-values measured within the zone of influence is 30.

Required

Allowable soil pressure for a settlement of 25 mm.

Solution

Assume that the foundation contact pressure (q) is greater than the preconstruction effective overburden pressure (p_0) at the foundation's base, in which case Eq. (7–27) applies.

$$S_c = B^{0.75} \frac{1.7}{\bar{N}^{1.4}} (q - 2p_0/3) \qquad \textbf{(7–27)}$$

Although settlement of 25 mm is required in this problem, as previously explained, it is recommended that a value of S_c of 25/1.6, or 16 mm, be used in Eq. (7–27). Hence,

$$S_c = 16 \text{ mm}$$

$$B = 3 \text{ m}$$

$$\bar{N} = 30$$

$$p_0 = (1.5 \text{ m})(18.30 \text{ kN/m}^3) = 27.45 \text{ kN/m}^2$$

$$16 \text{ mm} = (3 \text{ m})^{0.75} \frac{1.7}{30^{1.4}} [q - (2)(27.45 \text{ kN/m}^2)/3]$$

$$q = 501 \text{ kN/m}^2$$

Check the assumption that $q > p_0$.
Because $[q = 501 \text{ kN/m}^2] > [p_0 = 27.45 \text{ kN/m}^2]$, use of Eq. (7–27) is correct.

As an alternative solution, using Figure 7–24, with $B = 3.0$ m and $\bar{N} = 30$, obtain (from Figure 7–24) $Q = 30$ kN/m^2. Because $q > p_0$, use Eq. (7–32):

$$q = 16Q + 2p_0/3 \qquad \textbf{(7–32)}$$

$$q = (16)(30 \text{ kN/m}^2) + (2)(27.45 \text{ kN/m}^2)/3 = 498 \text{ kN/m}^2$$

7–9 PROBLEMS

7–1. A laboratory consolidation test was performed on a clayey soil specimen, which was drained on both top and bottom. The time for 50% consolidation was 6.2 min, and the specimen's thickness at 50% consolidation was 0.740 in. Two points on the field consolidation line have coordinates (p_1, e_1) and (p_2, e_2) of (1000 lb/ft^2, 1.167) and (2000 lb/ft^2, 1.108), respectively. Find the coefficient of permeability of the clay for the given loading range.

7–2. Determine the present effective overburden pressure at midheight of the compressible clay layer in the soil profile shown in Figure 7–25.

7–3. When the total pressure acting at midheight of a compressible clay layer is 100 kN/m^2, the corresponding void ratio is 1.09. When the total pressure increases to 400 kN/m^2, the corresponding void ratio decreases to 0.89. What would be the void ratio for a total pressure of 800 kN/m^2?

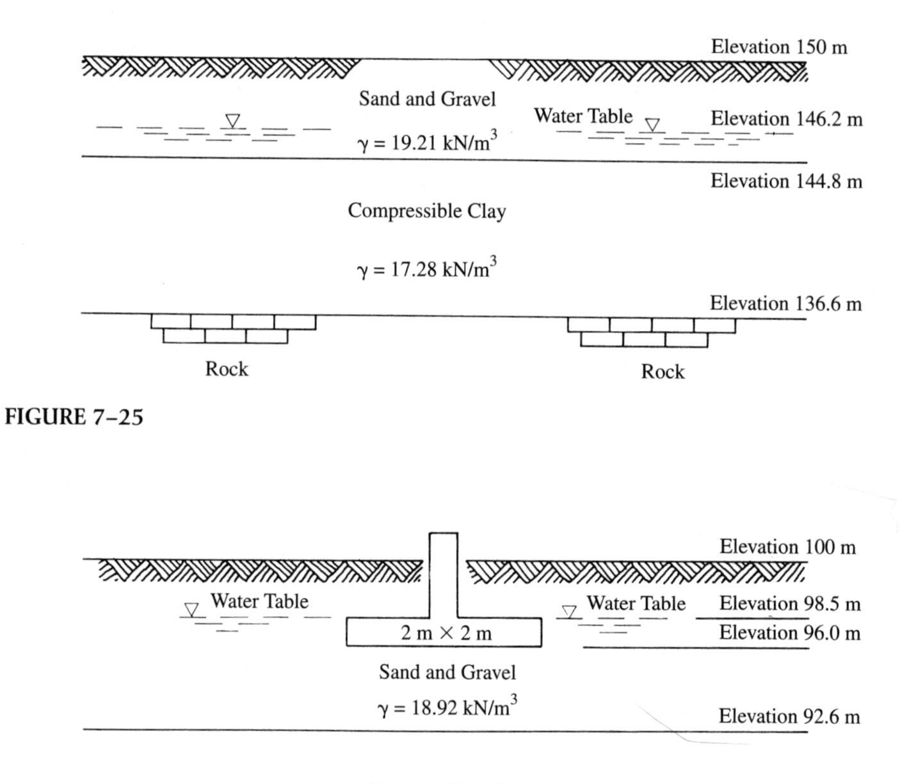

FIGURE 7–25

FIGURE 7–26

7–4. A compressible clay layer 10.0 m thick has an initial void ratio *in situ* of 1.026. Tests and computations show that the final void ratio of the clay layer after construction of a structure is 0.978. Determine the estimated primary consolidation settlement of the structure.

7–5. A foundation is to be constructed at a site where the soil profile is as shown in Figure 7–26. The base of the foundation, which is 2 m square, exerts a total load (weight of structure, foundation, and soil surcharge on the foundation) of 1000 kN. The initial void ratio *in situ* of the compressible clay layer is 1.058, and its compression index is 0.60. Find the estimated primary consolidation settlement for the clay layer.

7–6. Continuing Problem 7–5, tests and computations indicate that the coefficient of consolidation is 6.98×10^{-6} m^2/min. Compute the time required for 90% of the expected primary consolidation settlement to take place if the clay layer is underlain by (a) permeable sand and gravel, and (b) impermeable bedrock.

7–7. A sample of normally consolidated clay was obtained by a Shelby tube sampler from the midheight of a compressible clay layer (see Figure 7–27). A consolidation test was conducted on a portion of this sample, the results of which are given as follows:

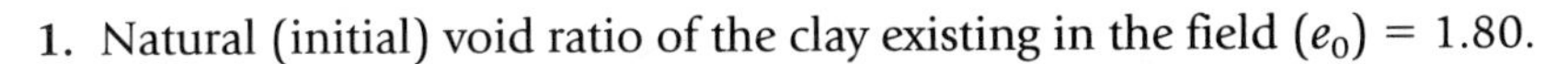

1. Natural (initial) void ratio of the clay existing in the field (e_0) = 1.80.

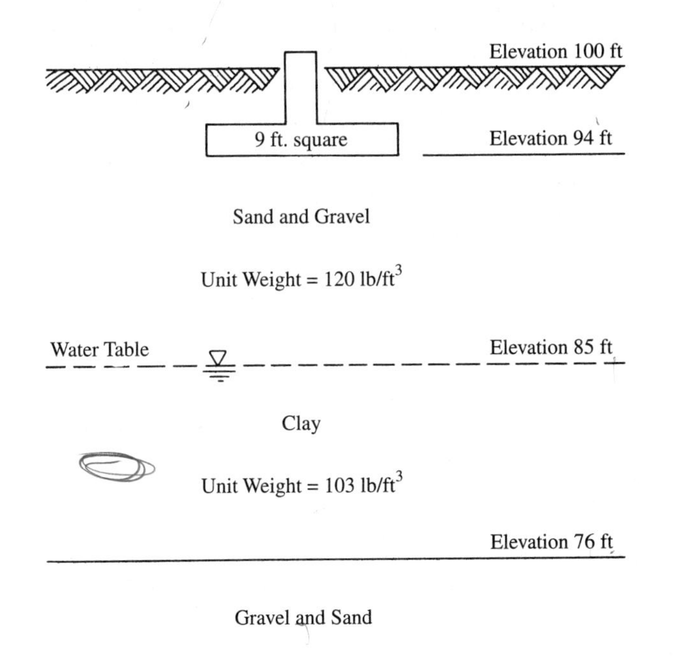

FIGURE 7–27

2. Pressure–void ratio relationships are as follows:

p (tons/ft^2)	e
0.250	1.72
0.500	1.70
1.00	1.64
2.00	1.51
4.00	1.34
8.00	1.15
16.00	0.95

A footing is to be constructed 6 ft below ground surface, as shown in Figure 7–27. The base of the footing is 9 ft by 9 ft, and it carries a total load of 200 tons, which includes the column load, weight of footing, and weight of soil surcharge on the footing.

a. From consolidation test results, prepare an e–log p curve and construct a field consolidation line, assuming that point f is located at $0.4e_0$.

b. Compute the total expected primary consolidation settlement of the compressible clay layer.

7–8. Continuing Problem 7–7, test results also indicated that the coefficient of consolidation (c_v) of the clay is 2.18×10^{-3} in.^{2}/min for the pressure increment from 1 to 2 tons/ft^2. Compute the time of primary consolidation settlement. Take U at 10% increments and plot these values on a settlement–log time curve.

7–9. A compressible 12-ft clay layer beneath a building is overlain by a stratum of sand and gravel and underlain by impermeable bedrock. The total expected primary consolidation settlement of the compressible clay layer due to the building load is 4.60 in. The coefficient of consolidation (c_v) is 9.04×10^{-4} in.^{2}/min.

1. How long will it take for 90% of the expected total primary consolidation settlement to take place?
2. Compute the amount of primary consolidation settlement that will occur in 1 yr.
3. How long will it take for primary consolidation settlement of 1 in. to take place?

7–10. Continuing Problem 7–5, assume that 100% primary consolidation will be complete in 14 yr. If the clay layer's natural water content is 35%, compute the estimated secondary compression settlement that would occur from 14 to 40 yr after construction.

7–11. A 9 ft by 9 ft square footing to carry a total load of 300 tons is to be installed 6 ft below ground surface on a sand stratum. Standard penetration tests were conducted on the site. Test results were corrected for overburden pressures, and the corrected N-values are listed as follows:

Depth (ft)	Corrected *N*-Values
2.5	25
5.0	28
7.5	27
10.0	30
12.5	28
15.0	23
17.5	24
20.0	28

No groundwater was encountered during subsurface exploration. Estimate the maximum expected settlement of the footing.

7–12. Assume the same conditions as in Problem 7–11, except that the groundwater table is located 8 ft below ground level and the sand's unit weight is 130 lb/ft^3. Estimate the maximum expected settlement of the footing.

7–13. A square footing 6 ft by 6 ft is to be installed 6 ft below ground level on a sand stratum. Standard penetration tests were conducted on the construction site. Test results were corrected for overburden pressures, and the lowest average corrected *N*-value was determined to be 18. Assuming that groundwater was not encountered, determine the allowable soil pressure for a maximum settlement of 1 in.

7–14. Assume the same conditions as in Problem 7–13, except that the groundwater table is located 8 ft below ground level and the sand's unit weight is 118 lb/ft^3. Determine the allowable soil pressure for a maximum settlement of 1 in.

7–15. A rectangular footing 3 m by 4 m located 2 m below ground level is to be constructed on sand having a unit weight of 18.8 kN/m^3. The footing is designed to take a total load of 6000 kN. If the arithmetic mean of SPT *N*-values measured within the zone of influence is 36, compute the settlement of the footing.

References

[1] Wayne C. Teng, *Foundation Design*, Prentice-Hall, Inc., Englewood Cliffs, N.J., 1962.

[2] Joseph E. Bowles, *Engineering Properties of Soils and Their Measurement*, 2nd ed., McGraw-Hill Book Company, New York, 1978.

[3] *1989 Annual Book of ASTM Standards*, ASTM, Philadelphia, 1989. Copyright, American Society for Testing and Materials, 1916 Race Street, Philadelphia, PA 19103. Reprinted with permission.

[4] Ralph B. Peck, Walter E. Hansen, and Thomas H. Thornburn, *Foundation Engineering*, 2nd ed., John Wiley & Sons, Inc., New York, 1974. Copyright © 1974, by John Wiley & Sons, Inc. Reprinted by permission of John Wiley & Sons, Inc.

[5] A. Casagrande, "The Determination of the Pre-consolidation Load and Its Practical Significance," *Proc. First Int. Conf. Soil Mech., Cambridge, Mass.*, **3**, 60–64 (1936).
[6] J. H. Schertmann, "The Undisturbed Consolidation Behavior of Clay," *Trans. ASCE*, **120**, 1201–1227 (1955).
[7] Karl Terzaghi and Ralph B. Peck, *Soil Mechanics in Engineering Practice*, John Wiley & Sons, Inc., New York, 1967. Copyright © 1967, by John Wiley & Sons, Inc. Reprinted by permission of John Wiley & Sons, Inc.
[8] J. H. Schertmann, "Estimating the True Consolidation Behavior of Clay from Laboratory Test Results," *Proc. ASCE*, **79**, Separate 311 (1953). 26 pp.
[9] A. W. Skempton, "Notes on the Compressibility of Clays," *Quart. J. Geol. Soc. Lond.*, **C**, 119–135 (1944).
[10] Karl Terzaghi, Ralph B. Peck, and Gholamreza Mesri, *Soil Mechanics in Engineering Practice*, 3rd ed., John Wiley & Sons, Inc., New York, 1996.
[11] *Design Manual: Soil Mechanics, Foundations, and Earth Structures*, NAVFAC DM-7, U.S. Department of the Navy, Naval Facilities Engineering Command, Alexandria, Va., 1971.
[12] Abdel Rahman Sadik Said Bazaraa, "Use of the Standard Penetration Test for Estimating Settlements of Shallow Foundations on Sand," Ph.D. thesis, University of Illinois, 1967.
[13] J. B. Burland and M. C. Burbidge, "Settlement of Foundations on Sand and Gravel," *Proc. Inst. Civil Eng. (London)*, **78** (Part 1), 1325–1381 (1985).

8

Shear Strength of Soil

8–1 INTRODUCTION

As a structural member, a piece of steel is capable of resisting compression, tension, and shear. Soil, however, like concrete and rock, is not capable of resisting high-tension stresses (nor is it required to do so). It is capable of resisting compression to some extent; but in the case of excessive (failure-producing) compression, failure usually occurs in the form of shearing along some internal surface within the soil. Thus, the structural strength of soil is primarily a function of its shear strength, where *shear strength* refers to the soil's ability to resist sliding along internal surfaces within a mass of the soil.

Because the ability of soil to support an imposed load is determined by its shear strength, the shear strength of soil is of great importance in foundation design (Chapter 9), lateral earth pressure calculations (Chapter 12), slope stability analysis (Chapter 14), and many other considerations. As a matter of fact, shear strength of soil is of such great importance that it is a factor in most soil problems. Determination of shear strength is one of the most frequent, important problems in soil mechanics.

As explained in Section 2–8, the shear strength of a given soil may be expressed by the Coulomb equation [1]:

$$s = c + \bar{\sigma} \tan \phi \qquad (2\text{–}16)$$

where s = shear strength
c = cohesion
$\bar{\sigma}$ = effective intergranular normal (perpendicular to the shear plane) pressure
ϕ = angle of internal friction
$\tan \phi$ = coefficient of friction

Cohesion (c) refers to strength gained from the ionic bond between grain particles and is predominant in clayey (cohesive) soils. The angle of internal friction (ϕ) refers to strength gained from internal frictional resistance (including sliding and rolling friction and the resistance offered by interlocking action among soil particles) and is predominant in granular (cohesionless) soils. Cohesion (c) and the angle of internal friction (ϕ) might be referred to as the *shear strength parameters*. They can be evaluated for a given soil by standard laboratory and/or field tests (Section 8–2), thereby defining the relationship for shear strength (s) as a function of effective intergranular normal pressure ($\bar{\sigma}$). The latter term ($\bar{\sigma}$) is not a soil property; it refers instead to the magnitude of the applied load.

As indicated in the preceding paragraph, the same two parameters affect shear strength of both cohesive and cohesionless soils. However, the predominant parameter differs depending on whether a cohesive soil or a cohesionless soil is being considered. Accordingly, study and analysis of shear strength of soil are normally done separately for cohesive and cohesionless soils.

Field and laboratory methods for determining shear strength parameters, from which shear strength can be evaluated, are presented in Section 8–2. Study and analysis of shear strength of cohesionless soils are presented in Section 8–4, and those of cohesive soils in Section 8–5.

8–2 METHODS OF INVESTIGATING SHEAR STRENGTH

There are several methods of investigating shear strength of soil. Some are laboratory methods; others are *in situ* (field) methods. Laboratory methods discussed here include the (1) unconfined compression test, (2) direct shear test, and (3) triaxial compression test. *In situ* methods discussed here include the (1) vane test, (2) standard penetration test, and (3) penetrometer test. The unconfined compression test can be used to investigate only cohesive soils, whereas the direct shear test and the triaxial compression test can be used to investigate both cohesive and cohesionless soils. The vane test can be used to investigate soft clays—particularly sensitive clays. The standard penetration test is limited primarily to cohesionless soils, whereas the penetrometer test is used mainly in fine-grained soils. The aforementioned methods for investigating shear strength of soil are discussed next. As done previously in this book, only generalized discussions of the various test procedures are presented here.

Laboratory Methods for Investigating Shear Strength

Unconfined Compression Test (ASTM D 2166). The unconfined compression test is perhaps the simplest, easiest, and least expensive test for investigating shear strength. It is quite similar to the usual determination of compressive strength of concrete, where crushing a concrete cylinder is carried out solely by measured increases in end loading. A cylindrical cohesive soil specimen is cut to a length of between 2 and 2½ times its diameter. It is then placed in a compression testing machine (see Figure 8–1) and subjected to an axial load. The axial load is applied

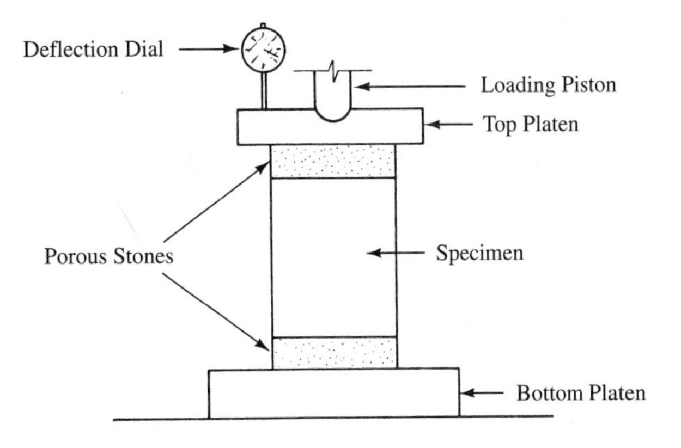

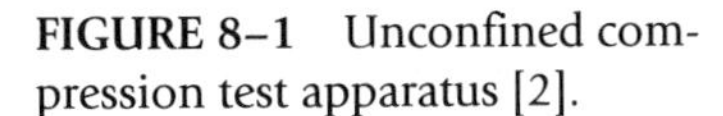
FIGURE 8–1 Unconfined compression test apparatus [2].

to produce axial strain at a rate of ½ to 2% per minute, and resulting stress and strain are measured.

As the load is applied to the specimen, its cross-sectional area will increase a small amount. For any applied load, the cross-sectional area, A, can be computed by the following equation:

$$A = \frac{A_0}{1 - \epsilon} \tag{8–1}$$

where A_0 is the specimen's initial area. The load itself, P, can be determined by multiplying the proving-ring dial reading by the proving-ring calibration factor, and the load per unit area can be found by dividing the load by the corresponding cross-sectional area. The axial unit strain, ϵ, can be computed by dividing the change in length of the specimen, ΔL, by its initial length, L_0. In equation form,

$$\epsilon = \frac{\Delta L}{L_0} \tag{8–2}$$

The value of ΔL is given by the deformation reading, provided the deflection dial is set to zero initially.

The largest value of the load per unit area or the load per unit area at 15% strain, whichever occurs first, is known as the *unconfined compressive strength*, q_u, and cohesion [c in Eq. (2–16)] is taken as one-half the unconfined compressive strength (i.e., $q_u/2$).

In the unconfined compression test, because there is no lateral support, the soil specimen must be able to stand alone in the shape of a cylinder. A cohesionless soil (such as sand) cannot generally stand alone in this manner without lateral support; hence, this test procedure is usually limited to cohesive soils.

EXAMPLE 8–1

Given

A clayey soil subjected to an unconfined compression test fails at a pressure of 2540 lb/ft^2 (i.e., q_u = 2540 lb/ft^2).

Required

Cohesion of this clayey soil.

Solution

$$\text{Cohesion} = \frac{\text{Unconfined compressive strength}}{2}$$

or

$$c = \frac{q_u}{2}$$

$$c = \frac{2540 \text{ lb/ft}^2}{2} = 1270 \text{ lb/ft}^2$$

Direct Shear Test (ASTM D3080). To carry out a direct shear test, one must place a soil specimen in a relatively flat box, which may be round or square (Figure 8–2). A normal load of specific (and constant) magnitude is applied. The box is "split" into two parts horizontally (Figure 8–2), and if half the box is held while the other half is pushed with sufficient force, the soil specimen will experience shear failure along horizontal surface *A*. This procedure is carried out in a direct shear apparatus (see Figure 8–3), and the particular normal load and shear stress that produced shear failure are recorded. The soil specimen is then removed from the shear box and discarded, and another specimen of the same soil sample is placed in the shear box. A normal load differing from (either higher or lower than) the one used in the first test is applied to the second specimen, and a shearing force is again applied with sufficient magnitude to cause shear failure. The normal load and shear stress that produced shear failure are recorded for the second test.

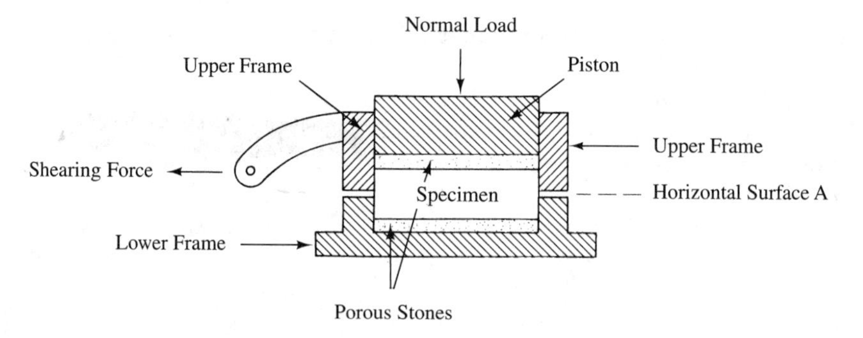

FIGURE 8–2 Typical direct shear box for single shear [3].

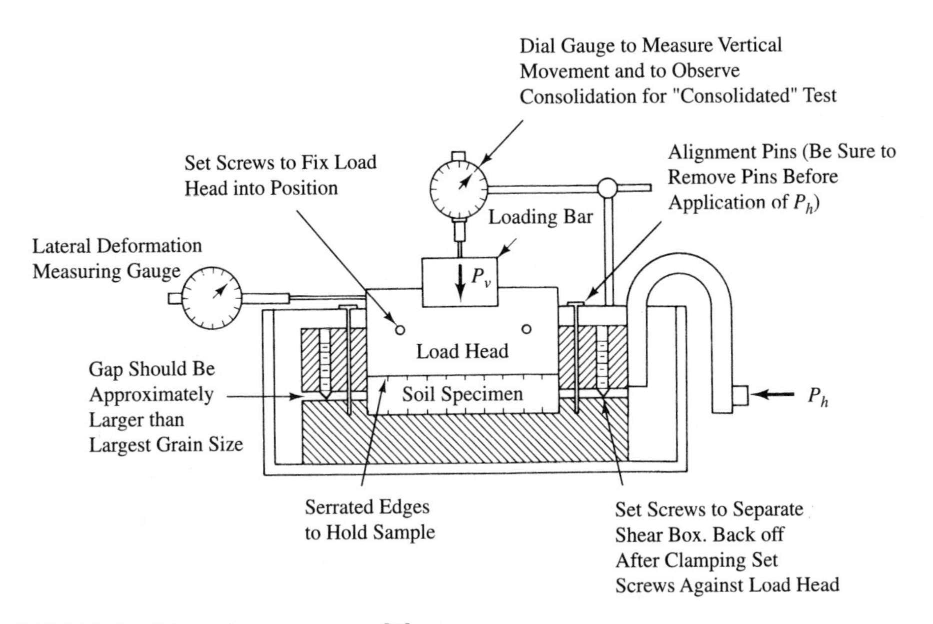

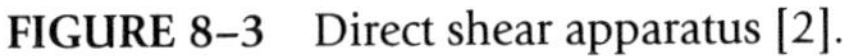
FIGURE 8–3 Direct shear apparatus [2].

The results of these two tests are plotted on a graph, with normal stress (which is the total normal load divided by the specimen's cross-sectional area) along the abscissa and the shear stress that produced failure of the specimen along the ordinate (see Figure 8–4). (The same scale must be used along both the abscissa and the ordinate.) A straight line drawn connecting these two plotted points is extended to intersect the ordinate. The angle between this straight line and a horizontal line (ϕ in Figure 8–4) is

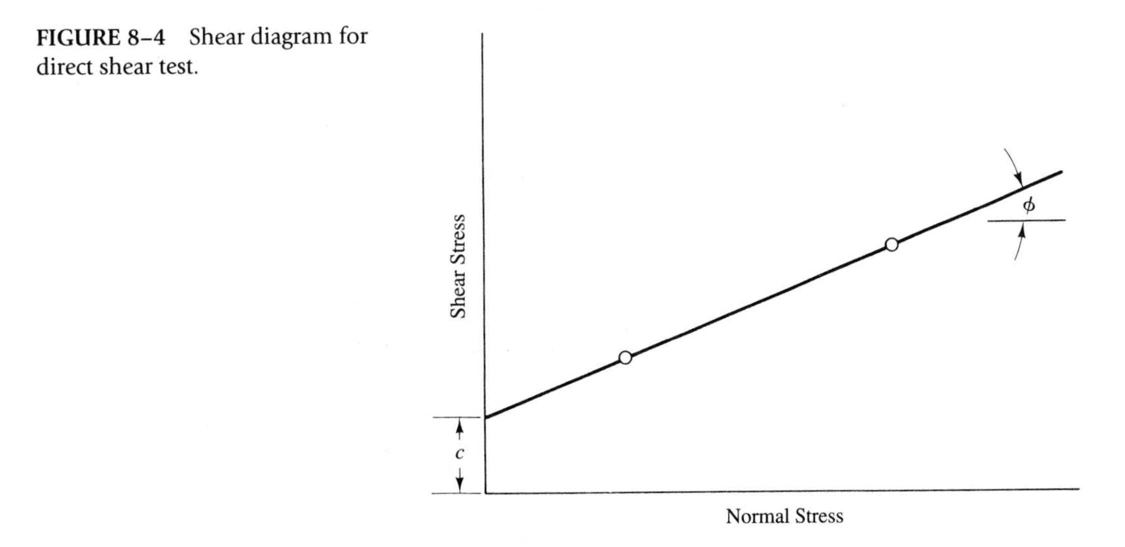

FIGURE 8–4 Shear diagram for direct shear test.

the angle of internal friction [ϕ in Eq. (2–16)], and the shear stress where the straight line intersects the ordinate (c in Figure 8–4) is the cohesion [c in Eq. (2–16)]. These values of ϕ and c can be used in Eq. (2–16) to determine the given soil's shear strength for any load (i.e., for any effective intergranular normal pressure, $\bar{\sigma}$).

In theory, it is adequate to have only two points to define the straight-line relationship of Figure 8–4. In practice, however, it is better to have three (or more) such points through which the best-fitting straight line can be drawn. This means, of course, that three (or more) separate tests must be made on three (or more) specimens from the same soil sample.

The direct shear test is a relatively simple means of determining shear strength parameters of soils. However, in this test shear failure is forced to occur along or across a predetermined plane (surface A in Figure 8–2), which is not necessarily the weakest plane of the soil specimen tested. Since development of the much better triaxial test (discussed subsequently), use of the direct shear test has decreased.

EXAMPLE 8–2

Given

A series of direct shear tests was performed on a soil sample. Each test was carried out until the soil specimen experienced shear failure. The test data are listed next.

Specimen Number	Normal Stress (lb/ft^2)	Shearing Stress (lb/ft^2)
1	604	1522
2	926	1605
3	1248	1720

Required

The soil's cohesion and angle of internal friction.

Solution

Given data are plotted on a shear diagram (see Figure 8–5). (Note that both the ordinate and abscissa scales are the same). Connect the plotted points by the best-fitting straight line and note that it makes an angle of 17° with the horizontal and intersects the ordinate at 1340 lb/ft^2. Therefore, cohesion (c) = 1340 lb/ft^2 and the angle of internal friction (ϕ) = 17°.

EXAMPLE 8–3

Given

A specimen of dry sand was subjected to a direct shear test that was carried out until the specimen sheared. A normal stress of 96.0 kN/m^2 was imposed for the test, and shear stress at failure was 65.0 kN/m^2.

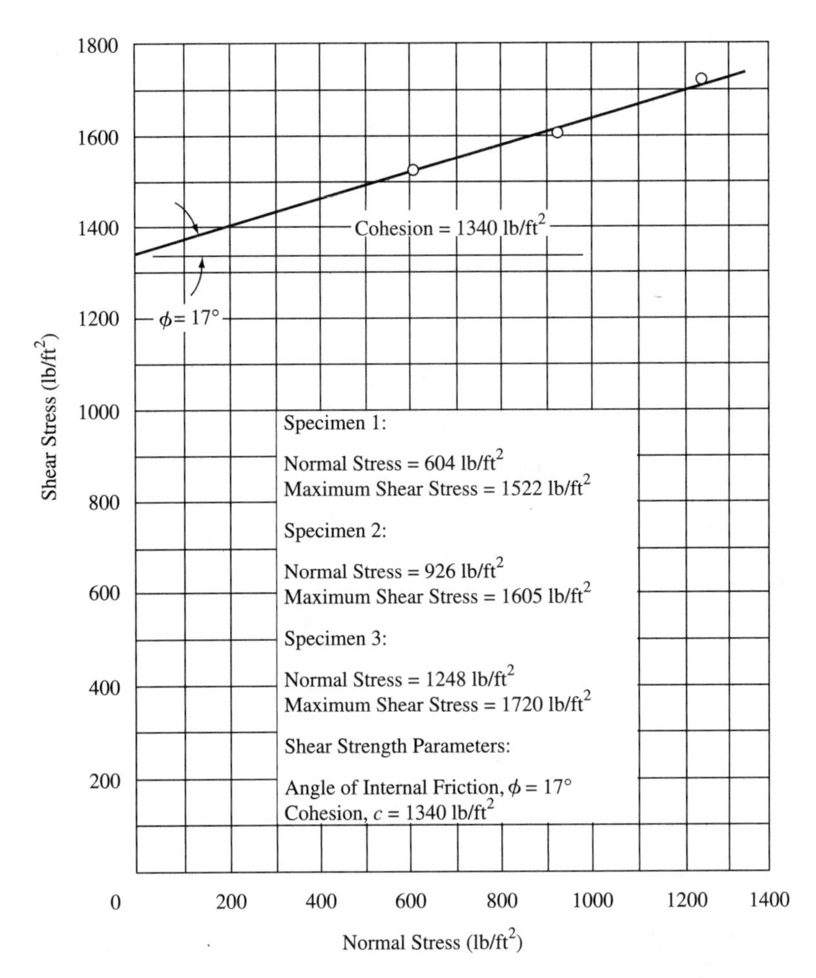

FIGURE 8–5 Maximum shear stress versus normal stress curve for Example 8–2.

Required

This sand's angle of internal friction.

Solution

Given data are plotted on a shear diagram (see Figure 8–6). (Note that both the ordinate and abscissa scales are the same.) Because cohesion is virtually zero for dry sand, the shear plot passes through the origin. Hence, draw a line through the plotted point and the origin. The angle between this line and the horizontal is measured to be 34°. Therefore, the sand's angle of internal friction (ϕ) is 34°. This value can also be determined by direct computation:

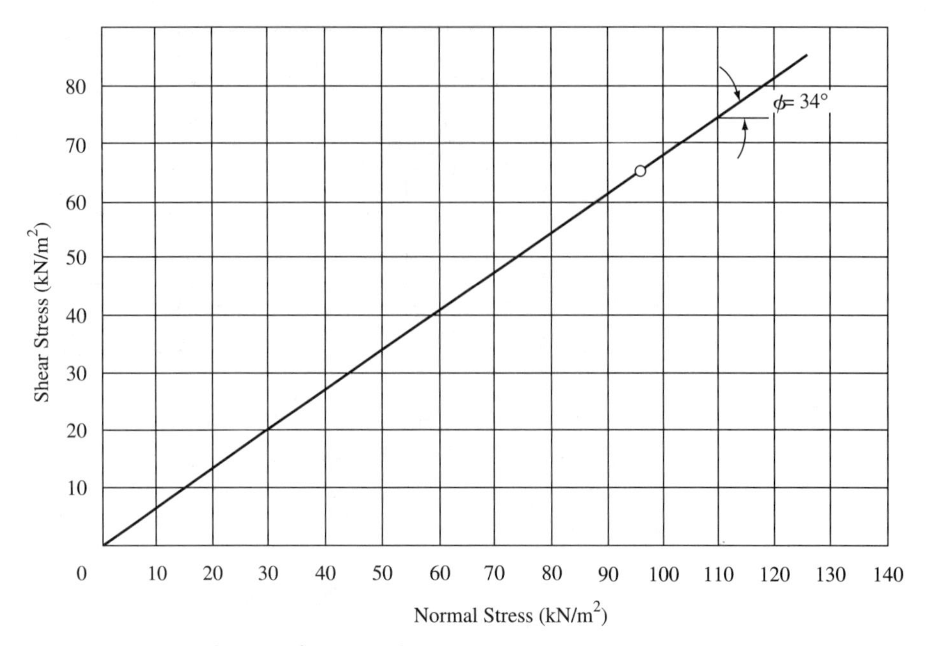

FIGURE 8–6 Shear diagram for Example 8–3.

$$\tan \phi = \frac{65.0\ \text{kN/m}^2}{96.0\ \text{kN/m}^2} = 0.6771$$

$$\phi = 34°$$

Triaxial Compression Test (ASTM D2850). The triaxial compression test is carried out in a manner somewhat similar to the unconfined compression test in that a cylindrical soil specimen is subjected to a vertical (axial) load. The major difference is that, unlike the unconfined compression test, where there is no confining (lateral) pressure, the triaxial test is carried out with confining (lateral) pressure present. Lateral pressure is made possible by enclosing the specimen in a chamber (see Figure 8–7) and introducing water or compressed air into the chamber to surround the soil specimen.

To carry out a test, one must wrap a cylindrical soil specimen having a length between 2 and 2½ times its diameter in a rubber membrane and must place the specimen in the triaxial chamber. Then, a specific (and constant) lateral pressure is applied by means of water or compressed air within the chamber. Next, a vertical (axial) load is applied externally and steadily increased until the specimen fails. The externally applied axial load that causes the specimen to fail and the lateral pressure are recorded. As in the direct shear test, it is necessary to remove the soil specimen and discard it, and then to place another specimen of the same soil sample in the triaxial chamber. The procedure is repeated for the new specimen for a different (either

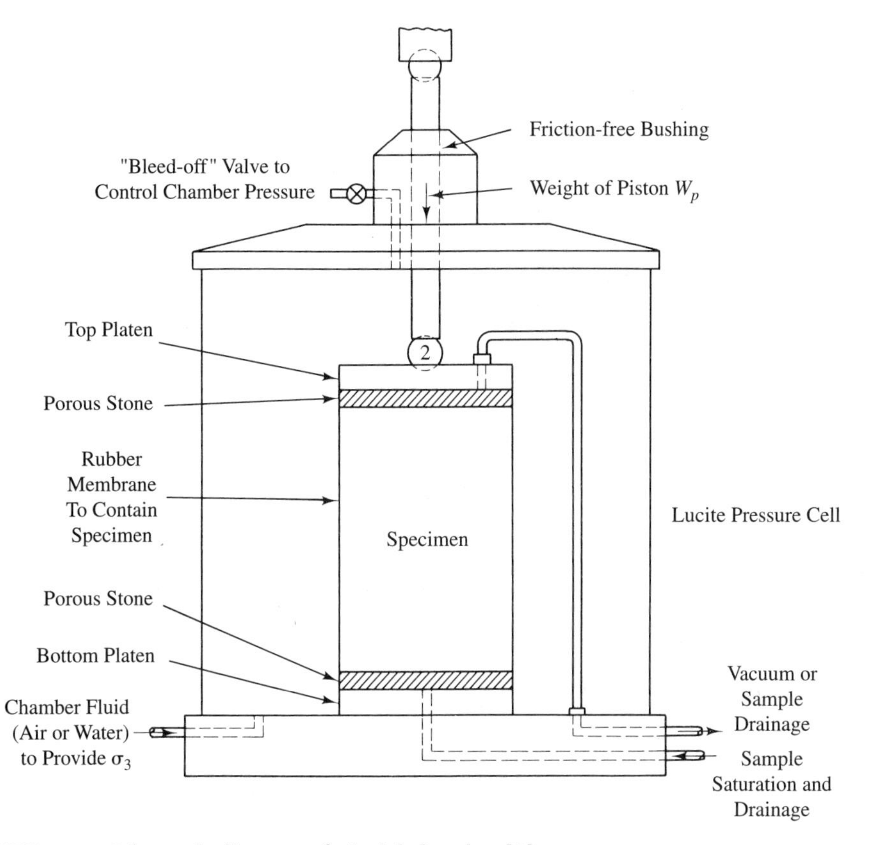

FIGURE 8–7 Schematic diagram of triaxial chamber [2].

higher or lower) lateral pressure. The axial load at failure and the lateral pressure are recorded for the second test.

Lateral pressure is designated as σ_3. However, it is applied not only to the specimen's sides, but also to its ends. This pressure is therefore called the *minor principal stress*. The externally applied axial load at failure divided by the cross-sectional area of the test specimen is designated as Δp and is called the *deviator stress at failure*. Total vertical (axial) pressure causing failure is the sum of the minor principal stress (σ_3) and the deviator stress at failure (Δp). This total vertical (axial) pressure at failure is designated as σ_1 and is called the *major principal stress*. In equation form,

$$\sigma_1 = \sigma_3 + \Delta p \tag{8-3}$$

The results of triaxial compression tests can be plotted in the following manner. Using the results of one of the triaxial tests, locate a point along the abscissa at distance σ_3 from the origin. This point is denoted by A in Figure 8–8, and it is indicated as being located along the abscissa at distance $(\sigma_3)_1$ from the origin. It is also necessary to locate another point along the abscissa at distance σ_1 from the origin.

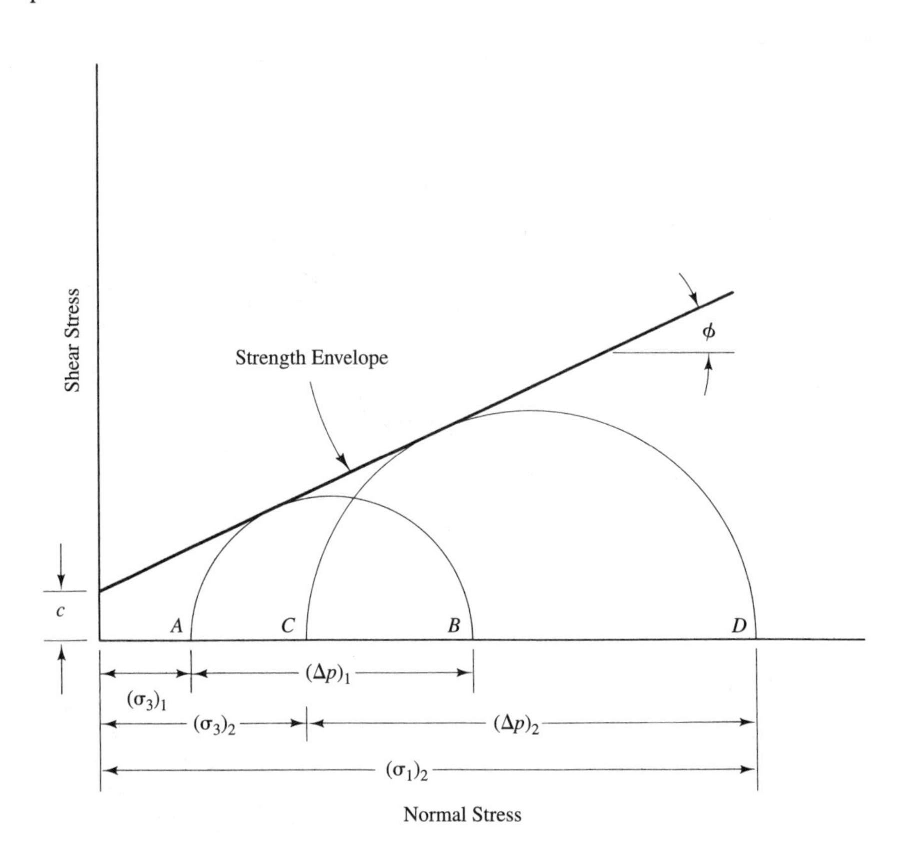

FIGURE 8–8 Shear diagram for triaxial compression test.

This point can be located by measuring either distance σ_1 from the origin or Δp from point A (the point located at distance σ_3 from the origin). This point is denoted by B in Figure 8–8 and is indicated as being located along the abscissa at distance $(\Delta p)_1$ from point A. Using AB as a diameter, construct a semicircle as shown in Figure 8–8. (This is known as a *Mohr's circle.*) The entire procedure is repeated using the data obtained from the triaxial test on the other specimen of the same soil sample. Thus, point C is located along the abscissa at distance $(\sigma_3)_2$ from the origin, and point D along the abscissa at distance $(\Delta p)_2$ from point C. Using CD as a diameter, construct another semicircle. The final step is to draw a straight line tangent to the semicircles, as shown in Figure 8–8. This straight line is called the *strength envelope, failure envelope,* or *Mohr's envelope.* As in the direct shear test (Figure 8–4), the angle between this straight line (the strength envelope) and a horizontal line (ϕ in Figure 8–8) is the angle of internal friction [ϕ in Eq. (2–16)], and the shear stress where the straight line intersects the ordinate (c in Figure 8–8) is the cohesion [c in Eq. (2–16)]. The same scale must be used along both the abscissa and the ordinate.

As in the direct shear test, it is adequate, in theory, to have only two Mohr's circles to define the straight-line relationship of Figure 8–8. In practice, however, it is better to have three (or more) Mohr's circles that can be used to draw the best strength envelope. This means, of course, that three (or more) separate tests must be performed on three (or more) specimens from the soil sample. In actuality, the strength envelope for both sand and clay will seldom be perfectly straight, except perhaps at low lateral pressures; therefore, it requires some interpretation to draw a best-fitting strength envelope of Mohr's circles.

EXAMPLE 8–4

Given

Triaxial compression tests on three specimens of a soil sample were performed. Each test was carried out until the specimen experienced shear failure. The test data are tabulated as follows:

Specimen Number	Minor Principal Stress, σ_3 (Confining Pressure) (kips/ft^2)	Deviator Stress at Failure, Δp (kips/ft^2)
1	1.44	5.76
2	2.88	6.85
3	4.32	7.50

Required

The soil's cohesion and angle of internal friction.

Solution

As shown in Figure 8–9, draw three Mohr's circles. Each one starts at a minor principal stress (σ_3) and has a diameter equal to the deviator stress at failure (Δp). Then draw the strength envelope tangent as nearly as possible to all three circles. The soil's cohesion is indicated by the intersection of the strength envelope and the ordinate, where a value of 1.8 kips/ft^2 is read. The soil's angle of internal friction, which is the angle between the strength envelope and the horizontal, is $17°$.

EXAMPLE 8–5

Given

A sample of dry, cohesionless soil was subjected to a triaxial compression test that was carried out until the specimen failed at a deviator stress of 105.4 kN/m^2. A confining pressure of 48.0 kN/m^2 was used for the test.

Required

This soil's angle of internal friction.

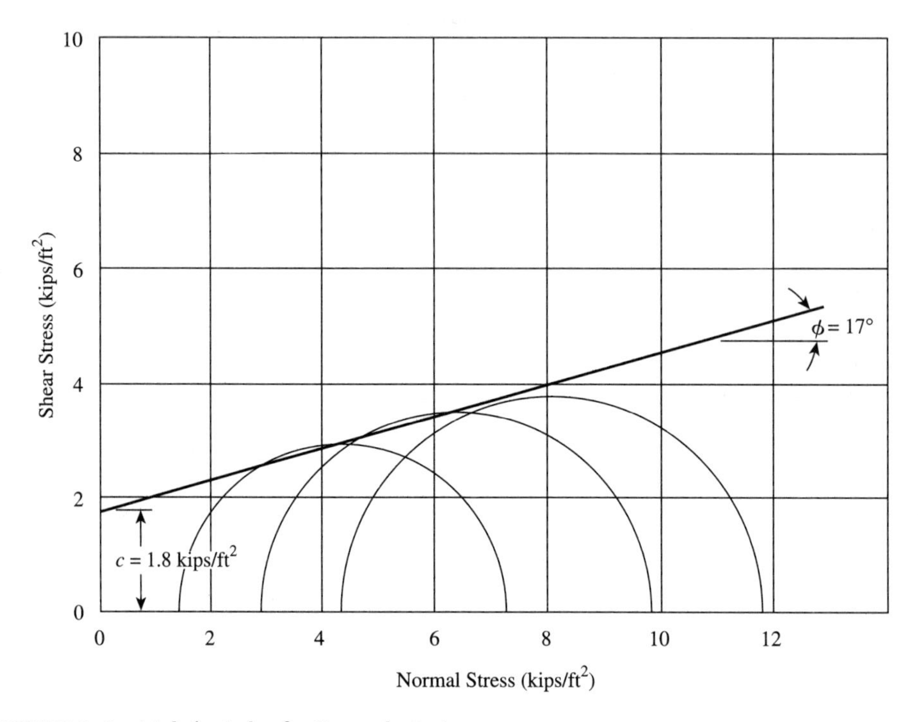

FIGURE 8–9 Mohr's circles for Example 8–4.

Solution

Given data are plotted on a shear diagram (see Figure 8–10). (Note that both the ordinate and abscissa scales are the same.) Point A is located along the abscissa at 48.0 kN/m^2 (the confining pressure—σ_3) and point B at 48.0 kN/m^2 + 105.4 kN/m^2, or 153.4 kN/m^2 (confining pressure plus deviator stress at failure—$\sigma_3 + \Delta p$). The Mohr's circle is drawn with a center along the abscissa at 100.7 kN/m^2 [i.e., 48.0 kN/m^2 + (105.4 kN/m^2)/2] and a radius of 52.7 kN/m^2. Because cohesion is virtually zero for dry, cohesionless soil, a line is drawn through the origin and tangent to the Mohr's circle. The angle between this line and the horizontal is measured to be 32°. Therefore, the soil's angle of internal friction (ϕ) is 32°.

EXAMPLE 8–6

Given

A sample of dry, cohesionless soil whose angle of internal friction is 37° is subjected to a triaxial test.

Required

If the minor principal stress (σ_3) is 14 lb/in.^{2}, at what values of deviator stress (Δp) and major principal stress (σ_1) will the test specimen fail?

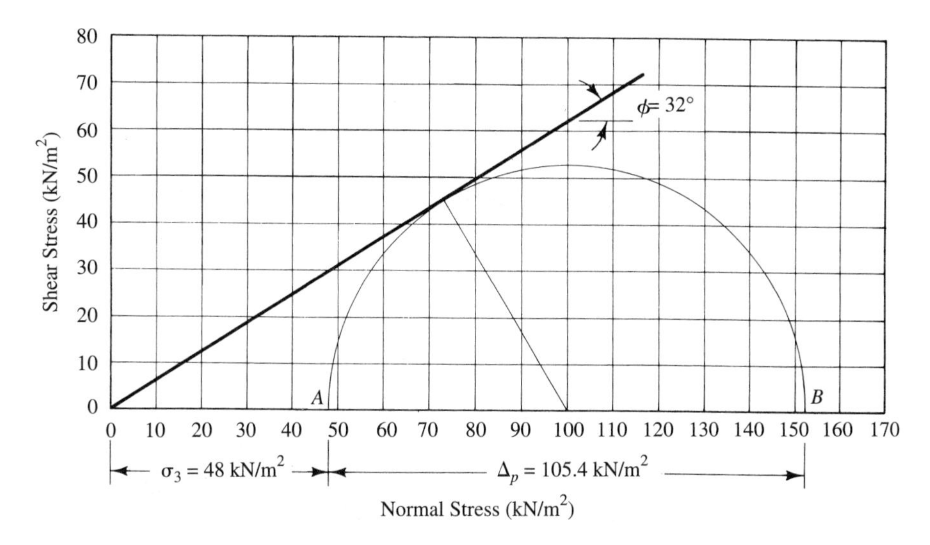

FIGURE 8–10 Mohr's circle for Example 8–5.

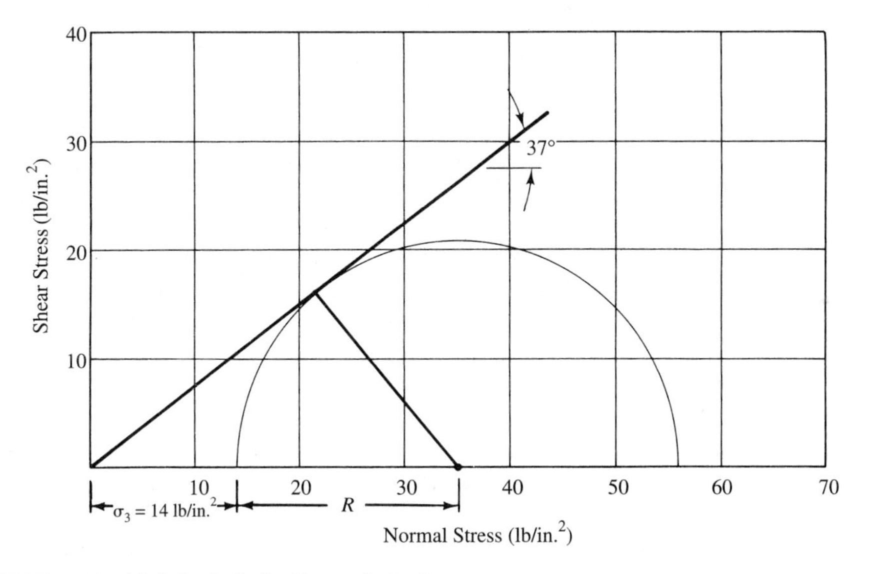

FIGURE 8–11 Mohr's circle for Example 8–6.

Solution

All samples of dry, cohesionless soils have cohesions of zero. Therefore, the Mohr's envelope must go through the origin. Draw a strength envelope starting at the origin for $\phi = 37°$. Then draw the Mohr's circle, starting at a minor principal stress (σ_3) of 14 lb/in.^{2} and tangent to the strength envelope (see Figure 8–11). It can now be determined that the deviator stress at failure (Δp) is 42.3 lb/in.^{2} (deviator stress at failure equals the diameter of the Mohr's circle), and the major principal stress at failure ($\sigma_1 = \Delta p + \sigma_3$) is 42.3 lb/in.^{2} + 14 lb/in.^{2}, or 56.3 lb/in.^{2}

This problem can also be solved analytically, using a sketch (i.e., drawing not made to scale). From Figure 8–11 with given values of ϕ of 37° and σ_3 of 14 lb/in.^{2},

$$\sin 37° = \frac{R}{14 + R} \text{ or } (14 + R) \sin 37° = R$$

$$8.4254 + 0.6018R = R$$

$$R = 21.16$$

$$\Delta p = 2R = (2)(21.16) = 42.3 \text{ lb/in.}^2$$

$$\sigma_1 = \Delta p + \sigma_3 = 42.3 \text{ lb/in.}^2 + 14 \text{ lb/in.}^2 = 56.3 \text{ lb/in.}^2$$

As is shown in Chapter 9, the angle of internal friction (ϕ) can be approximated for cohesionless soils, based on the results of a standard penetration test (SPT).

Variations in Shear Test Procedures

There are three basic types of shear test procedures as determined by the sample drainage condition: unconsolidated undrained (UU), consolidated undrained (CU), and consolidated drained (CD). These can be defined as follows. Although these three types apply to both direct shear and triaxial compression tests, they are explained for the triaxial test only.

The unconsolidated undrained (UU) test is carried out by placing the specimen in the chamber and introducing lateral (confining) pressure without allowing the specimen to consolidate (drain) under the lateral pressure. An axial load is then applied without allowing drainage of the sample. The UU test can be run rather quickly because the specimen is not required to consolidate under the lateral pressure or drain during application of the axial load. Because of the short time required to run this test, it is often referred to as the quick, or Q, test.

The consolidated undrained (CU) test is performed by placing the specimen in the chamber and introducing lateral pressure. The sample is then allowed to consolidate under the lateral pressure by leaving the drain lines open (Figure 8–7). The drain lines are then closed and axial stress is increased without allowing further drainage.

The consolidated drained (CD) test is similar to the CU test, except that the specimen is allowed to drain as the axial load is applied so that high excess pore pressures do not develop. Because the permeability of clayey soils is low, the axial load must be added very slowly during CD tests so that excess pore pressure can be dissipated. CD tests may take considerable time to run because of the time required for both consolidation under the lateral pressure and drainage during application of the axial load. Inasmuch as the time requirement is long for low-permeability soils, it is often referred to as the slow, or S, test.

The specific type of test (UU, CU, or CD) to be used in any given case depends largely on the field conditions to be simulated. For example, if field loading on a particular soil during construction of, say, an earthen dam is expected to be slow so that excess pore water will have drained by the end of construction, the slow (CD) test might be most appropriate. On the other hand, the quick (UU) test might be called for if loading during construction is to be very rapid. The CU test might be considered in practice as a compromise between the slow and quick tests.

In the final analysis, the type of test to be used may be based on the engineer's judgment of the problem at hand, the type of soil involved, and so on.

In Situ (Field) Methods for Investigating Shear Strength

Vane Test (ASTM D2573). The vane test, which was discussed in Section 3–7, can also be used to determine shear strength of cohesive soils. This test can be used in the field to determine *in situ* shear strength for soft, clayey soil—particularly for sensitive clays (those that lost part of their strength when disturbed). (The test can also be carried out in the laboratory on a cohesive soil sample.)

Standard Penetration Test (ASTM D1586). The standard penetration test (SPT) was discussed in Section 3–5. As noted there, through empirical testing, correlations between (corrected) SPT *N*-values and several soil parameters have been established. The correlation with shear strength was illustrated in Table 3–4.

Penetrometer Test. *In situ* bearing capacity of fine-grained soils at the surface can also be estimated by a penetrometer test. The test is performed by pushing the penetrometer steadily into the soil to the calibration mark at the penetrometer head and recording the (maximum) reading on the penetrometer scale as the penetrometer is pushed into the soil. This reading gives the pressure required to push the penetrometer into the soil to the calibration mark and is used as a guide to estimate the soil's bearing capacity. Use of the penetrometer test is limited to preliminary evaluations of the bearing capacity at the soil surface. It should also be noted that soil conditions present at the time of the test—particularly the water content—can influence the results of a penetrometer test. Figure 8–12 shows a hand penetrometer.

8–3 CHARACTERISTICS OF THE FAILURE PLANE

Whenever homogeneous soils are stressed to failure in unconfined and triaxial compression tests, failure tends to occur along a distinct plane, as shown in Figure 8–13a. The precise position of the failure plane is located at angle θ with the horizontal, which, as will be shown, is a function of the soil's angle of internal friction (ϕ). Figure 8–13b gives schematically the stresses acting on the failure plane, and Figure 8–13c shows the Mohr's circle and strength envelope for the given soil. Since the sum of the interior angles of a triangle is 180°,

FIGURE 8–12 Hand penetrometer.

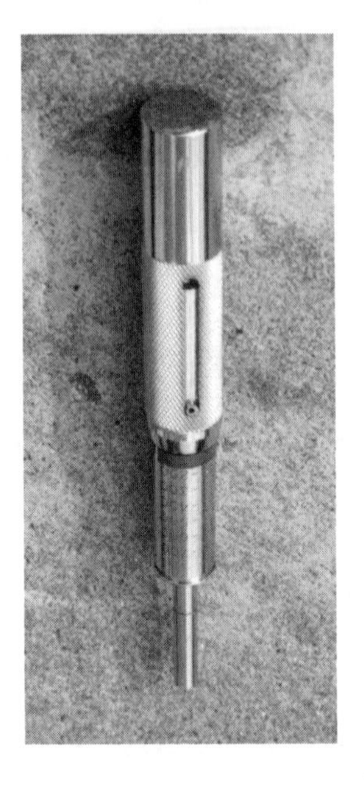

$$(180° - 2\theta) + 90° + \phi = 180° \tag{8–4}$$

Therefore,

$$\theta = 45° + \frac{\phi}{2} \tag{8–5}$$

From Figure 8–13c,

$$\sin \phi = \frac{\overline{DC}}{\overline{AC}} \tag{8–6}$$

$\overline{DC}$ and $\overline{AC}$ can be expressed as follows:

$$\overline{DC} = \frac{\sigma_1 - \sigma_3}{2} \tag{8–7}$$

$$\overline{AC} = \overline{AB} + \overline{BC} = c \cot \phi + \frac{\sigma_1 + \sigma_3}{2} \tag{8–8}$$

Substituting these values of $\overline{DC}$ and $\overline{AC}$ into Eq. (8–6) gives the following:

$$\sin \phi = \frac{(\sigma_1 - \sigma_3)/2}{c \cot \phi + (\sigma_1 + \sigma_3)/2} \tag{8–9}$$

Rearranging yields

$$\sigma_1 = \sigma_3\left(\frac{1 + \sin \phi}{1 - \sin \phi}\right) + 2c\left(\frac{\cos \phi}{1 - \sin \phi}\right) \tag{8–10}$$

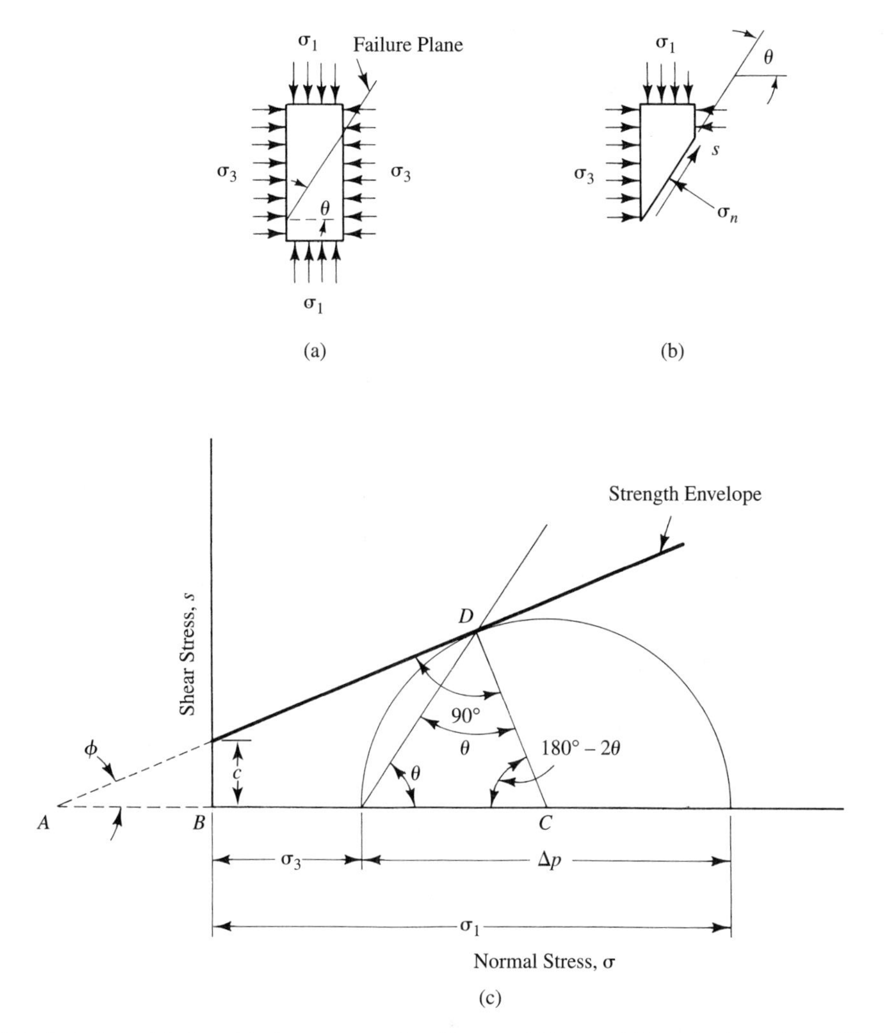

FIGURE 8–13 Relationship between angle of internal friction (ϕ) and orientation of failure plane (Θ): (a) failure plane; (b) stresses acting on the failure plane; (c) Mohr's circle.

However, from trigonometric identities,

$$\frac{1 + \sin \phi}{1 - \sin \phi} = \tan^2\left(45° + \frac{\phi}{2}\right) \quad \textbf{(8–11)}$$

$$\frac{\cos \phi}{1 - \sin \phi} = \tan\left(45° + \frac{\phi}{2}\right) \quad \textbf{(8–12)}$$

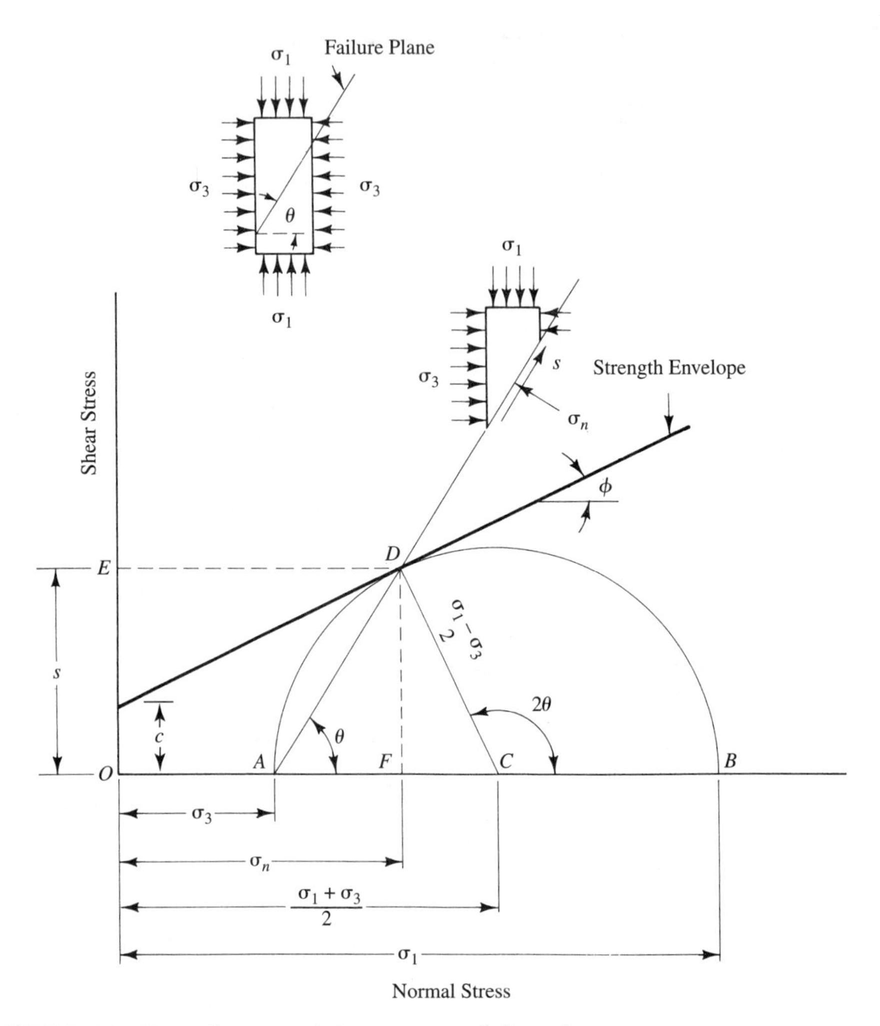

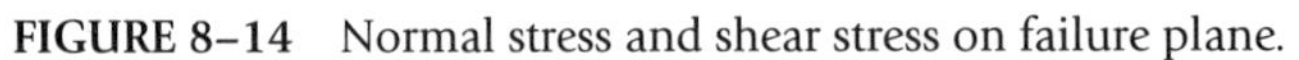

FIGURE 8–14 Normal stress and shear stress on failure plane.

Substituting these values into Eq. (8–10) gives the following:

$$\sigma_1 = \sigma_3 \tan^2\left(45° + \frac{\phi}{2}\right) + 2c \tan\left(45° + \frac{\phi}{2}\right) \tag{8–13}$$

The normal stress and shear stress on the failure plane (see Figure 8–14) can be calculated using the following equations, which result from the principles of solid mechanics.

$$\sigma_n = \frac{\sigma_1 + \sigma_3}{2} + \frac{\sigma_1 - \sigma_3}{2} \cos 2\theta \tag{8–14}$$

$$s = \frac{\sigma_1 - \sigma_3}{2} \sin 2\theta \tag{8–15}$$

where σ_n = normal stress on the failure plane
s = shear stress on the failure plane
σ_1 = major principal stress
σ_3 = minor principal stress
θ = angle between the failure plane and the horizontal plane (Figure 8–14)

Normal stress and shear stress can also be determined graphically. In Figure 8–14, points of tangency (e.g., point D in Figure 8–14) represent stress conditions on the failure plane in the test specimen. From the point where the strength envelope is tangent to the Mohr's circle (point D), a line drawn vertically downward intersects the abscissa at point F, and one drawn horizontally leftward intersects the ordinate at point E. With the coordinate system's origin denoted by O in Figure 8–14, OF is the normal stress (σ_n) on the failure plane, and OE is the shear stress (s). Furthermore, the angle between the abscissa and a line drawn from point A (the point located at distance σ_3 from the origin, see Figure 8–14) through the point of tangency (point D)—that is, angle DAB, or θ, in Figure 8–14—gives the orientation of the failure plane (i.e., angle θ in Figure 8–13a).

EXAMPLE 8–7

Given

The same conditions as given for Example 8–4.

Required

Angle of the failure plane and shear stress and normal stress on the failure plane for test specimen No. 1.

Solution

From Example 8–4, the following data are known:

$$c = 1.8 \text{ kips/ft}^2$$

$$\phi = 17°$$

$$\left.\begin{aligned} \sigma_3 &= 1.44 \text{ kips/ft}^2 \\ \Delta p &= 5.76 \text{ kips/ft}^2 \\ \sigma_1 &= 7.20 \text{ kips/ft}^2 \end{aligned}\right\} \text{ test specimen No. 1}$$

These data are plotted, as shown in Figure 8–15, according to the procedures described previously. Equation (8–5) may be used to find the angle of the failure plane (θ).

$$\theta = 45° + \frac{\phi}{2} \tag{8–5}$$

$$\theta = 45° + \frac{17°}{2} = 53.5°$$

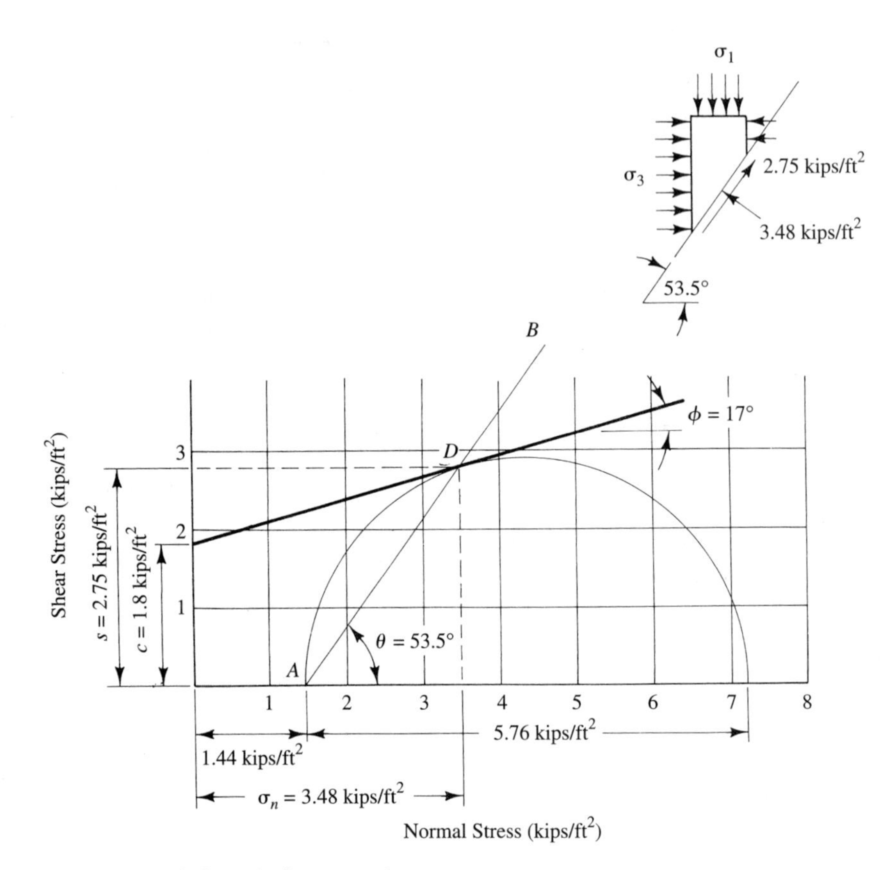

FIGURE 8–15 Mohr's circle for Example 8–7.

From point A, line AB is drawn at an angle of 53.5° (Figure 8–15), intersecting the Mohr's circle and the strength envelope at point D. The horizontal and vertical distances from the origin to point D are determined to be 3.48 kips/ft^2 and 2.75 kips/ft^2, respectively. Hence, the specimen's normal stress on the failure plane is 3.48 kips/ft^2, and its shear stress is 2.75 kips/ft^2.

These stresses can also be determined by computation. From Eq. (8–14),

$$\sigma_n = \frac{\sigma_1 + \sigma_3}{2} + \frac{\sigma_1 - \sigma_3}{2}\cos 2\theta \tag{8–14}$$

$$\sigma_n = \frac{7.20 \text{ kips/ft}^2 + 1.44 \text{ kips/ft}^2}{2}$$

$$+ \frac{7.20 \text{ kips/ft}^2 - 1.44 \text{ kips/ft}^2}{2} \cos [(2)(53.5°)] = 3.48 \text{ kips/ft}^2$$

From Eq. (8–15),

$$s = \frac{\sigma_1 - \sigma_3}{2} \sin 2\theta \tag{8–15}$$

$$s = \frac{7.20 \text{ kips/ft}^2 - 1.44 \text{ kips/ft}^2}{2} \sin [(2)(53.5°)] = 2.75 \text{ kips/ft}^2$$

8–4 SHEAR STRENGTH OF COHESIONLESS SOILS

Because of relatively large particle size, all mixtures of pure silt, sand, and gravel possess virtually no cohesion. This is because large particles have no tendency to stick together. Large particles do, however, develop significant frictional resistance, including sliding and rolling friction, as well as interlocking of the grains. This gives significant values of the angle of internal friction (ϕ); and with no cohesion ($c = 0$), Eq. (2–16) reverts to

$$s = \bar{\sigma} \tan \phi \tag{8–16}$$

Because most of a cohesionless soil's shear strength results from interlocking of grains, values of ϕ differ little whether the soil is wet or dry. Extrusion of water from void spaces is an extremely slow process for cohesive soils. Accordingly, the most critical condition with regard to shear strength usually occurs at construction time or upon application of a load. With cohesionless soils, any water contained in void spaces at construction time or upon application of a load will be driven out almost immediately, because of the high permeability of cohesionless soils. Thus, shear strength of cohesionless soils remains more or less constant throughout a structure's life.

The angle of internal friction (ϕ) of cohesionless soils can be obtained from laboratory or field tests (Section 8–2). However, ϕ can also be estimated based on the correlation between corrected SPT *N*-values and ϕ given by Peck et al. [4]. This correlation is shown in Figure 3–13. To use this graph, one enters at the upper right with the corrected *N*-value, moves horizontally to the curve marked *N*, then vertically downward to the abscissa, where the value of ϕ is read.

8–5 SHEAR STRENGTH OF COHESIVE SOILS

The shear strength of a given clay deposit is related to its water content and type of clay mineral, as well as the consolidation pressure experienced by the soil in the past (i.e., whether it is normally consolidated or overconsolidated clay). Shear strengths of clays may also differ enormously depending on whether a sample is undisturbed or remolded (as in fill).

Possible variation in a clay's shear strength is affected not only by the aforementioned factors, but also by pore water drainage that can occur during shearing deformation. Most clays in their natural state are at or near saturation; their relatively low permeabilities tend to inhibit pore water drainage that tries to occur during shearing. Thus, drainage considerations are important in the evaluation of shear strength of cohesive soils.

Normally Consolidated Clay

Strength in Drained Shear. If a saturated clay specimen is allowed to consolidate in a triaxial chamber under a lateral (confining) pressure equal to or greater than the maximum *in situ* pressure experienced by the clay, and if an axial load is slowly applied and increased and drainage is allowed at both ends of the sample, then a shear diagram similar to that shown in Figure 8–16 will be obtained. In the diagram, Mohr's circles are plotted for stress conditions at failure for three different lateral pressures, and the strength envelope is drawn tangent to the Mohr's circles.

The strength envelope shown in Figure 8–16 is sometimes referred to as the *effective stress strength envelope*, because it is based on effective stresses at failure. Because points of tangency represent stress conditions on the failure plane in each sample, the results of consolidated drained (CD) triaxial tests on normally consolidated clays can be expressed by Coulomb's equation [Eq. (2–16)], with $c = 0$. Thus,

$$s = \bar{\sigma} \tan \phi \qquad \textbf{(8–16)}$$

Consolidated Undrained Shear. The consolidated undrained (CU) test is performed by placing a saturated clay specimen in the chamber, introducing lateral

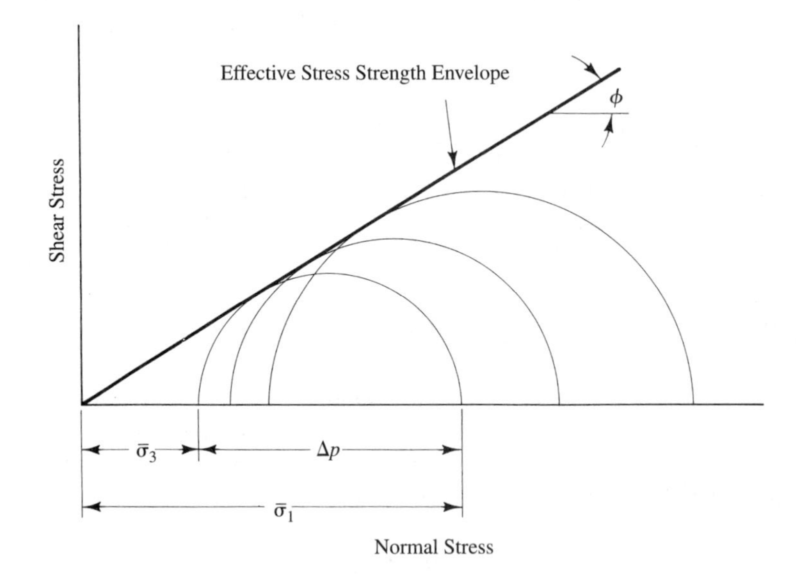

FIGURE 8–16 Results of consolidated drained (CD) triaxial tests on normally consolidated clay.

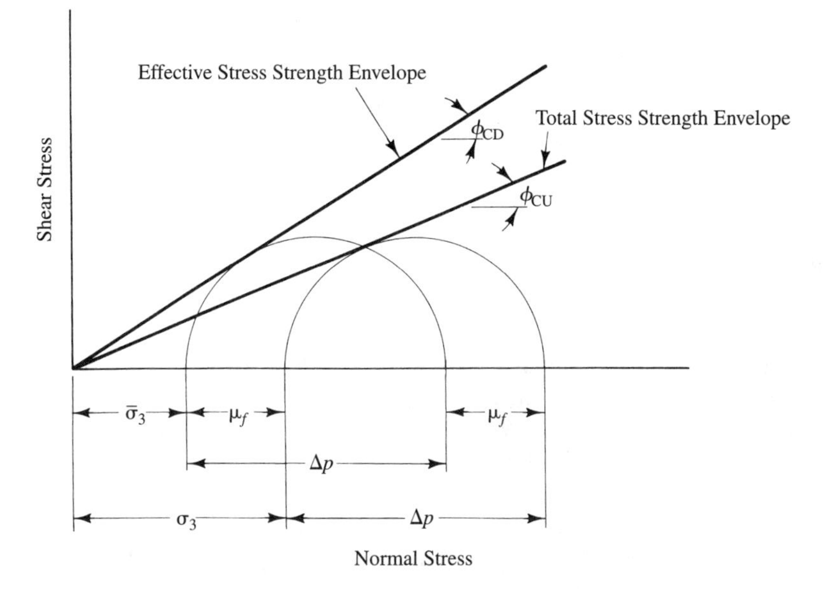

FIGURE 8–17 Results of consolidated undrained (CU) triaxial tests on normally consolidated clay.

(confining) pressure, and allowing the specimen to consolidate under the lateral pressure by leaving the drain lines open. Drain lines are then closed and an axial load is applied at a fairly rapid rate without allowing further drainage. With no drainage during axial load application, a buildup of excess pore pressure will result. [Initial excess pore pressure (μ_i) equals applied lateral pressure (σ_3) minus the pressure to which the sample had been consolidated ($\bar{\sigma}_c$)—that is, $\mu_i = \sigma_3 - \bar{\sigma}_c$. Hence, if σ_3 equals $\bar{\sigma}_c$, initial excess pore pressure in the specimen will be zero. If σ_3 is greater than $\bar{\sigma}_c$, initial pore pressure will be positive; if it is less than $\bar{\sigma}_c$, initial pore pressure will be negative.] The pore pressure (μ) during the test must be measured to obtain the effective stress needed to plot the Mohr's circle [effective stress ($\bar{\sigma}$) equals total pressure (σ) minus pore pressure (μ)—that is, $\bar{\sigma} = \sigma - \mu$] (see Figure 8–17). Pore pressure measurement can be accomplished by a pressure-measuring device connected to the drain lines at each end of the specimen.

The results of a CU test are also commonly presented with Mohr's circles plotted in terms of total stress (σ). The strength envelope in this case is referred to as the *total stress strength envelope*. Both the effective stress strength envelope and total stress strength envelope obtained from a CU test are shown in Figure 8–17. It can be noted that the Mohr's circle has equal diameters for total stresses and effective stresses, but the Mohr's circle for effective stresses is displaced leftward by an amount equal to the pore pressure at failure (μ_f) (Figure 8–17).

If several CU tests are performed on the same clay initially consolidated under different lateral pressures (σ_3), the total stress strength envelope is approximately a straight line passing through the origin (Figure 8–17). Hence, the results of CU triaxial tests on normally consolidated clays can be expressed by Coulomb's equation [Eq. (2–16)] as follows:

$$s = \sigma \tan \phi_{CU} \tag{8–17}$$

where ϕ_{CU} is known as the *consolidated undrained angle of internal friction.*

EXAMPLE 8–8

Given

A sample of normally consolidated clay was subjected to a CU triaxial compression test that was carried out until the specimen failed at a deviator stress of 50 kN/m^2. The pore water pressure at failure was recorded to be 18 kN/m^2, and a confining pressure of 48 kN/m^2 was used in the test.

Required

1. The consolidated undrained friction angle (ϕ_{CU}) for the total stress strength envelope.
2. The drained friction angle (ϕ_{CD}) for the effective stress strength envelope.

(See Figure 8–16.)

Solution

1. From Eq. (8–13),

$$\sigma_1 = \sigma_3 \tan^2\left(45° + \frac{\phi}{2}\right) + 2c \tan\left(45° + \frac{\phi}{2}\right) \tag{8–13}$$

$$\sigma_3 = 48 \text{ kN/m}^2 \quad \text{(given)}$$

$$\sigma_1 = \sigma_3 + \Delta p = 48 \text{ kN/m}^2 + 50 \text{ kN/m}^2 = 98 \text{ kN/m}^2$$

$$c = 0 \quad \text{(see Figure 8–17)}$$

$$98 \text{ kN/m}^2 = (48 \text{ kN/m}^2)\left[\tan^2\left(45° + \frac{\phi_{CU}}{2}\right)\right] + (2)(0)\left[\tan\left(45° + \frac{\phi_{CU}}{2}\right)\right]$$

$$\tan^2\left(45° + \frac{\phi_{CU}}{2}\right) = (98 \text{ kN/m}^2)/(48 \text{ kN/m}^2) = 2.042$$

$$45° + \frac{\phi_{CU}}{2} = 55.0°$$

$$\phi_{CU} = 20.0°$$

2. Use Eq. (8–13) again, but for this case,

$$\sigma_3 = 48 \text{ kN/m}^2 - 18 \text{ kN/m}^2 = 30 \text{ kN/m}^2$$

$$\sigma_1 = 30\ \text{kN/m}^2 + 50\ \text{kN/m}^2 = 80\ \text{kN/m}^2$$

Hence,

$$80\ \text{kN/m}^2 = (30\ \text{kN/m}^2)\left[\tan^2\left(45° + \frac{\phi_{CD}}{2}\right)\right] + (2)(0)\left[\tan\left(45° + \frac{\phi_{CD}}{2}\right)\right]$$

$$\tan^2\left(45° + \frac{\phi_{CD}}{2}\right) = (80\ \text{kN/m}^2)/(30\ \text{kN/m}^2) = 2.667$$

$$45° + \frac{\phi_{CD}}{2} = 58.5°$$

$$\phi_{CD} = 27.0°$$

Undrained Shear. The unconsolidated undrained (UU) shear test is performed by placing a specimen in the chamber and introducing lateral (confining) pressure without allowing the specimen to consolidate (drain) under the lateral pressure. An axial load is then applied without allowing drainage of the specimen.

Three Mohr's circles resulting from three UU tests run under different lateral pressures on an identical normally consolidated saturated clay are plotted in Figure 8–18 and labeled *A*, *B*, and *C*. It can be noted that all the circles have equal diameters; hence, the strength envelope is a horizontal line, which represents the undrained shear strength. Roughly the same effective stress at failure would result (see Mohr's circle *E* in Figure 8–18) for all three tests if pore pressures were measured and subtracted from total pressures (Figure 8–18). Hence, in terms of effective stresses, all undrained tests are represented by Mohr's circle *E* in Figure 8–18. When total stresses are plotted, the undrained test yields a series of Mohr's circles all having the same diameter, and the strength envelope for these forms a horizontal line (see Mohr's circles *A*, *B*, and *C* in Figure 8–18).

Mohr's circle *C* in Figure 8–18 is a special case of the UU test where the total minor stress (σ_3) is zero. In other words, this test was performed without any lateral pressure; hence, this special case of the UU test is the "unconfined compression test" that was discussed in Section 8–2. The diameter of Mohr's circle *C* is equal to the applied axial vertical stress at failure; it is referred to as the *unconfined compressive strength* (q_u). Because the Mohr's circle is tangent to a horizontal strength envelope, the undrained shear strength under $\phi = 0$ conditions may be evaluated on the basis of unconfined compression tests as follows:

$$s = c = \frac{q_u}{2} \tag{8–18}$$

[This phenomenon was stated previously (in Section 8–2) without detailed explanation at that point.]

When a load is applied to a saturated, or nearly saturated, normally consolidated cohesive soil (most clays in their natural condition are close to full saturation),

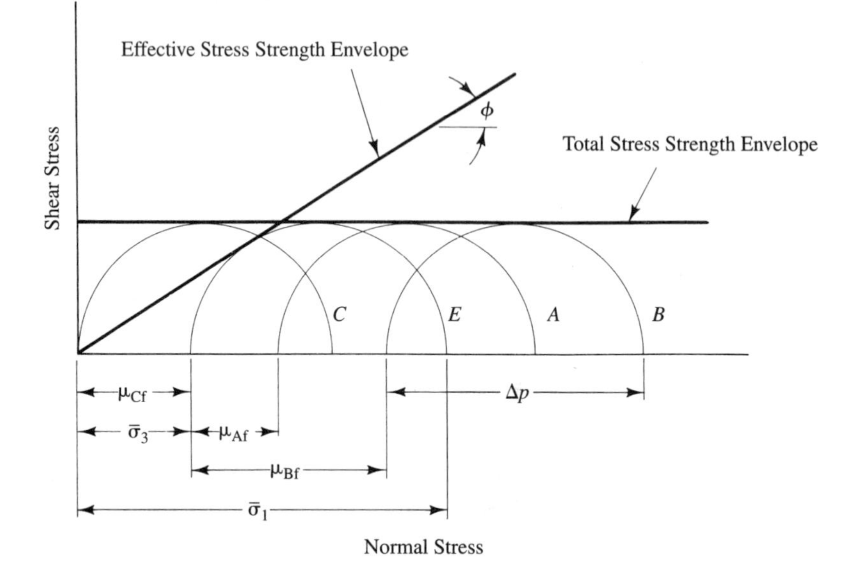

FIGURE 8–18 Results of unconsolidated undrained (UU) triaxial tests on normally consolidated clay.

water in the soil's voids carries the load first and consequently prevents the relatively small soil particles from coming into contact to develop frictional resistance. At that time, the soil's shear strength consists only of cohesion (i.e., $s = c$). As time goes on, water in the voids of cohesive soils is slowly expelled, and soil particles come together and offer frictional resistance. This increases the shear strength from $s = c$ to $s = c + \bar{\sigma} \tan \phi$ [see Eq. (2–16)]. Because the permeability of cohesive soil is very low, the process of water expulsion or extrusion from the voids is very slow, perhaps occurring over a period of years (i.e., the water content of clay does not change significantly for an appreciable time after application of a stress). What all this means is that immediately after a structure is built (i.e., immediately upon load application), the shear strength of a saturated normally consolidated cohesive soil consists of only cohesion. Therefore, in foundation design problems, the bearing capacity of normally consolidated cohesive soil should be estimated based on the assumption that soil behaves as if the angle of internal friction (ϕ) is equal to zero, and shear strength is equal to cohesion (the $\phi = 0$ concept). Such a design practice should be adequate at construction time, and any subsequent increase in shear strength should give an added factor of safety to the foundation [1].

For most normally consolidated cohesive soils, shear strength is estimated from the results of unconfined compression tests. Only for large projects and research work are the other types of shear tests generally justified. However, for soft and/or sensitive clays, shear strength is commonly obtained from the results of field or laboratory vane tests (Section 8–2).

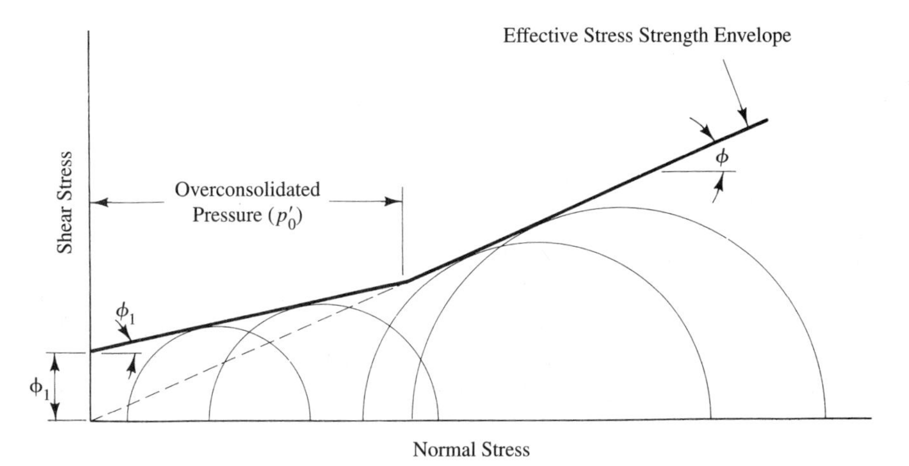

FIGURE 8–19 Results of consolidated drained (CD) triaxial tests on overconsolidated clay.

Overconsolidated Clay

As mentioned previously, overconsolidated clay has been subjected at some time in the past to pressure greater than that currently existing. If identical specimens of overconsolidated clay are sheared in a triaxial test under drained conditions, the resulting plots of data are as shown in Figure 8–19. The intersection of the first strength envelope with the ordinate is the cohesion, or the cohesive shear strength. The greater the overconsolidated pressure, the higher will be both the line labeled ϕ_1 and the cohesion.

The slope of this line (ϕ_1) represents the degree of relaxation of shear strength after removal of the overconsolidated pressure. No strength is retained in sands, so ϕ_1 is steep and equal to ϕ, and c (cohesion) is zero. Considerable strength may be retained in clays, however. Therefore, ϕ_1 is flatter and c may be quite high. For stress combinations up to the overconsolidated pressure (p_0'), a cohesion parameter (c_1) and reduced ϕ_1 are produced (Figure 8–19). Beyond p_0', the soil behaves as a normally consolidated clay.

Shear strength characteristics for an overconsolidated clay under drained conditions can be expressed by the following equations:

1. For effective normal pressure less than overconsolidated pressure (i.e., $\bar{\sigma} < p_0'$),

$$s = c_1 + \bar{\sigma} \tan \phi_1 \tag{8–19}$$

2. For effective normal pressure greater than overconsolidated pressure (i.e., $\bar{\sigma} > p_0'$),

$$s = \bar{\sigma} \tan \phi \tag{8–20}$$

The relative amount of overconsolidation is usually expressed as the *overconsolidation ratio* (OCR). As originally defined in Chapter 7, it is the ratio of overconsolidation pressure (p_0') to present overburden pressure (p_0). Hence,

$$\text{OCR} = \frac{p_0'}{p_0} \tag{7–2}$$

Under undrained conditions, the strength of an overconsolidated clay may be either smaller or larger than that under drained conditions, depending on the value of the OCR. If the OCR is in the range between 1 and about 4 to 8, the clay's volume tends to decrease during shear and, like that of normally consolidated clay, the undrained strength is less than the drained strength. If the OCR is greater than about 4 to 8, however, the clay's volume tends to increase during shear, pore water pressure decreases, and the undrained strength is greater than the drained strength.

For high OCRs, the undrained strength may be very high. However, clays with high OCRs exhibit strong negative pore pressures, which tend to draw water into the soil, causing it to swell and lose strength. Accordingly, undrained shear strength cannot be depended upon. Furthermore, in most practical problems, to apply the $\phi = 0$ concept for an overconsolidated clay would lead to results on the unsafe side, whereas for a normally consolidated clay the $\phi = 0$ and $c > 0$ concept would lead to errors in the conservative direction. Hence, except for OCRs as low as possibly 2 to 4, the $\phi = 0$ concept should not be used for overconsolidated clays [5].

Sensitivity

Cohesive soils often lose some of their shear strength if disturbed. A parameter known as *sensitivity* indicates the amount of strength lost by soil as a result of thorough disturbance. To determine a soil's sensitivity, one must perform unconfined compression tests on an undisturbed soil sample and on a remolded specimen of the same soil. Sensitivity (S_t) is the ratio of the unconfined compressive strength (q_u) of the undisturbed clay to that of the remolded clay. Hence,

$$S_t = \frac{(q_u)_{\text{undisturbed clay}}}{(q_u)_{\text{remolded clay}}} \tag{8–21}$$

Values of S_t for most clays range between 2 and about 4. For sensitive clays, they range from 4 to 8, and extrasensitive clays are encountered with values of S_t between 8 and 16. Clays with sensitivities greater than 16 are known as *quick clays* [5].

8–6 PROBLEMS

8–1. A specimen of dry sand was subjected to a direct shear test. A normal stress of 120.0 kN/m^2 was imposed on the specimen. The test was carried out until the specimen sheared, with a shear stress at failure of 75.0 kN/m^2. Determine the sand's angle of internal friction.

8–2. A series of direct shear tests was performed on a soil sample. Each test was carried out until the specimen sheared (failed). The laboratory data for the tests are tabulated as follows. Determine the soil's cohesion and angle of internal friction.

Specimen Number	Normal Stress (lb/ft^2)	Shearing Stress (lb/ft^2)
1	200	450
2	400	520
3	600	590
4	1000	740

8–3. The data shown in the following table were obtained in triaxial compression tests of three identical soil specimens. Find the soil's cohesion and angle of internal friction.

Specimen Number	Minor Principal Stress, σ_3 (lb/in.^{2})	Major Principal Stress, σ_1 (lb/ft^2)
1	5	23.0
2	10	38.5
3	15	53.6

8–4. A cohesionless soil sample was subjected to a triaxial test. The sample failed when the minor principal stress (confining pressure) was 1200 lb/ft^2 and the deviator stress was 3000 lb/ft^2. Find the angle of internal friction for this soil.

8–5. A triaxial test was performed on a dry, cohesionless soil under a confining pressure of 144.0 kN/m^2. If the sample failed when the deviator stress reached 395.8 kN/m^2, determine the soil's angle of internal friction.

8–6. A sample of dry, cohesionless soil has an angle of internal friction of 35°. If the minor principal stress is 15 lb/in.^{2}, at what values of deviator stress and major principal stress is the sample likely to fail?

8–7. Assume that both a triaxial shear test and a direct shear test are to be performed on a sample of dry sand. When the triaxial shear test is performed, the specimen fails when the major and minor principal stresses are 80 and 20 lb/in.^{2}, respectively. When the direct shear test is performed, what shear strength can be expected if the normal stress is 4000 lb/ft^2?

8–8. A cohesive soil sample is subjected to an unconfined compression test. The sample fails at a pressure of 3850 lb/ft^2 [i.e., unconfined compressive strength $(q_u) = 3850$ lb/ft^2]. Determine the soil's cohesion.

8–9. A triaxial shear test was performed on a clayey soil under unconsolidated undrained conditions. Find the cohesion of this soil if major and minor stresses at failure were 144 and 48 kN/m^2, respectively.

8–10. If an unconfined compression test is performed on the same clayey soil as described in Problem 8–9, what axial load can be expected at failure?

8–11. A triaxial compression test was performed under consolidated drained conditions on a normally consolidated clay. The test specimen failed at a confining pressure and deviator stress of 20 and 40 kN/m^2, respectively. Find the angle of internal friction for the effective stress strength envelope (i.e., ϕ_{CD}).

8–12. Determine the orientation (angle θ) of the failure plane and the shear stress and normal stress on the failure plane of specimen No. 2 of Problem 8–3.

References

[1] Wayne C. Teng, *Foundation Design,* Prentice-Hall, Inc., Englewood Cliffs, N.J., 1962.

[2] Joseph E. Bowles, *Engineering Properties of Soils and Their Measurement,* 2nd ed., McGraw-Hill Book Company, New York, 1978.

[3] *Standard Specifications for Transportation Materials and Methods of Sampling and Testing,* Part I, *Specifications,* 12th ed., AASHTO, 1978.

[4] Ralph B. Peck, Walter E. Hansen, and Thomas H. Thornburn, *Foundation Engineering,* 2nd ed., John Wiley & Sons, Inc., New York, 1974. Copyright © 1974, by John Wiley & Sons, Inc. Reprinted by permission of John Wiley & Sons, Inc.

[5] Karl Terzaghi and Ralph B. Peck, *Soil Mechanics in Engineering Practice,* John Wiley & Sons, Inc., New York, 1967. Copyright © 1967, by John Wiley & Sons, Inc. Reprinted by permission of John Wiley & Sons, Inc.

9 SHALLOW FOUNDATIONS

9–1 INTRODUCTION

The word *foundation* might be defined in general as "that which supports something." Many universities, for example, have an "athletic foundation," which supports in part the school's sports program. In the context of this book, *foundation* normally refers to something that supports a structure, such as a column or wall, along with the loads carried by the structure.

Foundations may be characterized as shallow or deep. *Shallow foundations* are located just below the lowest part of the superstructures they support; *deep foundations* extend considerably down into the earth. In the case of shallow foundations, the means of support is usually either a *footing,* which is often simply an enlargement of the base of the column or wall that it supports, or a *mat* or *raft foundation,* in which a number of columns are supported by a single slab. This chapter deals with shallow foundations—primarily footings. For deep foundations, the means of support is usually either a pier, caisson, or group of piles. These are covered in Chapters 10 and 11.

An individual footing is shown in Figure 9–1a. For purposes of analysis, a footing such as this may be thought of as a simple flat plate or slab, usually square in plan, acted on by a concentrated load (the column) and a distributed load (soil pressure) (Figure 9–1b). The enlarged size of the footing (compared with the column it supports) gives an increased contact area between the footing and the soil; the increased area serves to reduce pressure on the soil to an allowable amount, thereby preventing excessive settlement or bearing failure of the foundation.

Footings may be classified in several ways. For example, the footing depicted in Figure 9–1a is an *individual footing.* Sometimes one large footing may support two or more columns, as shown in Figure 9–2a. This is known as a *combined footing.* A footing extended in one direction to support a long structure such as a wall is called a *continuous footing,* or *wall footing* (Figure 9–2b). Two or more footings joined by a beam (called a *strap*) are called a *strap footing* (Figure 9–2c). A large slab supporting a number of columns not all of which are in a straight line is called a mat or raft foundation (Figure 9–2d).

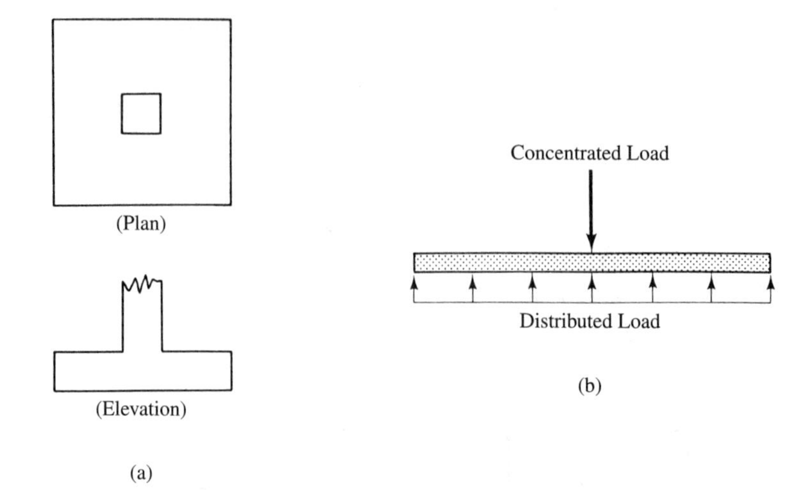

FIGURE 9–1 Individual footing.

Foundations must be designed to satisfy three general criteria:

1. They must be located properly (both vertical and horizontal orientation) so as not to be adversely affected by outside influences.
2. They must be safe from bearing capacity failure (collapse).
3. They must be safe from excessive settlement.

Specific procedures for designing footings are given in the remainder of this chapter. For initial orientation and future quick reference, the following steps are offered at this point:

1. Calculate the loads acting on the footing—Section 9–2.
2. Obtain soil profiles along with pertinent field and laboratory measurements and testing results—Chapter 3.
3. Determine the depth and location of the footing—Section 9–3.
4. Evaluate the bearing capacity of the supporting soil—Section 9–4.
5. Determine the size of the footing—Section 9–5.
6. Compute the footing's contact pressure and check its stability against sliding and overturning—Section 9–6.
7. Estimate the total and differential settlements—Chapter 7 and Section 9–7.
8. Design the footing structure—Section 9–8.

9–2 LOADS ON FOUNDATIONS [1]

When one is designing any structure, whether it is a steel beam or column, a floor slab, a foundation, or whatever, it is of basic and utmost importance that an accurate estimation (computation) of all loads acting on the structure be made. In general, a structure may be subjected upon construction or sometime in the future to some or all of the following loads, forces, and pressures: (1) dead load, (2) live load, (3) wind load, (4) snow load, (5) earth pressure, (6) water pressure, and (7) earthquake forces. These are discussed in this section.

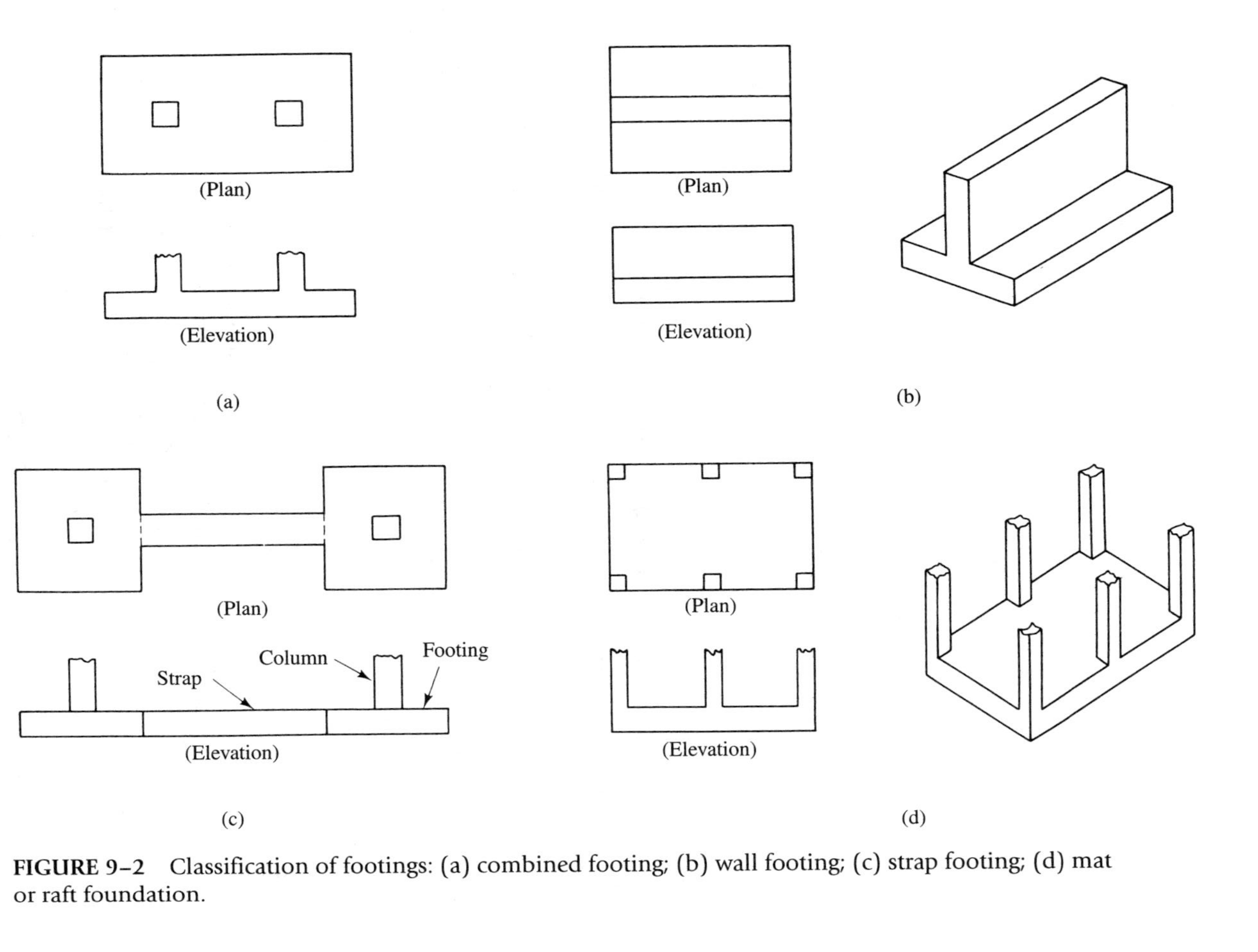

FIGURE 9–2 Classification of footings: (a) combined footing; (b) wall footing; (c) strap footing; (d) mat or raft foundation.

Dead Load

Dead load refers to the overall weight of a structure itself. It includes the weight of materials permanently attached to the structure (such as flooring) and fixed service equipment (such as air-conditioning equipment). Dead load can be calculated if sizes and types of structural material are known. This presents a problem, however, because a structure's weight is not known until its size is known, and its size cannot be known until it has been designed based (in part) on its weight. Normal procedure is to estimate dead load initially, use the estimated dead load (along with the live load, wind load, etc.) to size the structure, and then compare the sized structure's weight with the estimated dead load. If the sized structure's weight differs appreciably from the estimated dead load, the design procedure should be repeated, using a revised estimated weight.

Live Load

Live load refers to weights of applied bodies that are not permanent parts of a structure. These may be applied to the structure during part of its useful life (such as people, warehouse goods) or during its entire useful life (e.g., furniture). Because of the nature of live load, it is virtually impossible in most cases to calculate live load directly. Instead, live loads to be used in structural design are usually specified by local building codes. For example, a state building code might specify a minimum live loading of 100 lb/ft^2 for restaurants and 80 lb/ft^2 for office buildings.

Wind Load

Wind load, which is not considered as live load, may act on all exposed surfaces of structures. In addition, overhanging parts of buildings may be subject to uplift pressure as a result of wind. Like design live loads, design wind loads are usually calculated based on building codes. For example, a building code might specify a design wind loading for a particular locality of 15 lb/ft^2 for buildings less than 30 ft tall and 40 lb/ft^2 for buildings taller than 1200 ft, with a sliding scale in between.

Snow Load

Snow load results from accumulation of snow on roofs and exterior flat surfaces. The unit weight of snow varies, but it averages about 6 lb/ft^3. Thus, an accumulation of several feet of snow over a large roof area results in a very heavy load. (Two feet of snow over a 50-ft by 50-ft roof would be about 15 tons.) Design snow loads are also usually based on building codes. A building code might specify a minimum snow loading of 30 lb/ft^2 for a specific locality.

Earth Pressure

Earth pressure produces a lateral force that acts against the portion of substructure lying below ground or fill level (see Figure 9–3a). It is normally treated as dead load.

Water Pressure

Water pressure may produce a lateral force similar in nature to that produced by earth pressure. Water pressure may also produce a force that acts upward (hydrostatic uplift) on the bottom of a structure. These forces are illustrated in Figure 9–3b. Lateral water pressure is generally balanced, but hydrostatic uplift is not. It must be counteracted by the structure's dead load, or else some provision must be made to anchor the structure.

Earthquake Forces

Earthquake forces may act laterally, vertically, or torsionally on a structure in any direction. A building code should be consulted for the specification of earthquake forces to be used in design.

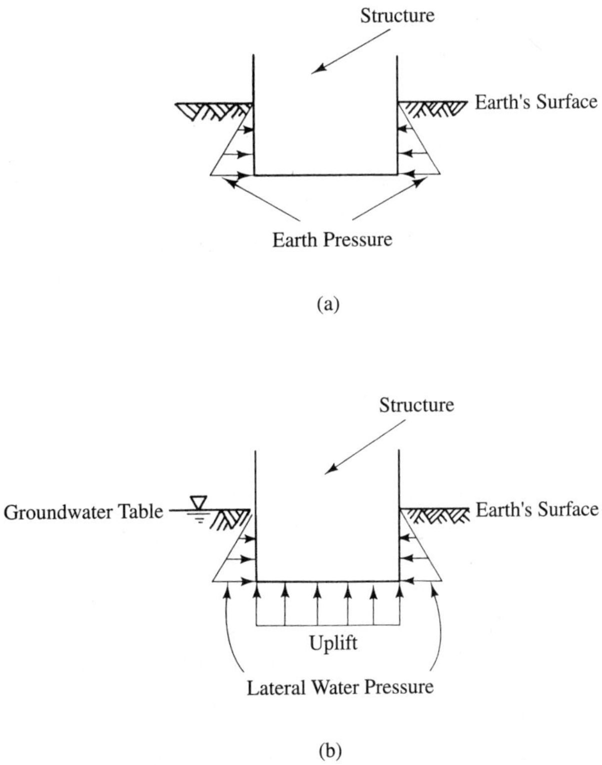

FIGURE 9–3 (a) Earth pressure; (b) water pressure.

9–3 DEPTH AND LOCATION OF FOUNDATIONS [2]

As related previously (Section 9–1), foundations must be located properly (both vertical and horizontal orientation) so as not to be adversely affected by outside influences. Outside influences include adjacent structures; water, including frost and groundwater; significant soil volume change; and underground defects (caves, for example). Thus, the depth and location of foundations are dependent on the following factors:

1. Frost action.
2. Significant soil volume change.
3. Adjacent structures and property lines.
4. Groundwater.
5. Underground defects.

These factors are discussed in this section.

Frost Action

In areas where air temperature falls below the freezing point, moisture in the soil near the ground surface will freeze. When the temperature subsequently rises above the freezing point, any frozen moisture will melt. As soil moisture freezes and melts, it alternately expands and contracts. Repeated expansion and contraction of soil moisture beneath a footing may cause it to be lifted during cold weather and dropped during warmer weather. Such a sequence generally cannot be tolerated by the structure.

Frost action on footings is prevented by placing the foundation below the maximum depth of soil that can be penetrated by frost. Depth of frost penetration varies from 4 ft (1.2 m) or more in some northern states (Maine, Minnesota) to zero in parts of some southern states (Florida, Texas). Because frost penetration varies with location, local building codes often dictate minimum depths of footings.

Significant Soil Volume Change

Some soils, particularly certain clays having high plasticity, shrink and swell significantly upon drying and wetting, respectively. This volume change is greatest near the ground surface and decreases with increasing depth. The specific depth and volume change relationship for a particular soil is dependent on the type of soil and level of groundwater. Volume change is usually insignificant below a depth of 5 to 10 ft (1.5 to 3.0 m) and does not occur below the groundwater table. In general, soil beneath the center of a structure is more protected from sun and precipitation; therefore, moisture change and resulting soil movement are smallest there. On the other hand, soil beneath the edges of a structure is less protected, and moisture change and consequent soil movement are greatest there.

As in the case of frost action, significant soil volume change beneath a footing may cause alternate lifting and dropping of the footing. Possible means of avoidance include placing the footing below all strata that are subject to significant volume changes (those soils with plasticity indices greater than 30%), placing it below the zone of volume change, and placing it below any objects that could affect moisture content unduly (such as roots, steam lines, etc.).

Adjacent Structures and Property Lines

Adjacent structures and property lines often affect the horizontal location of a footing. Existing structures may be damaged by construction of new foundations nearby, as a result of vibration, shock resulting from blasting, undermining by excavation, or lowering of the water table. After new foundations have been constructed, the (new) load they place on the soil may cause settlement of previously existing structures as a result of new stress patterns in the surrounding soil.

Because damage to existing structures by new construction may result in liability problems, new structures should be located and designed very carefully. In general, the deeper the new foundation and the closer to the old structure, the greater will be the potential for damage to the old structure. Accordingly, old and new foundations should be separated as much as is practical. This is particularly true if the new foundation will be lower than the old one. A general rule is that a straight line drawn downward and outward at a 45° angle from the end of the bottom of any new (or existing) higher footing should not intersect any existing (or new) lower footing (see Figure 9–4).

Special care must be exercised in placing a footing at or near a property line. One reason is that, because a footing is wider than the structure it supports, it is possible for part of the footing to extend across a property line and encroach on adjacent land, although the structure supported by the footing does not do so (see Figure 9–5). Also, excavation for a footing at or near a property line may have a harmful effect (cave-in, for example) on adjacent land. Either of these cases could result in liability problems; hence, much care should be exercised when footings are required near property lines.

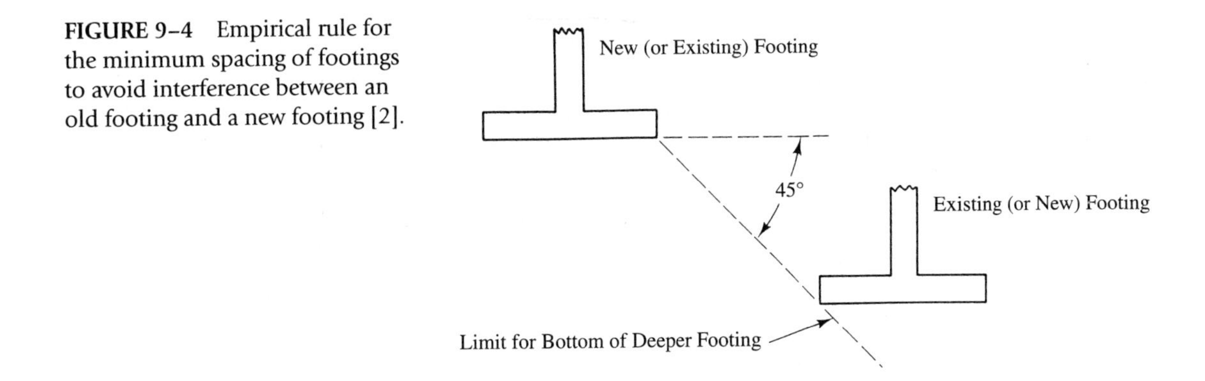

FIGURE 9–4 Empirical rule for the minimum spacing of footings to avoid interference between an old footing and a new footing [2].

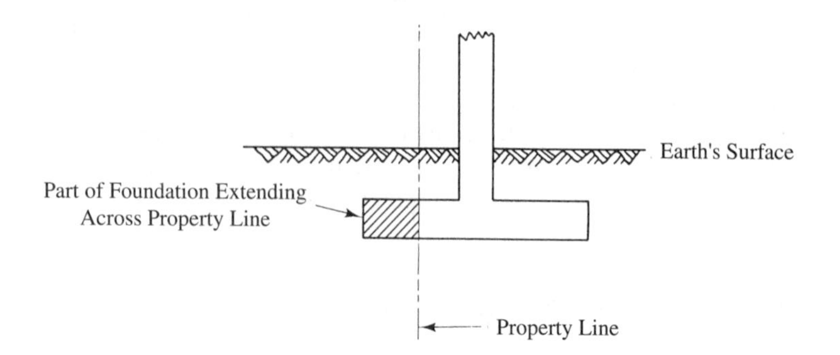

FIGURE 9–5 Sketch showing part of foundation extending across property line.

Groundwater

The presence of groundwater within soil immediately around a footing is undesirable for several reasons. First, footing construction below groundwater level is difficult and expensive. Generally, the area must be drained prior to construction. Second, groundwater around a footing can reduce the strength of soils by reducing their ability to carry foundation pressures. Third, groundwater around a footing may cause hydrostatic uplift problems; fourth, frost action may increase; and fifth, if groundwater reaches a structure's lowest floor, waterproofing problems are encountered. For these reasons, footings should be placed above the groundwater level whenever it is practical to do so.

Underground Defects

Footing location is also affected by the presence of underground defects, including faults, caves, and mines. In addition, human-made discontinuities such as sewer lines and underground cables and utilities must be considered when one is locating footings. Minor breaks in bedrock are seldom a problem unless they are active. Structures should never be built on or near tectonic faults that may slip. Certainly, foundations placed directly above a cave or mine should be avoided if at all possible. Human-made discontinuities are often encountered, and generally foundations should not be placed above them. When they are encountered where a footing is desired, either they or the footing should be relocated. As a matter of fact, a survey of underground utility lines should be made prior to excavation for a foundation in order to avoid damage to the utility lines (or even an explosion) during excavation.

9–4 BEARING CAPACITY ANALYSIS

The conventional method of designing foundations is based on the concept of bearing capacity. One meaning of the verb *to bear* is "to support or hold up." Generally, therefore, *bearing capacity* refers to the ability of a soil to support or hold up a foundation and structure. The *ultimate bearing capacity* of a soil refers to the loading per

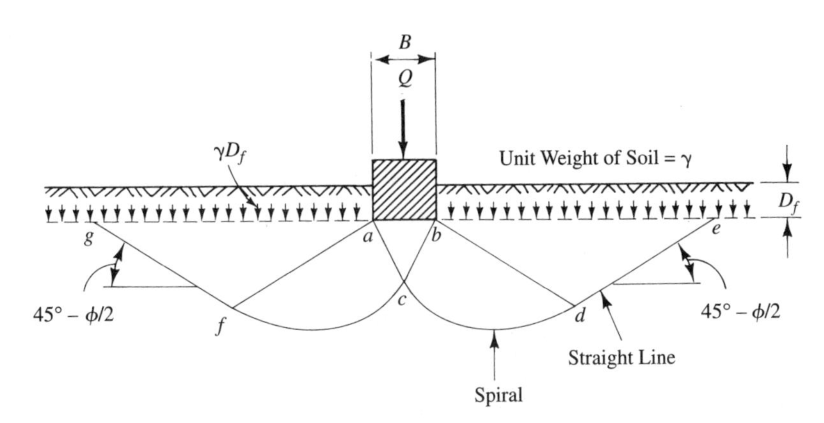

FIGURE 9–6 Plastic analysis of bearing capacity.

unit area that will just cause shear failure in the soil. It is given the symbol q_{ult}. The *allowable bearing capacity* (symbol q_a) refers to the loading per unit area that the soil is able to support without unsafe movement. It is the "design" bearing capacity. The allowable load is equal to allowable bearing capacity multiplied by area of contact between foundation and soil. The allowable bearing capacity is equal to the ultimate bearing capacity divided by the factor of safety. A factor of safety of 2.5 to 3 is commonly applied to the value of q_{ult}. Care must be taken to ensure that a footing design is safe with regard to (1) foundation failure (collapse) and (2) excessive settlement.

The basic principles governing bearing capacity theory as developed by Terzaghi [3] can be better followed by referring to Figure 9–6. As load (Q) is applied, the footing undergoes a certain amount of settlement as it is pushed downward, and a wedge of soil directly below the footing's base moves downward with the footing. The soil's downward movement is resisted by shear resistance of the foundation soil along slip surfaces *cde* and *cfg* and by the weight of the soil in sliding wedges *acfg* and *bcde.* For each set of assumed slip surfaces, the corresponding load Q that would cause failure can be determined. The set of slip surfaces giving the least applied load Q (that would cause failure) is the most critical; hence, the soil's ultimate bearing capacity (q_{ult}) is equal to the least load divided by the footing's area.

The following equations for calculating ultimate bearing capacity were developed by Terzaghi [3]:

Bearing Capacity factors p 272

Continuous footings (width B):

$$q_{\text{ult}} = cN_c + \gamma D_f N_q + 0.5\gamma B N_\gamma \tag{9–1}$$

Clay + Sand + Surcharge

Circular footings (radius R):

$$q_{\text{ult}} = 1.2cN_c + \gamma D_f N_q + 0.6\gamma R N_\gamma \tag{9–2}$$

Square footings (width B):

$$q_{\text{ult}} = 1.2cN_c + \gamma D_f N_q + 0.4\gamma B N_\gamma \tag{9–3}$$

The terms in these equations are as follows:

q_{ult} = ultimate bearing capacity
c = cohesion of soil
N_c, N_q, N_γ = Terzaghi's bearing capacity factors
γ = effective unit weight of soil
D_f = depth of footing, or distance from ground surface to base of footing
B = width of continuous or square footing
R = radius of a circular footing

The Terzaghi bearing capacity factors (N_c, N_q, N_γ) are functions of the soil's angle of internal friction (ϕ). The term in each equation containing N_c cites the influence of the soil's cohesion on its bearing capacity; the term containing N_q reflects the influence of surcharge; and that containing N_γ shows the influence of soil weight and foundation width or radius.

Values of the Terzaghi dimensionless bearing capacity factors for different values of ϕ can be obtained from Figure 9–7. The lines on Figure 9–7 representing N_q and N_c were drawn on the basis of the following equations [6]:

$$N_q = e^{\pi \tan \phi} \tan^2 \left(45^\circ + \frac{\phi}{2}\right) \quad \textbf{(9–4)}$$

$$N_c = \cot \phi \, (N_q - 1) \quad \textbf{(9–5)}$$

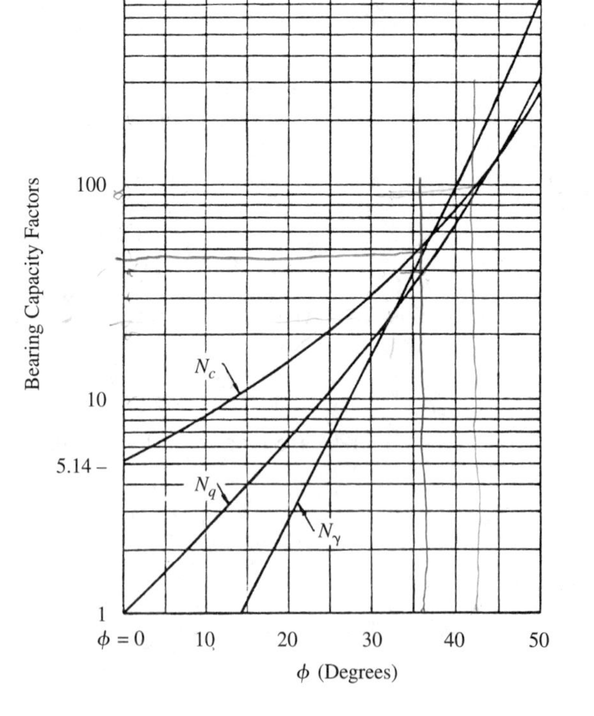

FIGURE 9–7 Chart showing relation between bearing capacity factors and ϕ [values of N_γ after Meyerhof (1955)] [4, 5].
Source: From Karl Terzaghi, Ralph B. Peck, and Gholamreza Mesri, *Soil Mechanics in Engineering Practice,* 3rd ed., John Wiley & Sons, Inc., New York, 1996. Copyright © 1996, by John Wiley & Sons, Inc. Reprinted by permission of John Wiley & Sons, Inc.

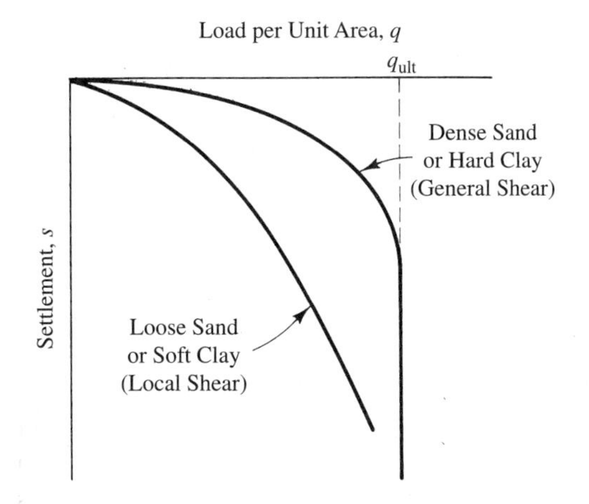

FIGURE 9–8 Relation between load and settlement of a footing on dense sand or hard clay and loose sand or soft clay.

The line on Figure 9–7 representing N_γ was found by plotting values determined in studies by Meyerhof [5].

Equations (9–1) through (9–3) are applicable for both cohesive and cohesionless soils. Dense sand and stiff clay produce what is called *general shear,* whereas loose sand and soft clay produce what is called *local shear* (see Figure 9–8). In the latter case (loose sand and soft clay), the term c (cohesion) in Eqs. (9–1) through (9–3) is replaced by c', which is equal to $\frac{2}{3}c$; in addition, the terms N_c, N_q, and N_γ are replaced by N_c', N_q', and N_γ', where the latter are obtained from Figure 9–7 using a modified ϕ value (ϕ') given by the following:

$$\phi' = \arctan\left(\tfrac{2}{3}\tan\phi\right) \tag{9–6}$$

Thus, for loose sand and soft clay, the terms c', N_c', N_q', and N_γ' are used in Eqs. (9–1) through (9–3) in place of the respective unprimed terms.

With cohesive soils, shear strength is most critical just after construction or as the load is first applied, at which time shear strength is assumed to consist of only cohesion. In this case, ϕ (angle of internal friction) is taken to be zero [1]. There are several means of evaluating cohesion [c terms in Eqs. (9–1) through (9–3)]. One is to use the unconfined compression test for ordinary sensitive or insensitive normally consolidated clay. In this test, c is equal to half the unconfined compressive strength (i.e., $\frac{1}{2}q_u$) (see Chapter 8). For sensitive clay, a field vane test may be used to evaluate cohesion (see Chapter 3).

In the case of cohesionless soils, the c term in Eqs. (9–1) through (9–3) is zero. The value of ϕ may be determined by several methods. One is to use corrected standard penetration test (SPT) values (see Chapter 3) and the curves shown in Figure 9–9. One enters the graph at the upper right with a corrected SPT N-value, moves horizontally to the curve marked N, then vertically downward to the abscissa to read the value of ϕ. This value of ϕ can be used with the curves in Figure 9–7 to determine the values of N_q and N_γ. Or, the values of N_q and N_γ may be determined using Figure 9–9 by projecting vertically downward from the curve marked N to the curves marked N_q and N_γ, then projecting horizontally over to the ordinate to read the values of N_q and N_γ, respectively. It is not necessary to determine a value of N_c because c is zero for

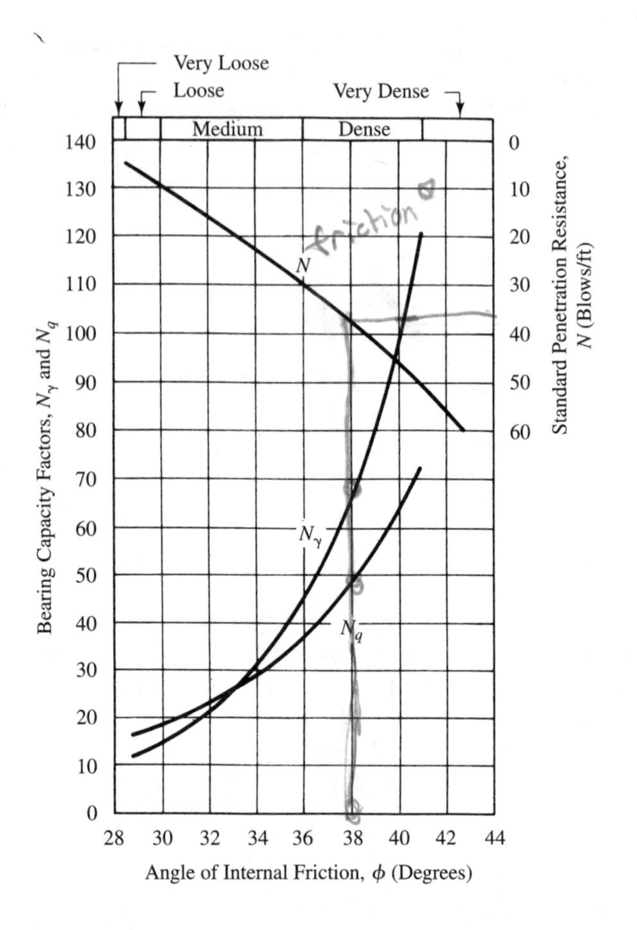

FIGURE 9–9 Curves showing the relationship between bearing capacity factors and ϕ, as determined by theory, and the rough empirical relationship between bearing capacity factors or ϕ and values of standard penetration resistance, N [7].

cohesionless soils; thus, the cN_c terms of Eqs. (9–1) through (9–3) are zero.

The four example problems that follow demonstrate the application of the Terzaghi bearing capacity formulas [that is, Eqs. (9–1) through (9–3)]. Example 9–1 deals with a wall footing in stiff clay. Example 9–2 involves a square footing in a stiff cohesive soil. A circular footing on a mixed soil is covered in Example 9–3, and a square footing in a dense cohesionless soil is considered in Example 9–4.

EXAMPLE 9–1

Given

1. A strip of wall footing 3.5 ft wide is supported in a uniform deposit of stiff clay (see Figure 9–10).
2. Unconfined compressive strength of this soil (q_u) = 2.8 kips/ft^2.
3. Unit weight of the soil (γ) = 130 lb/ft^3.
4. Groundwater was not encountered during subsurface soil exploration.
5. Depth of wall footing (D_f) = 2 ft.

FIGURE 9–10

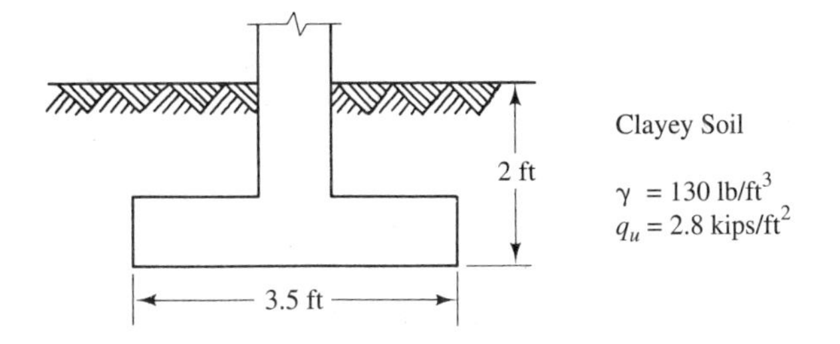

Required

1. Ultimate bearing capacity of this footing.
2. Allowable wall load, using a factor of safety of 3.

Solution

Because the supporting stratum is stiff clay, a general shear condition is evident in this case.

1. For a continuous wall footing,

$$q_{\text{ult}} = cN_c + \gamma D_f N_q + 0.5\gamma B N_\gamma \tag{9–1}$$

$$c = \frac{q_u}{2} = \frac{2.8 \text{ kips/ft}^2}{2} = 1.4 \text{ kips/ft}^2$$

$$\gamma = 0.130 \text{ kip/ft}^3$$
$$D_f = 2 \text{ ft}$$
$$B = 3.5 \text{ ft}$$

If we use $c > 0$, $\phi = 0$ analysis for cohesive soil, when $\phi = 0$, Figure 9–7 gives

$$N_c = 5.14$$
$$N_q = 1.0$$
$$N_\gamma = 0$$
$$q_{\text{ult}} = (1.4 \text{ kips/ft}^2)(5.14) + (0.130 \text{ kip/ft}^3)(2 \text{ ft})(1.0) + (0.5)(0.130 \text{ kip/ft}^3)(3.5 \text{ ft})(0) = 7.46 \text{ kips/ft}^2$$

2. $q_a = (7.46 \text{ kips/ft}^2)/3 = 2.49 \text{ kips/ft}^2$

Allowable wall loading $= q_a \times B = (2.49 \text{ kips/ft}^2)(3.5 \text{ ft})$

$= 8.72$ kips/ft of wall length

EXAMPLE 9–2

Given

1. A square footing with 5-ft sides is located 4 ft below the ground surface (see Figure 9–11).

FIGURE 9–11

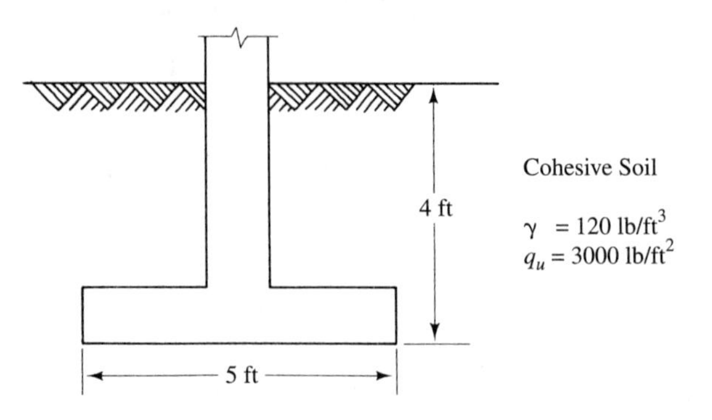

2. The groundwater table is at a great depth, and its effect can be ignored.
3. The subsoil consists of a thick deposit of stiff cohesive soil, with unconfined compressive strength (q_u) equal to 3000 lb/ft^2.
4. The unit weight (γ) of the soil is 120 lb/ft^3.

Required

Allowable bearing capacity, using a factor of safety of 3.0.

Solution

Because the supporting stratum is stiff clay, a general shear condition is evident in this case. For a square footing,

$$q_{\text{ult}} = 1.2cN_c + \gamma D_f N_q + 0.4\gamma B N_\gamma \tag{9-3}$$

$$c = \frac{q_u}{2} = \frac{3000 \text{ lb/ft}^2}{2} = 1500 \text{ lb/ft}^2$$

$$\gamma = 120 \text{ lb/ft}^3$$

$$D_f = 4 \text{ ft}$$

$$B = 5 \text{ ft}$$

If we use $c > 0$, $\phi = 0$ analysis for cohesive soil, when $\phi = 0$, Figure 9–7 gives

$$N_c = 5.14$$

$$N_q = 1.0$$

$$N_\gamma = 0$$

$$q_{\text{ult}} = (1.2)(1500 \text{ lb/ft}^2)(5.14) + (120 \text{ lb/ft}^3)(4 \text{ ft})(1.0) + (0.4)(120 \text{ lb/ft}^3)(5 \text{ ft})(0)$$
$$= 9732 \text{ lb/ft}^2$$

$$q_a = \frac{9732 \text{ lb/ft}^2}{3} = 3244 \text{ lb/ft}^2$$

FIGURE 9–12

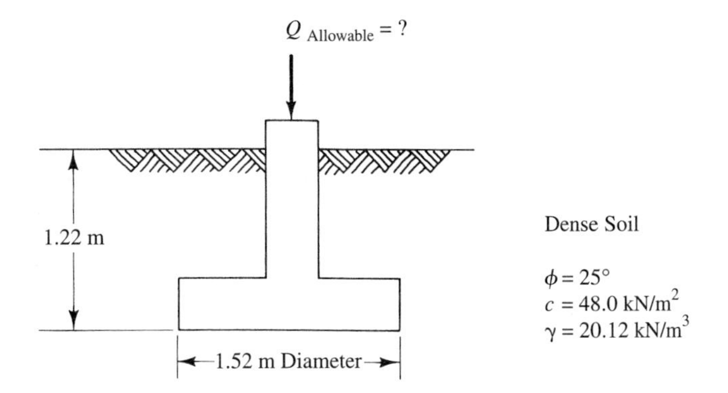

EXAMPLE 9–3

Given

1. A circular footing with a 1.52-m diameter is to be constructed 1.22 m below the ground surface (see Figure 9–12).
2. The subsoil consists of a uniform deposit of dense soil having the following strength parameters:

$$\text{Angle of internal friction} = 25°$$
$$\text{Cohesion} = 48.0\ \text{kN/m}^2$$

3. The groundwater table is at a great depth, and its effect can be ignored.

Required

The total allowable load (including column load, weight of footing, and weight of soil surcharge) that the footing can carry, using a factor of safety of 3.

Solution

Because the soil supporting the footing is dense soil, a general shear condition is evident. For a circular footing,

$$q_{\text{ult}} = 1.2cN_c + \gamma D_f N_q + 0.6\gamma R N_\gamma \tag{9–2}$$
$$c = 48.0\ \text{kN/m}^2$$
$$\gamma = 20.12\ \text{kN/m}^3$$
$$D_f = 1.22\ \text{m}$$
$$R = \frac{1.52\ \text{m}}{2} = 0.76\ \text{m}$$

From Figure 9–7, with $\phi = 25°$,

$$N_c = 21$$
$$N_q = 10$$
$$N_\gamma = 6$$

Therefore,

$$q_{\text{ult}} = (1.2)(48.0 \text{ kN/m}^2)(21) + (20.12 \text{ kN/m}^3)(1.22 \text{ m})(10) + (0.6)(20.12 \text{ kN/m}^3)(0.76 \text{ m})(6) = 1510 \text{ kN/m}^2$$

$$q_a = \frac{1510 \text{ kN/m}^2}{3} = 503 \text{ kN/m}^2$$

Therefore,

$$Q_{\text{allowable}} = A \times q_a = \frac{(\pi)(1.52 \text{ m})^2}{4}(503 \text{ kN/m}^2) = 913 \text{ kN}$$

EXAMPLE 9–4

Given

1. A column footing 6 ft by 6 ft is buried 5 ft below the ground surface in a dense cohesionless soil (see Figure 9–13).
2. The results of laboratory and field tests on the soil are as follows:
   a. Unit weight of soil (γ) = 128 lb/ft^3.
   b. Average corrected SPT N-value beneath the footing = 30.
   c. Groundwater was not encountered during subsurface soil exploration.
3. The footing is to carry a total load of 300 kips, including column load, weight of footing, and weight of soil surcharge.

Required

The factor of safety against bearing capacity failure.

Solution

Because the supporting stratum is dense cohesionless soil, a general shear condition is evident. Hence, the Terzaghi bearing capacity formula for a square footing is used, with $c = 0$, $\phi > 0$. For a square footing,

$$q_{\text{ult}} = 1.2cN_c + \gamma D_f N_q + 0.4\gamma B N_\gamma \qquad (9\text{–}3)$$

FIGURE 9–13

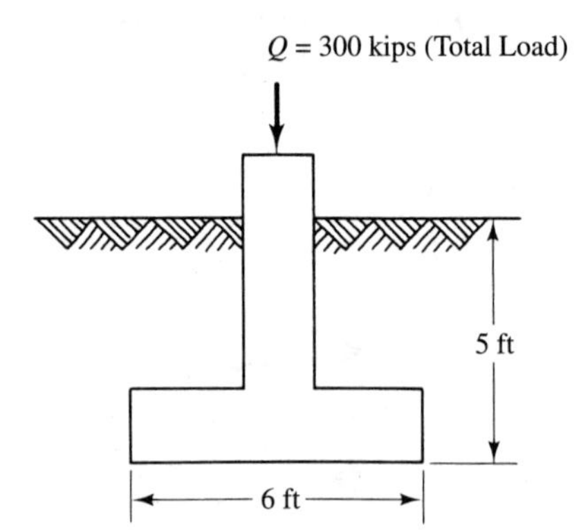

$$c = 0 \text{ (cohesionless soil)}$$
$$\gamma = 128 \text{ lb/ft}^3$$
$$D_f = 5 \text{ ft}$$
$$B = 6 \text{ ft}$$

From Figure 9–9, with the corrected N-value = 30, $\phi = 36°$. Then, from Figure 9–7, with $\phi = 36°$, the following bearing capacity factors are obtained:

$$N_q = 37$$
$$N_\gamma = 42$$
$$q_{\text{ult}} = (1.2)(0)(N_c) + (128 \text{ lb/ft}^3)(5 \text{ ft})(37) + (0.4)(128 \text{ lb/ft}^3)(6 \text{ ft})(42)$$
$$= 36{,}600 \text{ lb/ft}^2, \text{ or } 36.6 \text{ kips/ft}^2$$
$$q_{\text{actual}} = \frac{Q}{A} = \frac{300 \text{ kips}}{6 \text{ ft} \times 6 \text{ ft}} = 8.33 \text{ kips/ft}^2$$

$$\text{Factor of safety against bearing capacity failure} = \frac{q_{\text{ult}}}{q_{\text{actual}}} = \frac{36.6 \text{ kips/ft}^2}{8.33 \text{ kips/ft}^2}$$
$$= 4.4 > 3.0 \quad \therefore \text{ O.K.}$$

Effect of Water Table on Bearing Capacity [8]

Heretofore in this discussion of bearing capacity, it has been assumed that the water table was well below the footings and thus did not affect the soil's bearing capacity. This is not always the case, however. Depending on where the water table is located, two terms in Eqs. (9–1) through (9–3)—the γBN_γ (or γRN_γ) term and the $\gamma D_f N_q$ term—may require modification.

If the water table is at or above the footing's base, the soil's submerged unit weight (unit weight of soil minus unit weight of water) should be used in the γBN_γ (or γRN_γ) terms of Eqs. (9–1) through (9–3). If the water table is at distance B (note that B is the footing's width) or more below the footing's base (see Figure 9–14), the water table is assumed to have no effect, and the soil's full unit weight should be used. If the water table is below the base of the footing but less than distance B below the base, a linearly interpolated value of effective unit weight should be used in the γBN_γ (or γRN_γ) terms. (That is, the soil's effective unit weight is considered to vary linearly from the submerged unit weight at the footing's base to the full unit weight at distance B below the footing's base.)

If the water table is at the ground surface, the soil's submerged unit weight should be used in the $\gamma D_f N_q$ terms of Eqs. (9–1) through (9–3). If the water table is at or below the footing's base, the soil's full unit weight should be used in these terms. If the water table is between the footing's base and the ground surface, a linearly interpolated value of effective unit weight should be used in the $\gamma D_f N_q$ terms. (That is, the soil's effective unit weight is considered to vary linearly from submerged

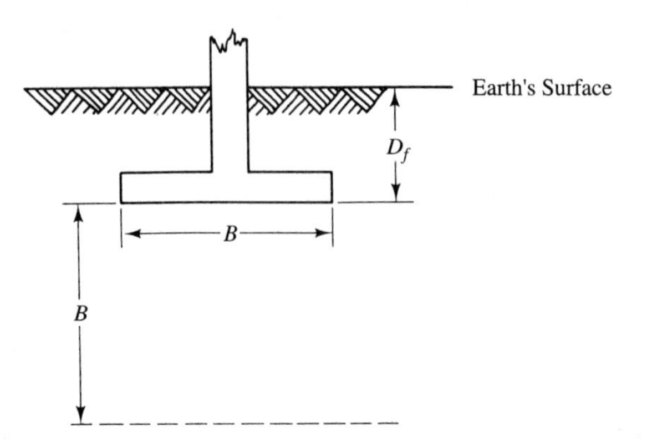

FIGURE 9–14 Sketch showing depth B (equal to footing width) below footing's base.

unit weight at the ground surface to the full unit weight at the footing's base.)

Example 9–5 deals with a square footing in soft, loose soil with the groundwater table located at the ground surface.

EXAMPLE 9–5

Given

1. A 7-ft by 7-ft square footing is located 6 ft below the ground surface (see Figure 9–15).
2. The groundwater table is located at the ground surface.
3. The subsoil consists of a uniform deposit of soft, loose soil. The laboratory test results are as follows:

$$\text{Angle of internal friction} = 20°$$
$$\text{Cohesion} = 300 \text{ lb/ft}^2$$
$$\text{Unit weight of soil} = 105 \text{ lb/ft}^3$$

Required

Allowable (design) load that can be imposed on this square footing, using a factor of safety of 3.

FIGURE 9–15

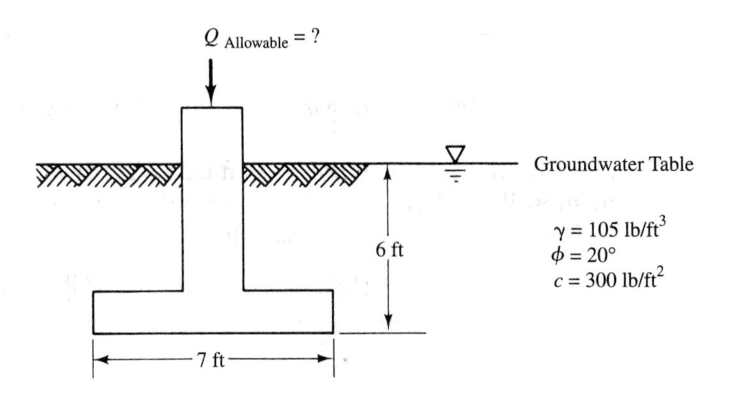

Solution

Because the footing is resting on soft, loose soil, Eq. (9–3) must be modified to reflect a local shear condition.

$$q_{\text{ult}} = 1.2c'N_c' + \gamma D_f N_q' + 0.4\gamma B N_\gamma'$$

$$c' = \tfrac{2}{3}c = \tfrac{2}{3} \times 300 \text{ lb/ft}^2 = 200 \text{ lb/ft}^2$$

From Eq. (9–6),

$$\phi' = \arctan\left(\tfrac{2}{3}\tan\phi\right) \qquad (9\text{–}6)$$

$$\phi' = \arctan\left(\tfrac{2}{3}\tan 20°\right) = 13.6°$$

With $\phi' = 13.6°$, Figure 9–7 gives

$$N_c' = 10$$

$$N_q' = 3$$

$$N_\gamma' = 1$$

$$B = 7 \text{ ft}$$

$$D_f = 6 \text{ ft}$$

$\gamma = 105 \text{ lb/ft}^3 - 62.4 \text{ lb/ft}^3 = 42.6 \text{ lb/ft}^3$ (with the water table at the ground surface, the soil's submerged unit weight must be used)

$$q_{\text{ult}} = (1.2)(200 \text{ lb/ft}^2)(10) + (42.6 \text{ lb/ft}^3)(6 \text{ ft})(3) + (0.4)(42.6 \text{ lb/ft}^3)(7 \text{ ft})(1)$$

$$= 3286 \text{ lb/ft}^2$$

$$q_a = \frac{3286 \text{ lb/ft}^2}{3} = 1095 \text{ lb/ft}^2$$

$$Q_{\text{allowable}} = q_a \times \text{Area of footing} = (1095 \text{ lb/ft}^2)(7 \text{ ft})(7 \text{ ft}) = 53{,}700 \text{ lb}$$

$$= 53.7 \text{ kips}$$

EXAMPLE 9–6

Given

1. A 6-ft by 6-ft square footing is located 5 ft below the ground surface (see Figure 9–16).
2. The groundwater table is located 7 ft below the ground level.
3. The subsoil consists of a uniform deposit of medium dense sand. The field and laboratory test results are as follows:

$$\text{Unit weight of soil} = 102 \text{ lb/ft}^3$$

$$\text{Angle of internal friction} = 32°$$

FIGURE 9–16

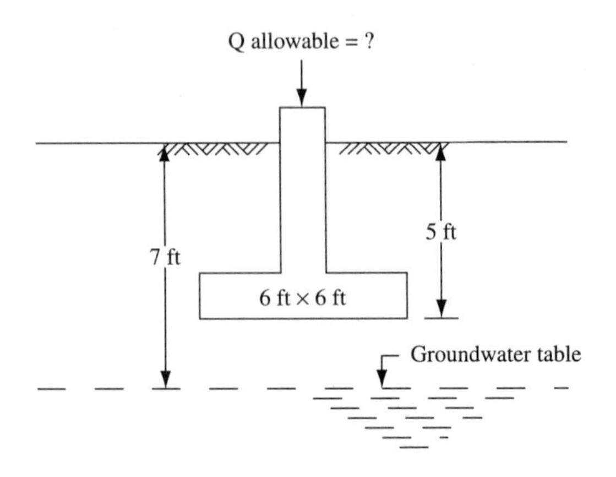

Required

Allowable (design) load that can be imposed on this square footing, using a factor of safety of 3.

Solution

Because the footing is resting on medium dense sand, a general shear condition prevails. For a square footing,

$$q_{\text{ult}} = 1.2cN_c + \gamma D_f N_q + 0.4\gamma B N_\gamma \tag{9–3}$$

Because the groundwater table is below the footing's base in this case, the soil's full unit weight ($\gamma = 102\ \text{lb/ft}^3$) should be used in the $\gamma D_f N_q$ term of Eq. (9–3). However, because the groundwater table is below the footing's base but less than distance B ($B = 5$ ft) below the base, a linearly interpolated value of effective unit weight should be used in the $\gamma B N_\gamma$ term of Eq. (9–3). Hence,

$$\gamma = (102\ \text{lb/ft}^3)(2\ \text{ft}/6\ \text{ft}) + (102\ \text{lb/ft}^3 - 62.4\ \text{lb/ft}^3)(4\ \text{ft}/6\ \text{ft})$$
$$= 60.4\ \text{lb/ft}^3$$

With $\phi = 32°$, Figure 9–7 gives

$$N_c = 34$$
$$N_q = 22$$
$$N_\gamma = 20$$

Because this soil is medium dense sand, $c = 0$.

$$q_{\text{ult}} = (1.2)(0)(34) + (102\ \text{lb/ft}^3)(5\ \text{ft})(22) + (0.4)(60.4\ \text{lb/ft}^3)(6\ \text{ft})(20)$$
$$= 14{,}119\ \text{lb/ft}^2$$
$$q_a = \frac{14{,}119\ \text{lb/ft}^2}{3} = 4706\ \text{lb/ft}^2$$
$$Q_{\text{allowable}} = q_a \times \text{Area of footing} = (4706\ \text{lb/ft}^2)(6\ \text{ft})(6\ \text{ft})$$
$$= 169{,}400\ \text{lb}$$
$$= 169.4\ \text{kips}$$

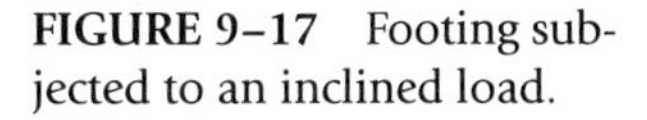

FIGURE 9–17 Footing subjected to an inclined load.

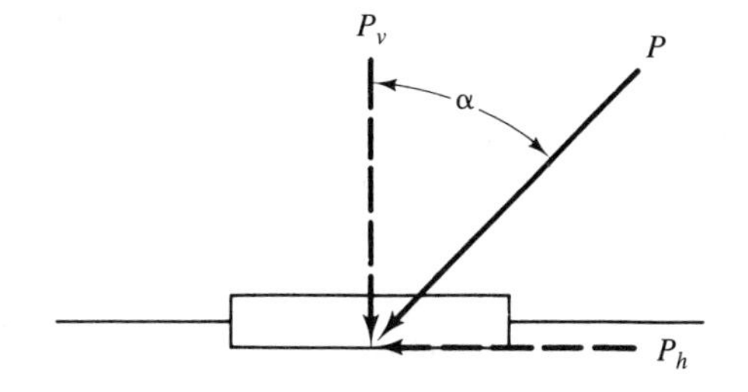

Inclined Load

If a footing is subjected to an inclined load (see Figure 9–17), the inclined load can be resolved into vertical and horizontal components. The vertical component can then be used for bearing capacity analysis in the same manner as described previously. After the bearing capacity has been computed by the normal procedure, it must be corrected by an R_i factor, which can be obtained from Figure 9–18. The footing's stability with regard to the inclined load's horizontal component must be checked by calculating the factor of safety against sliding (see Section 9–6).

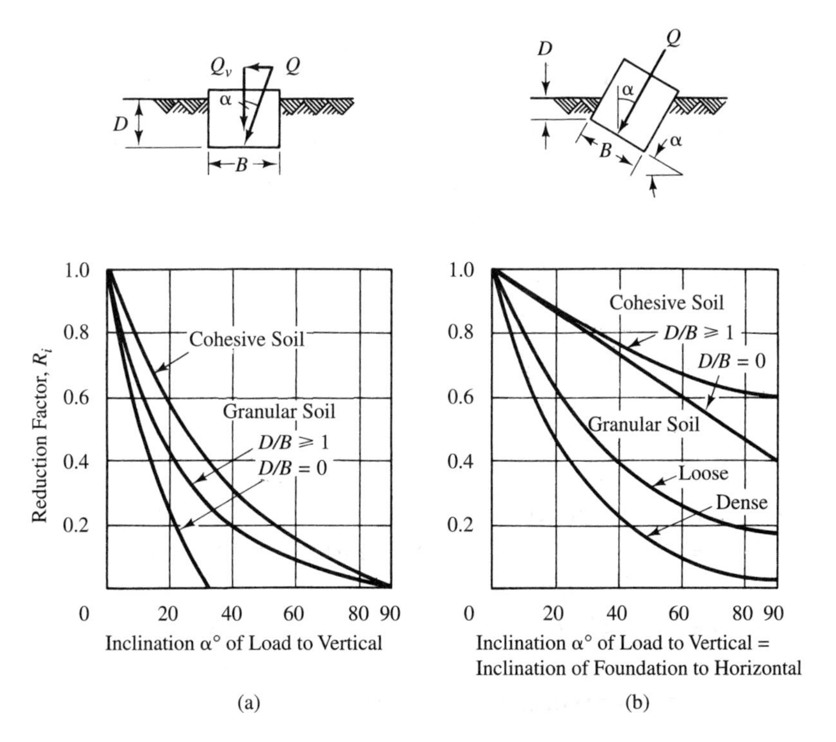

FIGURE 9–18 Inclined load reduction factors: (a) horizontal foundation [9]; (b) inclined foundation [10]. [1]

FIGURE 9–19

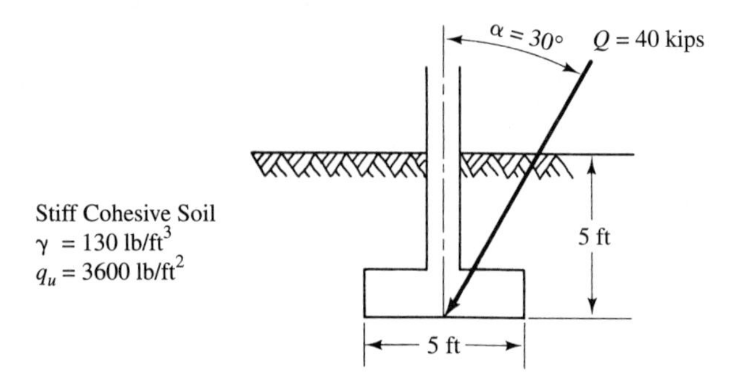

EXAMPLE 9–7

Given

A square footing (5 ft by 5 ft) is subjected to an inclined load as shown in Figure 9–19.

Required

The factor of safety against bearing capacity failure.

Solution

For a square footing,

$$q_{\text{ult}} = 1.2cN_c + \gamma D_f N_q + 0.4\gamma B N_\gamma \tag{9–3}$$

$$c = \frac{q_u}{2} = \frac{3600 \text{ lb/ft}^2}{2} = 1800 \text{ lb/ft}^2$$

$$\gamma = 130 \text{ lb/ft}^3$$

$$D_f = 5 \text{ ft}$$

$$B = 5 \text{ ft}$$

If we use $c > 0$, $\phi = 0$ analysis for cohesive soil, Figure 9–7 gives

$N_c = 5.14$

$N_q = 1.0$

$N_g = 0$

$$q_{\text{ult}} = (1.2)(1800 \text{ lb/ft}^2)(5.14) + (130 \text{ lb/ft}^3)(5 \text{ ft})(1.0) + (0.4)(130 \text{ lb/ft}^3)(5 \text{ ft})(0)$$
$$= 11{,}800 \text{ lb/ft}^2 = 11.8 \text{ kips/ft}^2$$

From Figure 9–18, with $\alpha = 30°$ and cohesive soil, the reduction factor for the inclined load is 0.42.

$$\text{Corrected } q_{\text{ult}} \text{ for inclined load} = (0.42)(11.8 \text{ kips/ft}^2) = 4.96 \text{ kips/ft}^2$$

$$Q_v = Q \cos 30° = (40 \text{ kips})(\cos 30°) = 34.6 \text{ kips}$$

$$\text{Factor of safety} = \frac{Q_{\text{ult}}}{Q_v} = \frac{(4.96 \text{ kips/ft}^2)(5 \text{ ft} \times 5 \text{ ft})}{34.6 \text{ kips}} = 3.6$$

Eccentric Load [1]

Design of a footing is somewhat more complicated if it must support an eccentric load. Eccentric loads result from loads applied somewhere other than the footing's centroid or from applied moments, such as those resulting at the base of a tall column from wind loads on the structure. Footings with eccentric loads may be analyzed for bearing capacity by two methods: (1) the concept of useful width and (2) application of reduction factors.

In the useful width method, only that part of the footing that is symmetrical with regard to the load is used to determine bearing capacity by the usual method, with the remainder of the footing being ignored. Thus, in Figure 9–20, with the (eccentric) load applied at the point indicated, the shaded area is symmetrical with regard to the load, and it is used to determine bearing capacity. That area is equal to $L \times (B - 2e_b)$ in this example.

Upon reflection, it can be observed that this method means mathematically that the bearing capacity decreases linearly as eccentricity (distance e_b in Figure 9–20) increases. This linear relationship has been confirmed in the case of cohesive soils. With cohesionless soils, however, a more nearly parabolic bearing capacity reduction

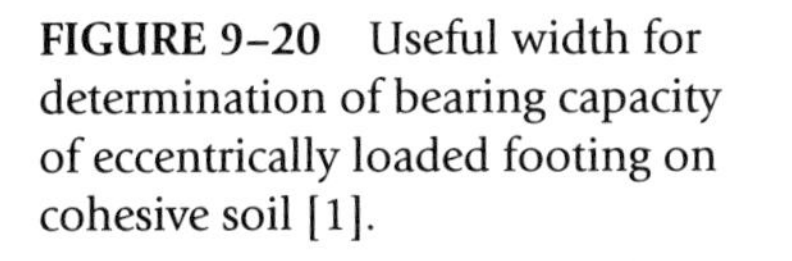
FIGURE 9–20 Useful width for determination of bearing capacity of eccentrically loaded footing on cohesive soil [1].

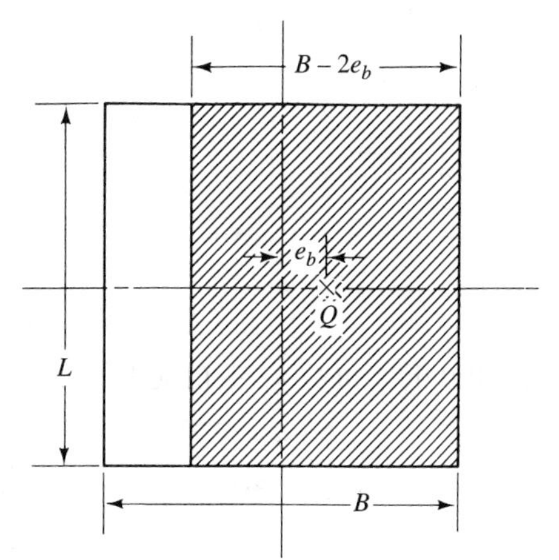

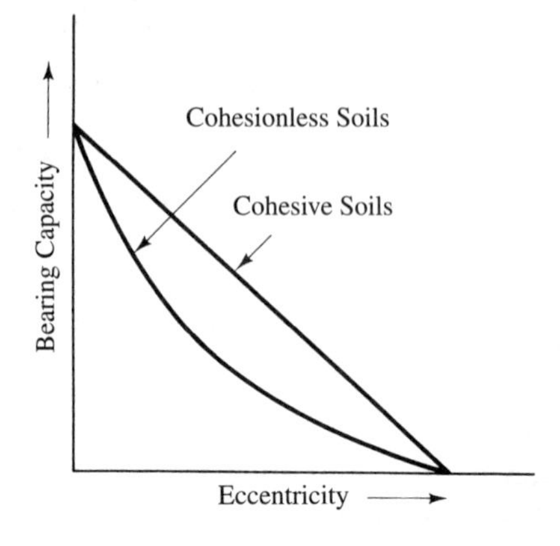

FIGURE 9–21 Relation between bearing capacity and eccentricity for cohesionless and cohesive soils.

has been determined [10]. The linear relationship for cohesive soils and the parabolic relationship for cohesionless soils are illustrated in Figure 9–21. Because the useful width method is based on a linear bearing capacity reduction, it is recommended that this method be used only with cohesive soils.

To use the reduction factors method, one first computes bearing capacity by the normal procedure, assuming that the load is applied at the centroid of the footing. The computed value of bearing capacity is then corrected for eccentricity by multiplying by a reduction factor (R_e) obtained from Figure 9–22.

Example 9–8 shows how bearing capacity can be calculated for an eccentric load in a cohesive soil by each of the two methods.

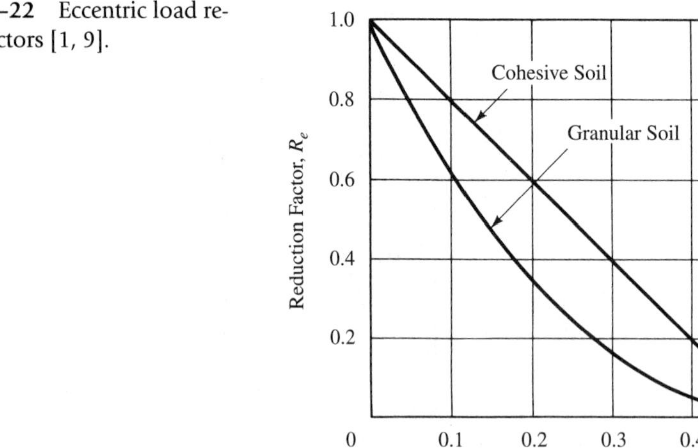

FIGURE 9–22 Eccentric load reduction factors [1, 9].

FIGURE 9–23

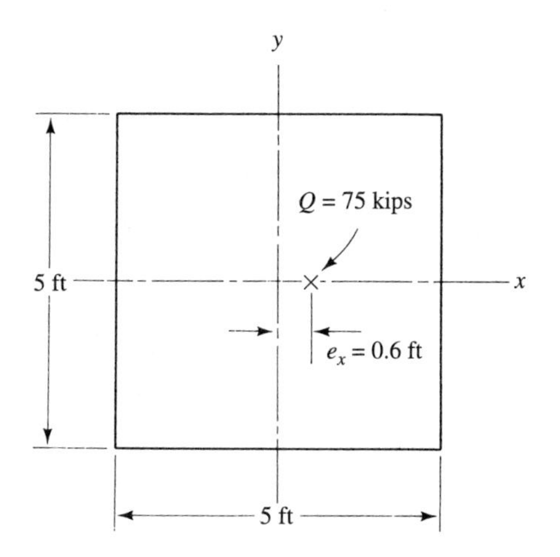

EXAMPLE 9–8

Given

1. A 5-ft by 5-ft square footing is located 4 ft below the ground surface.
2. The footing is subjected to an eccentric load of 75 kips (see Figure 9–23).
3. The subsoil consists of a thick deposit of cohesive soil with $q_u = 4.0\ \text{kips/ft}^2$ and $\gamma = 130\ \text{lb/ft}^3$.
4. The water table is at a great depth, and its effect on bearing capacity can be ignored.

Required

The factor of safety against bearing capacity failure:

1. By the concept of useful width.
2. Using a reduction factor from Figure 9–22.

Solution

1. *The concept of useful width:* From Figure 9–24, the useful width is 3.8 ft.

$$q_{\text{ult}} = 1.2cN_c + \gamma D_f N_q + 0.4\gamma BN_\gamma \qquad (9\text{–}3)$$

$$c = \frac{q_u}{2} = \frac{4.0\ \text{kips/ft}^2}{2} = 2.0\ \text{kips/ft}^2$$

FIGURE 9–24

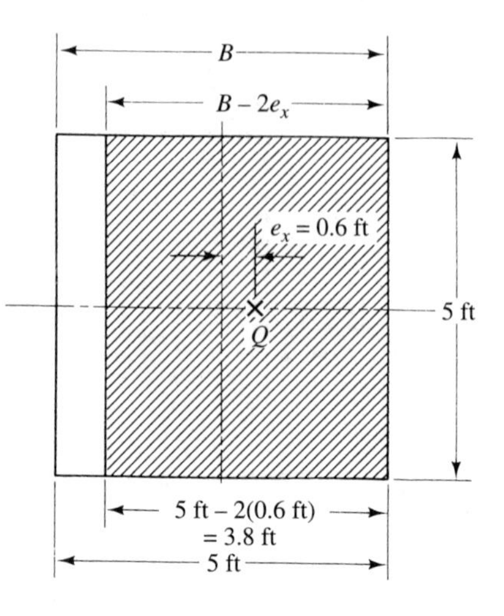

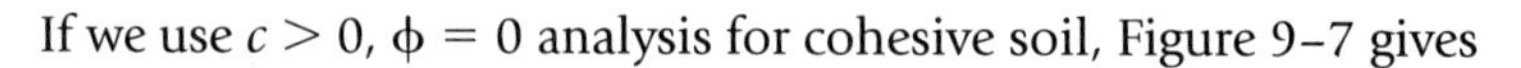

If we use $c > 0$, $\phi = 0$ analysis for cohesive soil, Figure 9–7 gives

$$N_c = 5.14$$

$$N_q = 1.0$$

$$N_\gamma = 0$$

$$\gamma = 0.130 \text{ kip/ft}^3$$

$$B = \text{Useful width} = 3.8 \text{ ft}$$

$$q_{\text{ult}} = (1.2)(2.0 \text{ kips/ft}^2)(5.14) + (0.130 \text{ kip/ft}^3)(4 \text{ ft})(1.0) + (0.4)(0.130 \text{ kip/ft}^3)(3.8 \text{ ft})(0) = 12.9 \text{ kips/ft}^2$$

$$\text{Factor of safety} = \frac{12.9 \text{ kips/ft}^2}{\left(\dfrac{75 \text{ kips}}{3.8 \text{ ft} \times 5 \text{ ft}}\right)} = 3.27$$

2. *Using a reduction factor from Figure 9–22:*

$$\text{Eccentricity ratio} = \frac{e_x}{B} = \frac{0.6 \text{ ft}}{5 \text{ ft}} = 0.12$$

For cohesive soil, Figure 9–22 gives $R_e = 0.76$. In this case, q_{ult} is computed based on the actual width: $B = 5$ ft.

$$q_{\text{ult}} = 1.2cN_c + \gamma D_f N_q + 0.4\gamma B N_\gamma \qquad \textbf{(9–3)}$$

$$q_{\text{ult}} = (1.2)(2.0 \text{ kips/ft}^2)(5.14) + (0.130 \text{ kip/ft}^3)(4 \text{ ft})(1.0) + (0.4)(0.130 \text{ kip/ft}^3)(5 \text{ ft})(0) = 12.9 \text{ kips/ft}^2$$

$$q_{\text{ult}} \text{ corrected for eccentricity} = q_{\text{ult}} \times R_e = (12.9 \text{ kips/ft}^2)(0.76)$$

$$= 9.80 \text{ kips/ft}^2$$

$$\text{Factor of safety} = \frac{9.80 \text{ kips/ft}^2}{\left(\dfrac{75 \text{ kips}}{5 \text{ ft} \times 5 \text{ ft}}\right)} = 3.27$$

Footings on Slopes

If footings are on slopes, their bearing capacities are less than if the footings were on level ground. In fact, bearing capacity of a footing is inversely proportional to ground slope.

Ultimate bearing capacity for continuous footings on slopes can be determined from the following equation [11]:

$$q_{\text{ult}} = cN_{cq} + \tfrac{1}{2}\gamma B N_{\gamma q} \tag{9–7}$$

where N_{cq} and $N_{\gamma q}$ are the bearing capacity factors for footings on slopes, and the other terms are as defined previously for Eqs. (9–1) through (9–3). Bearing capacity factors for use in Eq. (9–7) can be determined from Figure 9–25.

For circular or square footings on slopes, it is assumed that the ratios of their bearing capacities on the slope to their bearing capacities on level ground are in the same proportions as the ratio of bearing capacities of continuous footings on slopes to the bearing capacities of the continuous footings on level ground. Hence, their ultimate bearing capacities can be evaluated by first computing q_{ult} by Eq. (9–7) (i.e., as if the given footing on a slope were a continuous footing) and then multiplying that value by the ratio of q_{ult} computed from Eq. (9–2) or (9–3) (as if the given circular or square footing were on level ground) to q_{ult} determined from Eq. (9–1) (continuous footing on level ground). This may be expressed in equation form as follows [12]:

$$(q_{\text{ult}})_{c \text{ or } s \text{ footing on slope}} = (q_{\text{ult}})_{\text{continuous footing on slope}} \left[\frac{(q_{\text{ult}})_{c \text{ or } s \text{ footing on level ground}}}{(q_{\text{ult}})_{\text{continuous footing on level ground}}}\right] \tag{9–8}$$

Note: "c or s" footing denotes either circular or square footing. Examples 9–9 and 9–10 consider footings on slopes.

EXAMPLE 9–9

Given

A bearing wall for a building is to be located close to a slope as shown in Figure 9–26. The groundwater table is located at a great depth.

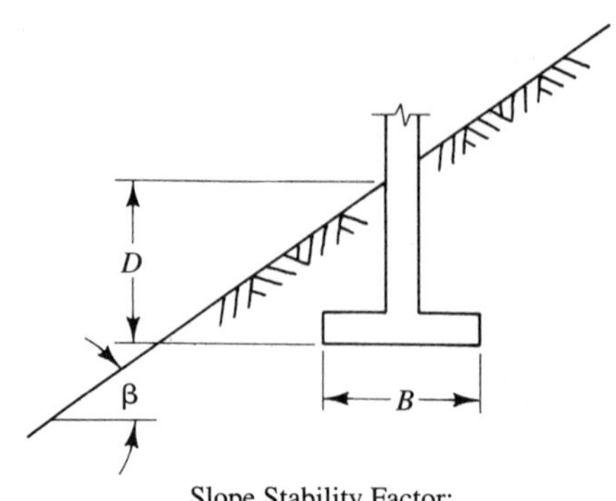

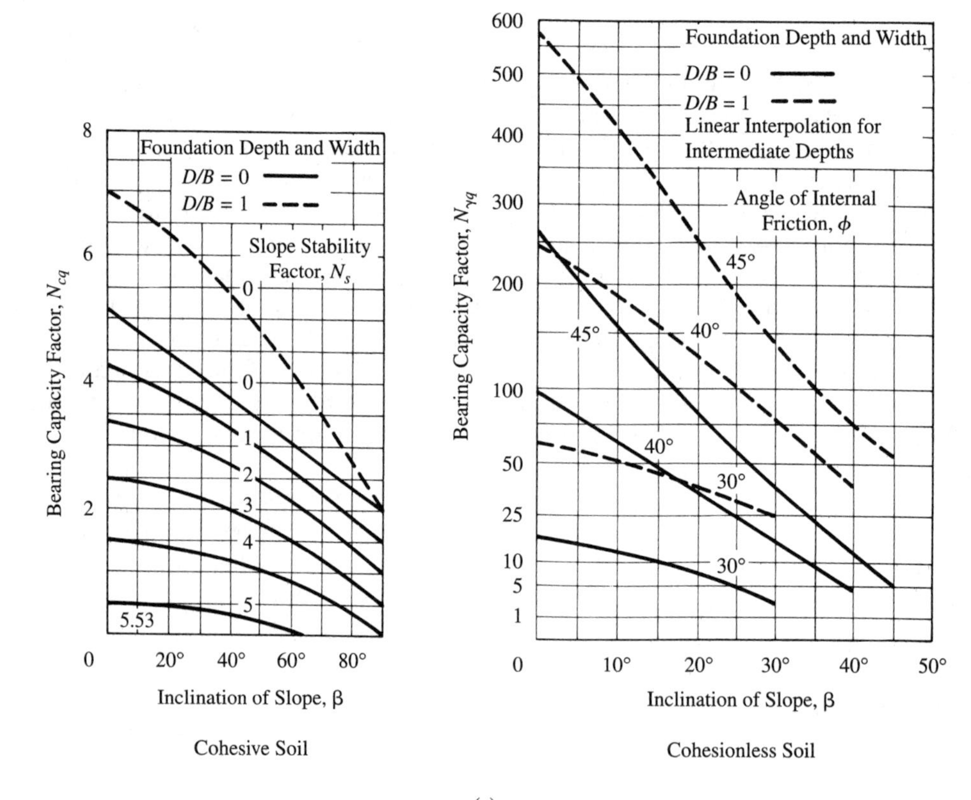

(a)

FIGURE 9–25 Bearing capacity factors for continuous footing on (a) face of slope and (b) top of slope [11].

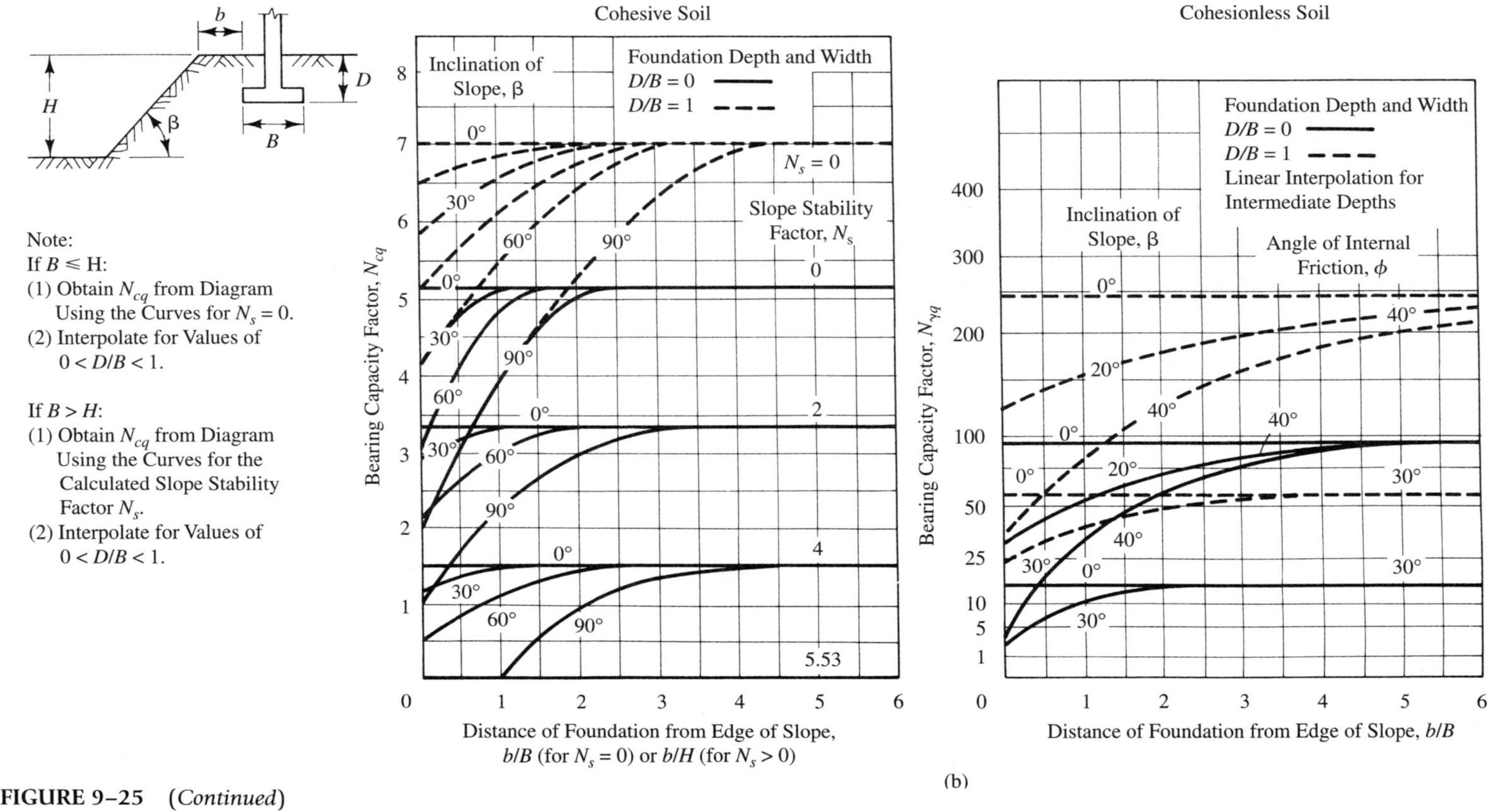

FIGURE 9–25 *(Continued)*

FIGURE 9–26

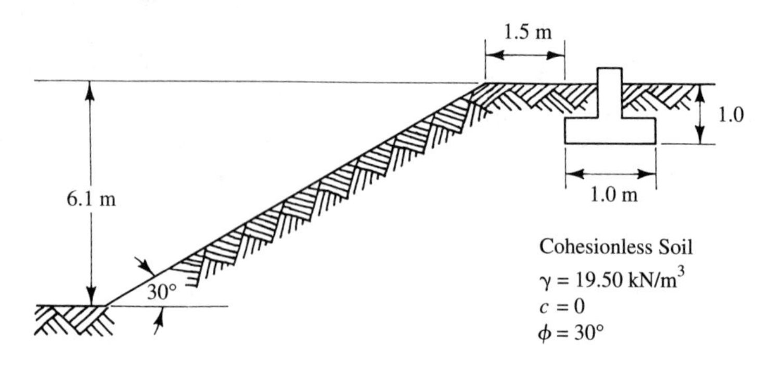

Required

Allowable bearing capacity, using a factor of safety of 3.

Solution

From Eq. (9–7),

$$q_{\text{ult}} = cN_{cq} + \tfrac{1}{2}\gamma B N_{\gamma q} \tag{9–7}$$

$$c = 0$$

$$\gamma = 19.50\ \text{kN/m}^3$$

$$B = 1.0\ \text{m}$$

From Figure 9–25b, with $\phi = 30°$,

$$\beta = 30°$$

$$\frac{b}{B} = \frac{1.5\ \text{m}}{1.0\ \text{m}} = 1.5$$

$$\frac{D_f}{B} = \frac{1.0\ \text{m}}{1.0\ \text{m}} = 1.0 \text{ (use the dashed line)}$$

$$N_{\gamma q} = 40$$

Therefore,

$$q_{\text{ult}} = (0)(N_{cq}) + (\tfrac{1}{2})(19.50\ \text{kN/m}^3)(1.0\ \text{m})(40) = 390\ \text{kN/m}^2$$

$$q_a = \frac{390\ \text{kN/m}^2}{3} = 130\ \text{kN/m}^2$$

EXAMPLE 9–10

Given

Same conditions as Example 9–9, except that a 1.0-m by 1.0-m square footing is to be constructed on the slope.

Required

Allowable bearing capacity, using a factor of safety of 3.

Solution

From Eq. (9–8),

$$(q_{\text{ult}})_{\text{square footing on slope}} = (q_{\text{ult}})_{\text{continuous footing on slope}}\left[\frac{(q_{\text{ult}})_{\text{square footing on level ground}}}{(q_{\text{ult}})_{\text{continuous footing on level ground}}}\right] \quad (9\text{–}8)$$

From Example 9–9,

$$(q_{\text{ult}})_{\text{continuous footing on slope}} = 390 \text{ kN/m}^2$$

From Eq. (9–3),

$$(q_{\text{ult}})_{\text{square footing on level ground}} = 1.2cN_c + \gamma D_f N_q + 0.4\gamma B N_\gamma \quad (9\text{–}3)$$

From Figure 9–7, with $\phi = 30°$,

$$N_c = 30$$
$$N_q = 18$$
$$N_\gamma = 16$$

$$(q_{\text{ult}})_{\text{square footing on level ground}} = (1.2)(0)(30) + (19.50 \text{ kN/m}^3)(1.0 \text{ m})(18) + (0.4)(19.50 \text{ kN/m}^3)(1.0 \text{ m})(16) = 475.8 \text{ kN/m}^2$$

From Eq. (9–1),

$$(q_{\text{ult}})_{\text{continuous footing on level ground}} = cN_c + \gamma D_f N_q + 0.5\gamma B N_\gamma \quad (9\text{–}1)$$

$$(q_{\text{ult}})_{\text{continuous footing on level ground}} = (0)(30) + (19.50 \text{ kN/m}^3)(1.0 \text{ m})(18) + (0.5)(19.50 \text{ kN/m}^3)(1.0 \text{ m})(16) = 507.0 \text{ kN/m}^2$$

Therefore, substituting into Eq. (9–8) yields the following:

$$(q_{\text{ult}})_{\text{square footing on slope}} = (390 \text{ kN/m}^2)\left(\frac{475.8 \text{ kN/m}^2}{507.0 \text{ kN/m}^2}\right) = 366 \text{ kN/m}^2$$

$$(q_a)_{\text{square footing on slope}} = \frac{366 \text{ kN/m}^2}{3} = 122 \text{ kN/m}^2$$

9–5 SIZE OF FOOTINGS

After the soil's allowable bearing capacity has been determined, the footing's required area can be determined by dividing the footing load by the allowable bearing capacity.

The following three examples illustrate the sizing of footings based on allowable bearing capacity.

EXAMPLE 9–11

Given

The footing shown in Figure 9–27 is to be constructed in a uniform deposit of stiff clay and must support a wall that imposes a loading of 152 kN/m of wall length.

Required

The width of the footing, using a factor of safety of 3.

Solution

From Eq. (9–1),

$$q_{\text{ult}} = cN_c + \gamma D_f N_q + 0.5\gamma B N_\gamma \tag{9-1}$$

$$c = \frac{q_u}{2} = \frac{145.8 \text{ kN/m}^2}{2} = 72.9 \text{ kN/m}^2$$

If we use $c > 0$, $\phi = 0$ analysis for cohesive soil, when $\phi = 0$, Figure 9–7 gives

$$N_c = 5.14$$

$$N_q = 1.0$$

$$N_\gamma = 0$$

$$\begin{aligned} q_{\text{ult}} &= (72.9 \text{ kN/m}^2)(5.14) + (18.82 \text{ kN/m}^3) \\ &\quad (1.20 \text{ m})(1.0) + (0.5)(18.82 \text{ kN/m}^3)(B)(0) \\ &= 397.3 \text{ kN/m}^2 \end{aligned}$$

$$q_a = \frac{397.3 \text{ kN/m}^2}{3} = 132.4 \text{ kN/m}^2$$

$$\text{Required width of wall} = \frac{152.0 \text{ kN/m}}{132.4 \text{ kN/m}^2} = 1.15 \text{ m}$$

FIGURE 9–27

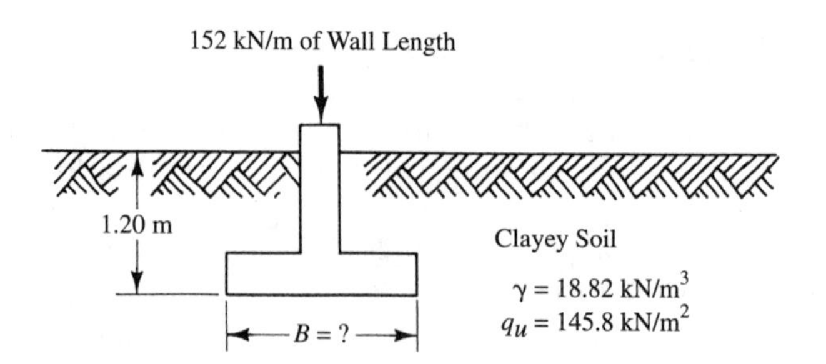

EXAMPLE 9–12

Given

1. A square footing rests on a uniform thick deposit of stiff clay with an unconfined compressive strength (q_u) of 2.4 kips/ft^2.
2. The footing is located 4 ft below the ground surface and is to carry a total load of 250 kips (see Figure 9–28).
3. The clay's unit weight is 125 lb/ft^3.
4. Groundwater is at a great depth.

Required

The necessary square footing dimension, using a factor of safety of 3. Also, find the necessary diameter of a circular footing, using a factor of safety of 3, if the footing is located 5 ft below the ground surface and is to carry a total load of 300 kips, and if $q_u = 2.6$ kips/ft^2.

Solution

Because the supporting stratum is stiff clay, a condition of general shear governs this case.

$$q_{\text{ult}} = 1.2cN_c + \gamma D_f N_q + 0.4\gamma B N_\gamma \qquad (9\text{–}3)$$

$$c = \frac{q_u}{2} = \frac{2.4 \text{ kips/ft}^2}{2} = 1.2 \text{ kips/ft}^2$$

Assuming $\phi = 0$, from Figure 9–7,

$$N_c = 5.14$$

$$N_q = 1.0$$

$$N_\gamma = 0$$

$$\gamma = 0.125 \text{ kip/ft}^3$$

$$D_f = 4 \text{ ft}$$

FIGURE 9–28

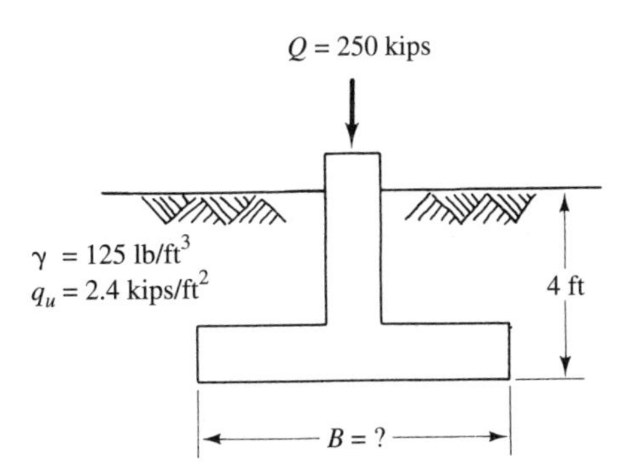

$$q_{\text{ult}} = (1.2)(1.2 \text{ kips/ft}^2)(5.14) + (0.125 \text{ kip/ft}^3)(4 \text{ ft})(1.0) + (0.4)(0.125 \text{ kip/ft}^3)(B)(0) = 7.90 \text{ kips/ft}^2$$

$$q_a = \frac{7.90 \text{ kips/ft}^2}{3} = 2.63 \text{ kips/ft}^2$$

$$\text{Required footing area} = \frac{250 \text{ kips}}{2.63 \text{ kips/ft}^2} = 95.1 \text{ ft}^2$$

Therefore,

$$B^2 = 95.1 \text{ ft}^2$$
$$B = 9.75 \text{ ft}$$

A 10-ft by 10-ft square footing would probably be specified.

For a circular footing,

$$q_{\text{ult}} = 1.2cN_c + \gamma D_f N_q + 0.6\gamma R N_\gamma \quad \textbf{(9–2)}$$

$$c = \frac{q_u}{2} = \frac{2.6 \text{ kips/ft}^2}{2} = 1.3 \text{ kips/ft}^2$$

Assuming $\phi = 0$, from Figure 9–7,

$$N_c = 5.14$$
$$N_q = 1.0$$
$$N_\gamma = 0$$
$$\gamma = 0.125 \text{ kip/ft}^3$$
$$D_f = 5 \text{ ft}$$

$$q_{\text{ult}} = (1.2)(1.3 \text{ kips/ft}^2)(5.14) + (0.125 \text{ kip/ft}^3)(5 \text{ ft})(1.0) + (0.6)(0.125 \text{ kip/ft}^3)(R)(0) = 8.64 \text{ kips/ft}^2$$

$$q_a = \frac{8.64 \text{ kips/ft}^2}{3} = 2.88 \text{ kips/ft}^2$$

$$\text{Required footing area} = \frac{300 \text{ kips}}{2.88 \text{ kips/ft}^2} = 104.2 \text{ ft}^2$$

Therefore,

$$\pi D^2/4 = 104.2 \text{ ft}^2$$
$$D = 11.5 \text{ ft}$$

EXAMPLE 9–13

Given

1. A uniform soil deposit has the following properties:

$$\gamma = 130 \text{ lb/ft}^3$$
$$\phi = 30°$$
$$c = 800 \text{ lb/ft}^2$$

FIGURE 9–29

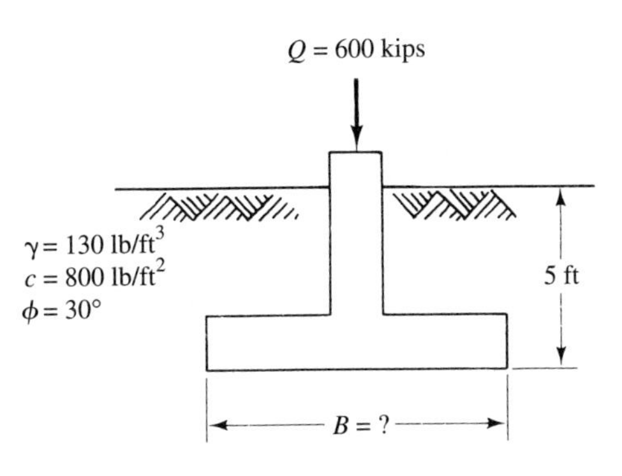

2. A proposed footing to be located 5 ft below the ground surface must carry a total load of 600 kips (see Figure 9–29).
3. The groundwater table is at a great depth, and its effect can be ignored.

Required

Determine the required dimension of a square footing to carry the proposed total load of 600 kips, using a general shear condition and a factor of safety of 3.

Solution

$$q_{ult} = 1.2cN_c + \gamma D_f N_q + 0.4\gamma B N_\gamma \qquad (9\text{–}3)$$

$$c = 800 \text{ lb/ft}^2$$

$$\gamma = 130 \text{ lb/ft}^3$$

$$D_f = 5 \text{ ft}$$

$$\phi = 30°$$

From Figure 9–7,

$$N_c = 30$$

$$N_q = 18$$

$$N_\gamma = 16$$

First Trial

Assume that $B = 10$ ft.

$$q_{ult} = (1.2)(800 \text{ lb/ft}^2)(30) + (130 \text{ lb/ft}^3)(5 \text{ ft})(18) + (0.4)(130 \text{ lb/ft}^3)(10 \text{ ft})(16)$$

$$= 48{,}820 \text{ lb/ft}^2$$

$$q_a = \frac{48{,}820 \text{ lb/ft}^2}{3} = 16{,}270 \text{ lb/ft}^2$$

$$\text{Required footing area} = \frac{600{,}000 \text{ lb}}{16{,}270 \text{ lb/ft}^2} = 36.9 \text{ ft}^2$$

$$B^2 = 36.9 \text{ ft}^2$$

$$B = 6.07 \text{ ft}$$

Second Trial

Assume that $B = 6$ ft.

$$q_{ult} = (1.2)(800 \text{ lb/ft}^2)(30) + (130 \text{ lb/ft}^3)(5 \text{ ft})(18) + (0.4)(130 \text{ lb/ft}^3)(6 \text{ ft})(16)$$
$$= 45{,}492 \text{ lb/ft}^2$$

$$q_a = \frac{45{,}492 \text{ lb/ft}^2}{3} = 15{,}164 \text{ lb/ft}^2$$

$$\text{Required footing area} = \frac{600{,}000 \text{ lb}}{15{,}164 \text{ lb/ft}^2} = 39.6 \text{ ft}^2$$

$$B^2 = 39.6 \text{ ft}^2$$

$$B = 6.29 \text{ ft}$$

A 6.5-ft by 6.5-ft square footing would probably be specified.

A footing sized in the manner just described and illustrated should be checked for settlement (see Chapter 7). If settlement is excessive (see Section 9–7), the size of the footing should be revised.

9–6 CONTACT PRESSURE

The pressure acting between a footing's base and the soil below is referred to as *contact pressure.* A knowledge of contact pressure and associated shear and moment distribution is important in footing design.

Contact pressure can be computed by using the flexural formula:

$$q = \frac{Q}{A} \pm \frac{M_x y}{I_x} \pm \frac{M_y x}{I_y} \tag{9–9}$$

where
q = contact pressure
Q = total axial vertical load
A = area of footing
M_x, M_y = total moment about respective x and y axes
I_x, I_y = moment of inertia about respective x and y axes
x, y = distance from centroid to the point at which the contact pressure is computed along respective x and y axes

In the special case where moments about both x and y axes are zero, contact pressure is simply equal to the total vertical load divided by the footing's area. In theory, contact pressure in this special case is uniform; in practice, however, it tends to vary somewhat because of distortion settlement. It is generally assumed to be uniform, however, for design purposes.

Use of the flexural formula to determine contact pressure is illustrated by the following examples. Example 9–14 illustrates computation of contact pressure when no moment is applied to either the x or y axis. Examples 9–15 and 9–16 illustrate the computation when moment is applied to one axis.

EXAMPLE 9–14

Given

1. A 5-ft by 5-ft square footing as shown in Figure 9–30.
2. Centric column load on the footing = 50 kips.
3. Unit weight of soil = 120 lb/ft^3.
4. Unit weight of concrete = 150 lb/ft^3.
5. Cohesive soil with unconfined compressive strength = 3000 lb/ft^2.

Required

1. Soil contact pressure.
2. Factor of safety against bearing capacity failure.

Solution

1. *Soil contact pressure:*

$$q = \frac{Q}{A} \pm \frac{M_x y}{I_x} \pm \frac{M_y x}{I_y} \tag{9-9}$$

Because the column load is imposed on the centroid of the footing, $M_x = 0$ and $M_y = 0$.

Q = Total axial vertical load on the footing's base

Q = Column load + Weight of footing's base pad + Weight of footing's pedestal + Weight of backfill soil

Column load = 50 kips (given)

Weight of footing's base = (5 ft)(5 ft)(1 ft)(0.150 kip/ft^3) = 3.75 kips

Weight of footing's pedestal = (1.5 ft)(1.5 ft)(3 ft)(0.150 kip/ft^3) = 1.01 kips

Weight of backfill soil = [(5 ft)(5 ft) − (1.5 ft)(1.5 ft)](3 ft) × (0.120 kip/ft^3) = 8.19 kips

$$Q = 50 \text{ kips} + 3.75 \text{ kips} + 1.01 \text{ kips} + 8.19 \text{ kips} = 62.95 \text{ kips}$$

$$A = (5 \text{ ft})(5 \text{ ft}) = 25 \text{ ft}^2$$

$$q = \frac{62.95 \text{ kips}}{25 \text{ ft}^2} = 2.52 \text{ kips/ft}^2$$

Thus, soil contact pressure = 2.52 kips/ft^2 (see Figure 9–31).

FIGURE 9–30

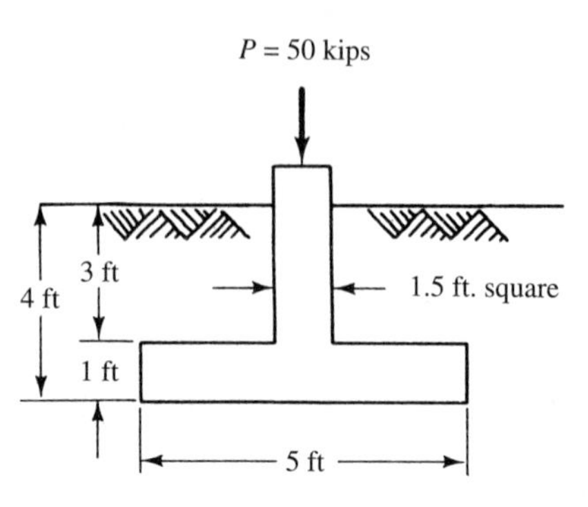

2. *Factor of safety against bearing capacity failure:* From Eq. (9–3),

$$q_{\text{ult}} = 1.2cN_c + \gamma D_f N_q + 0.4\gamma B N_\gamma \tag{9–3}$$

$$c = \frac{q_u}{2} = \frac{3000 \text{ lb/ft}^2}{2} = 1500 \text{ lb/ft}^2 = 1.50 \text{ kips/ft}^2$$

From Figure 9–7, if we use $c > 0$, $\phi = 0$ analysis,

$$N_c = 5.14$$
$$N_q = 1.0$$
$$N_\gamma = 0$$
$$D_f = 4 \text{ ft}$$

$$\begin{aligned} q_{\text{ult}} &= (1.2)(1.50 \text{ kips/ft}^2)(5.14) + (0.120 \text{ kip/ft}^3)(4 \text{ ft})(1.0) \\ &\quad + (0.4)(0.120 \text{ kip/ft}^3)(B)(0) \\ &= 9.73 \text{ kips/ft}^2 \end{aligned}$$

$$\text{Factor of safety} = \frac{9.73 \text{ kips/ft}^2}{2.52 \text{ kips/ft}^2} = 3.86$$

EXAMPLE 9–15

Given

1. A 6-ft by 6-ft square column footing as shown in Figure 9–32.
2. The column's base is hinged.
3. Load on the footing from the column (P) = 60 kips. Weight of concrete footing including pedestal and base pad (W_1) = 9.3 kips. Weight of backfill soil (W_2) = 11.2 kips.
4. Horizontal load acting on the base of the column = 4 kips.
5. Allowable bearing capacity of the supporting soil = 3.0 kips/ft^2.

Required

1. Contact pressure and soil pressure diagram.
2. Shear and moment at section A–A (Figure 9–32).

FIGURE 9–31

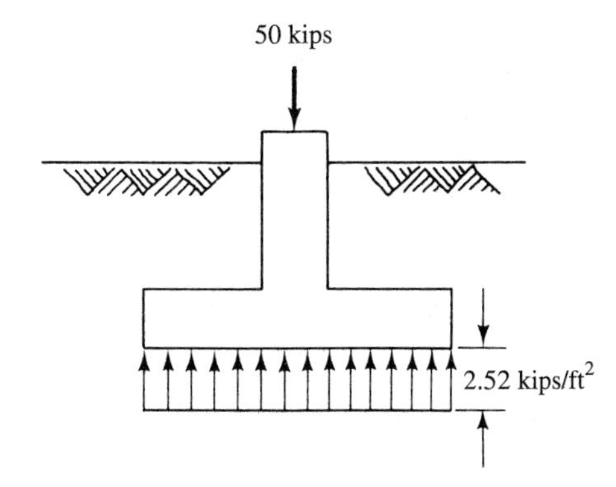

3. Factor of safety against sliding if the coefficient of friction between the footing base and the supporting soil is 0.40.
4. Factor of safety against overturning.

Solution

1. *Contact pressure and soil pressure diagram:*

$$q = \frac{Q}{A} \pm \frac{M_x y}{I_x} \pm \frac{M_y x}{I_y} \tag{9–9}$$

$$Q = P + W_1 + W_2 = 60 \text{ kips} + 9.3 \text{ kips} + 11.2 \text{ kips} = 80.5 \text{ kips}$$

$$A = 6 \text{ ft} \times 6 \text{ ft} = 36 \text{ ft}^2$$

$$M_y = 4 \text{ kips} \times 4.5 \text{ ft}$$

$= 18$ ft-kips (take moment at point C; see Figure 9–32)

$$x = \frac{6 \text{ ft}}{2} = 3 \text{ ft}$$

$$I_y = \frac{(6 \text{ ft})(6 \text{ ft})^3}{12} = 108 \text{ ft}^4$$

$$M_x = 0$$

$$\frac{M_x y}{I_x} = 0$$

$$q = \frac{80.5 \text{ kips}}{36 \text{ ft}^2} \pm \frac{(18 \text{ ft-kips})(3 \text{ ft})}{108 \text{ ft}^4} = 2.24 \text{ kips/ft}^2 \pm 0.50 \text{ kip/ft}^2$$

$$q_{\text{right}} = 2.24 \text{ kips/ft}^2 + 0.50 \text{ kip/ft}^2 = 2.74 \text{ kips/ft}^2 < 3.0 \text{ kips/ft}^2 \quad \therefore \text{ O.K.}$$

$$q_{\text{left}} = 2.24 \text{ kips/ft}^2 - 0.50 \text{ kip/ft}^2 = 1.74 \text{ kips/ft}^2 < 3.0 \text{ kips/ft}^2 \quad \therefore \text{ O.K.}$$

The pressure diagram is shown in Figure 9–33.

2. *Shear and moment at section A–A:* From Figure 9–34, $\triangle FDG$ and $\triangle EDH$ are similar triangles. Therefore,

FIGURE 9–32

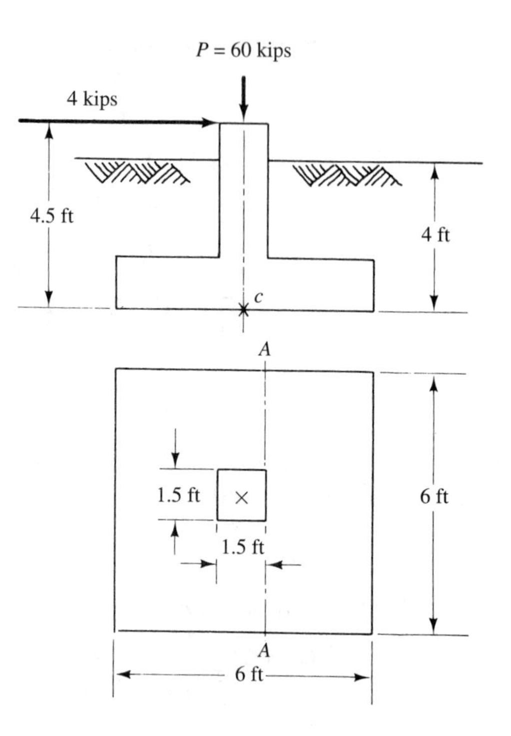

FIGURE 9–33

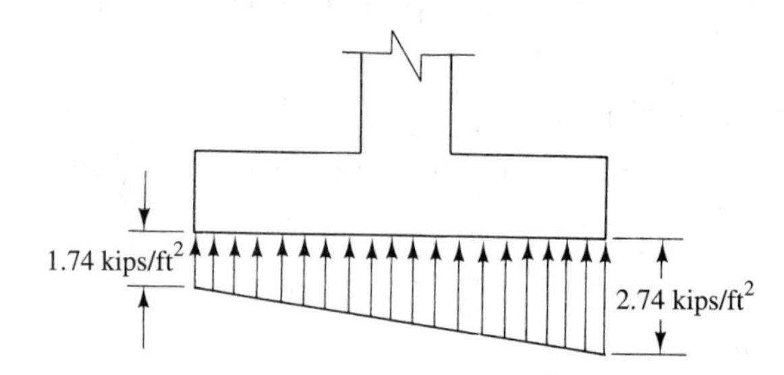

$$\frac{DE}{DF} = \frac{EH}{FG}$$

$$DF = 2.74 \text{ kips/ft}^2 - 1.74 \text{ kips/ft}^2 = 1.0 \text{ kip/ft}^2$$

$$EH = \frac{6 \text{ ft}}{2} - \frac{1.5 \text{ ft}}{2} = 2.25 \text{ ft} \quad \text{(see Figures 9–32 and 9–34)}$$

$$FG = 6 \text{ ft}$$

$$\frac{DE}{1.0 \text{ kip/ft}^2} = \frac{2.25 \text{ ft}}{6 \text{ ft}}$$

$$DE = 0.375 \text{ kip/ft}^2$$

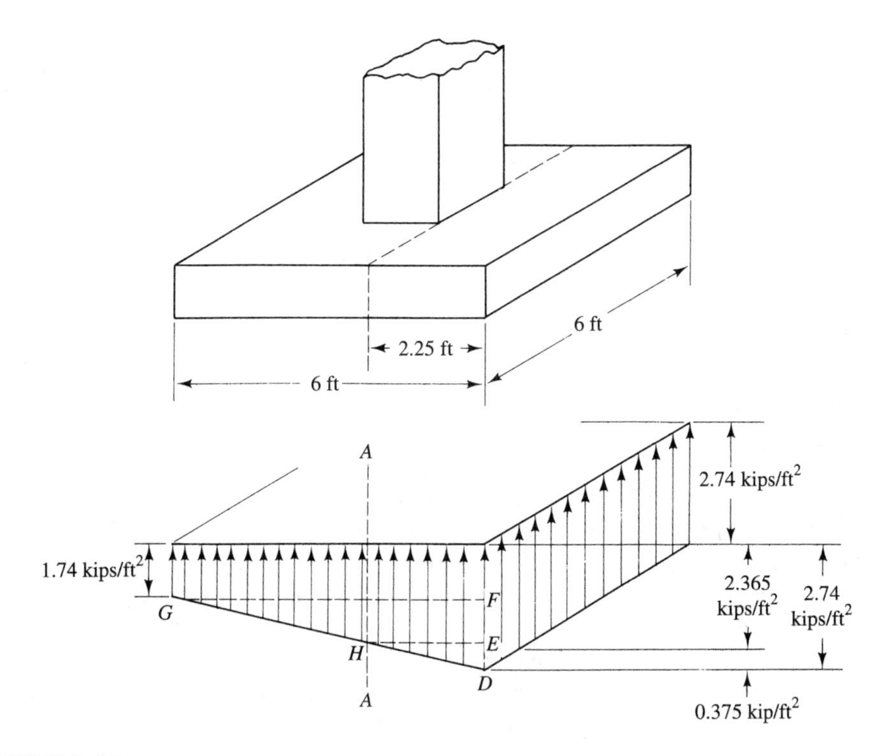

FIGURE 9–34

$$\text{Shear at } A\text{–}A = (2.25\text{ ft})(2.365\text{ kips/ft}^2)(6\text{ ft}) + (1/2)(2.25\text{ ft})(0.375\text{ kip/ft}^2)(6\text{ ft})$$

$$= 31.93\text{ kips} + 2.53\text{ kips} = 34.46\text{ kips}$$

$$\text{Moment at } A-A = (31.93\text{ kips})\left(\frac{2.25\text{ ft}}{2}\right) + (2.53\text{ kips})(^2\!/_3 \times 2.25\text{ ft})$$

$$= 39.7\text{ ft-kips}$$

3. *Factor of safety against sliding:*
   Factor of safety against sliding

$$= \frac{\text{Total vertical load} \times \text{Coefficient of friction between base and soil}}{\Sigma\text{ Horizontal forces}}$$

$$= \frac{(60\text{ kips} + 9.3\text{ kips} + 11.2\text{ kips})(0.40)}{4\text{ kips}} = 8.05$$

4. *Factor of safety against overturning:* See Figure 9–35. By taking moments at point K, one can compute the factor of safety against overturning as follows:

$$\text{Factor of safety} = \frac{\text{Moment to resist turning}}{\text{Turning moment}} = \frac{(80.5\text{ kips})(6\text{ ft}/2)}{(4\text{ kips})(4.5\text{ ft})} = 13.4$$

FIGURE 9–35

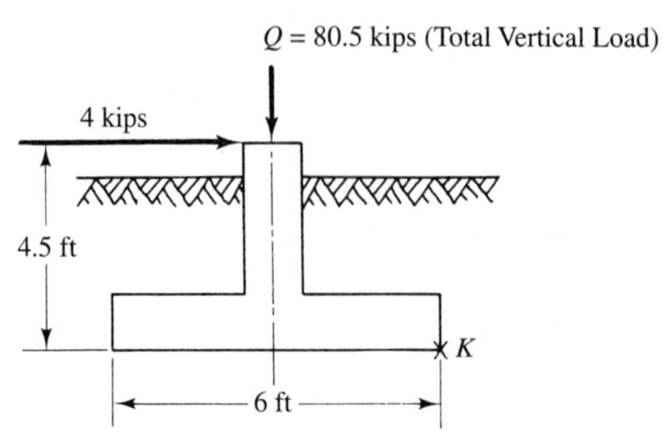

EXAMPLE 9–16

Given

1. A 7.5-ft by 10-ft rectangular column footing as shown in Figure 9–36.
2. The column's base is fixed into the foundation.
3. Load on the footing from the column (P) = 50 kips.
   Weight of the concrete footing and weight of the backfill soil (W) = 25 kips.
   Horizontal load acting on the column's base (H) = 3 kips.
   Moment acting on the foundation (M) = 30 ft-kips.
4. Allowable bearing capacity of the soil = 2 kips/ft^2.

Required

1. Contact pressure and soil pressure diagram.
2. Factor of safety against overturning.

Solution

1. *Contact pressure and soil pressure diagram:*

$$q = \frac{Q}{A} \pm \frac{M_x y}{I_x} \pm \frac{M_y x}{I_y} \tag{9–9}$$

$$Q = 50 \text{ kips} + 25 \text{ kips} = 75 \text{ kips}$$

$$A = 7.5 \text{ ft} \times 10 \text{ ft} = 75 \text{ ft}^2$$

$$M_y = (3 \text{ kips})(6 \text{ ft}) + 30 \text{ ft-kips}$$

$$= 48 \text{ ft-kips} \quad \text{(take moments at point } C\text{; see Figure 9–36)}$$

$$x = \frac{10 \text{ ft}}{2} = 5 \text{ ft}$$

$$I_y = \frac{(7.5 \text{ ft})(10 \text{ ft})^3}{12} = 625 \text{ ft}^4$$

$$M_x = 0$$

$$q = \frac{75 \text{ kips}}{75 \text{ ft}^2} \pm \frac{(48 \text{ ft-kips})(5 \text{ ft})}{625 \text{ ft}^4} = 1.00 \text{ kip/ft}^2 \pm 0.38 \text{ kip/ft}^2$$

FIGURE 9–36

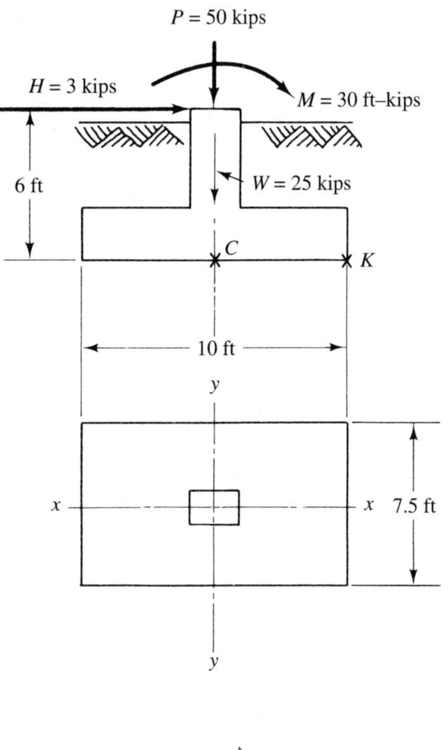

FIGURE 9–37

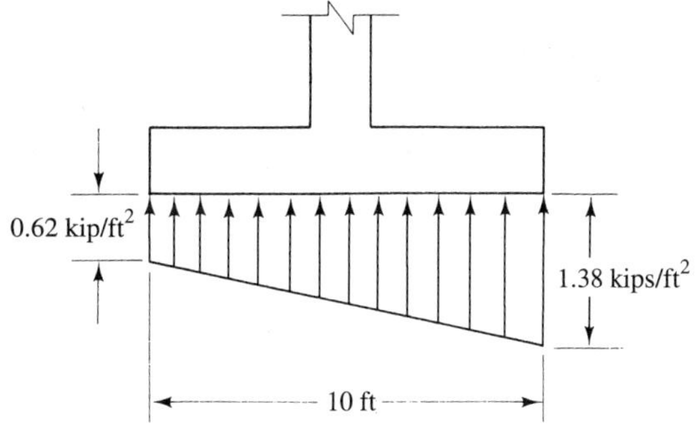

$$q_{\text{right}} = 1.38 \text{ kips/ft}^2 < 2 \text{ kips/ft}^2 \quad \therefore \text{ O.K.}$$

$$q_{\text{left}} = 0.62 \text{ kip/ft}^2 < 2 \text{ kips/ft}^2 \quad \therefore \text{ O.K.}$$

The pressure diagram is shown in Figure 9–37.

2. *Factor of safety against overturning:* By taking moments at point K (Figure 9–36), one finds that

$$\text{Factor of safety} = \frac{\text{Moment to resist turning}}{\text{Turning moment}}$$

$$= \frac{(50 \text{ kips} + 25 \text{ kips})(10 \text{ ft}/2)}{(3 \text{ kips})(6 \text{ ft}) + (30 \text{ ft-kips})} = 7.8$$

Under certain conditions, such as very large applied moments, Eq. (9–9) may give a negative value for the contact pressure. This implies tension between the footing and the soil. Soil cannot furnish any tensile resistance; hence, the flexural formula is not applicable in this situation. Instead, contact pressure may be calculated according to the basic equations of statics in the following manner.

Referring to Figure 9–38, by summing all forces in the vertical direction and all moments about point C and setting both sums equal to zero, one obtains the following two equations:

$$\Sigma V = 0 \uparrow +$$

$$\left(\frac{q}{2}\right)(d)(L) - P - W = 0 \qquad \textbf{(9–10)}$$

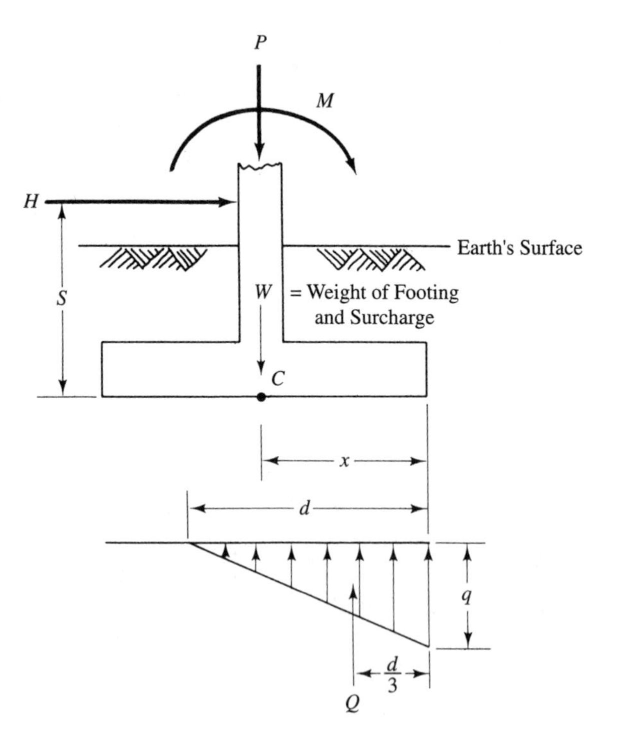

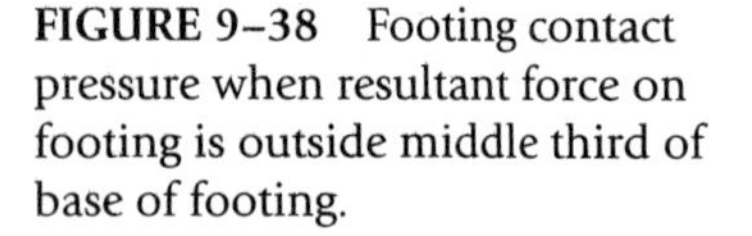
FIGURE 9–38 Footing contact pressure when resultant force on footing is outside middle third of base of footing.

$$\Sigma M_c = 0_{+}\curvearrowright$$

$$M + (H)(S) - \left(\frac{q}{2}\right)(d)(L)\left(x - \frac{d}{3}\right) = 0 \qquad (9\text{–}11)$$

Because all terms in Eqs. (9–10) and (9–11) are known except q and d, the two equations may be solved simultaneously to determine q and d. With q and d both known, the soil pressure diagram may be drawn. This technique is illustrated by Example 9–17.

EXAMPLE 9–17

Given

A rectangular footing 5 ft by 7.5 ft loaded as shown in Figure 9–39.

Required

Compute contact pressure and draw the soil pressure diagram.

Solution

By the flexural formula,

$$q = \frac{Q}{A} \pm \frac{M_x y}{I_x} \pm \frac{M_y x}{I_y} \qquad (9\text{–}9)$$

$$Q = 50 \text{ kips} + 20 \text{ kips} = 70 \text{ kips}$$

$$A = 5 \text{ ft} \times 7.5 \text{ ft} = 37.5 \text{ ft}^2$$

FIGURE 9–39

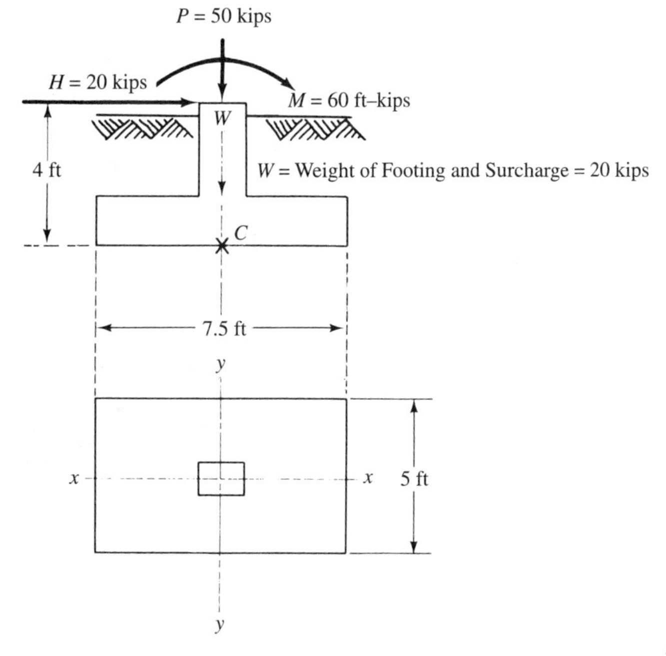

$$M_x = 0$$

$$M_y = (4 \text{ ft})(20 \text{ kips}) + 60 \text{ ft-kips} = 140 \text{ ft-kips}$$

(take moments at point C; see Figure 9–39)

$$x = \frac{7.5 \text{ ft}}{2} = 3.75 \text{ ft}$$

$$I_y = \frac{(5 \text{ ft})(7.5 \text{ ft})^3}{12} = 176 \text{ ft}^4$$

$$q = \frac{70 \text{ kips}}{37.5 \text{ ft}^2} \pm \frac{(140 \text{ ft-kips})(3.75 \text{ ft})}{176 \text{ ft}^4} = 1.87 \text{ kips/ft}^2 \pm 2.98 \text{ kips/ft}^2$$

$$q_{\text{right}} = +4.85 \text{ kips/ft}^2$$

$$q_{\text{left}} = -1.11 \text{ kips/ft}^2$$

Because q_{left} has a negative value, the flexural formula is not applicable in this case. Solve this problem by $\Sigma V = 0$ and $\Sigma M_c = 0$ [i.e., Eqs. (9–10) and (9–11)]. Referring to Figures 9–39 and 9–40, one finds that

$$\left(\frac{q}{2}\right)(d)(L) - P - W = 0 \qquad \textbf{(9–10)}$$

$$\left(\frac{qd}{2}\right)(5 \text{ ft}) = 70 \text{ kips} \qquad \textbf{(A)}$$

FIGURE 9–40

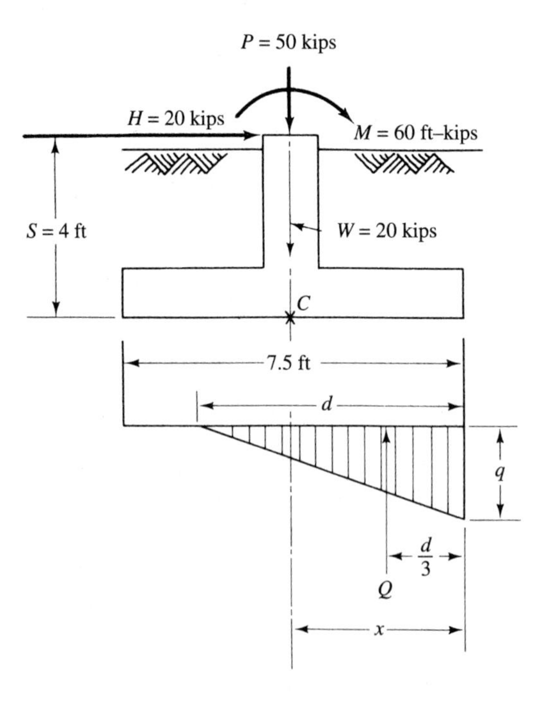

$$M + (H)(S) - \left(\frac{q}{2}\right)(d)(L)\left(x - \frac{d}{3}\right) = 0 \qquad \textbf{(9–11)}$$

$$60 \text{ ft-kips} + (20 \text{ kips})(4 \text{ ft}) - (70 \text{ kips})\left(\frac{7.5 \text{ ft}}{2} - \frac{d}{3}\right) = 0 \qquad \textbf{(B)}$$

[Note that $(qd/2)(L) = 70$ kips, from Eq. (A).] From Eq. (B),

$$60 \text{ ft-kips} + 80 \text{ ft-kips} - 262.5 \text{ ft-kips} + \frac{70 \text{ kips}}{3}d = 0$$

$$d = 5.25 \text{ ft}$$

Substitute $d = 5.25$ ft into Eq. (A):

$$\left(\frac{q}{2}\right)(5.25 \text{ ft})(5 \text{ ft}) = 70 \text{ kips}$$

$$q = 5.33 \text{ kips/ft}^2$$

The pressure diagram is shown in Figure 9–41.

9–7 TOTAL AND DIFFERENTIAL SETTLEMENT

Previous material in this chapter dealt primarily with bearing capacity analysis and prevention of bearing capacity failure of footings. Footings may also fail as a result of excessive settlement; thus, after the size of the footing has been determined by bearing capacity analysis, footing settlement should be calculated and the design revised if the calculated settlement is considered to be excessive.

Calculation of settlement has already been covered (Chapter 7). Maximum permissible settlement depends primarily on the nature of the superstructure. Some suggested maximum permissible settlement values are given in Table 9–1.

FIGURE 9–41

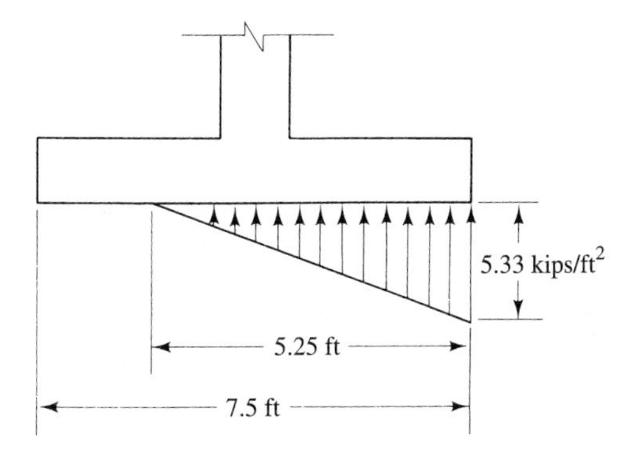

TABLE 9–1
Maximum Permissible Settlement [13]

Limiting Factor or Type of Structure	Maximum Permissible Settlement	
	Differential[1]	Total (in.)
Drainage of floors	0.01–0.02L	6–12
Stacking, warehouse lift trucks	0.01L	6
Tilting of smokestacks, silos	0.004B	3–12
Framed structure, simple	0.005L	2–4
Framed structure, continuous	0.002L	1–2
Framed structure with diagonals	0.0015L	1–2
Reinforced concrete structure	0.002–0.004L	1–3
Brick walls, one-story	0.001–0.002L	1–2
Brick walls, high	0.0005–0.001L	1
Cracking of panel walls	0.003L	1–2
Cracking of plaster	0.001L	1
Machine operation, noncritical	0.003L	1–2
Crane rails	0.003L	
Machines, critical	0.002L	

[1] L is the distance between adjacent columns; B is the width of the base.

9–8 STRUCTURAL DESIGN OF FOOTINGS

As was noted in Section 9–5, the required base area of a footing may be determined by dividing the column load by the allowable bearing capacity. Determining the thickness and shape of the footing and amount and location of reinforcing steel and performing other details of the actual structural design of footings are, however, ultimately the responsibility of a structural engineer.

In general, a soils engineer furnishes the contact pressure diagram and the shear and moment at a section (in the footing) at the face of the column, pedestal, or wall. This was demonstrated in Example 9–14 when the contact pressure diagram and the shear and moment at section A–A were determined. From this information, the structural engineer can do the actual structural design of the footing.

9–9 PROBLEMS

9–1. A strip of wall footing 3 ft wide is located 3.5 ft below the ground surface. Supporting soil has a unit weight of 125 lb/ft^3. The results of laboratory tests on the soil samples indicate that the supporting soil's cohesion and angle of internal friction are 1200 lb/ft^2 and 25°, respectively. Groundwater was not encountered during subsurface soil exploration. Determine the allowable bearing capacity, using a factor of safety of 3.

9–2. A square footing with a size of 10 ft by 10 ft is located 8 ft below the ground surface. The subsoil consists of a thick deposit of stiff cohesive soil with an unconfined compressive strength equal to 3600 lb/ft^2. The soil's unit weight is 128 lb/ft^3. Compute the ultimate bearing capacity.

9–3. A circular footing with a 1.22-m diameter is to be constructed 1.07 m below the ground surface. The subsoil consists of a uniform deposit of dense soil having a unit weight of 21.33 kN/m^3, an angle of internal friction of 20°, and a cohesion of 57.6 kN/m^2. The groundwater table is at a great depth, and its effect can be ignored. Determine the safe total load (including column load and weight of footing and soil surcharge), using a factor of safety of 3.

9–4. A footing 8 ft by 8 ft is buried 6 ft below the ground surface in a dense cohesionless soil. The results of laboratory and field tests on the supporting soil indicate that the soil's unit weight is 130 lb/ft^3, and the average corrected SPT N-value beneath the footing is 37. Compute the allowable (design) load that can be imposed onto this footing, using a factor of safety of 3.

9–5. A square footing with a size of 8 ft by 8 ft is to carry a total load of 40 kips. The depth of the footing is 5 ft below the ground surface, and groundwater is located at the ground surface. The subsoil consists of a uniform deposit of soft clay, the cohesion of which is 500 lb/ft^2. The soil's unit weight is 110 lb/ft^3. Compute the factor of safety against bearing capacity failure.

9–6. A square footing 0.3 m by 0.3 m is placed on the surface of a dense cohesionless sand (unit weight = 18.2 kN/m^3) and subjected to a load test. If the footing fails at a load of 13.8 kN, what is the value of ϕ for the sand?

9–7. A load test is performed on a 0.3-m by 0.3-m square footing on a dense cohesionless sand (unit weight = 18.0 kN/m^3). The footing's base is located 0.6 m below the ground surface. If the footing fails at a load of 82 kN, what is the failure load per unit area of the base of a square footing 2.0 m by 2.0 m loaded with its base at the same depth in the same materials?

9–8. A square footing 2 m by 2 m is to be constructed 1.22 m below the ground surface, as shown in Figure 9–42. The groundwater table is located 1.82 m below the ground surface. The subsoil consists of a uniform, medium dense, cohesionless soil with the following properties:

FIGURE 9–42

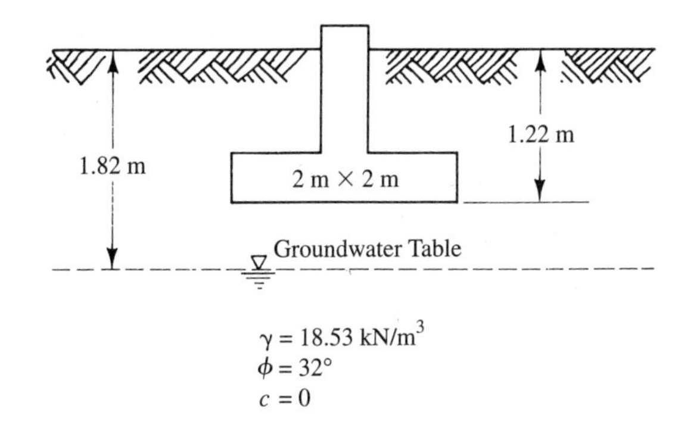

$$\text{Unit weight of soil} = 18.53 \text{ kN/m}^3$$
$$\text{Angle of internal friction} = 32°$$
$$\text{Cohesion} = 0$$

Determine the foundation soil's allowable bearing capacity if a factor of safety of 3 is used.

9–9. A square footing is to be constructed on a uniform thick deposit of clay with an unconfined compressive strength of 3 kips/ft^2. The footing will be located 5 ft below the ground surface and is designed to carry a total load of 300 kips. The unit weight of the supporting soil is 128 lb/ft^3. No groundwater was encountered during soil exploration. Considering general shear, determine the square footing dimension, using a factor of safety of 3.

9–10. A proposed square footing carrying a total load of 500 kips is to be constructed on a uniform thick deposit of dense cohesionless soil. The soil's unit weight is 135 lb/ft^3, and its angle of internal friction is 38°. The depth of the footing is to be 5 ft. Determine the dimension of this proposed footing, using a factor of safety of 3.

9–11. A bearing wall for a building is to be located close to a slope as shown in Figure 9–43. The groundwater table is at a great depth. Determine the foundation soil's allowable bearing capacity for the wall if a factor of safety of 3 is used.

9–12. Solve Problem 9–11 if the proposed footing is to be a 1.22-m by 1.22-m square footing (instead of a wall).

9–13. A wall footing is to be constructed on a uniform deposit of stiff clay, as shown in Figure 9–44. The footing is to support a wall that imposes 130 kN/m of wall length. Determine the required width of the footing if a factor of safety of 3 is used.

FIGURE 9–43

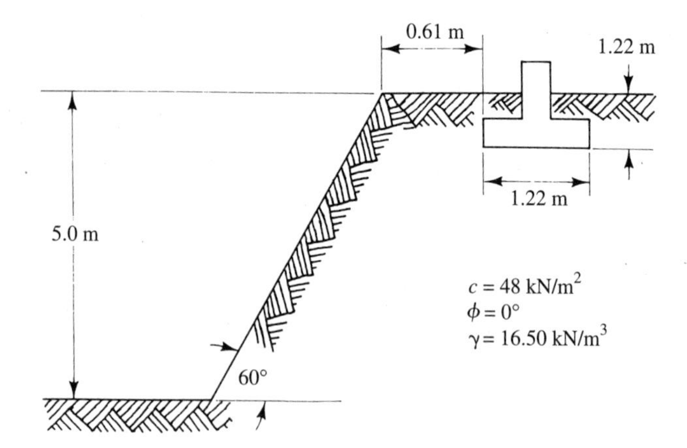

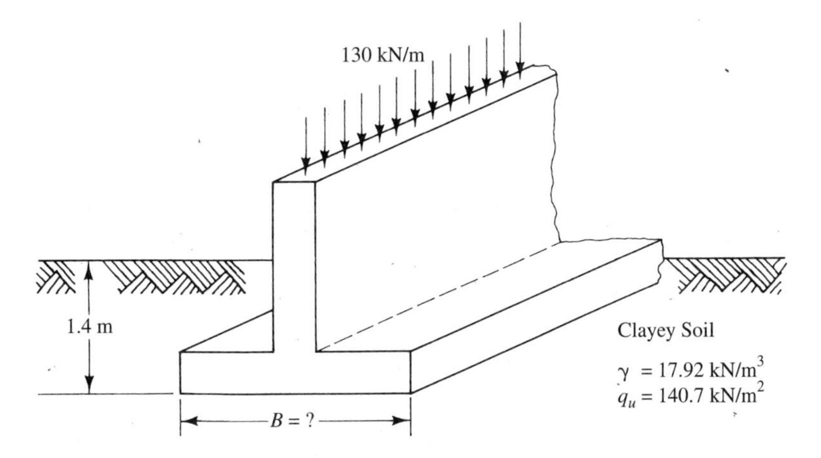

FIGURE 9–44 Clayey Soil

FIGURE 9–45

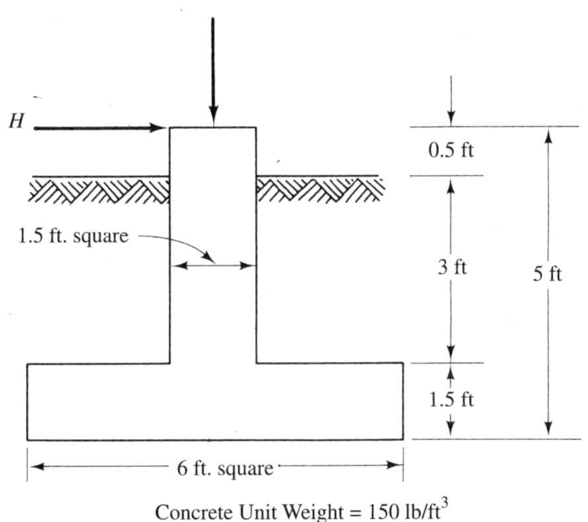

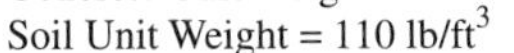

9–14. Compute and draw soil pressure diagrams for the footing shown in Figure 9–45 for the following loads:

1. $P = 70$ kips and $H = 20$ kips
2. $P = 70$ kips and $H = 10$ kips

9–15. Considering general shear, compute the safety factor against a bearing capacity failure for each of the two loadings in Problem 9–14 if the bearing soil is as follows:

FIGURE 9–46

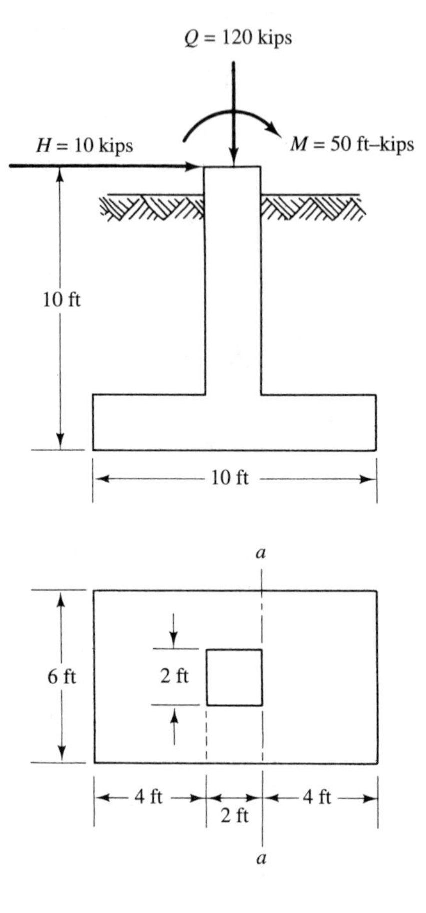

1. Cohesionless
   $\phi = 30°$
   $\gamma = 110\ \text{lb/ft}^3$
   $c = 0$

2. Cohesive
   $\phi = 0°$
   $\gamma = 110\ \text{lb/ft}^3$
   $c = 3000\ \text{lb/ft}^2$

In each case, groundwater is 10 ft below the base of the footing.

9–16. Same as Problem 9–15, except that groundwater is located at the ground surface.

9–17. For the footing shown in Figure 9–46, the vertical load, including the column load, surcharge weight, and weight of the footing, is 120 kips. The horizontal load is 10 kips, and a moment of 50 ft-kips (clockwise) is also imposed on the foundation.

1. Compute the soil contact pressure and draw the soil contact pressure diagram.
2. Compute the shear on section *a–a* (Figure 9–46).
3. Compute the moment on section *a–a* (Figure 9–46).
4. Compute the factor of safety against overturning.

5. Compute the factor of safety against sliding, if the coefficient of friction between the soil and the base of the footing is 0.60.
6. Compute the factor of safety against bearing capacity failure if the ultimate bearing capacity of the soil supporting the footing is 5.4 tons/ft^2.

9–18. A 6-ft by 6-ft square footing is buried 5 ft below the ground surface. The footing is subjected to an eccentric load of 200 kips. The eccentricity of the 200-kip load (e_x) is 0.8 ft. The supporting soil has values of $\phi = 38°$, $c = 0$, and $\gamma = 135$ lb/ft^3. Calculate the factor of safety against bearing capacity failure using a reduction factor from Figure 9–22.

References

[1] Wayne C. Teng, *Foundation Design,* Prentice-Hall, Inc., Englewood Cliffs, N.J., 1962.

[2] G. A. Leonards, Ed., *Foundation Engineering,* McGraw-Hill Book Company, New York, 1962.

[3] Karl Terzaghi and Ralph B. Peck, *Soil Mechanics in Engineering Practice,* John Wiley & Sons, Inc., New York, 1967. Copyright © 1967, by John Wiley & Sons, Inc. Reprinted by permission of John Wiley & Sons, Inc.

[4] Karl Terzaghi, Ralph B. Peck, and Gholamreza Mesri, *Soil Mechanics in Engineering Practice,* 3rd ed., John Wiley & Sons, Inc., New York, 1996.

[5] G. G. Meyerhof, "Influence of Roughness Base and Groundwater Conditions on the Ultimate Bearing Capacity of Foundations," *Geotechnique,* **5**, 227–242 (1955).

[6] H. Reissner, "Zum Erddruckproblem," *Proc. 1st Int. Conf. Appl. Mech.,* Delft, The Netherlands, 1924.

[7] Ralph B. Peck, Walter E. Hansen, and Thomas H. Thornburn, *Foundation Engineering,* 2nd ed., John Wiley & Sons, Inc., New York, 1974. Copyright © 1974, by John Wiley & Sons, Inc. Reprinted by permission of John Wiley & Sons, Inc.

[8] David F. McCarthy, *Essentials of Soil Mechanics and Foundations,* Reston Publishing Company, Inc., Reston, Va., 1977.

[9] *Manual of Recommended Practice,* Construction and Maintenance Section, Engineering Division, Association of American Railroads, Chicago, 1958.

[10] G. G. Meyerhof, "The Bearing Capacity of Foundations under Eccentric and Inclined Loads," *Proc. 3rd Int. Conf. Soil Mech. Found. Eng., Switzerland,* **1**, 440–445 (1953).

[11] G. G. Meyerhof, "The Ultimate Bearing Capacity of Foundations on Slopes," *Proc. 4th Int. Conf. Soil Mech. Found. Eng., London,* **1**, 385–386 (1957).

[12] *Design Manual: Soils Mechanics, Foundations, and Earth Structures,* NAVFAC DM-7, U.S. Department of the Navy, Naval Facilities Engineering Command, Alexandria, Va., 1971.

[13] Merlin G. Spangler and Richard L. Handy, *Soil Engineering,* 3rd ed., Intext Educational Publishers, New York, 1973. Copyright © 1951, 1960, 1973 by Harper & Row, Publishers, Inc. Reprinted by permission of the publisher.

10

PILE FOUNDATIONS

10–1 INTRODUCTION

Chapter 9 covered shallow foundations. Sometimes, however, the soil upon which a structure is to be built is of such poor quality that a shallow foundation would be subject to bearing capacity failure and/or excessive settlement. In such cases, *pile foundations* may be used to support the structure (i.e., to transmit the load of the structure to firmer soil, or rock, at a greater depth below the structure).

A pile foundation is a relatively long and slender member that is forced or driven into the soil, or it may be poured in place. If a pile is driven until it rests on a hard, impenetrable layer of soil or rock, the load of the structure is transmitted primarily axially through the pile to the impenetrable layer. This type of pile is called an *end-bearing pile.* With end-bearing piles, care must be exercised to ensure that the hard, impenetrable layer is adequate to support the load. If a pile cannot be driven to a hard stratum of soil or rock (e.g., if such a stratum is located too far below the ground surface), the load of the structure must be borne primarily by skin friction or adhesion between the surface of the pile and adjacent soil. Such a pile is known as a *friction pile.*

In addition to simply supporting the load of a structure, piles may perform other functions, such as densifying loose cohesionless soils, resisting horizontal loads, anchoring structures subject to uplift, and so on. The emphasis in this book, however, is on piles that support the load of a structure.

10–2 TYPES OF PILES

Piles may be classified according to the types of materials from which they are made. Virtually all piles are made of timber, concrete, or steel (or a combination of these). Each of these is discussed in general terms in this section.

Timber piles have been used for centuries and are still widely used. They are made relatively easily by delimbing tall, straight tree trunks. They generally make economical pile foundations. Timber piles have certain disadvantages, however. They have less capacity to carry a load than do concrete or steel piles. Also, the length of a timber pile is limited by the height of the tree available. Timber pile length is generally limited to around 60 ft (18 m), although longer timber piles are available in some locales. Timber piles may be damaged in the pile-driving process. In addition, they are subject to decay and attack by insects. This generally is not a problem if the pile is both in soil and always below the water table; if above the water table, timber piles can be treated chemically to increase their life.

Concrete piles can be either *precast* or *cast-in-place*. Precast concrete piles may be manufactured with circular, square, octagonal, or other cross-sectional shapes. They can be made of uniform cross section (with a pointed tip), or they may be tapered. Precast piles can be made of prestressed concrete. The main disadvantages of precast concrete piles have to do with problems of manufacturing and handling of the piles (space needed, time required for curing, heavy equipment necessary for handling and transporting, etc.).

Cast-in-place concrete piles may be *cased* or *uncased*. The cased type can be made by driving a shell containing a core into the soil, removing the core, and filling the shell with concrete. The uncased type can be made in a similar manner, except that the shell is withdrawn as concrete is poured. Cast-in-place concrete piles have several advantages over concrete piles that are precast. One is that, because the concrete is poured in place, damage due to pile driving is eliminated. Also, the length of the pile is known at the time the concrete is poured. (With a precast pile, the exact length of the pile to be cast must be known initially. If a given pile turns out to be too long or too short, extra cost is involved in cutting off the extra length of the pile or adding to it.)

Concrete piles generally have a somewhat larger capacity to carry load than do timber piles. They are usually not very susceptible to deterioration, except possibly by seawater and strong chemicals.

Steel piles are commonly either pipe-shaped or H-sections. Pipe-shaped steel piles may be filled with concrete after being driven. H-shaped steel piles are strong and capable of being driven to great depths through stiff layers. Steel piles are subject to damage by corrosion. They generally have a somewhat larger capacity to carry load than do timber piles or concrete piles.

Table 10–1 gives some customary design loads for different types of piles.

TABLE 10–1
Customary Design Loads for Piles [1]

Type of Pile	Allowable Load (tons)[1]
Wood	15–30
Composite	20–30
Cast-in-place concrete	30–50
Precast reinforced concrete	30–50
Steel pipe, concrete-filled	40–60
Steel H-section	30–60

[1] 1 ton = 8.896 kN.

TABLE 10–2
Available Lengths of Various Pile Types [2]

Pile Type	Comment, Available Maximum Length[1]
Timber	Depends on wood (tree) type. Lengths in the 50- to 60-ft range are usually available in most areas; lengths to about 75 ft are available but in limited quantity; lengths up to the 100-ft range are possible but very limited.
Steel H and pipe	Unlimited length; "short" sections are driven, and additional sections are field-welded to obtain a desired total length.
Steel shell, cast-in-place	Typically to between 100 and 125 ft, depending on shell type and manufacturer–contractor.
Precast concrete	Solid, small cross-section piles usually extend into the 50- to 60-ft length, depending on cross-sectional shape, dimensions, and manufacturer. Large-diameter cylinder piles can extend to about 200 ft long.
Drilled shaft, cast-in-place concrete	Usually in the 50- to 75-ft range, depending on contractor equipment.
Bulb-type, cast-in-place concrete	Up to about 100 ft.
Composite	Related to available lengths of material in the different sections. If steel and thin-shell cast-in-place concrete are used, the length can be unlimited; if timber and thin-shell cast-in-place concrete are used, lengths can be on the order of 150 ft.

[1] 1 ft = 0.3048 m.

10–3 LENGTH OF PILES

In the case of end-bearing piles, the required pile length can be found fairly accurately because it is the distance from the structure being supported by the pile to the hard, impenetrable layer of soil or rock on which the pile rests. This distance is established from soil boring tests.

With friction piles, the required pile length is determined indirectly. Friction piles must be driven to such a depth that adequate lateral surface area of the pile is in contact with soil in order that sufficient skin friction or adhesion can be developed.

Table 10–2 gives available lengths of various types of piles.

10–4 PILE CAPACITY

The capacity of a single pile may be evaluated by the structural strength of the pile and by the supporting strength of the soil.

Pile Capacity as Evaluated by the Structural Strength of the Pile

Obviously, a pile must be strong enough structurally to carry the load imposed upon it. A pile's structural strength depends on its size and shape, as well as the type of material from which it is made.

Allowable structural strengths of different types of piles are specified by a number of building codes. Table 10–3 shows allowable stress in various types of pile, according to one code.

Pile Capacity as Evaluated by the Supporting Strength of the Soil

In addition to the strength of the pile itself, pile capacity is limited by the soil's supporting strength. As mentioned previously, the load carried by a pile is ultimately

TABLE 10–3
Allowable Stress in Piles [3]

(a) Timber Piles

Species	Compression Parallel to Grain (lb/in.^{2})	Bending (lb/in.^{2})	Shear Horizontal (lb/in.^{2})	Compression Perpendicular to Grain (lb/in.^{2})	Modulus of Elasticity (lb/in.^{2})
Douglas fir (all varieties)	1150	2300	110	225	1,500,000
Southern yellow pine (market weighted averages)	1150	2300	110	225	1,400,000
Southern red oak	950	2400	110	325	1,100,000

(b) Steel Piles

The design load shall not cause a stress in the steel greater than 12,600 lb/in.^{2} and a stress in any concrete used to fill piles, driven either open or closed end, greater than 25% of its ultimate 28-day compressive strength.

(c) Concrete Piles

Cast-in-place piles: The stress in concrete shall not exceed 25% of the ultimate 28-day strength of the concrete.

Prestressed concrete piles: The maximum allowable compressible stress in precast piles due to an externally applied load shall not exceed.

$$f_c = 0.33\, f_c' - 0.27 f_{pe}$$

where f_c' is the 28-day compression strength of concrete, and f_{pe} is the effective prestress stress on the gross section.

borne by either or both of two ways. The load is transmitted to the soil surrounding the pile by friction or adhesion between the soil and the sides of the pile, and/or the load is transmitted directly to the soil just below the pile's tip. This can be expressed in equation form as follows:

$$Q_{\text{ultimate}} = Q_{\text{friction}} + Q_{\text{tip}} \tag{10–1}$$

where Q_{ultimate} = ultimate (at failure) bearing capacity of a single pile
Q_{friction} = bearing capacity furnished by friction or adhesion between the soil and the sides of the pile
Q_{tip} = bearing capacity furnished by the soil just below the pile's tip

The term Q_{friction} in Eq. (10–1) can be evaluated by multiplying the unit skin friction or adhesion between the soil and the sides of the pile (f) by the pile's surface (skin) area (A_{surface}). The term Q_{tip} can be evaluated by multiplying the ultimate bearing capacity of the soil at the tip of the pile (q) by the area of the tip (A_{tip}). Hence, Eq. (10–1) can be expressed as follows:

$$Q_{\text{ultimate}} = f \cdot A_{\text{surface}} + q \cdot A_{\text{tip}} \tag{10–2}$$

In the case of end-bearing piles, the term Q_{tip} of Eq. (10–1) or $q \cdot A_{\text{tip}}$ of Eq. (10–2) will be predominant, whereas with friction piles, the term Q_{friction} of Eq. (10–1) or $f \cdot A_{\text{surface}}$ of Eq. (10–2) will be predominant.

Equations (10–1) and (10–2) are generalized and therefore applicable for all soils. The manner in which some of the terms of Eq. (10–2) are evaluated differs, however, depending on whether the pile is driven in sand or clay. It is convenient, therefore, to consider separately piles driven in sand and those driven in clay.

Piles Driven in Sand. In the case of piles driven in sand, skin friction between the soil and the sides of the pile [$f \cdot A_{\text{surface}}$ in Eq. (10–2)] can be evaluated by multiplying the coefficient of friction between sand and pile surface ($\tan \delta$) by the total horizontal soil pressure acting on the pile. The coefficient of friction between sand and pile surface can be obtained from Table 10–4. The total horizontal soil pressure acting on the pile is a function of effective vertical (overburden) pressure of soil adjacent to the pile. Soil pressure normally increases as depth increases. In the special case of piles driven in sand, however, it has been determined that the effective vertical (overburden) pressure of soil adjacent to a pile does not increase without limit as depth increases. Instead, effective vertical pressure increases as depth increases until a certain depth of penetration is reached. Below this depth, which is called the *critical depth* and denoted D_c, effective vertical pressure remains more or less constant. The critical depth is dependent on the field condition of the sand and the pile's size. Tests indicate that critical depth ranges from about 10 pile diameters for loose sand to about 20 pile diameters for dense compact sand [2]. Thus, effective vertical pressure of soil adjacent to a pile varies with depth as illustrated in Figure 10–1.

The term $f \cdot A_{\text{surface}}$ of Eq. (10–2) can now be determined for a pile by multiplying the pile's circumference by the area under the p_v versus depth curve (Figure 10–1)

TABLE 10–4
Coefficient of Friction between Sand and Pile Materials [2]

Material	Tan δ
Concrete	0.45
Wood	0.4
Steel (smooth)	0.2
Steel (rough, rusted)	0.4
Steel (corrugated)	Use tan φ of sand

by the coefficient of lateral earth pressure (K) by the coefficient of friction between sand and pile surface (tan δ). The coefficient of lateral earth pressure is assumed to vary between 0.60 and 1.25, with lower values used for silty sands and higher values for other deposits [7].

The bearing capacity at the pile tip [q in Eq. (10–2)] can be calculated by using bearing capacity equations for cohesionless soil, which were developed by Terzaghi and Peck [1]:

$$q_{\text{tip}} = \gamma D_f N_q + 0.6\gamma R N_\gamma \qquad \text{(for circular piles)} \tag{10–3}$$

$$q_{\text{tip}} = \gamma D_f N_q + 0.4\gamma B N_\gamma \qquad \text{(for square piles)} \tag{10–4}$$

where q_{tip} = bearing capacity at pile tip
γ = unit weight of soil
D_f = embedded length of pile
N_γ, N_q = bearing capacity factors (see Figure 9–7)
R = radius of pile tip (for circular piles)
B = width of pile tip (for square piles)

It can be noted that these equations have the same general form as the bearing capacity equations given in Chapter 9 for shallow foundations. However, as indicated previously, the magnitude of effective vertical (overburden) pressure of soil adjacent to a pile is more or less constant below the critical depth. Thus, for design purposes, the term $\gamma D_f N_q$ in Eqs. (10–3) and (10–4) should be replaced by the term $p_v N_q$, where p_v is the effective vertical pressure adjacent to the pile at the pile tip (Figure 10–1) [2].

In most cases, driven piles are relatively small in cross section; therefore, the terms in Eqs. (10–3) and (10–4) involving R and B are small compared with the other term in the equations. Thus, for many cases, Eqs. (10–3) and (10–4) may be approximated as follows:

$$q_{\text{tip}} = p_v N_q \tag{10–5}$$

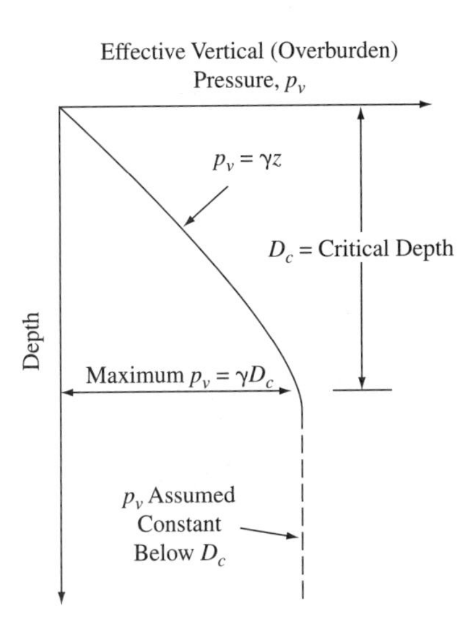

FIGURE 10–1 Variation of effective vertical (overburden) pressure of soil adjacent to a pile with depth [2, 4–6].

As piles are driven through soil, they wedge their way downward, displacing adjacent soil. The soil just below pile tips is displaced outward and upward, causing shearing stresses to be induced in the soil above the tip. These stresses alter shear patterns below the tip as compared with those for shallow footings. Hence, the value of N_q (from Figure 9–7, for shallow foundations) should be increased to N_q^*, the bearing capacity factor for piles driven in sand. Therefore, Eq. (10–5) becomes

$$q_{\text{tip}} = p_v N_q^* \tag{10–6}$$

The value of N_q^* is related to the angle of internal friction (ϕ) of the sand, and it should, of course, be based on the value of the angle of internal friction of the sand located in the general vicinity of where the pile tip will ultimately rest. The angle of internal friction of the sand at this location can be determined by laboratory tests on a sample taken from the specified location or by correlation with penetration resistance tests in a boring hole [i.e., corrected standard penetration test (SPT) N-value] (see Figures 3–9 and 9–9). Values of N_q^* can then be obtained from Fig. 10–2. [N_q^* should also be used in lieu of N_q if Eq. (10–3) or (10–4) is used.]

To summarize the method described in this section for computing pile capacity for piles driven in sand, Eq. (10–2) is used, with the term $f \cdot A_{\text{surface}}$ evaluated by multiplying the pile's circumference by the area under the p_v versus depth curve (Figure 10–1) by the coefficient of lateral earth pressure (K) by the coefficient of friction between sand and pile surface ($\tan \delta$), and the term $q \cdot A_{\text{tip}}$ evaluated by multiplying the value of q_{tip} obtained from Eq. (10–6) by the area of the pile tip. Pile capacity thus determined represents the ultimate load that can be applied to the pile. In practice, it is common to apply a factor of safety of 2 to determine the (downward) design load for the pile [2].

Examples 10–1 and 10–2 illustrate the procedure for calculating pile capacity for piles driven in sand.

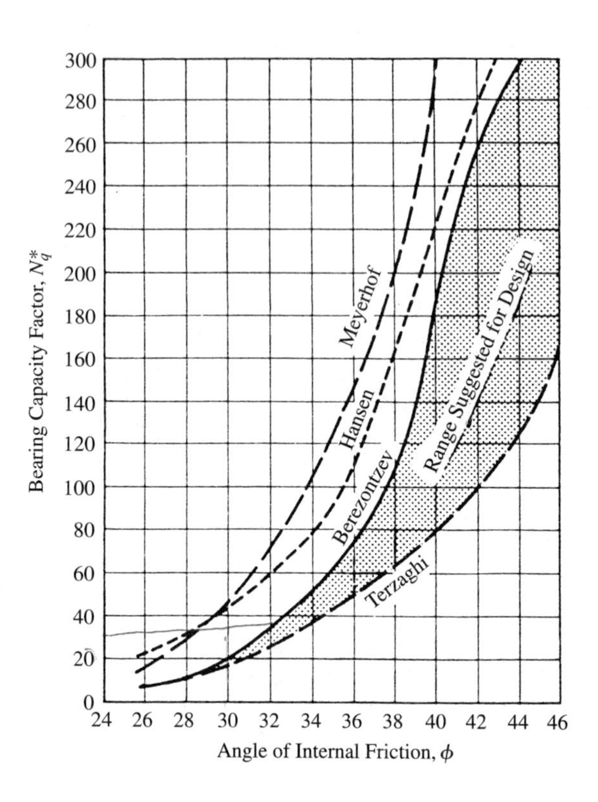

FIGURE 10–2 Bearing capacity factor, N_q^*, for piles penetrating into sand [8].

EXAMPLE 10–1

Given

1. A concrete pile is to be driven into a medium dense to dense sand.
2. The pile's diameter is 12 in. and its embedded length is 25 ft.
3. Soil conditions are shown in Figure 10–3.
4. No groundwater was encountered, and the groundwater table is not expected to rise during the life of the structure.

Required

The pile's axial capacity if the coefficient of lateral earth pressure (K) is assumed to be 0.95, and the factor of safety (F.S.) is 2.

Solution

For dense sand,

$$D_c = 20 \times \text{Pile's diameter} = 20 \times 1 \text{ ft} = 20 \text{ ft} \qquad \text{(see Figure 10–4)}$$

From Eq. (10–2),

$$Q_{\text{ultimate}} = f \cdot A_{\text{surface}} + q \cdot A_{\text{tip}} \tag{10–2}$$

$$f \cdot A_{\text{surface}} = (\text{Circumference of pile})(\text{Area of } p_v \text{ diagram})(K)(\tan \delta)$$

FIGURE 10–3

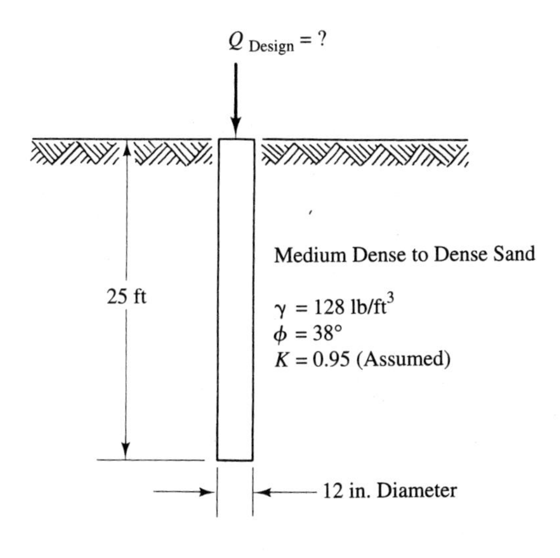

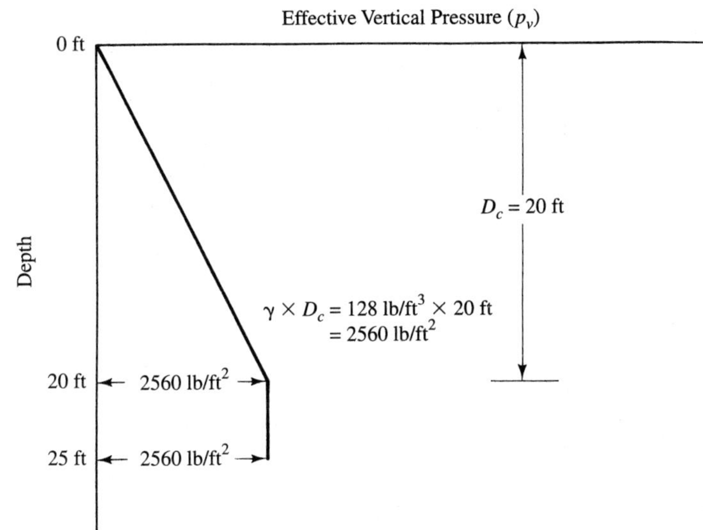

FIGURE 10–4

$$\text{Circumference of pile} = \pi d = (\pi)(1\text{ ft}) = 3.14\text{ ft}$$

$$\text{Area of } p_v \text{ diagram} = (\tfrac{1}{2})(2560\text{ lb/ft}^2)(20\text{ ft}) + (2560\text{ lb/ft}^2)(25\text{ ft} - 20\text{ ft})$$

$$= 38{,}400\text{ lb/ft}$$

$$K = 0.95 \qquad \text{(given)}$$

$$\tan\delta = 0.45 \qquad \text{(see Table 10–4 for concrete pile)}$$

$$f \cdot A_{\text{surface}} = (3.14\text{ ft})(38{,}400\text{ lb/ft})(0.95)(0.45)$$

$$= 51{,}500\text{ lb} = 51.5\text{ kips}$$

From Eq. (10–6),

$$q_{\text{tip}} = p_v N_q^* \tag{10–6}$$

$$p_v = 2560 \text{ lb/ft}^2 \qquad \text{(see Figure 10–4)}$$

$$N_q^* = 80 \qquad \text{(from Figure 10–2 for } \phi = 38°\text{, using the mid-area of the "Range suggested for design")}$$

$$q_{\text{tip}} = (2560 \text{ lb/ft}^2)(80) = 204{,}800 \text{ lb/ft}^2$$

$$A_{\text{tip}} = \frac{\pi d^2}{4} = \left(\frac{\pi}{4}\right)(1 \text{ ft})^2 = 0.785 \text{ ft}^2$$

$$q \cdot A_{\text{tip}} = (204{,}800 \text{ lb/ft}^2)(0.785 \text{ ft}^2) = 160{,}800 \text{ lb} = 160.8 \text{ kips}$$

$$Q_{\text{ultimate}} = 51.5 \text{ kips} + 160.8 \text{ kips} = 212.3 \text{ kips}$$

$$Q_{\text{design}} = \frac{Q_{\text{ultimate}}}{\text{F. S.}} = \frac{212.3 \text{ kips}}{2} = 106.2 \text{ kips}$$

EXAMPLE 10–2

Given

The same conditions as in Example 10–1, except that groundwater is located 10 ft below the ground surface (see Figure 10–5).

Required

The pile's axial capacity if K is 0.95, and a factor of safety of 2 is used.

Solution

$$D_c = 20 \times 1 \text{ ft} = 20 \text{ ft} \qquad \text{(see Figure 10-6)}$$

$$f \cdot A_{\text{surface}} = (\text{Circumference of pile})(\text{Area of } p_v \text{ diagram})(K)(\tan \delta)$$

$$\text{Circumference of pile} = \pi d = (\pi)(1 \text{ ft}) = 3.14 \text{ ft}$$

$$\text{Area of } p_v \text{ diagram} = (\tfrac{1}{2})(1280 \text{ lb/ft}^2)(10 \text{ ft}) + (\tfrac{1}{2})(1280 \text{ lb/ft}^2 + 1936 \text{ lb/ft}^2)(10 \text{ ft}) + (1936 \text{ lb/ft}^2)(5 \text{ ft})$$

$$= 32{,}200 \text{ lb/ft}$$

$$K = 0.95$$

$$\tan \delta = 0.45$$

$$f \cdot A_{\text{surface}} = (3.14 \text{ ft})(32{,}200 \text{ lb/ft})(0.95)(0.45) = 43{,}200 \text{ lb} = 43.2 \text{ kips}$$

$$q_{\text{tip}} = p_v N_q^* \tag{10–6}$$

$$N_q^* = 80$$

$$q_{\text{tip}} = (1936 \text{ lb/ft}^2)(80) = 154{,}900 \text{ lb/ft}^2$$

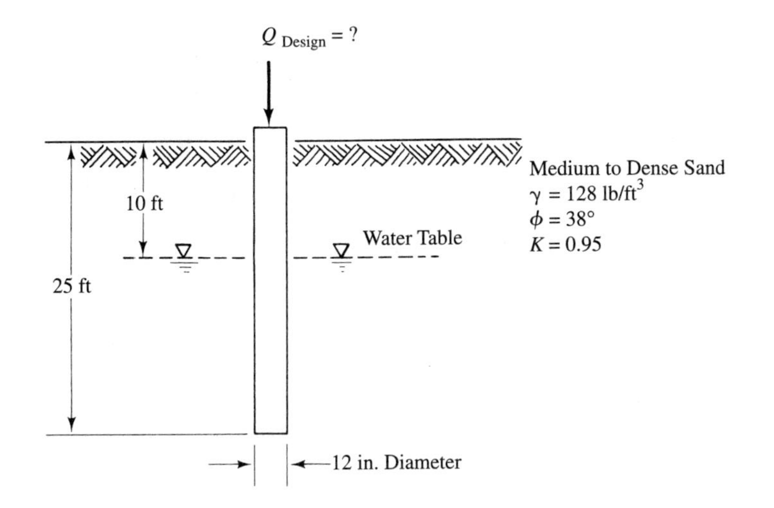

FIGURE 10–5

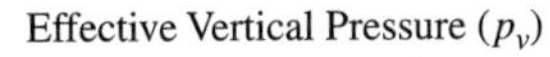

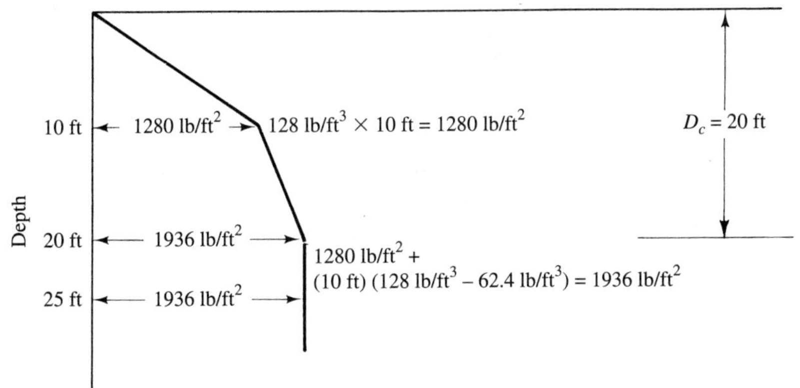

FIGURE 10–6

$$A_{\text{tip}} = 0.785 \text{ ft}^2$$

$$q \cdot A_{\text{tip}} = (154{,}900 \text{ lb/ft}^2)(0.785 \text{ ft}^2) = 121{,}600 \text{ lb} = 121.6 \text{ kips}$$

$$Q_{\text{ultimate}} = 43.2 \text{ kips} + 121.6 \text{ kips} = 164.8 \text{ kips}$$

$$Q_{\text{design}} = \frac{164.8 \text{ kips}}{2} = 82.4 \text{ kips}$$

Dennis and Olson [9, 10] studied the results of a number of load tests on piles carried to ultimate failure in sand. In their statistical analyses, they subdivided the data on the basis of the description of the sands and types of piles; established sets of values for the coefficient of lateral pressure (K), friction angle between sand and pile surface (δ), and bearing capacity factor (N_q); and set upper limiting values for skin friction and end-point resistance. They then substituted various values into the load equation [Eq. (10–2)] until they found the combinations that gave the best answers corresponding to the respective results of actual load tests. In other words, they found the combinations that gave mean ratios of computed bearing capacities to measured load capacities nearest to 1.0 for all tests and produced the least scatter.*

Taking K equal to 0.8 and assigning reasonable values to δ and N_q in accordance with the standard penetration resistance values, Olson [11] developed a table of soil properties for use in Eq. (10–2) that gives mean ratios of computed bearing capacities to measured load capacities nearest to 1.0 for all tests and produces the least scatter (see Table 10–5 and Figure 10–7). Such a semiempirical approach, with load-test data correlating to a static equation such as Eq. (10–2), seems to be a logical way to improve practice. However, a large number of good-quality load-test data will be needed for further statistical studies [9].*

If detailed information for computing pile bearing capacity is unavailable, rough estimates of unit skin friction can be obtained from Figure 10–8 as a function of depth in pile diameters. This is an empirical relationship based on the fact that the skin friction developed between the sand and the pile is strongly influenced by the condition of the sand around the pile.

In a similar vein, Figure 10–9 gives unit tip resistance as a function of depth in pile diameters. This is also an empirical relationship based on a number of pile load tests.

Piles Driven in Clay. For piles driven in clay, unit adhesion between the soil and the sides of the pile [f in Eq. (10–2)] can be evaluated by multiplying the cohesion of the clay (c) by the adhesion factor (α). The adhesion factor can be determined by using Figure 10–10. The term $f \cdot A_{\text{surface}}$ of Eq. (10–2) can thus be evaluated by multiplying the (undisturbed) cohesion of the clay (c) by the adhesion factor (α) by the surface (skin) area of the pile (A_{surface}).

With soft clays, there is a tendency for the clay to come in close contact with the pile, in which case adhesion is assumed to be equal to cohesion (meaning $\alpha = 1.0$). In the case of stiff clays, pile driving disturbs surrounding soil and may cause a small open space to develop between the clay and the pile. Thus, adhesion is smaller than cohesion (meaning $\alpha < 1.0$).

The bearing capacity [q in Eq. (10–2)] at the pile tip can be calculated by using the following equation [2]:

$$q_{\text{tip}} = cN_c \tag{10–7}$$

*From Karl Terzaghi, Ralph B. Peck, and Gholamreza Mesri, *Soil Mechanics in Engineering Practice,* 3rd ed., John Wiley & Sons, Inc., New York, 1996. Copyright © 1996, by John Wiley & Sons, Inc. Reprinted by permission of John Wiley & Sons, Inc.

TABLE 10–5
Soil Properties Used in Olson's Final Analyses [9, 11][1,2]

Soil Type	Range in N-Values	δ (deg)	f_{lim} (kN/m^2)	N_q	q_{lim} (MN/m^2)
Gravel	0–4	(20)	(70)	(12)	(3)
	5–10	(25)	(85)	(20)	(5)
	11–30	(30)	(100)	(40)	(10)
	over 30	(35)	(120)	(60)	(12.5)
Sand/gravel	0–4	(20)	(70)	(12)	(3)
	5–10	(25)	(85)	(20)	(5)
	11–30	(30)	(100)	(40)	(10)
	over 30	(35)	(120)	(60)	(12.5)
Sand	0–4	(20)	(50)	(50)	(2)
	5–10	30	55	120	6
	11–30	35	95	120	95
	31–50	40	130	120	9.5
	51–100	40	165	130	10
	over 100	40	190	220	26.5
Sand/silt	0–4	10	(50)	(10)	(0.5)
	5–10	10	(50)	(20)	(2)
	11–30	15	(70)	50	5.5
	31–50	20	100	100	8
	51–100	(30)	(100)	(100)	(10)
	over 100	(34)	(1000)	(100)	(10)
Silt	0–4	(10)	(50)	(10)	(2)
	5–10	15	(50)	(10)	(2)
	11–30	20	(70)	(10)	(2)
	31–50	20	(70)	(12)	(3)
	over 50	(25)	(70)	(12)	(3)

[1] Numbers in parentheses were not used in the analyses.
[2] From Karl Terzaghi, Ralph B. Peck, and Gholamreza Mesri, *Soil Mechanics in Engineering Practice,* 3rd ed., John Wiley & Sons, Inc., New York, 1996. Copyright © 1996, by John Wiley & Sons, Inc. Reprinted by permission of John Wiley & Sons, Inc.

where q_{tip} = bearing capacity at pile tip
c = cohesion of the clay located in the general vicinity of where the pile tip will ultimately rest
N_c = bearing capacity factor and has a value of about 9 [2]

Thus, the term $q \cdot A_{\text{tip}}$ of Eq. (10–2) can be evaluated by multiplying the value of q_{tip} from Eq. (10–7) by the area of the pile tip.

To summarize the method described in this section for computing pile capacity for piles driven in clay, Eq. (10–2) is used, with the term $f \cdot A_{\text{surface}}$ evaluated by multiplying the cohesion of the clay (c) by the adhesion factor (α) by the surface (skin) area of the pile, and the term $q \cdot A_{\text{tip}}$ evaluated by multiplying the value of q_{tip}

FIGURE 10–7 Comparison of measured and calculated axial load capacities of driven steel piles [after Olson (1990)] [9, 11]. *Source:* From Karl Terzaghi, Ralph B. Peck, and Gholamreza Mesri, *Soil Mechanics in Engineering Practice*, 3rd ed., John Wiley & Sons, Inc., New York, 1996. Copyright © 1996, by John Wiley & Sons, Inc. Reprinted by permission of John Wiley & Sons, Inc.

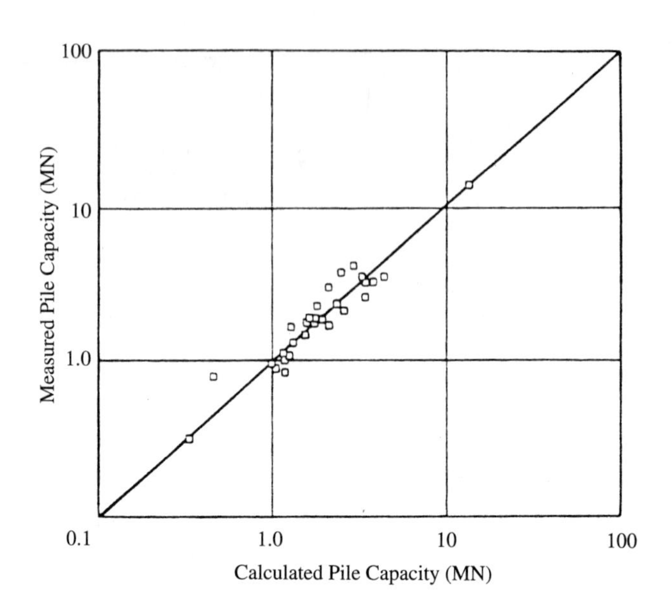

FIGURE 10–8 Ultimate unit side resistance, f_s, versus D/B—compression/tension (1 tsf = 95.76 kN/m^2) [12].

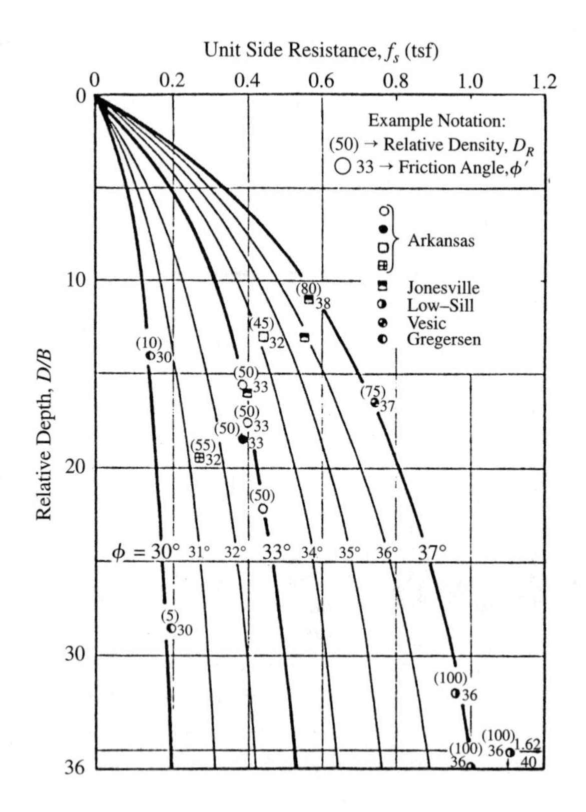

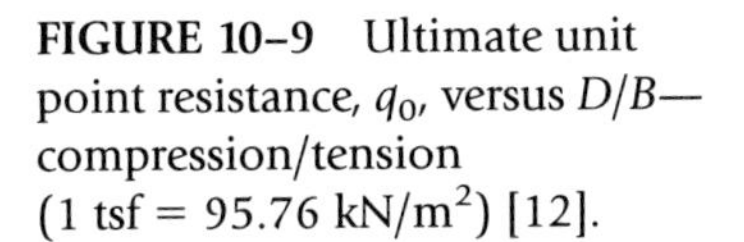

FIGURE 10–9 Ultimate unit point resistance, q_0, versus D/B—compression/tension ($1 \text{ tsf} = 95.76 \text{ kN/m}^2$) [12].

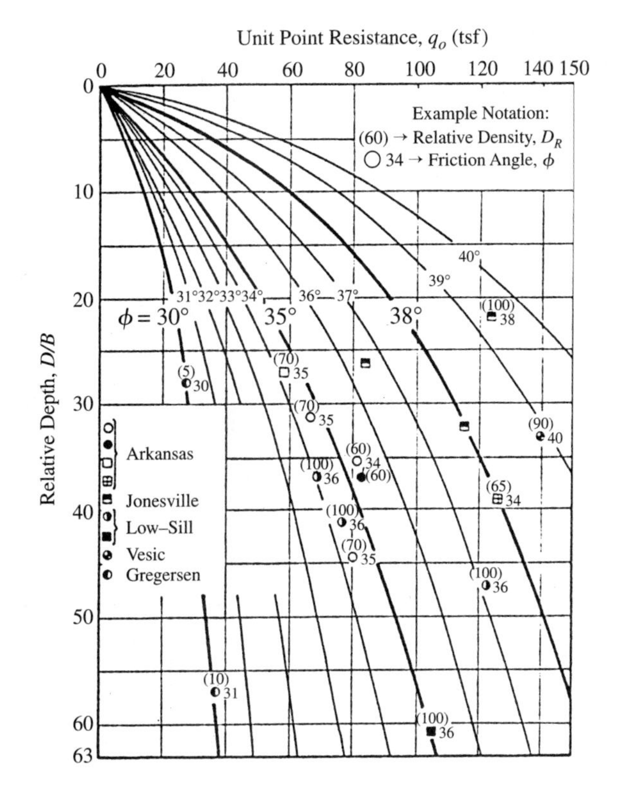

obtained from Eq. (10–7) by the area of the pile tip. Pile capacity thus determined represents the ultimate load that can be applied to the pile. In practice, it is common to apply a factor of safety of 2 to determine the (downward) design load for a pile [2].

Examples 10–3 through 10–5 illustrate the procedure for calculating pile capacity for piles driven in clay.

EXAMPLE 10–3

Given

1. A 12-in.-diameter concrete pile is driven at a site as shown in Figure 10-11.
2. The embedded length of the pile is 35 ft.

Required

Design capacity of the pile, using a factor of safety of 2.

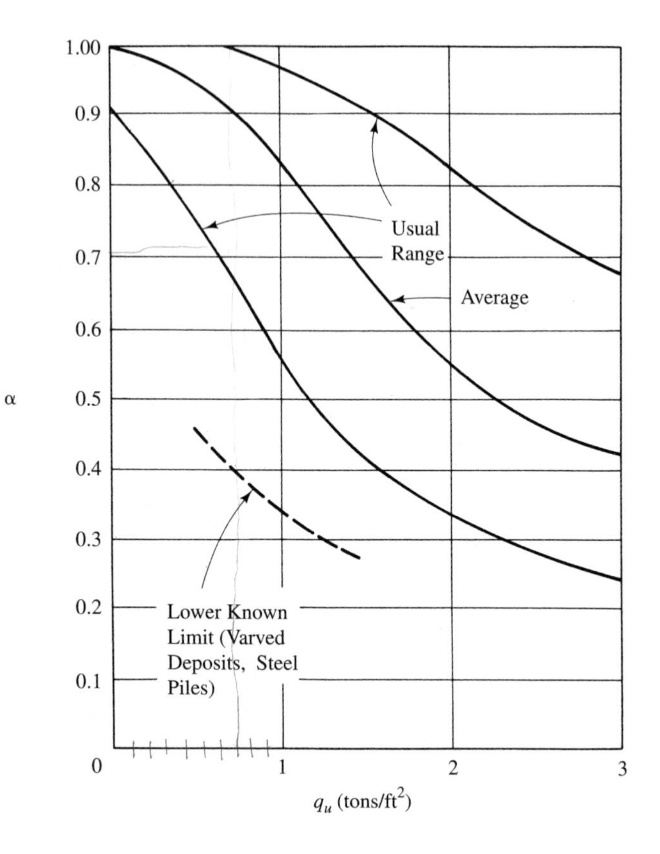

FIGURE 10–10 Relationship between adhesion factor, α, and unconfined compressive strength, q_u (1 ton/ft^2 = 95.76 kN/m^2) [13].

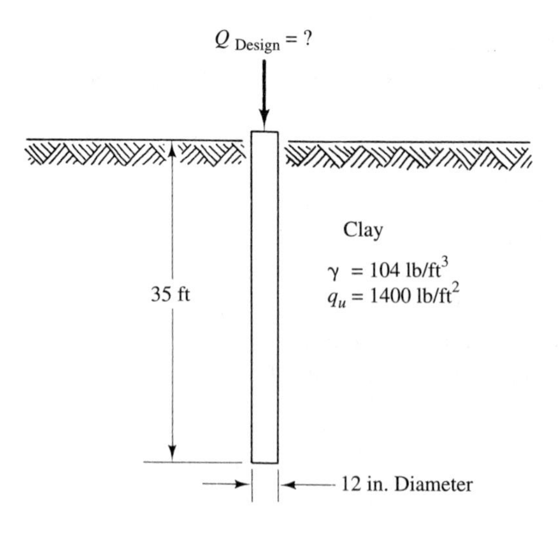

FIGURE 10–11

Solution

From Eq. (10–2),

$$Q_{\text{ultimate}} = f \cdot A_{\text{surface}} + q \cdot A_{\text{tip}} \qquad (10\text{–}2)$$

$$f = \text{Adhesion} = \alpha c$$

$$q_u = 1400\ \text{lb/ft}^2 = 0.7\ \text{ton/ft}^2$$

$$\alpha = 0.9 \qquad (\text{see Figure 10–10 with } q_u = 0.7\ \text{ton/ft}^2)$$

$$c = \frac{q_u}{2} = \frac{1400\ \text{lb/ft}^2}{2} = 700\ \text{lb/ft}^2$$

$$f = (0.9)(700\ \text{lb/ft}^2) = 630\ \text{lb/ft}^2$$

$$A_{\text{surface}} = (\pi d)(L) = (\pi)(1\ \text{ft})(35\ \text{ft}) = 110\ \text{ft}^2$$

$$q_{\text{tip}} = cN_c \qquad (10\text{–}7)$$

$$q_{\text{tip}} = (700\ \text{lb/ft}^2)(9) = 6300\ \text{lb/ft}^2$$

$$A_{\text{tip}} = \frac{\pi d^2}{4} = \frac{\pi}{4}(1\ \text{ft})^2 = 0.785\ \text{ft}^2$$

$$Q_{\text{ultimate}} = (630\ \text{lb/ft}^2)(110\ \text{ft}^2) + (6300\ \text{lb/ft}^2)(0.785\ \text{ft}^2) = 74{,}200\ \text{lb}$$

$$= 74.2\ \text{kips}$$

$$Q_{\text{design}} = \frac{74.2\ \text{kips}}{2} = 37.1\ \text{kips}$$

EXAMPLE 10–4

Given

A 12-in.-diameter concrete pile is driven at a site as shown in Figure 10–12.

Required

Design capacity of the pile, using a factor of safety of 2.

Solution

From Eq. (10–1),

$$Q_{\text{ultimate}} = Q_{\text{friction}} + Q_{\text{tip}} \qquad (10\text{–}1)$$

$$Q_{\text{friction}} = f \cdot A_{\text{surface}} = f_1 \cdot A_{\text{surface}_1} + f_2 \cdot A_{\text{surface}_2}$$

From Figure 10–10, with $q_{u_1} = 1400\ \text{lb/ft}^2 = 0.7\ \text{ton/ft}^2$, $\alpha_1 = 0.9$.

$$c_1 = \frac{q_{u_1}}{2} = \frac{1400\ \text{lb/ft}^2}{2} = 700\ \text{lb/ft}^2$$

$$f_1 = c_1\alpha_1 = (700\ \text{lb/ft}^2)(0.9) = 630\ \text{lb/ft}^2$$

$$A_{\text{surface}_1} = (\pi d)(L_1) = (\pi)(1\ \text{ft})(20\ \text{ft}) = 62.8\ \text{ft}^2$$

FIGURE 10–12

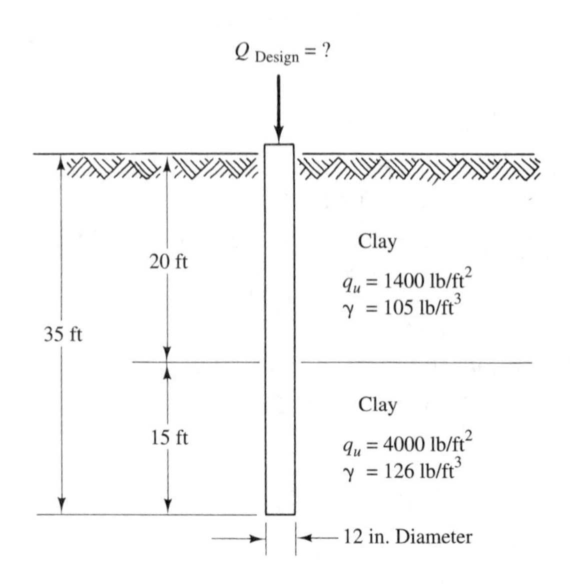

From Figure 10–10, with $q_{u_2} = 4000 \text{ lb/ft}^2 = 2.0 \text{ tons/ft}^2$, $\alpha_2 = 0.56$.

$$c_2 = \frac{q_{u_2}}{2} = \frac{4000 \text{ lb/ft}^2}{2} = 2000 \text{ lb/ft}^2$$

$$f_2 = c_2\alpha_2 = (2000 \text{ lb/ft}^2)(0.56) = 1120 \text{ lb/ft}^2$$

$$A_{\text{surface}_2} = (\pi d)(L_2) = (\pi)(1 \text{ ft})(15 \text{ ft}) = 47.1 \text{ ft}^2$$

$$Q_{\text{friction}} = (630 \text{ lb/ft}^2)(62.8 \text{ ft}^2) + (1120 \text{ lb/ft}^2)(47.1 \text{ ft}^2) = 92{,}300 \text{ lb}$$

$$= 92.3 \text{ kips}$$

$$q_{\text{tip}} = cN_c \tag{10-7}$$

$$q_{\text{tip}} = (2000 \text{ lb/ft}^2)(9) = 18{,}000 \text{ lb/ft}^2$$

$$A_{\text{tip}} = \frac{\pi}{4}d^2 = \frac{\pi}{4}(1 \text{ ft})^2 = 0.785 \text{ ft}^2$$

$$Q_{\text{tip}} = (18{,}000 \text{ lb/ft}^2)(0.785 \text{ ft}^2) = 14{,}100 \text{ lb} = 14.1 \text{ kips}$$

$$Q_{\text{ultimate}} = 92.3 \text{ kips} + 14.1 \text{ kips} = 106.4 \text{ kips}$$

$$Q_{\text{design}} = \frac{106.4 \text{ kips}}{2} = 53.2 \text{ kips}$$

EXAMPLE 10–5

Given

1. A 0.36-m square prestressed concrete pile is to be driven in a clayey soil (see Figure 10–13).
2. The design capacity of the pile is 360 kN.

FIGURE 10–13

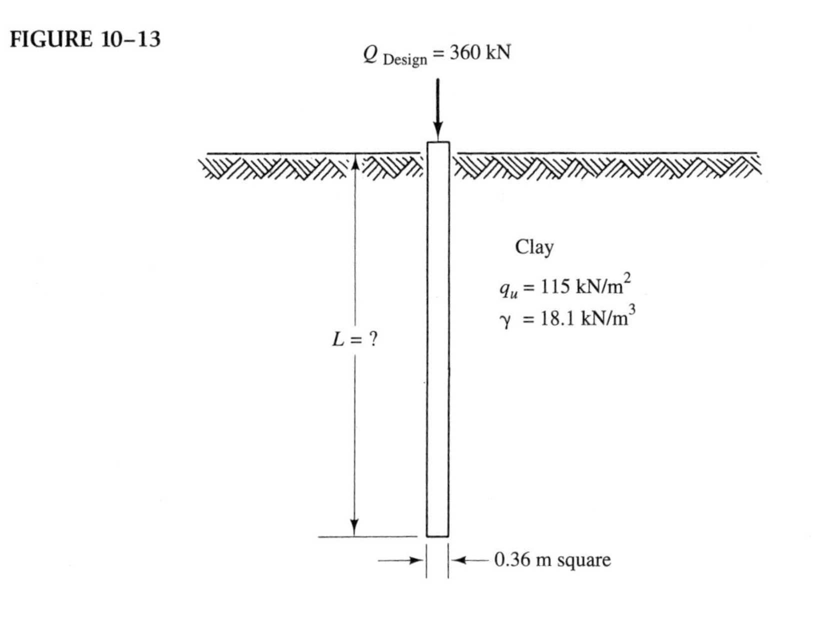

Required

The necessary length of the pile if the factor of safety is 2.

Solution

$$Q_{\text{design}} = 360 \text{ kN}$$
$$Q_{\text{ultimate}} = \text{F.S.} \times Q_{\text{design}} = (2)(360 \text{ kN}) = 720 \text{ kN}$$
$$c = \frac{115 \text{ kN/m}^2}{2} = 57.5 \text{ kN/m}^2$$

$$q_{\text{tip}} = cN_c = (57.5 \text{ kN/m}^2)(9) = 518 \text{ kN/m}^2$$
$$Q_{\text{tip}} = (518 \text{ kN/m}^2)(0.36 \text{ m})(0.36 \text{ m}) = 67.1 \text{ kN}$$

From Eq. (10–1),

$$Q_{\text{ultimate}} = Q_{\text{friction}} + Q_{\text{tip}} \qquad \textbf{(10–1)}$$
$$Q_{\text{friction}} = Q_{\text{ultimate}} - Q_{\text{tip}}$$
$$Q_{\text{friction}} = 720 \text{ kN} - 67.1 \text{ kN} = 652.9 \text{ kN}$$
$$Q_{\text{friction}} = f \cdot A_{\text{surface}} = \alpha c A_{\text{surface}}$$

From Figure 10–10, with $q_u = 115 \text{ kN/m}^2$,

$$\alpha = 0.76$$

$$652.9 \text{ kN} = (0.76)(57.5 \text{ kN/m}^2)(4 \times 0.36 \text{ m})(L)$$

$$L = 10.4 \text{ m}$$

The required length of the 0.36-m square pile is 10.4 m.

Soft clays adjacent to piles may lose a large portion of their strength as a result of being disturbed by pile driving. Propitiously, the disturbed clay gains strength rapidly after driving stops. The original clay's full strength is usually regained within a month or so after pile driving has terminated. Ordinarily, this is not a problem because piles are not usually loaded immediately after driving; thus, the clay has time to regain its original strength prior to being loaded. In cases where piles are to be loaded immediately after driving, however, the effect of decreased strength must be taken into account by performing laboratory tests to determine the extent of strength reduction and rate of strength recovery [14].

Slender piles driven in soft clay have a tendency to buckle when loaded. The ultimate load for buckling of slender steel piles in soft clay can be estimated by using the following equation [15]:

$$Q_{ult} = \lambda\sqrt{cEI} \tag{10–8}$$

where Q_{ult} = ultimate bearing capacity of a single slender pile for buckling in soft clay
λ = 8 for very soft clay; 10 for soft clay
c = cohesion of the soil
E = modulus of elasticity of the steel
I = moment of inertia of the cross section of the pile

Heavy steel, timber, and concrete piles do not tend to buckle if embedded in the soil for their entire lengths.

10–5 PILE-DRIVING FORMULAS

In theory, it seems possible to calculate pile capacity based on the amount of energy delivered to a pile by the hammer and resulting penetration of the pile. Intuitively, the greater the resistance required to drive a pile, the greater will be the capacity of the pile to carry load. Hence, many attempts have been made to develop *pile-driving formulas* by equating energy delivered by the hammer to work done by the pile as it penetrates a certain distance against a certain resistance, with an allowance made for energy losses.

Generally, no pile-driving formula has been developed that gives accurate results for pile capacity. Soil resistance does not remain constant during and after the pile-driving operation. In addition, pile-driving formulas give varying results. Although pile-driving formulas are not generally used to determine pile capacity, they may be used to determine when to stop driving a pile so that its bearing capacity will be the same as that of a test pile or of other piles driven in the same subsoil. To accomplish this, one should drive piles until the number of blows required to drive the last inch is the same as that of the test piles that furnished the information for evaluating the design load. However, piles driven in soft silt or clay should all be driven to the same depth, rather than driven a certain number of blows [1]. Penetration resistance can also be used to prevent pile damage due to overdriving.

One simple and widely used pile-driving formula is known as the *Engineering-News formula.* It is given as follows [16]:

$$Q_a = \frac{2W_r H}{S + C} \tag{10–9}$$

where Q_a = allowable pile capacity, lb
W_r = weight of ram, lb
H = height of fall of ram, ft
S = amount of pile penetration per blow, in./blow
C = 1.0 for drop hammer
C = 0.1 for steam hammer

For use with SI units, Eq. (10–9) may be expressed as

$$Q_a = \frac{1000W_r H}{6(S + C)} \tag{10–10}$$

with Q_a computed in kN if W_r is in kN, H in m, S in mm/blow, and $C = 25$ for drop hammers and 2.5 for steam hammers. The Engineering-News formula has a built-in factor of safety of 6. Tests have shown that this formula is not reliable for computing pile loads, and it should be avoided except as a rough guide [2].

EXAMPLE 10–6

Given

The design capacity of a 0.3-m-diameter concrete pile is 160 kN. The pile is driven by a drop hammer with a manufacturer's hammer energy rating of 40 kN · m.

Required

Average penetration of the pile from the last few driving blows.

Solution

From Eq. (10–10),

$$Q_a = \frac{1000W_r H}{6(S + C)} \tag{10–10}$$

$$Q_a = 160 \text{ kN}$$

$$W_r H = 40 \text{ kN} \cdot \text{m}$$

$$C = 25 \text{ (for a drop hammer)}$$

Therefore,

$$160 \text{ kN} = \frac{(1000)(40 \text{ kN} \cdot \text{m})}{(6)(S + 25)}$$

$$S = 17 \text{ mm/blow}$$

TABLE 10–6
Pile Hammer Efficiency [7]

Type of Hammer	Efficiency, e_h
Drop hammer	0.75–1.00
Single-acting hammer	0.75–0.85
Double-acting hammer	0.85
Diesel hammer	0.85–1.00

Another pile-driving formula is known as the *Danish formula.* It is given as follows [2]:

$$Q_{\text{ultimate}} = \frac{e_h(E_h)}{S + \frac{1}{2}S_0} \qquad (10\text{–}11)$$

where Q_{ultimate} = ultimate capacity of the pile
e_h = efficiency of pile hammer (see Table 10–6)
E_h = manufacturer's hammer energy rating (see Table 10–7)
S = average penetration of the pile from the last few driving blows
S_0 = elastic compression of the pile
$S_0 = [(2e_hE_hL)/(AE)]^{1/2}$
L = length of pile
A = cross-sectional area of pile
E = modulus of elasticity of pile material

Statistical studies indicate that a factor of safety of 3 should be used with the Danish formula.

Example 10–7 demonstrates how the Danish formula can be used as a field control during pile driving to indicate when the desired pile capacity has been obtained.

EXAMPLE 10–7

Given

1. The design capacity of a 12-in. steel-pipe pile is 100 kips.
2. The pile's modulus of elasticity is 29,000 kips/in.^{2}
3. The pile's length is 40 ft.
4. The pile's cross-sectional area is 16 in.^{2}
5. The hammer is a Vulcan 140C with a weight of pile hammer ram of 14,000 lb and manufacturer's hammer energy rating of 36,000 ft-lb.
6. Hammer efficiency is assumed to be 0.80.

Required

1. What should be the average penetration of the pile from the last few driving blows?

TABLE 10–7
Properties of Selected Impact Pile Hammers [13][1]

Rated Energy (ft-lb)	Make	Model	Type[2]	Blows per Minute[3]	Stroke at Rated Energy (in.)	Weight Striking Parts (lb)
7,260	Vulcan	2	S	70	29	300
8,750	MKT[4]	9B3	DB	145	17	1600
13,100	MKT	10B3	DB	105	19	3000
15,000	Vulcan	1	S	60	36	5000
15,100	Vulcan	50C	DF	120	15½	5000
16,000	MKT	DE–20	DE	48	96	2000
18,200	Link-Belt	440	DE	86–90	36⅞	4000
19,150	MKT	11B3	DB	96	19	5000
19,500	Raymond	65C	DF	100–110	16	6500
19,500	Vulcan	06	S	60	36	6500
22,400	MKT	DE–30	DE	48	96	2800
22,500	Delmag	D–12	DE	42–60		2750
24,375	Vulcan	0	S	50	39	7500
24,400	Kobe	K13	DE	45–60	102	2870
24,450	Vulcan	80C	DF	111	16	8000
26,000	Vulcan	08	S	50	39	8000
26,300	Link-Belt	520	DE	80–84	43⅙	5070
32,000	MKT	DE–40	DE	48	96	4000
32,500	MKT	S10	S	55	39	10,000
32,500	Vulcan	010	S	50	39	10,000
32,500	Raymond	00	S	50	39	10,000
36,000	Vulcan	140C	DF	103	15½	14,000
39,700	Delmag	D–22	DE	42–60		4850
40,600	Raymond	000	S	50	39	12,500
41,300	Kobe	K–22	DE	45–60	102	4850
42,000	Vulcan	014	S	60	36	14,000
48,750	Vulcan	016	S	60	36	16,250

[1] 1 ft-lb = 1.356 N · m; 1 in. = 25.4 mm; 1 lb = 4.448 N.
[2] S, single-acting steam; DB, double-acting steam; DF, differential-acting steam; DE, diesel.
[3] After development of significant driving resistance.
[4] For many years known as McKiernan–Terry.

2. How many blows/ft for the last foot of penetration are required for the design capacity, using the Danish formula?

Solution

1. From Eq. (10–11),

$$Q_{\text{ultimate}} = \frac{e_h(E_h)}{S + \frac{1}{2}S_0} \tag{10–11}$$

$$S + \frac{1}{2}S_0 = \frac{e_h(E_h)}{Q_{\text{ultimate}}}$$

$$S = \frac{e_h(E_h)}{Q_{\text{ultimate}}} - \tfrac{1}{2}S_0$$

$$Q_{\text{design}} = \frac{Q_{\text{ultimate}}}{F.S.} = \frac{Q_{\text{ultimate}}}{3}$$

$$Q_{\text{ultimate}} = 3 \times Q_{\text{design}} = 3 \times 100 \text{ kips} = 300 \text{ kips}$$

$$S_0 = [(2e_hE_hL)/(AE)]^{1/2}$$

$$e_h = 0.80$$

$$E_h = 36{,}000 \text{ ft-lb} = 36 \text{ ft-kips}$$

$$L = 40 \text{ ft}$$

$$A = 16 \text{ in.}^2$$

$$E = 29{,}000 \text{ kips/in.}^2$$

$$S_0 = \left[\frac{(2)(0.80)(36 \text{ ft-kips})(40 \text{ ft})}{(16 \text{ in.}^2)(29{,}000 \text{ kips/in.}^2)}\right]^{1/2} = 0.070 \text{ ft} = 0.84 \text{ in.}$$

$$S = \frac{(0.80)(36 \text{ ft-kips})(12 \text{ in./ft})}{300 \text{ kips}} - (\tfrac{1}{2})(0.84 \text{ in.}) = 0.73 \text{ in./blow}$$

2. Number of blows required for last foot of penetration

$$= \frac{12 \text{ in./ft}}{0.73 \text{ in./blow}} = 16 \text{ blows/ft}$$

10–6 PILE LOAD TESTS

Load tests are performed on-site on test piles to determine or verify the design capacity of piles. Normally, piles are designed initially by analytic or other methods, based on estimated loads and soil characteristics. Pile load tests are performed on test piles during the design stage to check the design capacity. Should load test results indicate possible bearing failure or excessive settlement, the pile design should be revised accordingly. Also, data collected from pile load tests are used in the development of criteria for the foundation installation.

To carry out pile load tests, one must first drive test piles. They should be driven at a location where soil conditions are known (such as near a borehole) and where soil conditions are relatively poor. Both test piles and the method of driving them should be exactly the same as will be used in the construction project. A penetration record should be kept as each test pile is driven.

The next step is to load the test piles. For reasons explained previously in this chapter, test piles in clays should not be loaded until some time (at least several weeks) has passed after the piles are driven. Test piles in sands, however, may be loaded several days after they are driven. Test piles may be loaded by adding dead weight or by hydraulic jacking (against a fixed platform, for example). (Figure 10–14 illustrates

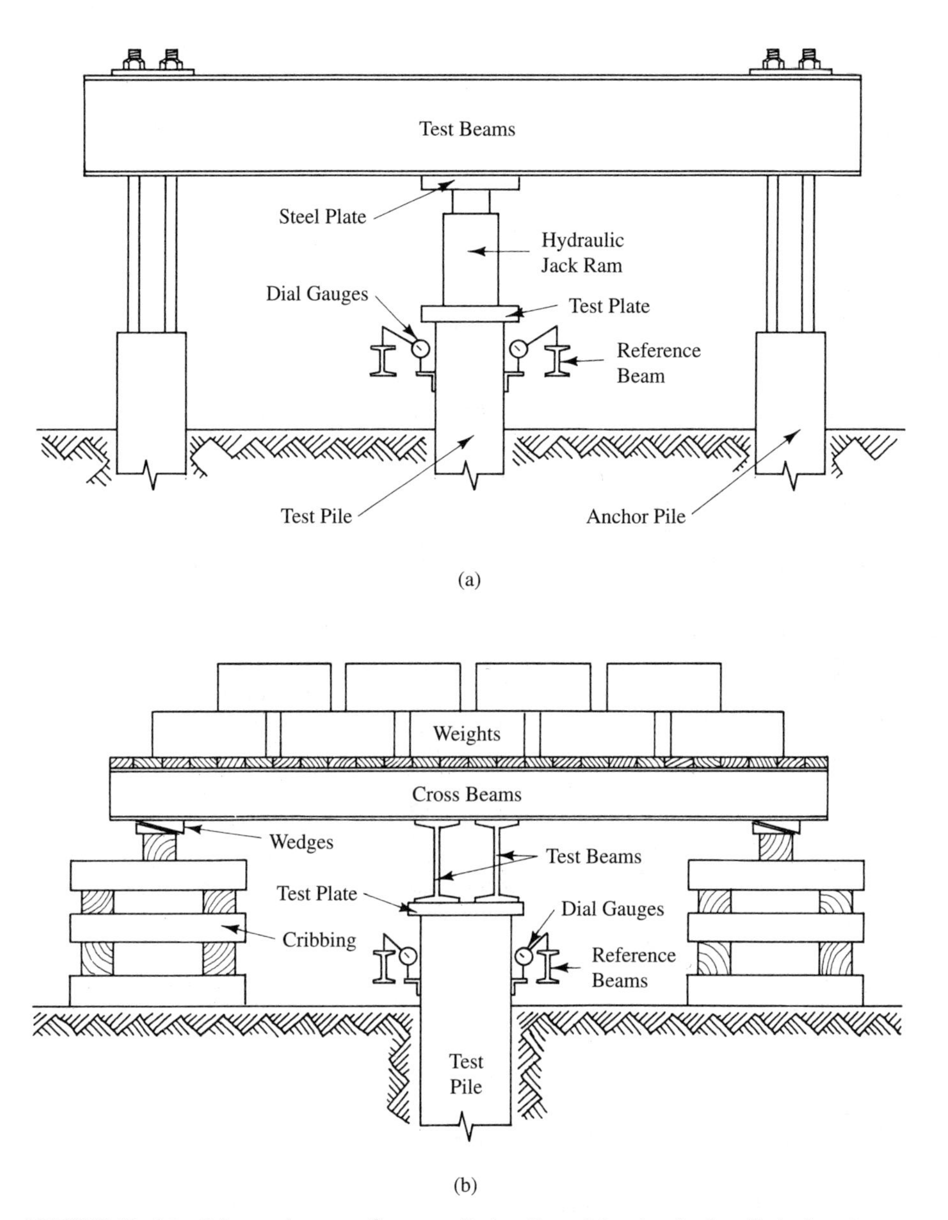

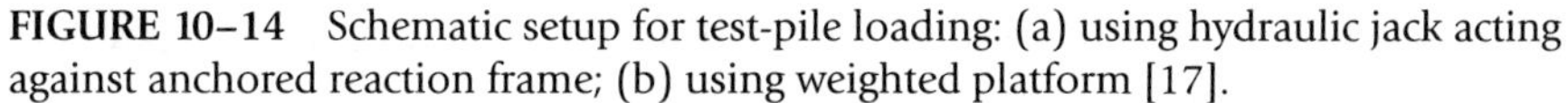

FIGURE 10–14 Schematic setup for test-pile loading: (a) using hydraulic jack acting against anchored reaction frame; (b) using weighted platform [17].

schematically how test piles can be loaded by these methods.) The total load on test piles should be 200% of the proposed design load. The load should be applied to the pile in increments of 25% of the total test load. For specific details regarding loading, the reader is referred to the *ASTM Book of Standards.* In any event, a record of the load and corresponding settlement must be kept as each test pile is loaded and unloaded.

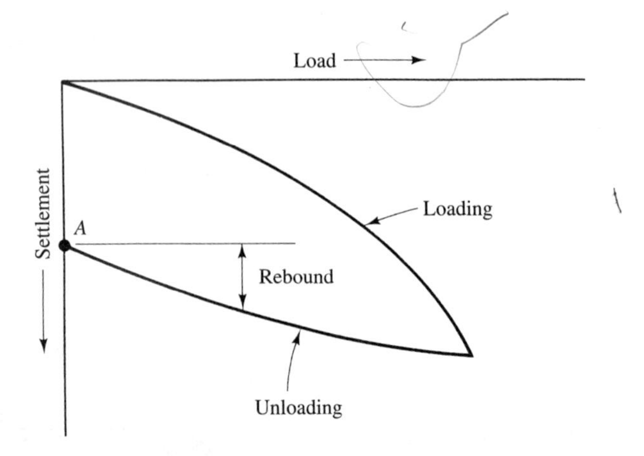

FIGURE 10–15 Typical load versus settlement graph.

The next step is to plot a load versus settlement graph, as shown in Figure 10–15. From this graph, the relationship between the load and net settlement can be obtained. Ordinates along the loading curve of Figure 10–15 give gross settlement. Subtracting the final settlement upon unloading (point *A* in Figure 10–15) from ordinates along the unloading curve gives the rebound. Net settlement can then be determined by subtracting the rebound from the corresponding gross settlement.

The allowable pile load is generally determined based on criteria specified by applicable building codes. There are many building codes and therefore many criteria for determining allowable pile loads based on pile tests. It is, of course, the responsibility of soils engineers to follow criteria specified by the applicable building code. Examples 10–8 and 10–9, in addition to illustrating the determination of allowable pile loads, give two possible building code criteria for determining pile capacity by the pile load test.

EXAMPLE 10–8

Given

1. A 12-in.-diameter pipe pile with a length of 50 ft was subjected to a pile load test.
2. The test results were plotted and the load–settlement curve is shown in Figure 10–16.
3. The local building code states that the allowable pile load is taken as one-half of that load that produces a net settlement of not more than 0.01 in./ton, but in no case more than 0.75 in.

Required

Allowable pile load.

Solution

$$\text{Net settlement} = \text{Gross settlement} - \text{Rebound}$$

Test Load (kips)	Test Load (tons)	Gross Settlement (in.)	Rebound (in.)	Net Settlement (in.)	Building Code Maximum Allowable Settlement (in.)
100	50	0.20	2.39 − 2.20 = 0.19	0.20 − 0.19 = 0.01	<0.5
200	100	0.45	2.54 − 2.20 = 0.34	0.45 − 0.34 = 0.11	<~~1.0~~ (use 0.75)
300	150	0.76	2.64 − 2.20 = 0.44	0.76 − 0.44 = 0.32	<~~1.5~~ (use 0.75)
400	200	1.25	2.73 − 2.20 = 0.53	1.25 − 0.53 = 0.72	<~~2.0~~ (use 0.75)
500	250	2.80	2.80 − 2.20 = 0.60	2.80 − 0.60 = 2.20	>~~2.5~~ (use 0.75)

Because a test load of 200 tons produces a net settlement of 0.72 in. and the maximum allowable settlement is 0.75 in.,

$$\text{Allowable pile load} = \frac{200 \text{ tons}}{2} = 100 \text{ tons}$$

EXAMPLE 10–9

Given

The same conditions as in Example 10–8, except that another local building code is to be applied as follows: "The allowable pile load shall be not more than one-half of that test load that produces a net settlement per ton of test load of not more than 0.01 in., but in no case more than 0.5 inch."

FIGURE 10–16

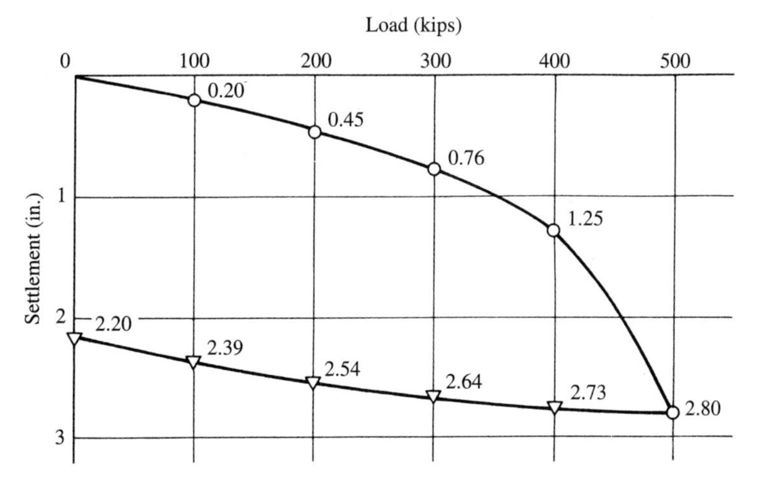

Required

Allowable pile load.

Solution

From Example 10–8,

Test Load (tons)	Net Settlement (in.)	Building Code Maximum Allowable Settlement (in.)
50	0.01	<0.5
100	0.11	<~~1.0~~ (use 0.5)
150	0.32	<~~1.5~~ (use 0.5)
200	0.72	>~~2.0~~ (use 0.5)
250	2.20	>~~2.5~~ (use 0.5)

Because a test load of 150 tons produces a net settlement of 0.32 in. and the maximum allowable settlement is 0.5 in.,

$$\text{Allowable pile load} = \frac{150 \text{ tons}}{2} = 75 \text{ tons}$$

Some building codes use a "breaking in the curve" or the point defined by tangents drawn on either side of a break of a load–settlement graph. One building code [3] states that

> the design load on piles may be determined by the designer based on an analysis of the results of pile load tests performed in accordance with ASTM D-1143. The allowable pile load shall be determined by the application of a safety factor of 2 to the ultimate pile capacity as determined by the intersection of the initial and final tangents to a curve fitted to the plotted results of the pile load test. The fitted curve shall not extend to any point at which the pile continued to move under the applied load. . . .

EXAMPLE 10–10

Given

The results of a pile load test are as follows:

Load (kN)	Settlement (mm)
250	2.7
500	5.8
750	9.3
1000	12.5
1250	16.2
1500	20.0
1750	44.0
2000	80.0

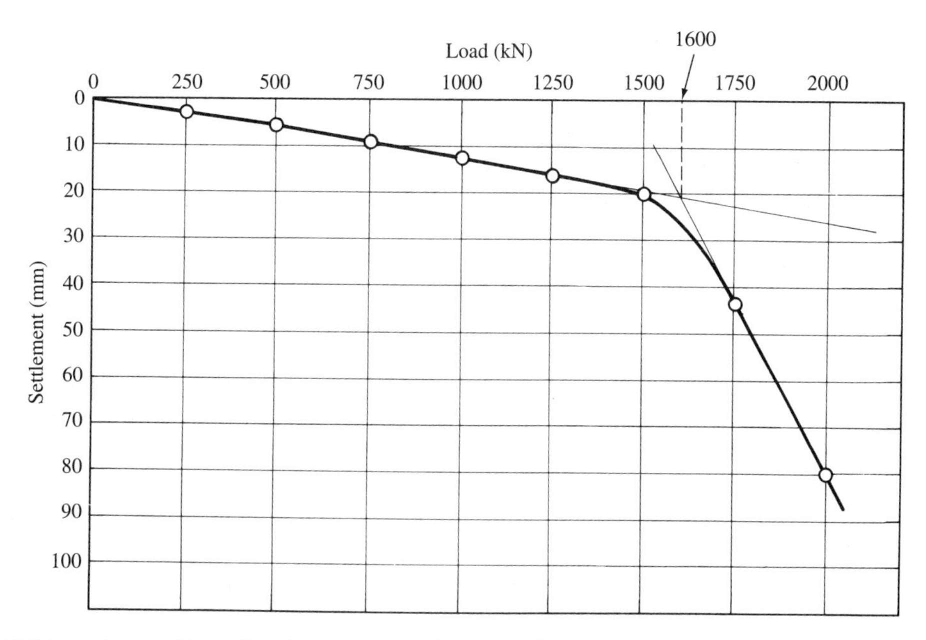

FIGURE 10–17 Plot of load-test data for Example 10–10.

Required

Assuming that the building code given just prior to this example is applicable, find the allowable load on the pile.

Solution

Load-test data are shown plotted in Figure 10–17. Initial and final tangents to the plotted curve intersect at a load of 1600 kN. Hence, according to the code, the allowable load on the pile is (1600 kN)/2, or 800 kN.

10–7 NEGATIVE SKIN FRICTION (DOWN DRAG)

As related throughout this chapter, piles depend, in part at least, on skin friction for support. Under certain conditions, however, skin friction may develop that causes down drag on a pile rather than support. Skin friction that causes down drag is known as *negative skin friction.*

Negative skin friction may occur if soil adjacent to a pile settles more than the pile itself. This is most likely to happen when a pile is driven through compressible soil, such as soft to medium clay or soft silt. Subsequent consolidation of the soil (caused by newly placed fill, for example) can cause negative skin friction as soil adjacent to the pile moves downward while the pile, restrained at the tip, remains fixed. A similar phenomenon may occur as a result of lowering the water table at the site.

Negative skin friction is, of course, detrimental with regard to a pile's ability to carry load. Hence, if conditions at a particular site suggest that negative skin friction may occur, its magnitude should be determined and subtracted from the pile's load-carrying ability.

10–8 PILE GROUPS AND SPACING OF PILES

Heretofore in this chapter, discussion has pertained to a single pile. In reality, however, piles are almost always arranged in groups of three or more. Furthermore, the group of piles is commonly tied together by a pile cap, which is attached to the head of individual piles and causes the several piles to act together as a pile foundation. Figure 10–18 illustrates some typical pile grouping patterns.

If two piles are driven close together, soil stresses caused by the piles tend to overlap, and the bearing capacity of the pile group consisting of two piles is less than the sum of the individual capacities. If the two piles are moved farther apart, so that individual stresses do not overlap, the bearing capacity of the pile group is not reduced significantly from the sum of the individual capacities. Thus, it would appear that piles should be spaced relatively far apart. This consideration is offset, however, by the unduly large pile caps that would be required for the wider spacing.

Minimum allowable pile spacing is often specified by applicable building codes. For example, a building code may state that "the minimum center-to-center spacing of piles not driven to rock shall be not less than twice the average diameter of a round pile, nor less than 1.75 times the diagonal dimension of a rectangular or rolled structural steel pile, nor less than 2 ft 6 in. (0.76 m). For piles driven to rock, the minimum center-to-center spacing of piles shall be not less than twice the average diameter of a round pile, nor less than 1.75 times the diagonal dimension of a rectangular or rolled structural steel pile, nor less than 2 ft 0 in. (0.61 m)" [3].

10–9 EFFICIENCY OF PILE GROUPS

As related in the last section, the capacity of a pile group may be less than the sum of the individual capacities of the piles making up the group. Inasmuch as it would be convenient to estimate the capacity of a group of piles based on the capacity of a single pile, attempts have been made to determine the efficiency of pile groups. (Efficiency of a pile group is the capacity of a pile group divided by the sum of the individual capacities of the piles making up the group.)

In the case where a pile group is comprised of end-bearing piles resting on bedrock (or on a layer of dense sand and gravel overlying bedrock), an efficiency of 1.0 may be assumed [18]. (In other words, the group of n piles will carry n times the capacity of a single pile.) An efficiency of 1.0 is also often assumed by designers for friction piles driven in cohesionless soil. For a pile group composed of friction piles driven in cohesive soil, an efficiency of less than 1.0 is to be expected because stresses from individual piles build up and reduce the capacity of the pile group.

One equation that has been used to compute pile-group efficiency is known as the *Converse–Labarre equation* [18]:

$$E_g = 1 - \theta \frac{(n-1)m + (m-1)n}{90mn} \tag{10–12}$$

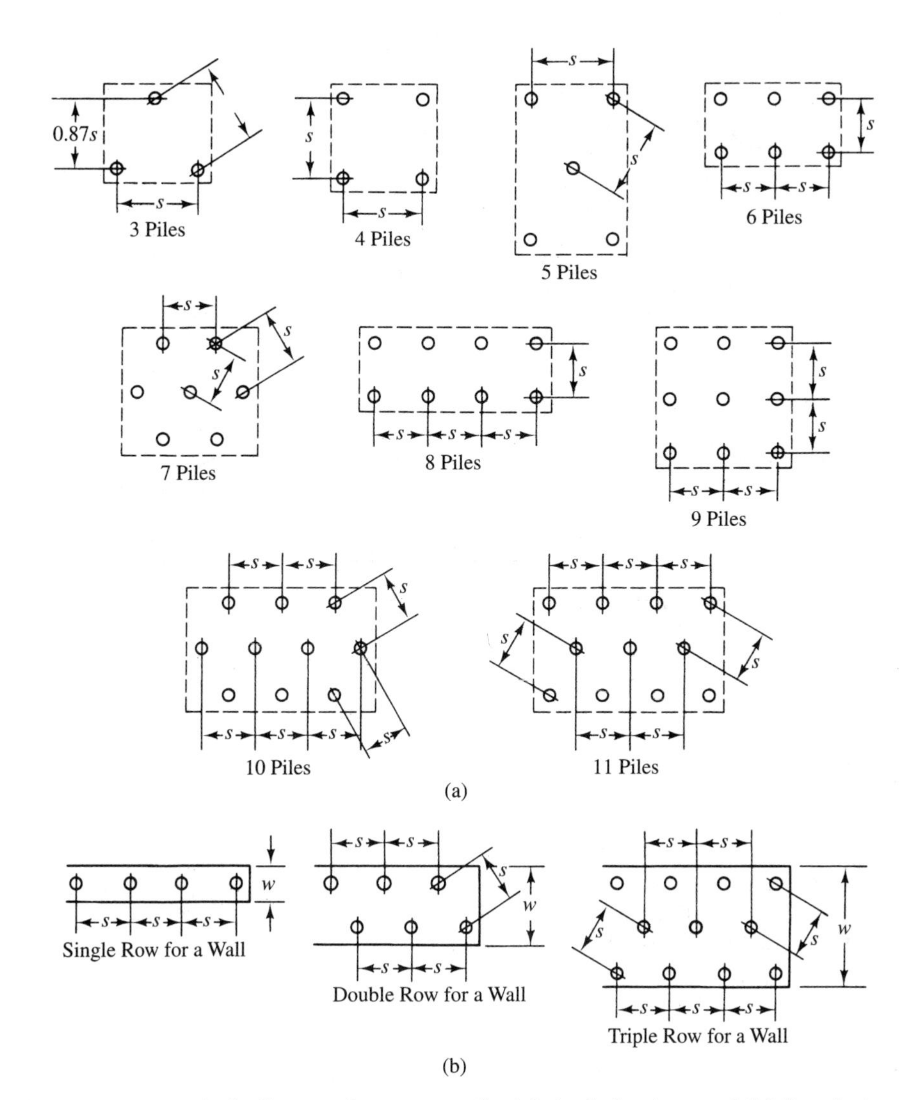

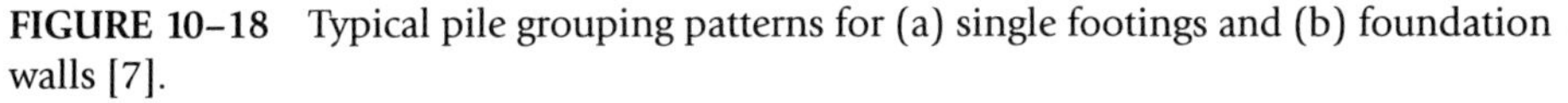

FIGURE 10–18 Typical pile grouping patterns for (a) single footings and (b) foundation walls [7].

where
E_g = pile-group efficiency
θ = arctan d/s, deg
n = number of piles in a row
m = number of rows of piles
d = diameter of piles
s = spacing of piles, center to center, in same units as pile diameter

Example 10–11 illustrates the application of the Converse–Labarre equation.

FIGURE 10–19

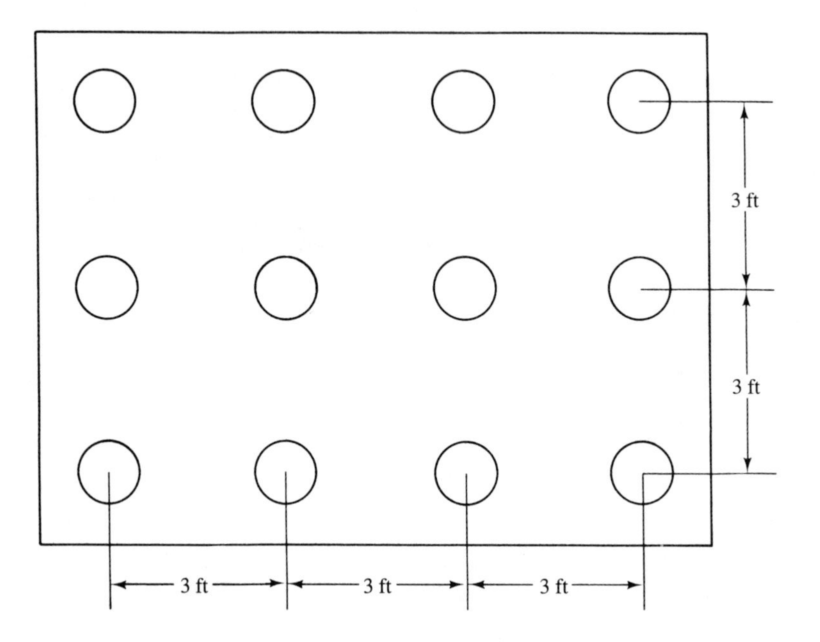

EXAMPLE 10–11

Given

1. A pile group consists of 12 friction piles in cohesive soil (see Figure 10–19).
2. Each pile's diameter is 12 in. and center-to-center spacing is 3 ft.
3. By means of a load test, the ultimate load of a single pile was found to be 100 kips.

Required

Design capacity of the pile group, using the Converse–Labarre equation.

Solution

$$E_g = 1 - \theta \frac{(n-1)m + (m-1)n}{90mn} \qquad \textbf{(10–12)}$$

$$\theta = \arctan \frac{d}{s} = \arctan \frac{1}{3} = 18.4°$$

$$E_g = 1 - (18.4)\frac{(4-1)(3) + (3-1)(4)}{(90)(3)(4)} = 0.710$$

$$\text{Allowable bearing capacity of a single pile} = \frac{100 \text{ kips}}{2} = 50 \text{ kips}$$

Design capacity of the pile group = (0.710)(12)(50 kips) = 426 kips

---

For friction piles driven in cohesive soil, Coyle and Sulaiman suggested that pile-group efficiency may be assumed to vary linearly from a value of 0.7 at a pile spacing of three times the pile diameter to a value of 1.0 at a pile spacing of eight times the pile diameter [19, 20]. For pile spacings less than three times the pile diameter, group capacity may be considered as block capacity, and total capacity can be estimated by treating the group as a pier and applying the following equation [1, 19]:

$$Q_g = 2D(W + L)f + 1.3 \times c \times N_c \times W \times L \tag{10–13}$$

where Q_g = ultimate bearing capacity of pile group
D = depth of pile group
W = width of pile group
L = length of pile group
f = unit adhesion developed between cohesive soil and pile surface (equal to αc)
α = ratio of adhesion to cohesion (see Figure 10–10)
c = cohesion
N_c = bearing capacity factor for a shallow rectangular footing (see Figure 9–7)

A pile group can be considered safe against block failure if the total design load (i.e., "safe design load" per pile multiplied by the number of piles) does not exceed $Q_g/3$. If the total design load exceeds $Q_g/3$, the foundation design must be revised.

Figure 10–20 gives a summary of criteria for pile-group capacity.

EXAMPLE 10–12

Given

1. A pile group consists of four friction piles in cohesive soil (see Figure 10–21).
2. Each pile's diameter is 12 in. and center-to-center spacing is 2.5 ft.

Required

1. Block capacity of the pile group. Use a factor of safety of 3.
2. Allowable group capacity based on individual pile failure. Use a factor of safety of 2, along with the Converse–Labarre equation for pile-group efficiency.
3. Design capacity of the pile group.

Solution

1. *Block capacity:* Because center-to-center spacing of the piles is 2.5 ft, which is less than 3 ft (i.e., 3 diameters), according to the criteria suggested by Coyle and Sulaiman [20], the block capacity of the pile group can be estimated by Eq. (10–13):

FIGURE 10–20 Summary of criteria for pile-group capacity. (a) Individual pile failure in cohesionless soils: $Q_g = n \times Q_u$; individual pile failure in cohesive soils; for $S \geq 3$ diameters, $Q_g = E_g \times n \times Q_u$, E_g varies linearly from 0.7 at $S = 3$ diameters to 1.0 at $S \geq 8$ diameters. (b) Block failure in cohesive soils: for $S < 3.0$ diameters, $Q_g = 2D(W + L)f + 1.3 \times c \times N_c \times W \times L$ [19, 20].

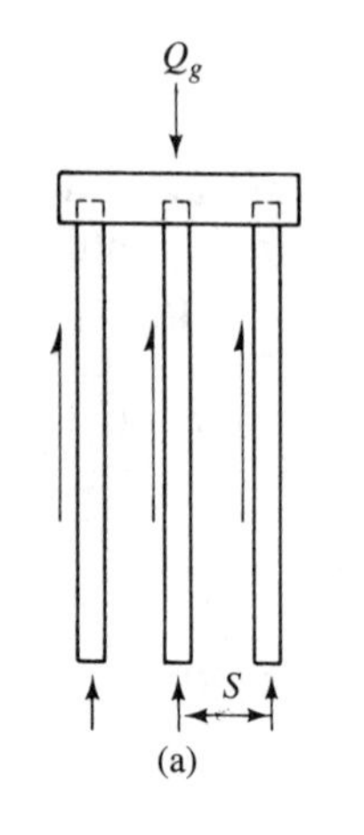

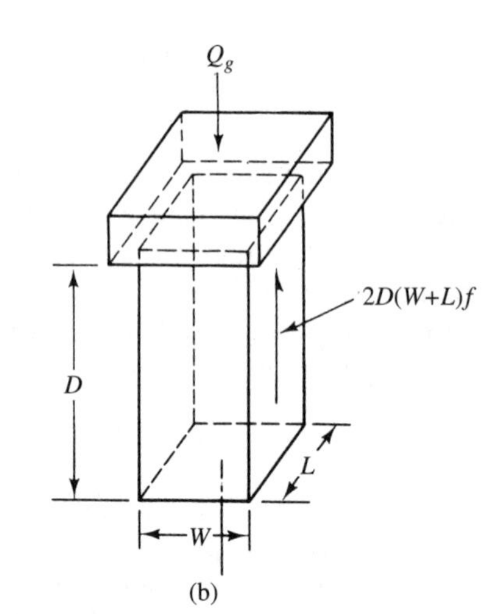

In Cohesionless Soils:
$Q_g = n \times Q_u$

In Cohesive Soils:
For $S < 3.0$ Diameters,
$Q_g = 2D(W+L)f + 1.3 \times c \times N_c \times W \times L$

In Cohesive Soils:
For $S \geqslant 3$ Diameters,
$Q_g = E_g \times n \times Q_u$
E_g Varies Linearly from 0.7 at $S = 3$ Diameters to 1.0 at $S \geqslant 8$ Diameters

$$Q_g = 2D(W + L)f + 1.3 \times c \times N_c \times W \times L \qquad (10\text{–}13)$$

$$D = 35 \text{ ft}$$

$$W = 2.5 \text{ ft} + 0.5 \text{ ft} + 0.5 \text{ ft} = 3.5 \text{ ft}$$

$$L = 2.5 \text{ ft} + 0.5 \text{ ft} + 0.5 \text{ ft} = 3.5 \text{ ft}$$

$$f = \alpha c$$

$$q_u = 4000 \text{ lb/ft}^2 = 2.0 \text{ tons/ft}^2$$

$$c = \frac{4000 \text{ lb/ft}^2}{2} = 2000 \text{ lb/ft}^2 = 2 \text{ kips/ft}^2$$

From Figure 10–10, with $q_u = 2.0 \text{ tons/ft}^2$,

$$\alpha = 0.56$$

$$f = (0.56)(2000 \text{ lb/ft}^2) = 1120 \text{ lb/ft}^2 = 1.12 \text{ kips/ft}^2$$

$$N_c = 5.14 \qquad \text{(from Figure 9–7 for } \phi = 0° \text{ for clay)}$$

$$Q_g = (2)(35 \text{ ft})(3.5 \text{ ft} + 3.5 \text{ ft})(1.12 \text{ kips/ft}^2) + (1.3)(2 \text{ kips/ft}^2)(5.14)(3.5 \text{ ft})(3.5 \text{ ft}) = 713 \text{ kips}$$

FIGURE 10–21

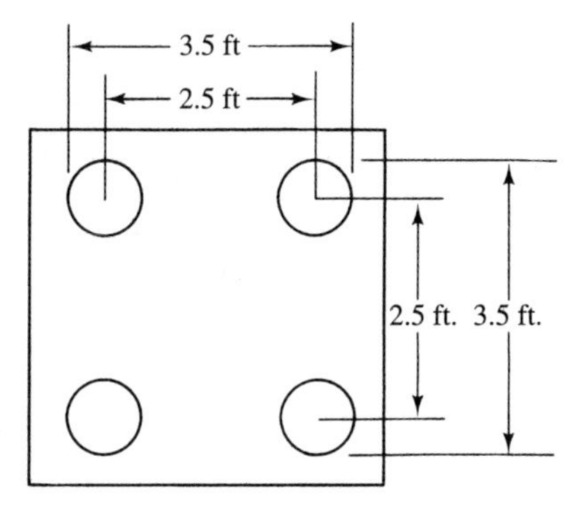

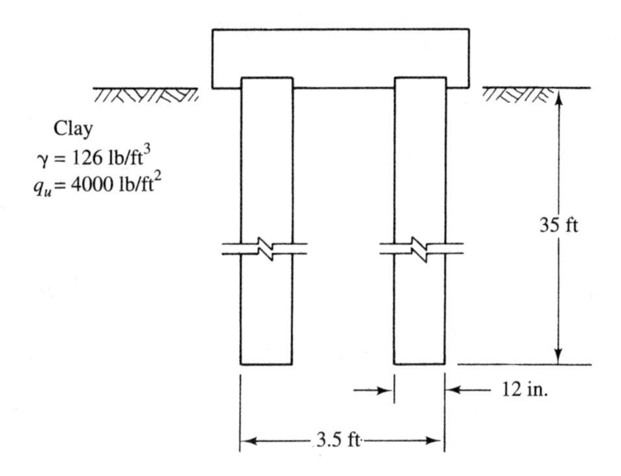

$$\text{Allowable block capacity} = \frac{713 \text{ kips}}{3} = 238 \text{ kips}$$

2. *Group capacity based on individual pile:*

$$Q_{\text{ultimate}} = Q_{\text{friction}} + Q_{\text{tip}} \tag{10–1}$$

$$Q_{\text{friction}} = f \cdot A_{\text{surface}}$$

$$f = 1.12 \text{ kips/ft}^2 \qquad [\text{from (1) above}]$$

$$A_{\text{surface}} = (\pi d)(L) = (\pi)(1 \text{ ft})(35 \text{ ft}) = 110.0 \text{ ft}^2$$

$$Q_{\text{friction}} = (1.12 \text{ kips/ft}^2)(110.0 \text{ ft}^2) = 123 \text{ kips}$$

$$Q_{\text{tip}} = cN_cA_{\text{tip}} = (2 \text{ kips/ft}^2)(9)\left(\frac{\pi}{4}\right)(1 \text{ ft})^2 = 14 \text{ kips}$$

$$Q_{\text{ultimate}} = 123 \text{ kips} + 14 \text{ kips} = 137 \text{ kips}$$

$$Q_a = \frac{137 \text{ kips}}{2}$$

$$= 68.5 \text{ kips} \quad \text{(allowable load for an individual pile)}$$

$$E_g = 1 - \theta \frac{(n-1)m + (m-1)n}{90mn} \qquad \textbf{(10–12)}$$

$$\theta = \arctan \frac{d}{s} = \arctan \frac{1}{2.5} = 21.8°$$

$$n = 2$$

$$m = 2$$

$$E_g = 1 - (21.8)\frac{(2-1)(2) + (2-1)(2)}{(90)(2)(2)} = 0.758$$

$$\text{Allowable } Q = (68.5 \text{ kips})(4)(0.758)$$

$$= 208 \text{ kips (allowable load for pile group)}$$

3. *Design capacity of the pile group:* This is the smaller group capacity of (1) and (2), which is 208 kips.

10–10 DISTRIBUTION OF LOADS IN PILE GROUPS

The load on any particular pile within a pile group may be computed by using the elastic equation [14]:

$$Q_m = \frac{Q}{n} \pm \frac{M_y x}{\Sigma(x^2)} \pm \frac{M_x y}{\Sigma(y^2)} \qquad \textbf{(10–14)}$$

where Q_m = axial load on any pile m
Q = total vertical load acting at the centroid of the pile group
n = number of piles
M_x, M_y = moment with respect to x and y axes, respectively
x, y = distance from pile to y and x axes, respectively

(Both x and y axes pass through the centroid of the pile group and are perpendicular to each other.) It should be noted that shears and bending moments can be determined for any section of pile cap by using elastic and static equations.

EXAMPLE 10–13

Given

1. A pile group consists of nine piles as shown in Figure 10–22.

2. A column load of 450 kips acts vertically on point A.

Required

Load on piles 1, 6, and 8.

Solution

From Eq. (10–14)

$$Q_m = \frac{Q}{n} \pm \frac{M_y x}{\Sigma(x^2)} \pm \frac{M_x y}{\Sigma(y^2)} \qquad (10\text{–}14)$$

$$Q = 450 \text{ kips}$$

$$n = 9$$

$$\Sigma(x^2) = (6)(3 \text{ ft})^2 = 54 \text{ ft}^2$$

$$\Sigma(y^2) = (6)(3 \text{ ft})^2 = 54 \text{ ft}^2$$

$$M_x = (450 \text{ kips})\left(\frac{15 \text{ in.}}{12 \text{ in./ft}}\right) = 562.5 \text{ kip-ft}$$

$$M_y = (450 \text{ kips})\left(\frac{9 \text{ in.}}{12 \text{ in./ft}}\right) = 337.5 \text{ kip-ft}$$

FIGURE 10–22

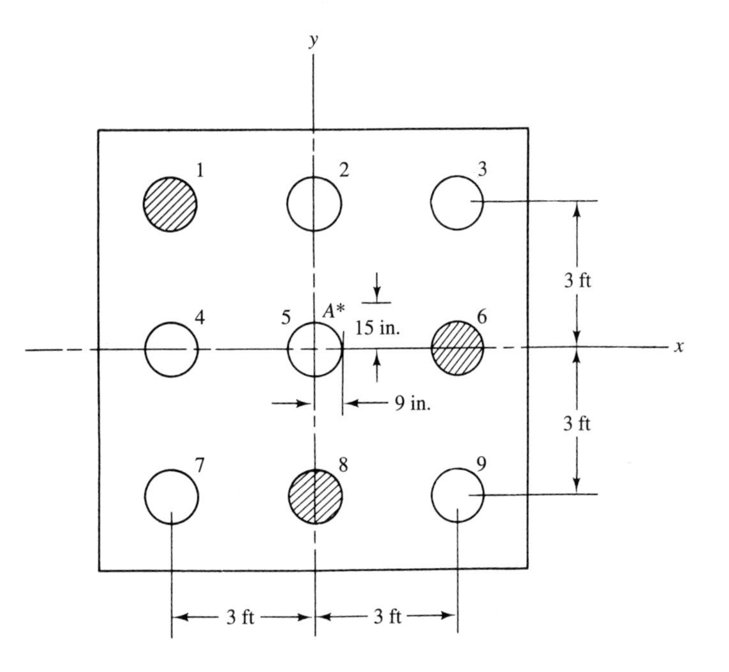

Load on Pile No. 1

$$Q_1 = \frac{450 \text{ kips}}{9} + \frac{(337.5 \text{ kip-ft})(-3 \text{ ft})}{54 \text{ ft}^2} + \frac{(562.5 \text{ kip-ft})(+3 \text{ ft})}{54 \text{ ft}^2} = 62.5 \text{ kips}$$

Load on Pile No. 6

$$Q_6 = \frac{450 \text{ kips}}{9} + \frac{(337.5 \text{ kip-ft})(+3 \text{ ft})}{54 \text{ ft}^2} + \frac{(562.5 \text{ kip-ft})(0)}{54 \text{ ft}^2} = 68.8 \text{ kips}$$

Load on Pile No. 8

$$Q_8 = \frac{450 \text{ kips}}{9} + \frac{(337.5 \text{ kip-ft})(0)}{54 \text{ ft}^2} + \frac{(562.5 \text{ kip-ft})(-3 \text{ ft})}{54 \text{ ft}^2} = 18.8 \text{ kips}$$

EXAMPLE 10–14

Given

1. Figure 10–23 shows a pile foundation consisting of five piles.
2. The pile foundation is subjected to a 200-kip vertical load and a moment with respect to the y axis of 140 kip-ft (Figure 10–23).

Required

Shear and bending moment on section a–a due to the pile reacting under the pile cap.

Solution

From Eq. (10–14),

$$Q_m = \frac{Q}{n} \pm \frac{M_y x}{\Sigma(x^2)} \pm \frac{M_x y}{\Sigma(y^2)} \qquad \textbf{(10–14)}$$

$$Q = 200 \text{ kips}$$

$$n = 5$$

$$M_y = 140 \text{ kip-ft}$$

$$M_x = 0$$

$$\Sigma(x^2) = (4)(3.5 \text{ ft})^2 = 49 \text{ ft}^2$$

$$Q_2 = Q_4 = \frac{200 \text{ kips}}{5} + \frac{(140 \text{ kip-ft})(3.5 \text{ ft})}{49 \text{ ft}^2} + \frac{(0)y}{\Sigma(y^2)} = 50 \text{ kips}$$

Shear at section a–a = (50 kips)(2) = 100 kips

Moment at section a–a = (2)(50 kips)(3.5 ft − 1 ft) = 250 kip-ft

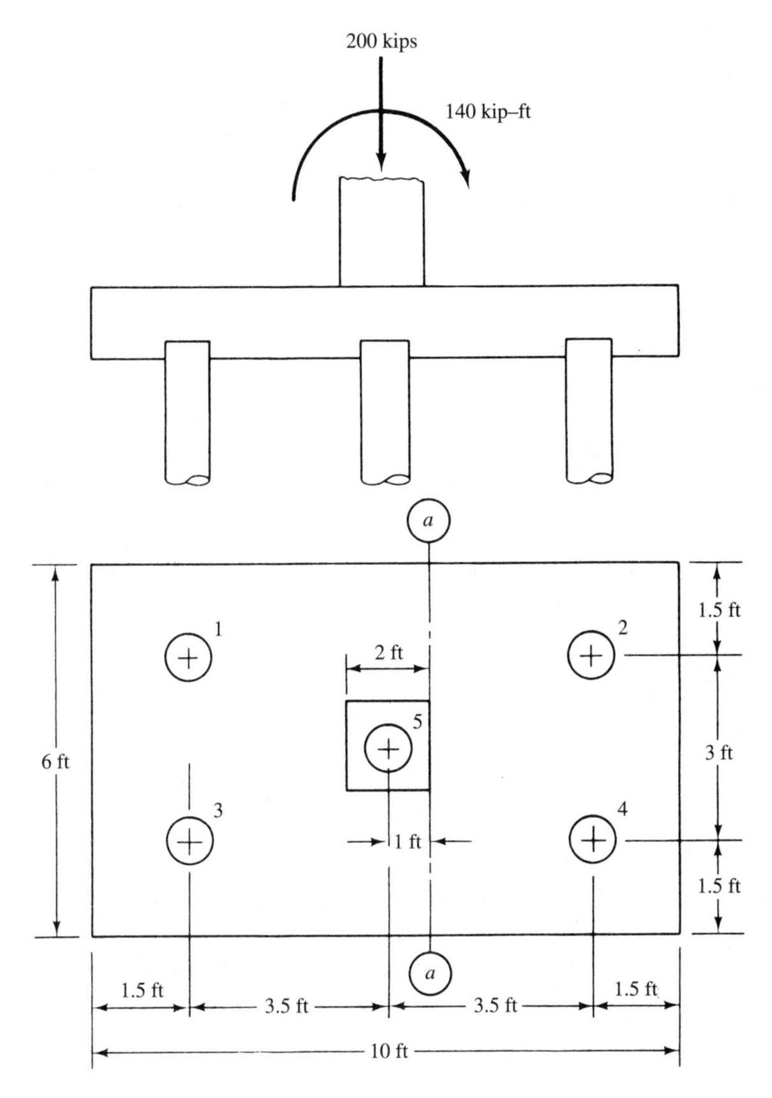

FIGURE 10–23

10–11 SETTLEMENT OF PILE FOUNDATIONS

Like shallow foundations, pile foundations must be analyzed to predict their settlement to ensure that it is tolerable. Unfortunately, universally accepted methods for predicting pile settlements are not available today. The following give some possible methods for predicting pile settlement for end-bearing piles on bedrock, piles in sand, and piles in clay.

Settlement of End-Bearing Piles on Bedrock

A well-designed and constructed pile foundation on hard bedrock generally will not experience an objectionable amount of settlement. The amount of settlement of pile foundations on soft bedrock is very difficult to predict accurately and can be estimated only by judging from the characteristics of rock core samples. Local experience, if available, should be employed as guidance [14].

Settlement of Piles in Sand

Settlement of a pile group is substantially larger than that of a single test pile. In fact, group settlement can be two to 10 times that of a single pile or even greater. Also, the larger the pile group, the greater the settlement will generally be. For sandy soils, the settlement for a pile group can be estimated based on the settlement of a single test pile (from a field load test) using the equation [21]

$$S = S_0 \, [\bar{B}/B]^{1/2} \tag{10–15}$$

where S = group settlement
S_0 = settlement of a single pile (from a field load test)
$\bar{B}$ = smallest dimension of the pile group
B = diameter of the tested pile

All terms in Eq. (10–15) are in length units, but S and S_0 must be in the same units (ordinarily in. or cm) and $\bar{B}$ and B must be in the same units (usually ft or m). For a 12-in.-diameter test pile, Eq. (10–15) shows that a 16-ft-wide pile group would settle about four times as much as that of the test pile.

Settlement of a pile group in sand can also be estimated by using Eq. (10–16). In this case, settlement is a function of net foundation pressure, pile-group width, and the SPT N-value. For saturated sand and gravel in a homogeneous deposit not underlain by more compressible soil at a greater depth [4],

$$S = \frac{2p\sqrt{B}I}{N} \tag{10–16}$$

where S = settlement, in.
p = net foundation pressure, tons/ft^2
B = pile-group width, ft
I = $1 - D/8B \geq 0.5$, where D = effective depth of pile group, ft
N = average corrected SPT N-value (at a depth of roughly one pile-group width in homogeneous soil), blows/ft

Settlement of Piles in Clay

Prediction of pile settlements in deep clay requires first an estimate of load distribution in the soil, followed by settlement calculation in accordance with consolidation

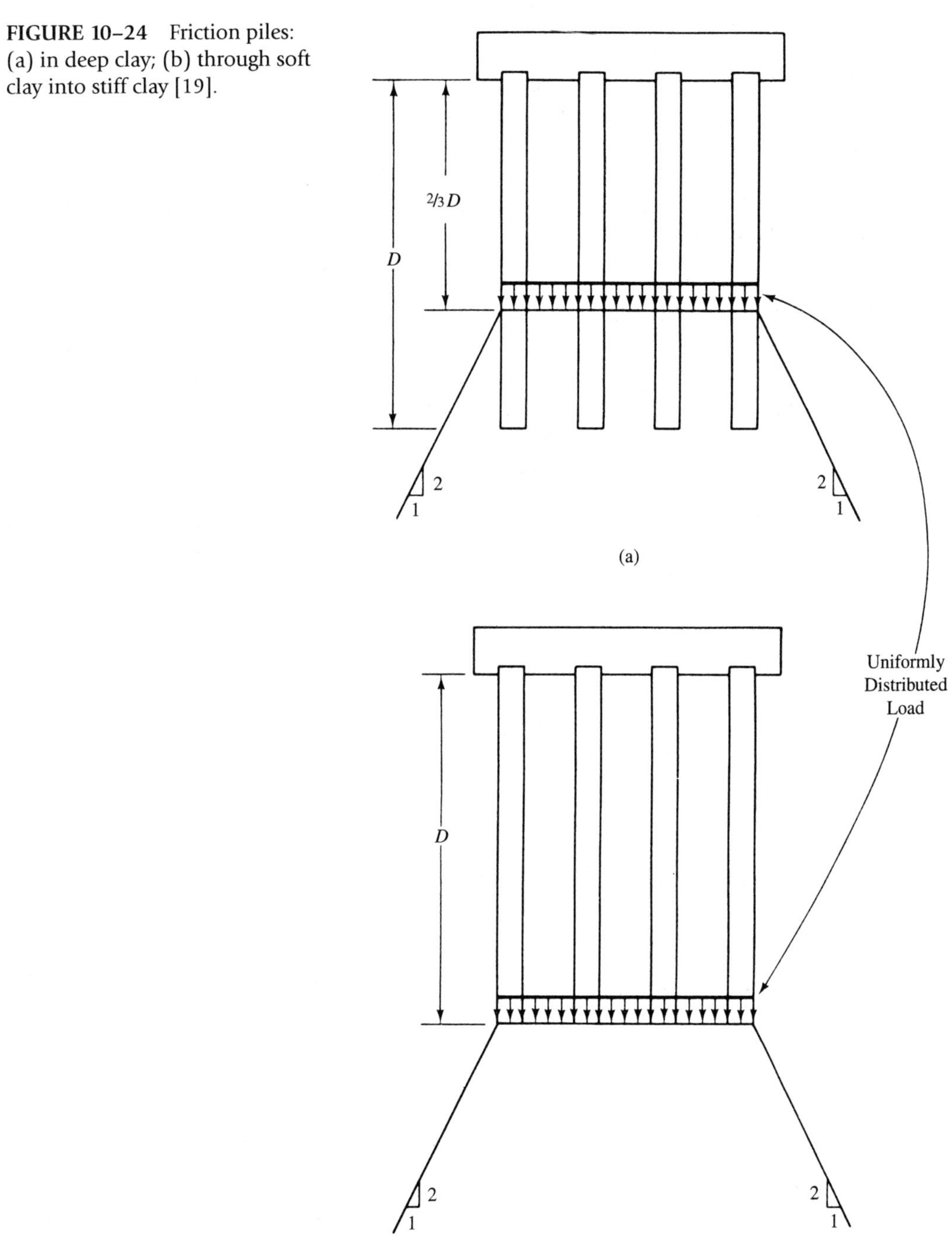

FIGURE 10–24 Friction piles: (a) in deep clay; (b) through soft clay into stiff clay [19].

theory. One method of estimating load distribution is to assume that the load is applied to an equivalent flexible mat (i.e., an imaginary mat) at some selected level and then to compute the distribution of the load from that imaginary mat. For friction piles in deep clay, the equivalent (imaginary) mat may be assumed at a plane located at two-thirds the pile depth [1, 19] (see Figure 10–24a). Consolidation of soil below that plane is then computed as if the piles are no longer present. If piles pass through a layer of very soft clay to a firm bearing in a layer of stiff clay, an equivalent mat may be placed at the level of the pile tips, assuming eventual concentration of the load at that level (Figure 10–24b).

Settlement analysis is then performed, based on consolidation test results, to predict the expected, approximate settlement that would occur for an ordinary (unpiled) foundation as if the foundation were a mat of the same depth and dimensions at the same plane. In such cases, the method of settlement analysis of pile-supported foundations is the same as that used for shallow foundations. From Chapter 7, based on consolidation test results, the amount of settlement due to consolidation can be calculated for a layer of compressible soil by the following equation [1, 14]:

$$S = \frac{e_0 - e}{1 + e_0}(H) \qquad \textbf{(7–14)}$$

or

$$S = C_c\left(\frac{H}{1 + e_0}\right) \log \frac{p_0 + \Delta p}{p_0} \qquad \textbf{(7–18)}$$

where S = consolidation settlement
e_0 = initial void ratio (void ratio *in situ*)
e = final void ratio
H = thickness of layer of compressible soil
C_c = compression index (slope of field e–log p curve)
p_0 = effective overburden pressure (effective weight of soil above midheight of the consolidating layer)
Δp = consolidation pressure (net additional pressure)

Example 10–15 illustrates computation of approximate total settlement of a pile foundation in deep clay.

EXAMPLE 10–15

Given

1. A group of friction piles in deep clay is shown in Figure 10–25.
2. The total load on the piles reduced by the weight of soil displaced by the foundation is 300 kips.

Required

Approximate total settlement of the pile foundation.

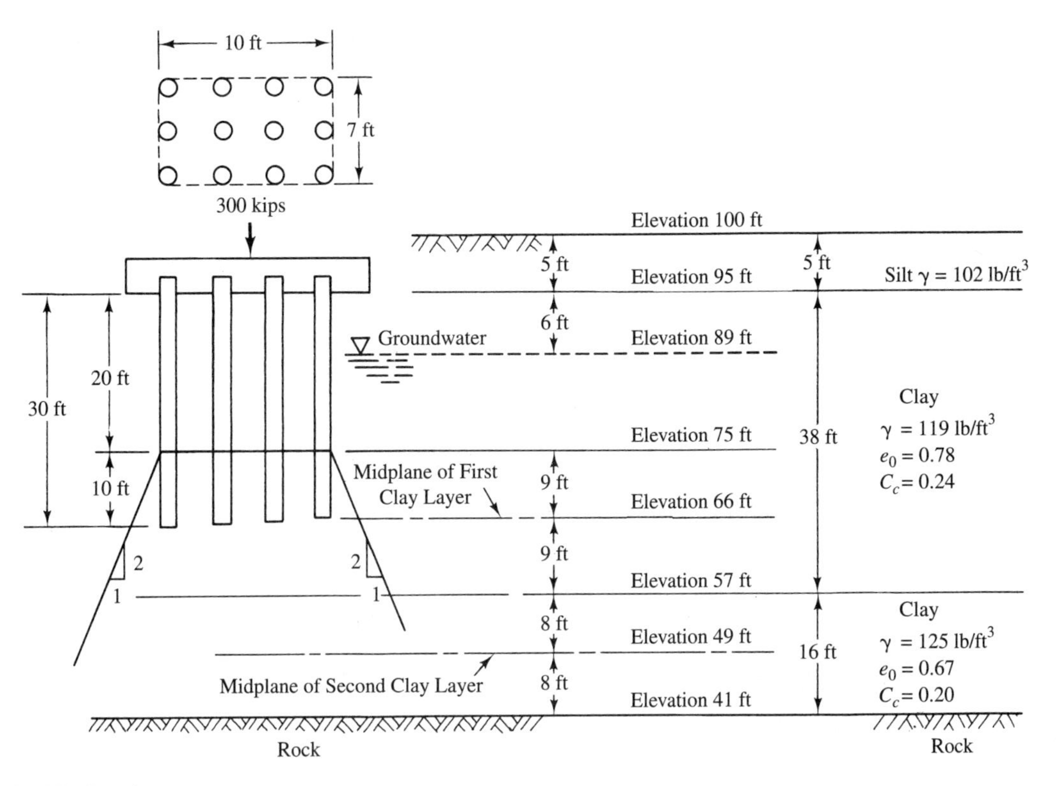

FIGURE 10–25

Solution

Computation of Effective Overburden Pressures (p_0)

$$\begin{aligned} p_0 \text{ at elev. 66 ft} &= (100\text{ ft} - 95\text{ ft})(102\text{ lb/ft}^3) + (95\text{ ft} - 89\text{ ft})(119\text{ lb/ft}^3) \\ &\quad + (89\text{ ft} - 66\text{ ft})(119\text{ lb/ft}^3 - 62.4\text{ lb/ft}^3) \\ &= 2530\text{ lb/ft}^2\text{, or } 2.53\text{ kips/ft}^2 \end{aligned}$$

$$\begin{aligned} p_0 \text{ at elev. 49 ft} &= (100\text{ ft} - 95\text{ ft})(102\text{ lb/ft}^3) + (95\text{ ft} - 89\text{ ft})(119\text{ lb/ft}^3) \\ &\quad + (89\text{ ft} - 57\text{ ft})(119\text{ lb/ft}^3 - 62.4\text{ lb/ft}^3) \\ &\quad + (57\text{ ft} - 49\text{ ft})(125\text{ lb/ft}^3 - 62.4\text{ lb/ft}^3) \\ &= 3540\text{ lb/ft}^2\text{, or } 3.54\text{ kips/ft}^2 \end{aligned}$$

Computation of Δp

$$\text{Area at elev. 66 ft} = [10\text{ ft} + (2)(75\text{ ft} - 66\text{ ft})(\tfrac{1}{2})] \cdot [7\text{ ft} + (2)(75\text{ ft} - 66\text{ ft})(\tfrac{1}{2})] = 304\text{ ft}^2$$

$$\Delta p \text{ at elev. 66 ft} = \frac{300\text{ kips}}{304\text{ ft}^2} = 0.99\text{ kip/ft}^2$$

$$\text{Area at elev. 49 ft} = [10\text{ ft} + (2)(75\text{ ft} - 49\text{ ft})(\tfrac{1}{2})] \cdot [7\text{ ft} + (2)(75\text{ ft} - 49\text{ ft})(\tfrac{1}{2})] = 1188\text{ ft}^2$$

$$\Delta p \text{ at elev. 49 ft} = \frac{300\text{ kips}}{1188\text{ ft}^2} = 0.25\text{ kip/ft}^2$$

Settlement Computations

From Eq. (7–18),

$$S = C_c\left(\frac{H}{1 + e_0}\right)\log\frac{p_0 + \Delta p}{p_0} \tag{7–18}$$

Elev. 75 to 57 ft:

$$S = (0.24)\left(\frac{18\text{ ft}}{1 + 0.78}\right)\log\frac{2.53\text{ kips/ft}^2 + 0.99\text{ kip/ft}^2}{2.53\text{ kips/ft}^2} = 0.35\text{ ft}$$

Elev. 57 to 41 ft:

$$S = (0.20)\left(\frac{16\text{ ft}}{1 + 0.67}\right)\log\frac{3.54\text{ kips/ft}^2 + 0.25\text{ kip/ft}^2}{3.54\text{ kips/ft}^2} = 0.06\text{ ft}$$

Approximate total settlement = 0.35 ft + 0.06 ft = 0.41 ft = 4.9 in.

10–12 CONSTRUCTION OF PILE FOUNDATIONS

Construction of pile foundations consists of installing the piles (see Figure 10–26) (usually by driving) and constructing pile caps. Pile caps are often made of concrete, and their construction is usually a relatively simple structural problem.

With regard to pile installation, most piles are driven by a device called a *pile hammer.* Simply speaking, a pile hammer is a weight that is alternately raised and dropped onto the top of a pile to drive the pile into the soil. Hammer weights vary considerably. As a general rule, a hammer's weight should be at least half the weight of the pile being driven, and the driving energy should be at least 1 ft-lb for each pound of pile weight [14]. The hammer itself is contained within a larger device, with the hammer operated between a pair of parallel steel members known as *leads.*

Several types of pile hammers are available. *Drop hammers* consist of a heavy ram that is raised by a cable and hoisting drum and dropped onto the pile. For *single-acting hammers,* the ram is raised by steam or compressed air and dropped onto the pile. With *double-acting hammers,* the ram is both raised and accelerated downward by steam or air. *Differential-acting hammers* are similar to double-acting hammers. *Diesel hammers* use gasoline for fuel, which causes an explosion that advances the pile and lifts the ram. The total driving energy delivered to the pile includes both the impact of the ram and the energy delivered by the explosion. Table 10–7 (in Section 10–5) gives more specific infofmation on various pile hammers.

FIGURE 10–26 Pile installation.
Source: Courtesy of Associated Pile & Fitting Corporation of New Jersey.

Selection of a pile hammer for a specific job depends on a number of factors. Table 10–8 gives data for selection of pile hammers for various conditions.

Repeated striking of a pile by a pile hammer's heavy ram can damage the pile. A wood pile's fibers at its head (top) may be crushed by the ram (an action known as *brooming*), causing the pile to split near its top end. Brooming and splitting can be minimized by putting a heavy steel ring over the pile's head while it is being driven into the soil. Any damaged part of the pile must be cut off and removed prior to loading the pile. (Hence, a somewhat longer wood pile than is ultimately needed should be used at the beginning to allow for the length of pile that must be cut off.) Precast concrete piles may be protected by placing a metal cap over the pile's head with laminated layers of wood beneath the cap (i.e., between the cap and the pile's head) and a block of hardwood above the cap—all of this to help protect the pile as it is being driven by cushioning the ram's blow.

The other end of a pile—the tip—also needs protection—particularly if the pile is being driven through very hard soil or boulders. Such protection is provided by *driving points* (sometimes referred to as *driving shoes*). Figure 10–27 illustrates some

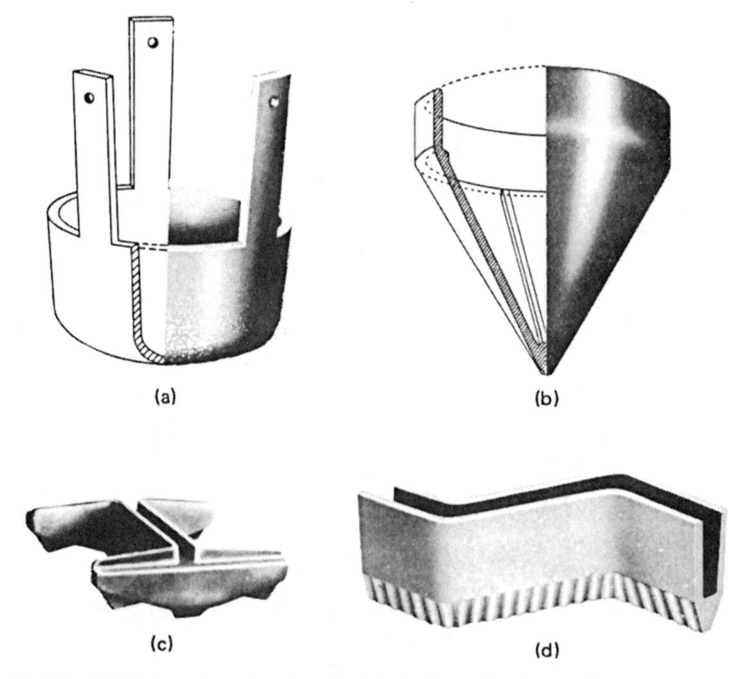

FIGURE 10–27 Driving points (or shoes): (a) timber pile shoes; (b) pipe pile point; (c) H-pile point; (d) sheet pile protector.
Source: Courtesy of Associated Pile & Fitting Corporation of New Jersey.

commercially available driving points (or shoes) for various types of piles. If hard driving is anticipated for precast concrete piles, driving points (or shoes) may be cast at the tips of the piles (see Figure 10–28).

10–13 PROBLEMS

10–1. A 12-in. square concrete pile is driven into loose sand to a depth of 30 ft. Soil conditions are shown in Figure 10–29. Find the pile's axial capacity if K is assumed to be 0.7 and the factor of safety is 2.

10-2. Rework Problem 10–1, assuming that the groundwater table is located 5 ft below the ground surface.

10-3. A 0.5-m-diameter steel pile is driven into dense sand. The pile is driven with the tip closed by a flat plate. The closed-end, steel-pipe pile is filled with concrete after driving. The embedded length of the pile is 20 m. Soil conditions are as shown in Figure 10–30. Determine the design capacity of the pile, using a factor of safety of 2.

TABLE 10–8

Data for Selection of Pile Hammers for Driving Concrete, Timber, and Steel Sheetpiling under Average and Heavy Driving Conditions [14][1,2]

Length of Pile (ft)	Depth of Penetration (%)	Sheet Pile[3]			Timber Pile		Concrete Pile	
		Light	*Medium* (ft-lb per blow)	*Heavy*	*Light* (ft-lb per blow)	*Heavy*	*Light* (ft-lb per blow)	*Heavy*
Driving through earth, sand, loose gravel—normal frictional resistance								
25	50	1000–1800	1000–1800	1800–2500	3600–4200	3600–7250	7250–8750	8750–15,000
	100	1000–3600	1800–3600	1800–3600	3600–7250	3600–8750	7250–8750	13,000–15,000
50	50	1800–3600	1800–3600	3600–4200	3600–8750	7250–8750	8750–15,000	13,000–25,000
	100	3600–4200	3600–4200	3600–7500	7250–8750	7250–15,000	13,000–15,000	15,000–25,000
75	50		3600–7500	3600–8750		13,000–15,000		19,000–36,000
	100			3600–8750		15,000–19,000		19,000–36,000
Driving through stiff clay, compacted gravel—very resistant								
25	50	1800–2500	1800–2500	1800–4200	7250–8750	7250–8750	7250–8750	8750–15,000
	100	1800–3600	1800–3600	1800–4200	7250–8750	7250–8750	7250–15,000	13,000–15,000
50	50	1800–4200	3600–4200	3600–8750	7250–15,000	7250–15,000	13,000–15,000	13,000–25,000
	100		3600–8750	3600–13,000		13,000–15,000		19,000–36,000
75	50		3600–8750	3600–13,000		13,000–15,000		19,000–36,000
				7500–19,000		15,000–25,000		19,000–36,000
Weight (per lin. ft)		20 lb	30 lb	40 lb	30 lb	60 lb	150 lb	400 lb
Pile size (approx.)		15 in.	15 in.	15 in.	13-in. diam	18-in. diam	12 in.^{2}	20 in.

[1] Tennessee Valley Authority.
[2] 1 ft-lb = 1.356 N · m; 1 in. = 25.4 mm; 1 lb = 4.448 N.
[3] Energy required in driving a single sheet pile. Double these when driving two piles at a time.

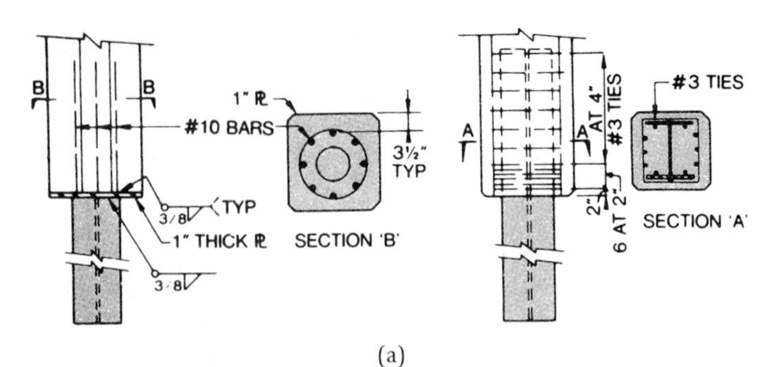

(a)

(b)

FIGURE 10–28 Driving points (or shoes) cast at the tips of the piles. (a) The Prestressed Concrete Institute Standard for 10- to 36-in. piles has these details. HARD-BITE™ or Pluyn Points will protect the vulnerable corners of the H and assure penetration into dense and boulder-filled soils. The H and points prevent damage to the tip of the precast concrete pile. (b) H extends 4 ft into concrete; plate in web of H adds to bond strength. PILE-TIPS July–August 1983.
Source: Courtesy of Associated Pile & Fitting Corporation of New Jersey.

FIGURE 10–29

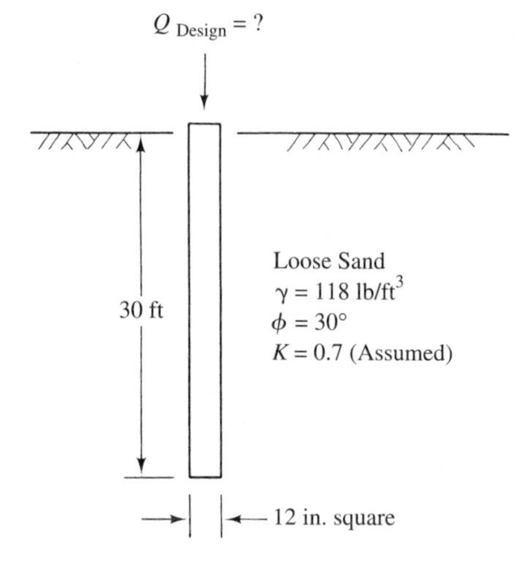

10–4. A 14-in. square concrete pile is driven at a site as shown in Figure 10–31. The embedded length of the pile is 40 ft. Determine the pile's design capacity, using a factor of safety of 2.

FIGURE 10–30

FIGURE 10–31

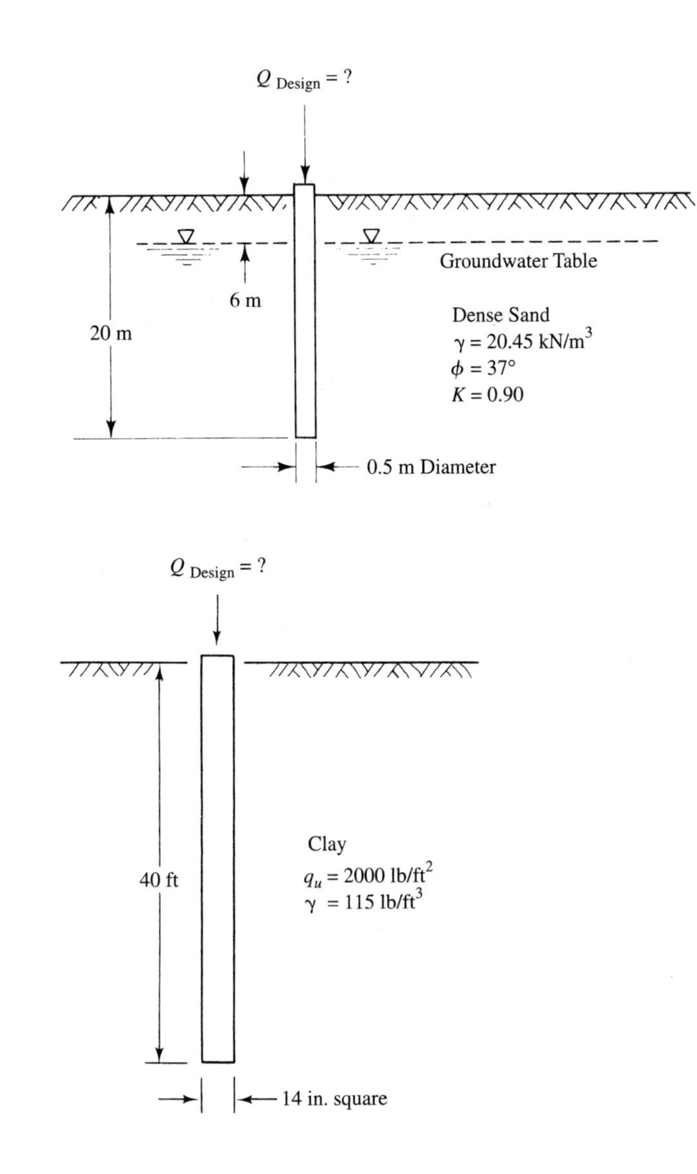

10–5. A 12-in.-diameter concrete pile is driven at a site as shown in Figure 10–32. What is the pile's design capacity if the factor of safety is 2?

10–6. A 0.5-m-diameter steel pile is driven into a varved clay deposit. The pile is driven with the tip closed by a flat plate. The closed-end, steel-pipe pile is filled with concrete after driving. The embedded length of the pile is 15 m. The clay deposit has a unit weight of 17.92 kN/m^3 and an unconfined compressive strength of 120 kN/m^2. Determine the design capacity of the pile, using a factor of safety of 2.

FIGURE 10–32

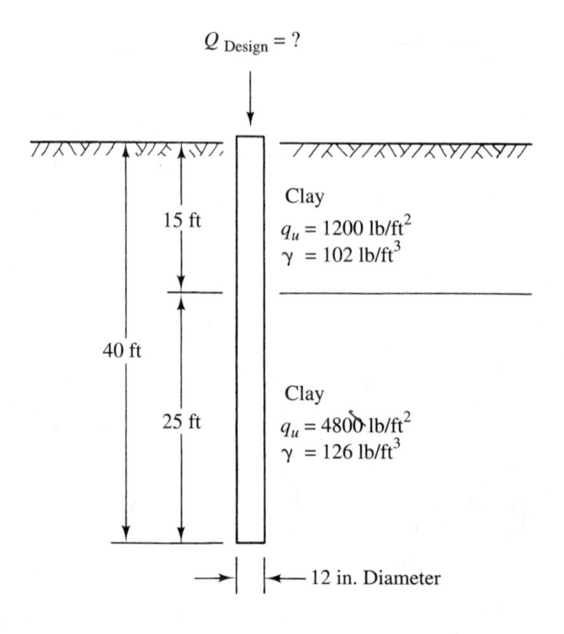

FIGURE 10–33

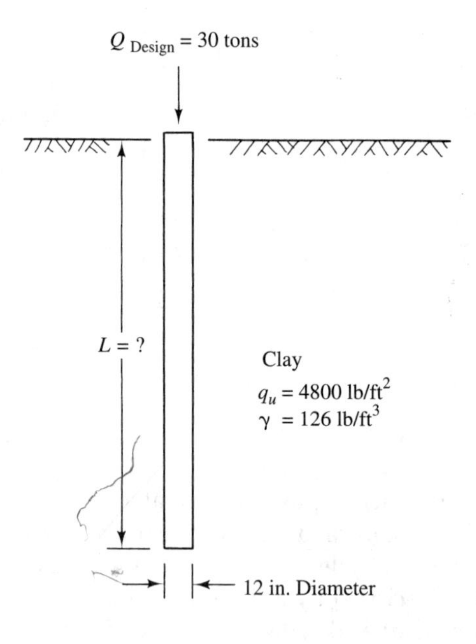

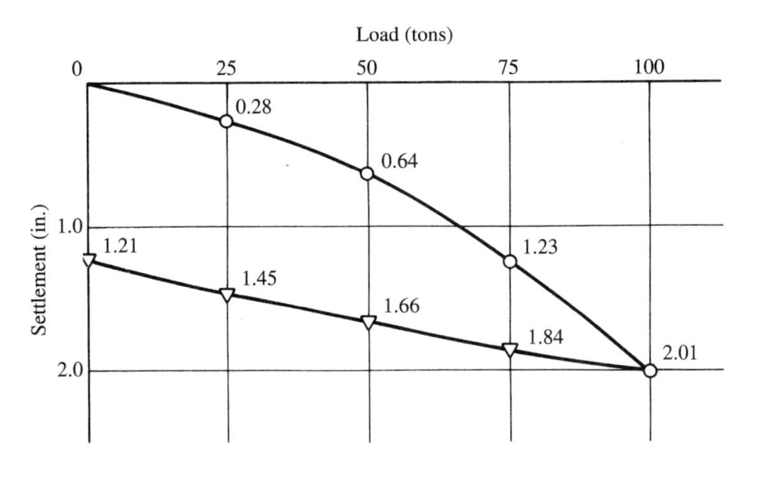

FIGURE 10–34

10–7. Rework Problem 10–6 if the embedded length of the pile is 20 m and the clay deposit's unit weight and unconfined compressive strength are 17.29 kN/m^3 and 96 kN/m^2, respectively.

10–8. A 12-in.-diameter concrete pile is to be driven into a clay soil as shown in Figure 10–33. The pile's design capacity is 30 tons. Determine the pile's required length if the factor of safety is 2.

10–9. The design capacity of a steel pile is 250 kN. The pile is driven by a steam hammer with a manufacturer's hammer energy rating of 36 kN · m. Determine the average penetration of the pile from the last few driving blows. Use the Engineering-News formula.

10–10. A steel-pipe pile is to be driven to an allowable load (design load) of 35-tons capacity by an MKT-11B3 double-acting steam hammer. The steel pipe has a net cross-sectional area of 17.12 in.^{2} and a length of 45 ft. The Danish pile-driving formula is to be used to control field installation of the piles. How many blows per foot are required for the last foot of penetration?

10–11. Rework Problem 10–10 using the Engineering-News formula.

10–12. A pile load test produced the settlement and rebound curves given in Figure 10–34. The pile has a 12-in. diameter and is 25 ft long. Determine the allowable load for this pile using a local building code that states the following: "The allowable load shall not be more than one-half of that test load that produces a net settlement per ton of test load of not more than 0.01 in., but in no case more than 0.75 in."

10–13. Rework Problem 10–12, except that the local building code is changed to read as follows: "The allowable pile load is taken as one-half of that load that produces a net settlement of not more than 0.01 in./ton of test load, but in no case more than 0.5 in."

FIGURE 10–35

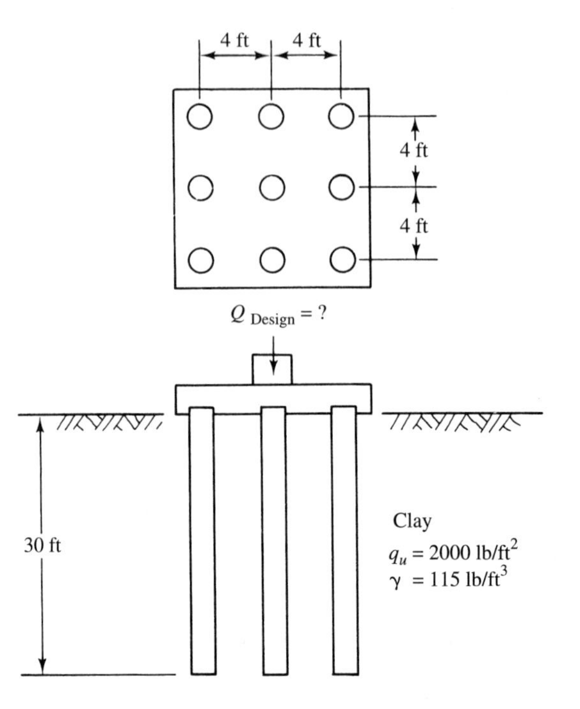

10–14. A pile group consists of nine friction piles in clay soil (see Figure 10–35). The diameter of each pile is 16 in. and the embedded length is 30 ft each. Center-to-center pile spacing is 4 ft. Soil conditions are shown in Figure 10–35. Find the pile group's design capacity if the factor of safety is 2. Use the Converse–Labarre equation.

10–15. A concrete pile with a diameter of 0.3 m and length of 20 m was subjected to a pile load test, with the following results:

Load (kN)	Settlement (mm)
250	5.0
500	9.1
750	12.6
1000	16.2
1250	20.0
1500	32.0
1750	48.0
2000	67.1

Determine the allowable load for this pile using the building code cited on page 344.

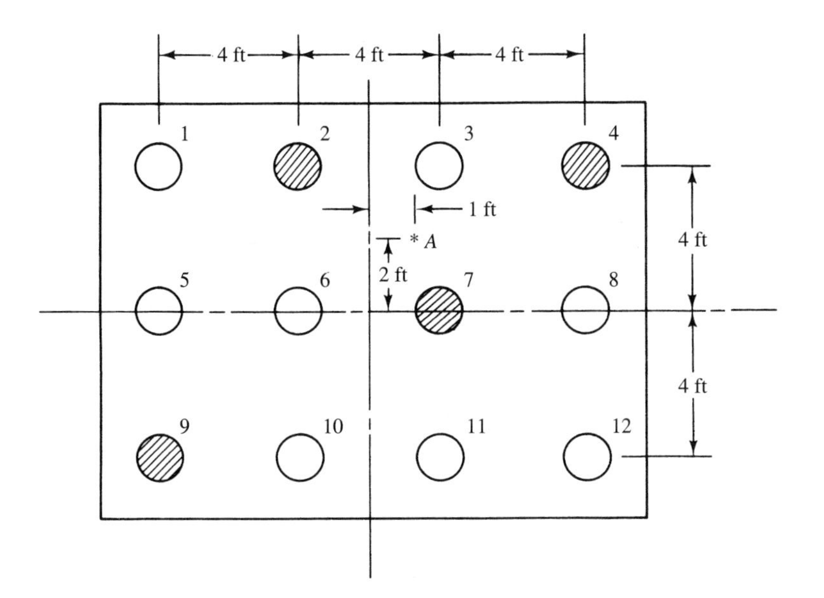

FIGURE 10–36

10–16. A nine-pile group consists of 12-in.-diameter friction concrete piles 30 ft long. The piles are driven into clay, the unconfined compressive strength of which is 6000 lb/ft^2 and the unit weight of which is 125 lb/ft^3. Pile spacing is 2½ diameters. Find (a) the block capacity of the pile group, using a factor of safety of 3; (b) the allowable group capacity based on individual pile failure, using a factor of safety of 2 along with the Converse–Labarre equation for pile-group efficiency; and (c) the design capacity of the pile group.

10–17. A pile group consists of 12 piles as shown in Figure 10–36. A vertical load of 480 kips acts vertically on point *A*. Determine the load on piles 2, 4, 7, and 9.

10–18. A pile group consists of four friction piles in cohesive soil. Each pile's diameter is 0.4 m, and center-to-center spacing is 1.5 m. The ultimate capacity of each pile is 453 kN. Estimate the design capacity of the pile group, using a factor of safety of 2 and the criteria suggested by Coyle and Sulaiman (Figure 10–20).

10–19. A pile group consists of nine friction piles in cohesive soil. Each pile's diameter is 0.3 m, and center-to-center spacing is 1.2 m. The ultimate capacity of each pile is 300 kN. Estimate the design capacity of the pile group, using a factor of safety of 2 and the criteria suggested by Coyle and Sulaiman (Figure 10–20).

10–20. The tower shown in Figure 10–37 is subjected to a wind pressure of 25 lb/ft^2 on its projected area. The tower and foundation weigh 320 kips. Determine the maximum and minimum pile reactions for the layout shown.

FIGURE 10–37

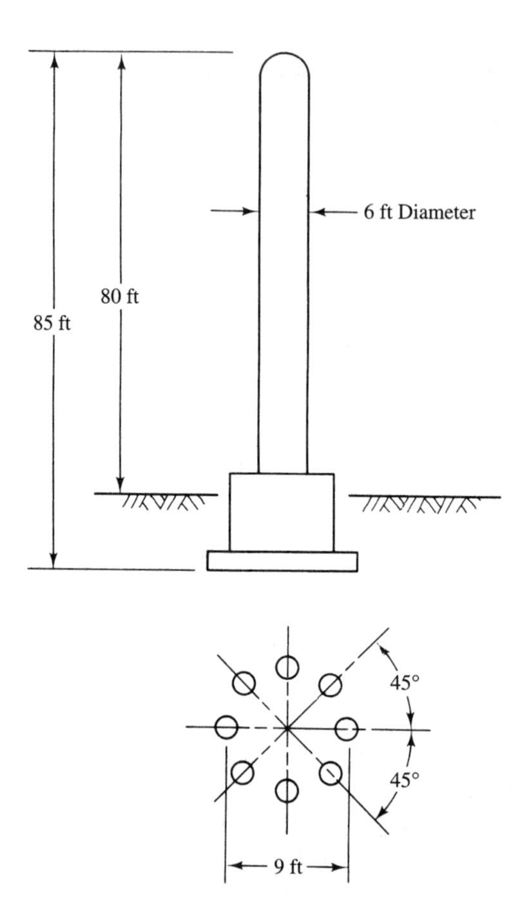

10–21. A group of friction piles in deep clay is shown in Figure 10–38. The total load on the piles reduced by the weight of soil displaced by the foundation is 400 kips. Find the expected total settlement of the pile foundation.

References

[1] Karl Terzaghi and Ralph B. Peck, *Soil Mechanics in Engineering Practice,* John Wiley & Sons, Inc., New York, 1967. Copyright © 1967, by John Wiley & Sons, Inc. Reprinted by permission of John Wiley & Sons, Inc.

[2] David F. McCarthy, *Essentials of Soil Mechanics and Foundations,* Reston Publishing Company, Inc., Reston, Va., 1977.

[3] *North Carolina State Building Code,* Vol. I, *General Construction,* 1978 ed.

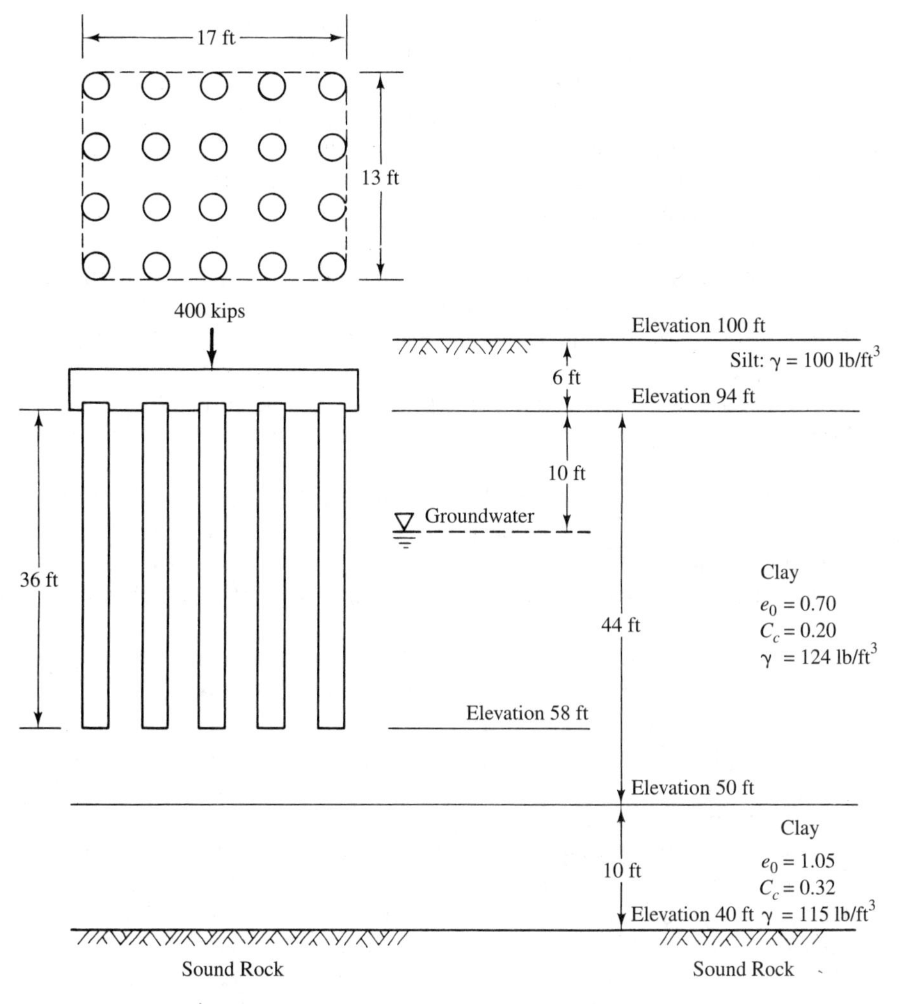

FIGURE 10–38

[4] G. G. Meyerhof, "Bearing Capacity and Settlement of Pile Foundations," *J. Geotech. Eng. Div. ASCE*, **102**(GT3) 197–228 (1976).

[5] A. S. Vesic, "Ultimate Loads and Settlement of Deep Foundations in Sand," *Proc. Bearing Capacity Settlement Found. Symp.*, Duke University, Durham, N.C., 1967.

[6] A. S. Vesic, "Test on Instrumental Piles, Ogeechee River Site," *J. Soil Mech. Found. Eng. Div., Proc. ASCE*, **96**(SM2) 561–584 (1970).

[7] Joseph E. Bowles, *Foundation Analysis and Design*, 2nd ed., McGraw-Hill Book Company, New York, 1977.

[8] David F. McCarthy, *Essentials of Soil Mechanics and Foundations,* 4th ed., Regents/Prentice Hall, Englewood Cliffs, N.J., 1993.

[9] Karl Terzaghi, Ralph B. Peck, and Gholamreza Mesri, *Soil Mechanics in Engineering Practice,* 3rd ed., John Wiley & Sons, Inc., New York, 1996.

[10] N. D. Dennis and R. E. Olson, "Axial Capacity of Steel Piles in Sand," *Proc. ASCE Conf. Geotech. Practice Offshore Eng.,* Austin, Tex., 1983.

[11] R. E. Olson, "Axial Load Capacity of Steel Pipe Piles in Sand," *Proc. Offshore Tech. Conf.,* Houston, Tex., 1990.

[12] H. M. Coyle and R. R. Castello, "New Design Correlations for Piles in Sand," *J. Geotech. Eng. Div. ASCE,* **107**(GT7) 965–986 (1981).

[13] Ralph B. Peck, Walter E. Hansen, and Thomas H. Thornburn, *Foundation Engineering,* 2nd ed., John Wiley & Sons, Inc., New York, 1974. Copyright © 1974, by John Wiley & Sons, Inc. Reprinted by permission of John Wiley & Sons, Inc.

[14] Wayne C. Teng, *Foundation Design,* Prentice-Hall, Inc., Englewood Cliffs, N.J., 1962.

[15] *Design Manual: Soils Mechanics, Foundations, and Earth Structures,* NAVFAC DM-7, U.S. Department of the Navy, Naval Facilities Engineering Command, Alexandria, Va., 1971.

[16] R. H. Karol, *Soils and Soil Engineering,* Prentice-Hall, Inc., Englewood Cliffs, N.J., 1960.

[17] *1989 Annual Book of ASTM Standards,* ASTM, Philadelphia, 1981. Copyright, American Society for Testing and Materials, 1916 Race Street, Philadelphia, PA 19103. Reprinted with permission.

[18] Alfreds R. Jumikis, *Foundation Engineering,* Intext Educational Publishers, Scranton, Pa., 1971. Copyright © 1971 by Harper & Row, Publishers, Inc. Reprinted by permission of the publisher.

[19] Bramlett McClelland, "Design and Performance of Deep Foundations," *Proc. Specialty Conf. Perform. Earth Earth-Supported Struct., ASCE, 2*(June 1972).

[20] Harry M. Coyle and Ibrahim H. Sulaiman, *Bearing Capacity of Foundation Piles: State of the Art,* Highway Res. Board, Record N, 333, 1970.

[21] *Foundations and Earth Structures* (NAVFAC DM-7.2), U.S. Department of the Navy, Naval Facilities Engineering Command, Alexandria, Va., 1982.

11

DRILLED CAISSONS

11–1 INTRODUCTION

A drilled caisson is a type of deep foundation that is constructed in place by drilling a hole into the soil, often to bedrock or a hard stratum, and subsequently placing concrete in the hole. The concrete may or may not contain reinforcing steel. Some drilled caissons have straight sides throughout (straight-shaft caissons); others are constructed with enlarged bases (belled caissons) (see Figure 11–1). The enlarged base area results in a decreased contact pressure (soil pressure) at the caisson's base.

The purpose of a drilled caisson is to transmit a structural load to the caisson's base, which may be bedrock or another hard stratum. In essence, a drilled caisson is primarily a compression member with an axial load applied at its top, a reaction at its bottom, and lateral support along its sides.

Drilled caissons are constructed by using auger drill equipment to form the hole in the soil. Soil is removed from the hole during drilling, in contrast to the driven pile, which only compresses soil aside. Thus, such problems as shifting and lifting of driven piles do not occur with drilled caissons. In some cases, such as in dry, strong cohesive soil, the hole may be drilled dry and without any side support. In this case, concrete placed in the hole makes direct contact with the soil forming the sides of the hole. If cohesionless soil and/or groundwater is encountered, a bentonite slurry may be introduced during drilling to prevent the soil from caving in. (Protective casing may also be used to prevent a cave-in.) In this case, concrete is placed from the bottom up so as to displace the slurry. If a casing is used, it is slowly removed as concrete is placed, and the operator makes sure that soil does not fall into the excavated hole and mix with the concrete.

Drilled caissons are a popular type of deep foundation for several reasons. Drilling equipment is relatively light and easy to use compared with pile-driving equipment. Drilling equipment is much quieter than pile drivers and does not cause massive ground vibrations that can adversely affect adjacent piles. Finally, drilled holes afford better (visual) inspection of the subsoil encountered.

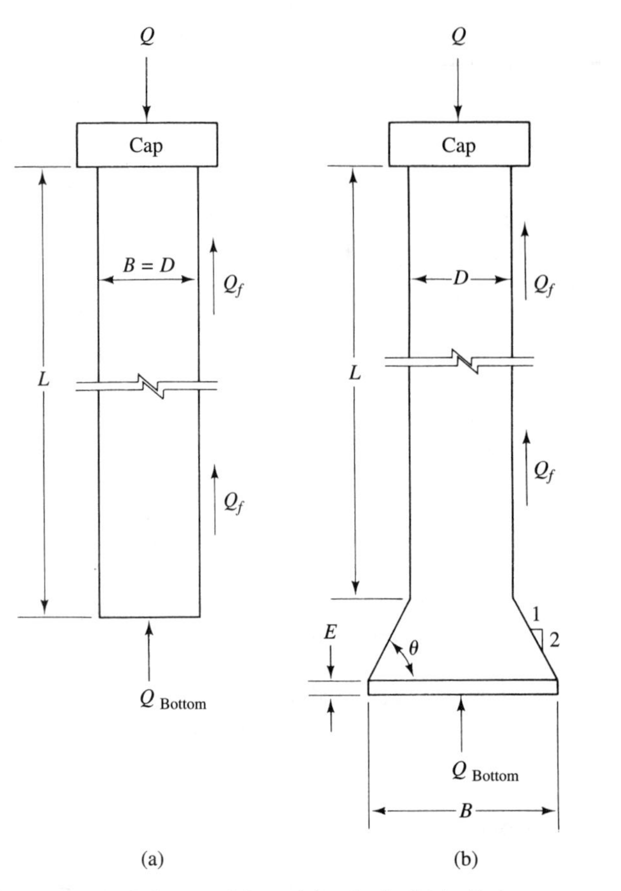

FIGURE 11–1 Caissons: (a) straight-shaft; (b) belled.

11–2 BEARING CAPACITY OF DRILLED CAISSONS

As with a pile, a caisson gets its supporting power from two sources—skin friction and bearing capacity at the caisson's base. Thus, at failure, the load on a drilled caisson may be expressed (as for a pile) by Eq. (10–1), which is reproduced as follows:

$$Q_{\text{ultimate}} = Q_{\text{friction}} + Q_{\text{tip}} \tag{10–1}$$

Because caissons are not driven (as piles are), they do not make tight contact with surrounding soil (as piles do). Consequently, the supporting power for caissons provided by skin friction is relatively small.

To evaluate bearing capacity, it is helpful to consider separately drilled caissons in cohesive soils, in sands, and on bedrock.

Drilled Caissons in Cohesive Soils

The analysis of drilled caissons in cohesive soils is similar to that of piles, in that the caisson's total bearing capacity results from resistance provided by its end bearing and skin friction, in accordance with Eq. (10–1). The term Q_{tip} of Eq. (10–1) can be evaluated by multiplying the cohesion (c) of the soil at the caisson's bottom by the bearing capacity factor (N_c), and this by the area of the caisson's bottom. The term Q_{friction} of Eq. (10–1) can be evaluated by multiplying the unit adhesion or skin friction developed between the shaft's surface and the soil (f) by the shaft's surface area (A_{shaft}) (obtained by multiplying the circumference of the caisson shaft by the depth of the caisson from the ground surface to the top of the bell). Making these substitutions in Eq. (10–1) gives [1]

$$Q_{\text{total}} = cN_c A_{\text{bottom}} + fA_{\text{shaft}} \tag{11–1}$$

Thus far in this discussion, the analysis has been approximately the same as that for piles driven in clay. There is a significant difference between the two, however, and that is in determining the bearing capacity factor (N_c) and the unit adhesion or skin friction (f) of Eq. (11–1). For drilled caissons, the value of the bearing capacity factor (N_c) can be obtained from Table 11–1, and the value of the unit adhesion or skin friction (f) can be obtained from Table 11–2. Adhesion or skin friction that develops along the shaft is related to the clay's cohesion and the manner in which the caisson is drilled.

A factor of safety of 3 is recommended for Q_{bottom} [i.e., the term $cN_c A_{\text{bottom}}$ in Eq. (11–1)] [1]. Thus,

$$Q_{\text{allowable}} = \tfrac{1}{3} cN_c A_{\text{bottom}} + fA_{\text{shaft}} \tag{11–2}$$

TABLE 11–1
Bearing Capacity Factors for Drilled Caissons [2]

Ratio of Depth of Caisson to Diameter of Caisson Bottom	N_c
0	6.2
0.5	7.1
1.0	7.7
1.5	8.1
2.0	8.4
2.5	8.6
3.0	8.8
4.0 and over	9.0

TABLE 11–2
Adhesion or Skin Friction Values for Drilled Caisson Foundations in Clay [1]

Foundation Type and Drilling Method Utilized	Adhesion or Skin Friction, f*	Upper Limit on f Value (kips/ft^2)†
Straight shaft, excavation drilled dry	$0.5c$	1.8
Straight shaft, drilled with slurry	$0.3c$	0.8
Belled, drilled dry	$0.3c$	0.8
Belled, drilled with slurry	$0.15c$	0.5

* c is soil cohesion determined from triaxial testing, not *in situ* vane shear tests.
† 1 kip/ft^2 = 47.88 kN/m^2.

EXAMPLE 11–1

Given

1. A 3-ft-diameter plain-concrete drilled caisson is constructed in clay.
2. Soil conditions and a sketch of the caisson are shown in Figure 11–2.
3. The excavation is drilled dry.
4. A local building code states: "The shafts of caissons shall be designed as concrete columns with continuous lateral support. The unit compressive stress in the concrete shall not exceed 33% of its ultimate 28-day compressive strength nor 1200 lb/in.^{2} No steel reinforcement is required in concrete-filled, drilled piers or caissons unless required by the load imposed thereon."

FIGURE 11–2

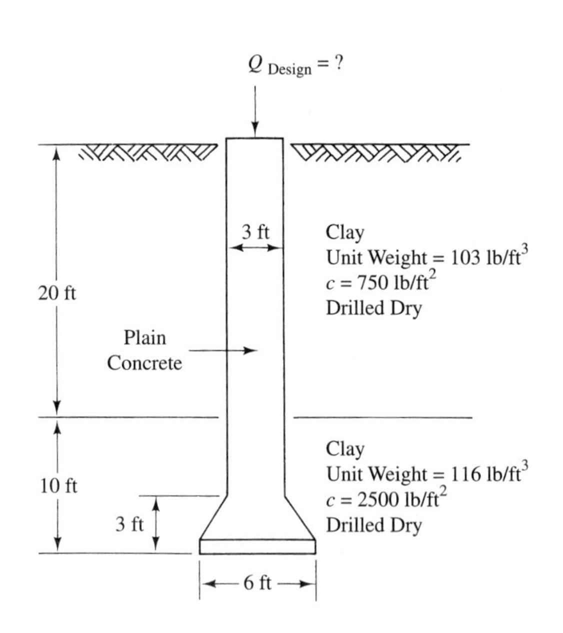

5. The caisson will be made of concrete with f_c' of 4000 lb/in.^{2}

Required

Design capacity of the drilled caisson.

Solution

Supporting Strength of Soil

From Eq. (11–2),

$$Q_{\text{allowable}} = \tfrac{1}{3}cN_c A_{\text{bottom}} + fA_{\text{shaft}} \tag{11–2}$$

$$c = 2500 \text{ lb/ft}^2 = 2.5 \text{ kips/ft}^2$$

$$\frac{\text{Depth of caisson}}{\text{Diameter of caisson bottom}} = \frac{30\ ft}{6\ ft} = 5$$

From Table 11–1, $N_c = 9.0$.

$$A_{\text{bottom}} = \frac{\pi(6 \text{ ft})^2}{4} = 28.27 \text{ ft}^2$$

$$\tfrac{1}{3}cN_c A_{\text{bottom}} = \frac{(2.5 \text{ kips/ft}^2)(9.0)(28.27 \text{ ft}^2)}{3} = 212 \text{ kips}$$

From Table 11–2, for a belled caisson, drilled dry, adhesion or skin friction (f) is $0.3c$ but not more than 0.8 kip/ft^2.

$$f_1 = (0.3)(750 \text{ lb/ft}^2) = 225 \text{ lb/ft}^2 = 0.225 \text{ kip/ft}^2 < 0.8 \text{ kip/ft}^2 \quad \therefore \text{O.K.}$$

$$f_2 = (0.3)(2500 \text{ lb/ft}^2) = 750 \text{ lb/ft}^2 = 0.750 \text{ kip/ft}^2 < 0.8 \text{ kip/ft}^2 \quad \therefore \text{O.K.}$$

A_{shaft} = Circumference of caisson shaft × Effective length of shaft in developing skin friction

$$A_{\text{shaft}_1} = (\pi)(3 \text{ ft})(20 \text{ ft}) = 188.5 \text{ ft}^2$$

$$A_{\text{shaft}_2} = (\pi)(3 \text{ ft})(10 \text{ ft} - 3 \text{ ft}) = 66.0 \text{ ft}^2$$

$$fA_{\text{shaft}} = (0.225 \text{ kip/ft}^2)(188.5 \text{ ft}^2) + (0.750 \text{ kip/ft}^2)(66.0 \text{ ft}^2) = 92 \text{ kips}$$

$$Q_{\text{allowable}} = 212 \text{ kips} + 92 \text{ kips} = 304 \text{ kips}$$

Supporting Strength of Concrete Shaft

According to the local building code given,

$$(4000 \text{ lb/in.}^2)(0.33) = 1320 \text{ lb/in.}^2 > 1200 \text{ lb/in.}^2$$

Therefore, use $q_a = 1200$ lb/in.^{2}

$$Q_{\text{allowable}} = 1200 \text{ lb/in.}^2 \times \frac{\pi(36\ in.)^2}{4} = 1{,}221{,}000 \text{ lb} = 1221 \text{ kips}$$

The design capacity of the caisson is the smaller of the allowable capacities, which is 304 kips.

EXAMPLE 11–2

Given

The excavation for a caisson to carry a total load of 1200 kN is to be drilled with slurry in the soil profile shown in Figure 11–3. The caisson will be made of concrete with f_c' of 27,500 kN/m^2. It is specified that the unit compressive stress of the caisson shaft shall not exceed 33% of its ultimate 28-day compressive strength nor 8274 kN/m^2.

Required

The diameter of the plain-concrete straight-shaft caisson.

Solution

Diameter of Caisson as Determined by the Soil's Supporting Strength

From Eq. (11–2),

$$Q_{\text{allowable}} = \tfrac{1}{3}cN_c A_{\text{bottom}} + fA_{\text{shaft}} \tag{11–2}$$

$$c = 120 \text{ kN/m}^2$$

Assume that the ratio of depth of caisson to diameter of caisson bottom is greater than 4. From Table 11–1, $N_c = 9.0$.

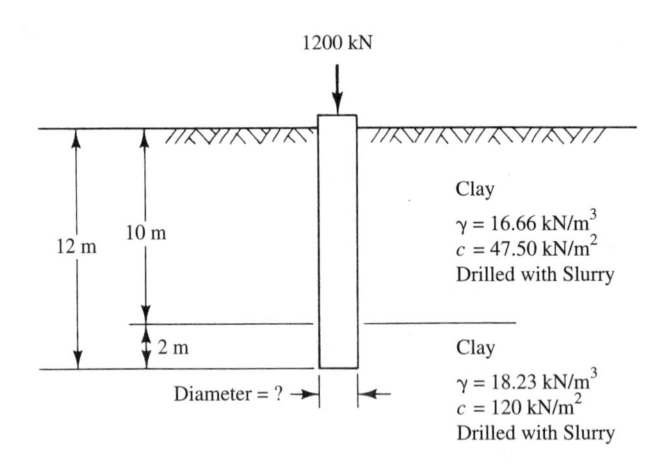

FIGURE 11–3

$$A_{\text{bottom}} = \frac{\pi d^2}{4}$$

$$\tfrac{1}{3}cN_c\,A_{\text{bottom}} = (\tfrac{1}{3})(120\ \text{kN/m}^2)(9.0)\left(\frac{\pi d^2}{4}\right) = 282.7d^2\ \text{kN/m}^2$$

From Table 11–2, for a straight-shaft caisson, drilled with slurry, adhesion or skin friction (f) is $0.3c$ but not more than $(0.8)(47.88\ \text{kN/m}^2)$, or $38.3\ \text{kN/m}^2$.

$$f_1 = (0.3)(47.5\ \text{kN/m}^2) = 14.2\ \text{kN/m}^2 < 38.3\ \text{kN/m}^2 \quad \therefore \text{O.K.}$$

$$f_2 = (0.3)(120\ \text{kN/m}^2) = 36.0\ \text{kN/m}^2 < 38.3\ \text{kN/m}^2 \quad \therefore \text{O.K.}$$

$$(A_{\text{shaft}})_1 = (\pi d)(10\ \text{m}) = 31.42d\ \text{m}$$

$$(A_{\text{shaft}})_2 = (\pi d)(2\ \text{m}) = 6.283d\ \text{m}$$

$$fA_{\text{shaft}} = (14.2\ \text{kN/m}^2)(31.42d\ \text{m}) + (36.0\ \text{kN/m}^2)(6.283d\ \text{m})$$

$$= 672.4d\ \text{kN/m}$$

Because $Q_{\text{allowable}} = 1200$ kN, substituting into Eq. (11–2) gives

$$1200\ \text{kN} = 282.7d^2\ \text{kN/m}^2 + 672.4d\ \text{kN/m}$$

Solving this quadratic equation gives

$$d = 1.19\ \text{m}$$

Diameter of Caisson as Determined by the Concrete Shaft's Strength

$$(0.33)(27{,}500\ \text{kN/m}^2) = 9075\ \text{kN/m}^2 > 8274\ \text{kN/m}^2$$

Therefore, use $q_a = 8274\ \text{kN/m}^2$.

$$Q_{\text{allowable}} = Aq_a$$

$$1200\ \text{kN} = \left(\frac{\pi d^2}{4}\right)(8274\ \text{kN/m}^2)$$

$$d = 0.430\ \text{m}$$

The required caisson diameter is the larger of the two values, which is 1.19 m. (Because the ratio of depth of caisson to diameter of bottom is greater than 4, the use of $N_c = 9.0$ is valid.)

Drilled Caissons in Sands

The analysis of drilled caissons in sands is also similar to that of piles, in accordance with Eq. (10–1). The term Q_{tip} of Eq. (10–1) can be evaluated by multiplying effective vertical pressure (p_v) considering the limits imposed by the concept of critical depth by the bearing capacity factor (N_q), and this by the area of the bottom of the caisson (A_{bottom}). The term Q_{friction} of Eq. (10–1) can be evaluated by multiplying the coefficient of lateral earth pressure of the soil at rest (K_0) by effective vertical pressure (p_v) by the coefficient of friction between sand and concrete (tan δ) by the skin area of the caisson shaft (A_{shaft}). Making these substitutions into Eq. (10–1) gives [1]

$$Q_{\text{ultimate}} = p_v N_q A_{\text{bottom}} + (K_0 p_v \tan \delta) A_{\text{shaft}} \tag{11–3}$$

The value of the coefficient of lateral earth pressure at rest (K_0) ranges from about 0.4 for dense sand to 0.5 for loose sand [3]. The value of the coefficient of friction between sand and concrete (tan d) can be taken to be the value of the coefficient of friction among sand particles (tan φ) if the excavation has been drilled dry. If the excavation is drilled using a slurry, some reduction should be applied to the value of tan φ used [1].

When a drilled caisson in sand is designed by the preceding procedure, a factor of safety of 2 to 3 is recommended.

Example 11–3 illustrates the computation of allowable bearing capacity for a drilled caisson in sand.

EXAMPLE 11–3

Given

1. A 3-ft-diameter straight-shaft caisson is constructed in sand.
2. Soil conditions and a sketch of the caisson are shown in Figure 11–4.
3. The excavation is drilled dry.

Required

Allowable bearing capacity of the caisson as determined by the soil's supporting strength.

FIGURE 11–4

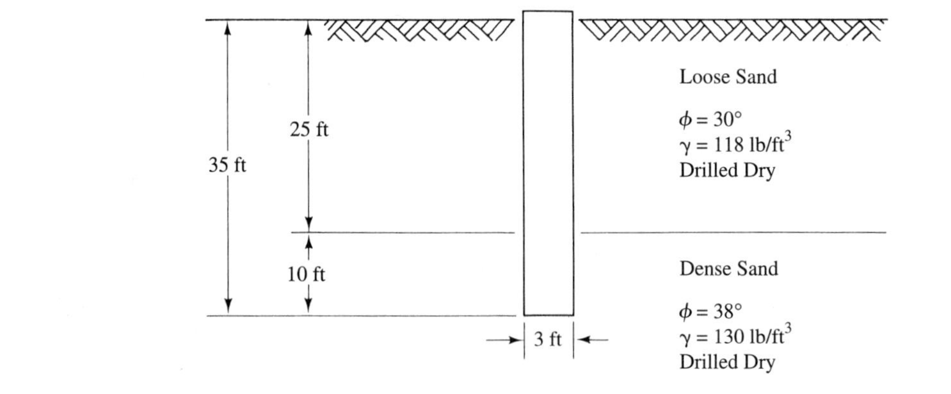

Solution

$$D_c = \text{Critical depth} = 10 \times \text{Caisson's diameter (for loose sand)}$$

$$D_c = 10 \times 3 \text{ ft} = 30 \text{ ft} \qquad \text{(see Figure 11–5)}$$

From Eq. (11–3),

$$Q_{\text{ultimate}} = p_v N_q A_{\text{bottom}} + (K_0 p_v \tan \delta) A_{\text{shaft}} \tag{11–3}$$

$$p_v = 3600 \text{ lb/ft}^2 \qquad \text{(see Figure 11–5)}$$

$$N_q = 50 \qquad \text{(from Figure 9–7 for } \phi = 38°\text{)}$$

$$A_{\text{bottom}} = \frac{\pi}{4}(3 \text{ ft})^2 = 7.07 \text{ ft}^2$$

$$Q_{\text{bottom}} = (3600 \text{ lb/ft}^2)(50)(7.07 \text{ ft}^2) = 1{,}273{,}000 \text{ lb} = 1273 \text{ kips}$$

$$Q_{\text{friction}} = (K_0)(\text{Area of } p_v \text{ diagram})(\text{Circumference of caisson shaft})(\tan \delta)$$

$$K_0 = 0.5 \text{ (for loose sand)}$$

$$K_0 = 0.4 \text{ (for dense sand)}$$

$$\tan \delta = \tan 30° \text{ (for upper layer of loose sand)}$$

$$\tan \delta = \tan 38° \text{ (for lower layer of dense sand)}$$

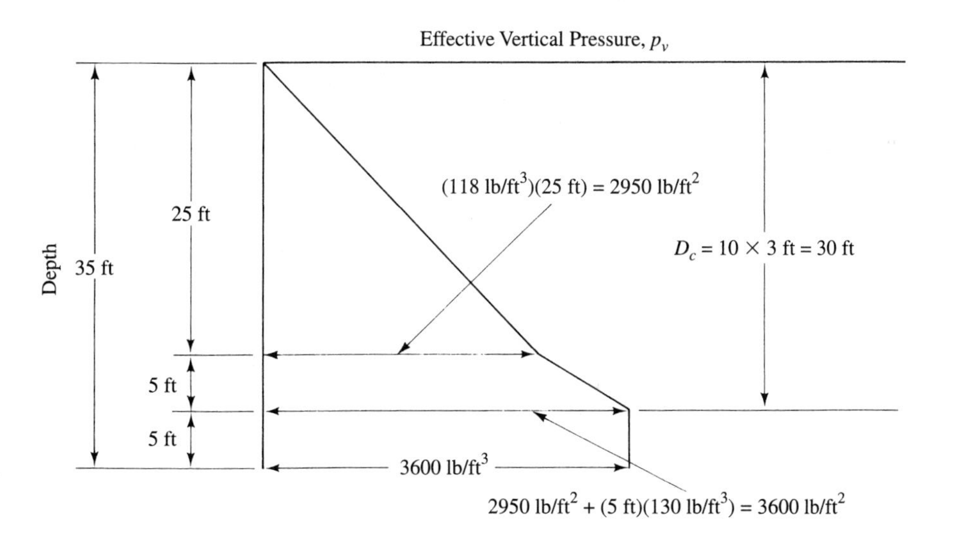

FIGURE 11–5

$$
\begin{aligned}
Q_{\text{friction}} &= (0.5)(\tfrac{1}{2} \times 2950 \text{ lb/ft}^2 \times 25 \text{ ft})(\pi \times 3 \text{ ft})(\tan 30°) \\
&\quad + (0.4)[(2950 \text{ lb/ft}^2)(5 \text{ ft}) + (\tfrac{1}{2})(5 \text{ ft})(3600 \text{ lb/ft}^2 \\
&\quad - 2950 \text{ lb/ft}^2)] \times (\pi \times 3 \text{ ft})(\tan 38°) + (0.4)(5 \text{ ft} \\
&\quad \times 3600 \text{ lb/ft}^2)(\pi \times 3 \text{ ft})(\tan 38°) \\
&= 202{,}000 \text{ lb} = 202 \text{ kips} \\
Q_{\text{ultimate}} &= 1273 \text{ kips} + 202 \text{ kips} = 1475 \text{ kips} \\
Q_{\text{allowable}} &= \frac{1475 \text{ kips}}{3} = 492 \text{ kips}
\end{aligned}
$$

Reese and O'Neill [4, 5] suggested the following equation for approximating base capacity*:

$$q_p = 0.06N \tag{11–4}$$

where q_p = ultimate base capacity (capacity at settlement equal to 5% of base diameter), MN/m^2

N = corrected standard penetration test (SPT) N-value

Equation (11–4) gives values of q_p in MN/m^2, with an upper limit of 4.3 MN/m^2. In the tests that led to Eq. (11–4), base diameters did not exceed approximately 1.5 m; hence, Reese and O'Neill suggested that where larger diameters are encountered, values of q_p determined from this equation should be reduced by the ratio of 1.5 to the base diameter [i.e., 1.5/(Base diameter)]. Making this reduction will restrict settlement to that corresponding to a 1.5-m base diameter.

* From Karl Terzaghi, Ralph B. Peck, and Gholamreza Mesri, *Soil Mechanics in Engineering Practice,* 3rd ed., John Wiley & Sons, Inc., New York, 1996. Copyright © 1996, by John Wiley & Sons, Inc. Reprinted by permission of John Wiley & Sons, Inc.

FIGURE 11–6

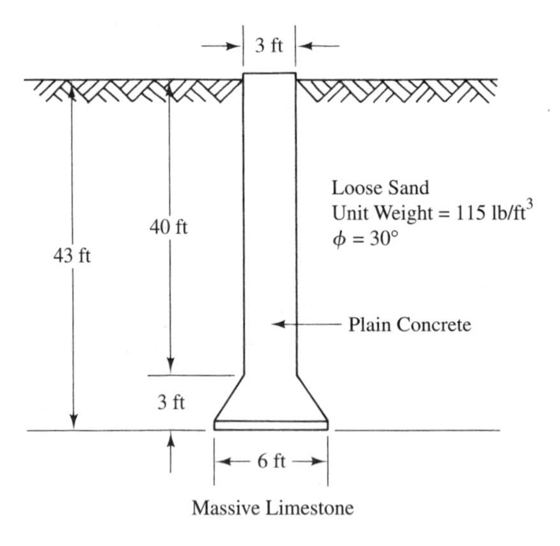

In the skin-friction term in Eq. (11–3) [i.e., $(K_0 p_v \tan\delta)A_{shaft}$], the first three terms (K_0, p_v, and $\tan\delta$) are all influenced significantly by the construction of the shaft; hence, accurate evaluation of each of these terms is difficult. O'Neill and Reese [4, 5] attempted to combine all these terms except p_v into a single variable, β, as follows*:

$$\beta = K \tan\delta \tag{11-5}$$

where*

$$\beta = 1.5 - 0.246z^{0.5} \qquad 1.2 \geq \beta \geq 0.25 \tag{11-6}$$

where z = depth, m.

O'Neill and Reese recommend using the results of Eqs. (11–4) and (11–5) with caution, suggesting an upper limiting value of unit skin friction of 0.2 MN/m². In addition, they suggest that friction resistance should be considered constant for depths below 26 m, which is the depth at which β becomes constant according to Eq. (11–6). Less-conservative designs should be validated by the results of full-scale loading [4].

Drilled Caissons on Bedrock

Clear-cut procedures do not exist for determining the design capacity for drilled caissons on bedrock. The designer usually goes by the applicable local building code, which is often based on past experience in the area. Such codes may give criteria with regard to the concrete's structural strength and/or the bedrock's supporting strength. Unconfined compression tests may be performed on rock samples. Using a factor of safety of 5 to 8, one can then evaluate the allowable bearing pressure [2]. However, if the allowable bearing capacity of rock specified by a local building code is less than the allowable bearing pressure evaluated by unconfined compression tests, the allowable bearing capacity specified by the building code should be used.

Example 11–4 illustrates computation of the design capacity for a drilled caisson on bedrock. This example gives a sample of a possible local building code's specification regarding drilled caissons on rock.

EXAMPLE 11–4

Given

1. A 3-ft-diameter plain-concrete drilled caisson is constructed on massive limestone, the unconfined compressive strength of which is found to be 2500 lb/in.²
2. Soil conditions and a sketch of the caisson are shown in Figure 11–6.
3. The excavation is drilled dry.
4. A local building code states the following:

* From Karl Terzaghi, Ralph B. Peck, and Gholamreza Mesri, *Soil Mechanics in Engineering Practice*, 3rd ed., John Wiley & Sons, Inc., New York, 1996. Copyright © 1996, by John Wiley & Sons, Inc. Reprinted by permission of John Wiley & Sons, Inc.

a. "The shafts of caissons shall be designed as concrete columns with continuous lateral support. The unit compressive stress in the concrete shall not exceed 33% of its ultimate 28-day compressive strength nor 1200 lb/in.^{2} No steel reinforcement is required in concrete-filled, drilled piers or caissons unless required by the load imposed thereon."

b. "Allowable bearing capacity of rock shall not exceed the following:

Massive igneous or metamorphic rock	100 tons/ft^2
Massive sedimentary rock	20 tons/ft^2"

5. The caisson will be made of concrete with f_c' of 4000 lb/in.^{2}

Required

Design capacity of the caisson (neglect skin friction of the caisson shaft).

Solution

Allowable Bearing Capacity of the Caisson as Determined by the Structural Strength of the Concrete

$$(4000\ \text{lb/in.}^2)(0.33) = 1320\ \text{lb/in.}^2 > 1200\ \text{lb/in.}^2$$

Therefore, use $q_a = 1200$ lb/in.^{2}

$$Q_{\text{allowable}} = \frac{(\pi)(3\ \text{ft})^2}{4}(1200\ \text{lb/in.}^2)(144\ \text{in.}^2/\text{ft}^2)$$

$$= 1{,}221{,}450\ \text{lb} = 611\ \text{tons}$$

Allowable Bearing Capacity of the Caisson as Determined by the Supporting Strength of the Rock

$$q_a = \left(\frac{2500\ \text{lb/in.}^2}{8}\right)\left(\frac{144\ \text{in.}^2/\text{ft}^2}{2000\ \text{lb/ton}}\right) = 22.5\ \text{tons/ft}^2$$

(This uses a factor of safety of 8 for the unconfined compressive strength of the rock.) However, the local building code specifies that q_a shall not exceed 20 tons/ft^2. Therefore, q_a of 20 tons/ft^2 should be used.

$$Q_{\text{allowable}} = \frac{\pi}{4}(6\ \text{ft})^2(20\ \text{tons/ft}^2) = 565\ \text{tons}$$

The design capacity of the caisson is the smaller of the allowable capacities, which is 565 tons.

11–3 SETTLEMENT OF DRILLED CAISSONS

Settlement of drilled caissons in clay depends largely on the load history of the clay. This is similar to settlement of footings. Because drilled caissons are uneconomical in normally consolidated clays and settlement thereon is excessive, drilled caissons should be used only in overconsolidated clays. Long-term settlement analysis in clay soils can be performed by using consolidation theory and assuming the drilled caisson's bottom to be a footing [3].

Settlement of drilled caissons in sand "at any depth is likely to be about one-half the settlement of an equally loaded footing covering the same area on sand of the same characteristics" [3]. Generally, such settlement will not be detrimental because the caisson will normally be found on dense sand and settlement will be small. Settlement in sand can be computed by using the procedures given in Chapter 7 for footings on sand. It should be kept in mind, however, that settlement of the caisson should be about one-half the settlement computed for the equivalent footing.

Settlement of drilled caissons on bedrock should be very small if the rock is dry. However, water may be found at the bottom of some caissons, and it can cause some settlement—sometimes large settlement if soft rocks disintegrate upon soaking. Therefore, it is desirable that the water be pumped out and the caisson thoroughly cleaned during the last stage of drilling [2].

11–4 CONSTRUCTION AND INSPECTION OF DRILLED CAISSONS

Construction of drilled caissons consists for the most part of excavation of soil and placement of concrete (perhaps with reinforcing steel). As related in Section 11–1, drilled caissons generally are excavated by using an auger drill or another type of drilling equipment. An auger is a screwlike device (see Figure 11–7) that is attached to a shaft and rotated under power. The rotating action digs into the soil and raises it to the surface. If a caisson is to have a bell at the bottom, the bell is made by using a reamer.

While excavation is being done, soil is exposed in the walls. Soil at the caisson's bottom and exposed in the walls should be examined (and records kept) whenever possible to check the adequacy of the supporting soil at the caisson's bottom and to determine the depth to, and thickness of, different soil strata. Sometimes a person may be able to descend in the shaft for inspection.

After excavation of the soil, the concrete must be of acceptable quality and properly placed. It is preferable that concrete not strike the sides of the hole as it is

FIGURE 11–7 Large auger used in drilled caisson construction.

FIGURE 11–8 Steel casing used in drilled caisson construction.

FIGURE 11–9 Reinforcing steel used in drilled caisson construction.

being poured. A casing (see Figure 11–8), if used, is generally removed as the concrete is poured. Normally, only the concrete in the upper part of the shaft is vibrated. It is always best to pour concrete in the dry, but if water is present, concrete can be placed underwater. Installation of reinforcing steel (see Figure 11–9) (if specified) should be carefully checked prior to placing concrete.

One final aspect of the overall construction process is inspection. A drilled caisson should be inspected for accuracy of the caisson's alignment and dimensions, for bearing capacity of the soil at the caisson's bottom, for proper placement of reinforcing steel and concrete, and so on. Normally, the owner's representative should be present during construction of the caisson to ensure that it is done properly and according to specifications. Figure 11–10 shows a setup for a caisson field inspection.

11–5 PROBLEMS

11–1. A plain-concrete drilled caisson is to be constructed in a clayey soil. The caisson shaft's diameter is 4 ft, and the belled bottom is 8 ft in diameter. The

FIGURE 11–10 Set-up for a caisson field inspection.

FIGURE 11–11

4 ft
Clay
Unit Weight = 104 lb/ft^3
c = 900 lb/ft^2
Drilled with Slurry
28 ft
36 ft
Plain Concrete
4 ft
4 ft
Clay
Unit Weight = 118 lb/ft^3
c = 2700 lb/ft^2
Drilled with Slurry
8 ft

drilled caisson extends to a total depth of 36 ft. Soil conditions are illustrated in Figure 11–11. Compute the caisson's allowable bearing capacity if (a) the excavation is drilled dry, and (b) the foundation is to be drilled with bentonite slurry, and a factor of safety of 3 is to be used. (*Note:* The maximum allowable compressive stress of plain concrete is assumed to be 1200 lb/in.^{2})

11–2. The excavation for a caisson to carry a total load of 1500 kN is to be drilled dry in the soil profile shown in Figure 11–12. The caisson will be made of concrete with f_c' of 27,500 kN/m^2. It is specified that the unit compressive stress of the caisson shaft shall not exceed 33% of its ultimate 28-day compressive strength nor 8274 kN/m^2. Find the required diameter of the plain-concrete straight-shaft caisson.

FIGURE 11–12

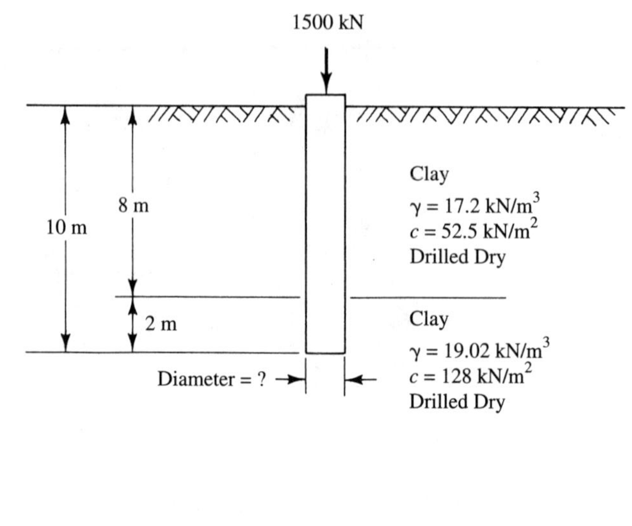

FIGURE 11–13

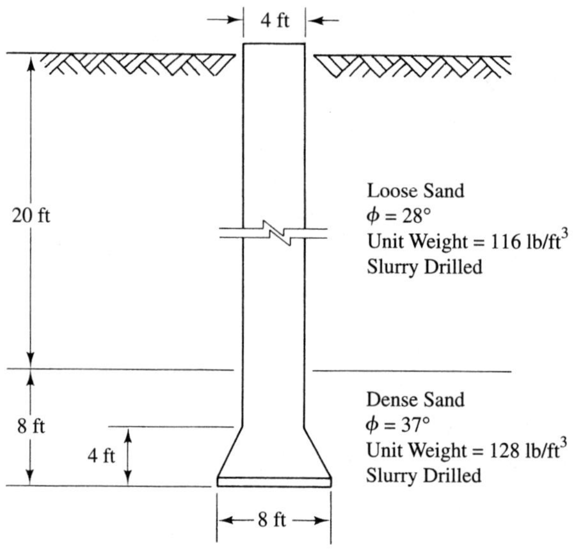

11–3. A drilled caisson 4 ft in diameter and supported by a bell end is to be constructed of plain concrete in sand. Soil conditions and a sketch of the caisson are shown in Figure 11–13. Compute the caisson's design capacity if the excavation is slurry drilled and the factor of safety is 3. Assume that the coefficient of friction between sand and concrete is tan $\frac{2}{3}\phi$ for this bentonite-slurry-drilled caisson. The maximum allowable compressive stress of plain concrete is assumed to be 1200 lb/in.^{2}

11–4. A straight drilled caisson, 4 ft in diameter and made of reinforced concrete, rests on horizontal bedded granite (massive igneous rock). The unconfined

FIGURE 11–14

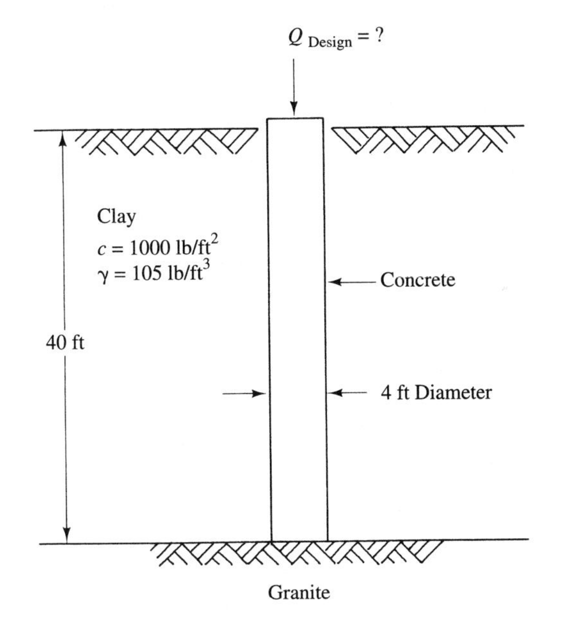

compressive strength of the intact granite sample is 20,000 lb/in.^{2} Determine the safe design load on the caisson if the skin friction of the caisson shaft is neglected (see Figure 11–14). A local building code states the following:

1. "The unit allowable compressive stress shall not exceed 1200 lb/in.^{2} for plain concrete."
2. "The allowable bearing value of the rock shall not exceed the following:
   Massive igneous or metamorphic rock — 100 tons/ft^2
   Massive sedimentary rock — 20 tons/ft^2"

References

[1] David F. McCarthy, *Essentials of Soil Mechanics and Foundations,* Reston Publishing Company, Inc., Reston, Va., 1977.

[2] Wayne C. Teng, *Foundation Design,* Prentice-Hall, Inc., Englewood Cliffs, N.J., 1962.

[3] Karl Terzaghi and Ralph B. Peck, *Soil Mechanics in Engineering Practice,* John Wiley & Sons, Inc., New York, 1967. Copyright © 1967, by John Wiley & Sons, Inc. Reprinted by permission of John Wiley & Sons, Inc.

[4] Karl Terzaghi, Ralph B. Peck, and Gholamreza Mesri, *Soil Mechanics in Engineering Practice,* 3rd ed., John Wiley & Sons, Inc., New York, 1996.

[5] L. C. Reese and M. W. O'Neill, *Drilled Shafts: Construction Procedures and Design Methods,* Fed. Hwy. Adm. Publ. FHWA-HI-88-042, McLean, Va., 1988.

12 LATERAL EARTH PRESSURE

12–1 INTRODUCTION

The word *lateral* means "to the side" or "sideways." Thus, *lateral earth pressure* means "pressure to the side," or "sideways pressure." Analysis and determination of lateral earth pressure are necessary to design retaining walls and other earth retaining structures, such as bulkheads, abutments, and the like. Obviously, the magnitude and location of lateral earth pressure must be known in order to design a retaining wall or other retaining structure that can withstand applied pressure with an adequate safety margin. Almost always, engineers calculate earth pressures and forces on a unit (1-ft or 1-m) section of the retaining wall.

There are three categories of earth pressure—*earth pressure at rest, active earth pressure,* and *passive earth pressure.* Earth pressure at rest (P_0) refers to lateral pressure caused by earth that is prevented from lateral movement by an unyielding wall. In actuality, however, some retaining-wall movement often occurs, resulting in either active or passive earth pressure as explained next.

If a wall moves away from soil, as sketched in Figure 12–1, the earth surface will tend to be lowered, and lateral pressure on the wall will be decreased. If the wall moves far enough away, shear failure of the soil will occur, and a sliding soil wedge will tend to move forward and downward. The earth pressure exerted on the wall at this state of failure is known as active earth pressure (P_a), and it is at minimum value.

If, on the other hand, a wall moves toward soil, as shown in Figure 12–2, the earth surface will tend to be raised, and lateral pressure on the wall will be increased. If the wall moves far enough toward the soil, shear failure of the soil will occur, and a sliding soil wedge will tend to move backward and upward. The earth pressure exerted on the wall at this state of failure is known as passive earth pressure (P_p), and it is at maximum value. Figure 12–3 illustrates the relationship between wall movement and variation in lateral earth pressure.

Section 12–2 discusses earth pressure at rest, whereas Sections 12–3 and 12–4 cover determination of active and passive earth pressures according to Rankine and Coulomb theory, respectively. The effects of a surcharge load on active thrust are discussed in Section 12–5. Culmann's graphic solution for finding active earth

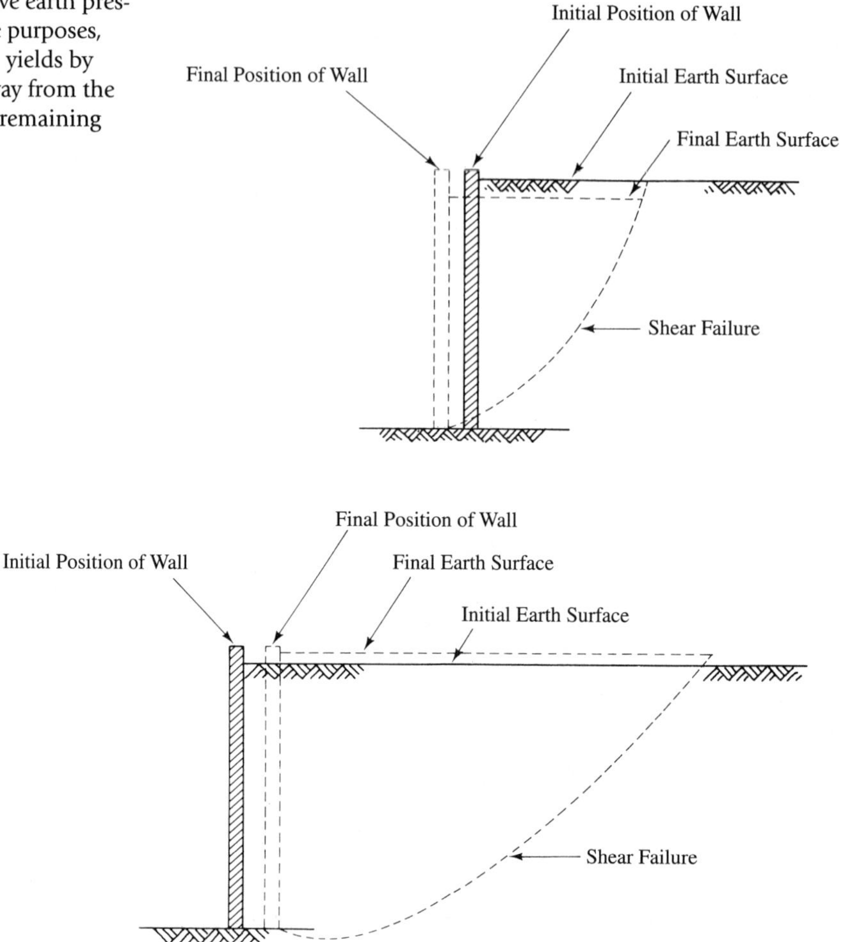

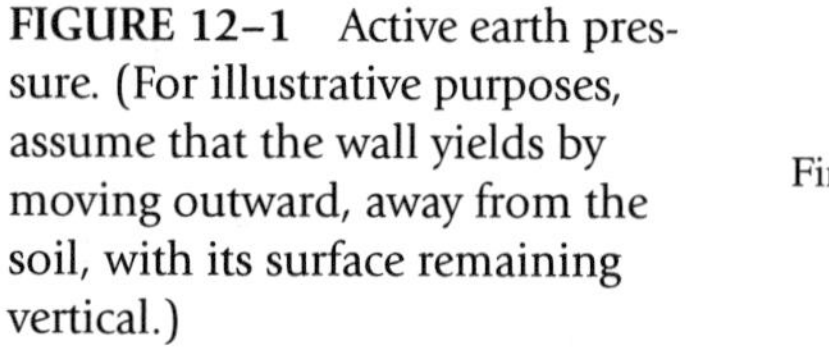
FIGURE 12–1 Active earth pressure. (For illustrative purposes, assume that the wall yields by moving outward, away from the soil, with its surface remaining vertical.)

FIGURE 12–2 Passive earth pressure. (For illustrative purposes, assume that the wall moves backward, toward the soil, with its surface remaining vertical.)

pressure is presented in Section 12–6. Section 12–7 gives some design considerations for retaining walls. Lateral earth pressure on braced sheetings is considered in Section 12–8.

12–2 EARTH PRESSURE AT REST

As noted in Section 12–1, earth pressure at rest refers to lateral pressure caused by earth that is prevented from lateral movement by an unyielding wall. Such a condition can occur, for example, when earth rests against the outer sides of a building's

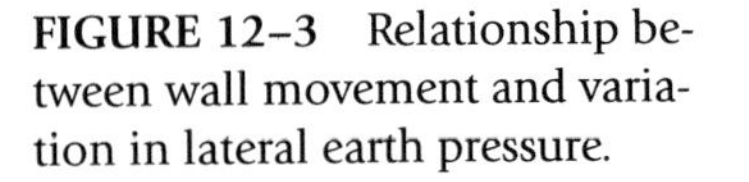

FIGURE 12–3 Relationship between wall movement and variation in lateral earth pressure.

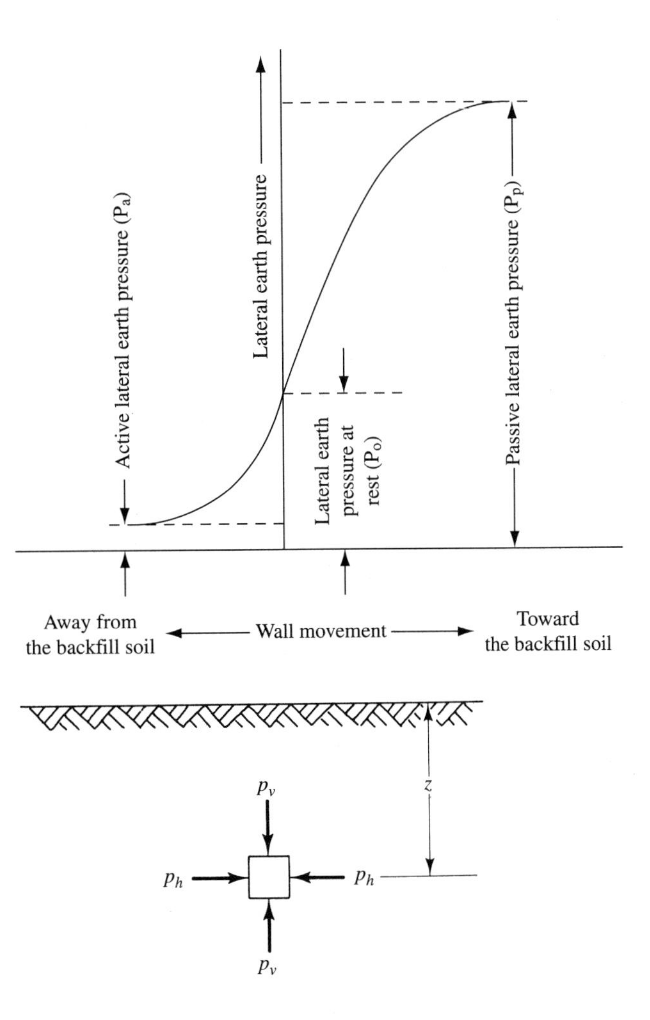

FIGURE 12–4 Subsurface stresses in a soil mass at depth z.

basement walls. With virtually no wall movement, soil in contact with the wall does not undergo lateral strain and does not therefore develop its full shearing resistance. In this case, the magnitude of earth pressure on the wall (i.e., the earth pressure at rest) falls somewhere between the active and passive pressures.

To analyze earth pressure at rest, consider the stress conditions on an element of soil at depth z (see Figure 12–4). Although the element can deform vertically when loaded, it cannot deform laterally because the element is confined by the same soil under identical loading conditions. This configuration is equivalent to soil resting against a smooth, immovable wall (see Figure 12-5), and the soil is in a state of elastic equilibrium. In this case, pressure at the base of the wall and the resultant force per unit length of wall can be determined for dry soil by using the following equations:

FIGURE 12–5 Earth pressure at rest for dry soil.

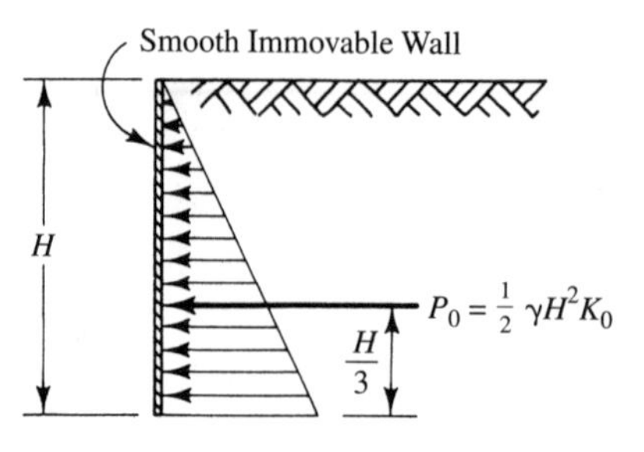

$$p_0 = K_0 \gamma H \tag{12–1}$$

$$P_0 = \tfrac{1}{2} K_0 \gamma H^2 \tag{12–2}$$

where p_0 = pressure at base of the wall
P_0 = resultant force per unit length of wall for earth pressure at rest
K_0 = coefficient of earth pressure at rest (defined in the following paragraph)
γ = unit weight of the soil
H = height of the wall

For the zero lateral strain condition, lateral and vertical stresses (p_h and p_v in Figure 12–4) are related by Poisson's ratio, μ, as follows:

$$\frac{p_h}{p_v} = \frac{\mu}{1 - \mu} \tag{12–3}$$

The ratio of p_h to p_v in a soil mass is known as the *coefficient of earth pressure at rest* and is denoted by K_0. Hence,

$$K_0 = \frac{p_h}{p_v} \tag{12–4}$$

K_0 has been observed in experiments to be dependent on a soil's angle of internal friction (ϕ) and plasticity index, as well as its stress history.

As stated in Chapter 11, for granular soils the coefficient of lateral earth pressure at rest ranges from about 0.4 for dense sand to 0.5 for loose sand. K_0 can also be determined for sands by the following empirical relationship [1]:

$$K_0 = 1 - \sin \phi \tag{12–5}$$

For normally consolidated clays, the following empirical equation can be used to estimate K_0 [2]:

$$K_0 = 0.19 + 0.233 \log (PI) \tag{12–6}$$

where *PI* is the soil's plasticity index. For overconsolidated clays, values of K_0 tend to be larger than those of normally consolidated clays. Figure 12–6 gives an empirical relationship for determining K_0 as a function of the overconsolidation ratio (OCR) (see Section 8–5).

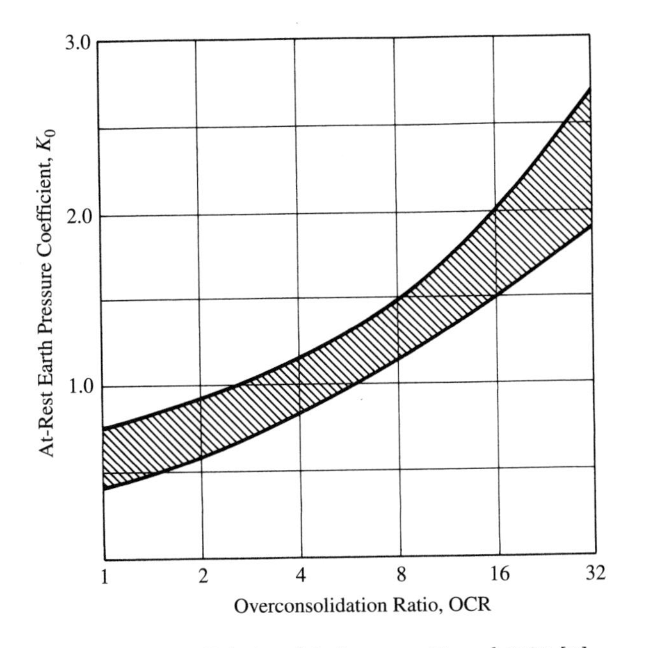

FIGURE 12–6 Relationship between K_0 and OCR [3].

When some or all of the wall in question is below the groundwater table, hydrostatic pressure acting against the submerged section of wall must be added to the effective lateral soil pressure. From Figure 12–7, it can be observed that the lateral earth pressure at rest at the water table (p_1) is given by

$$p_1 = K_0 \gamma z_1 \tag{12–7}$$

whereas that at the base of the wall (p_2) is

$$p_2 = K_0 \gamma z_1 + K_0 \gamma_{sub} z_2 + \gamma_w z_2 \tag{12–8}$$

(γ, γ_{sub}, and γ_w represent the unit weight of soil, submerged unit weight of soil, and unit weight of water, respectively.) The resultant force per unit length of wall (P_0) can be determined by finding the area under the lateral earth pressure diagram:

$$P_0 = \frac{p_1 z_1}{2} + \frac{p_1 + p_2}{2}(z_2) \tag{12–9}$$

EXAMPLE 12–1

Given

A smooth, unyielding wall retains a dense cohesionless soil with no lateral movement of soil (i.e., "at-rest condition" is assumed), as shown in Figure 12–8.

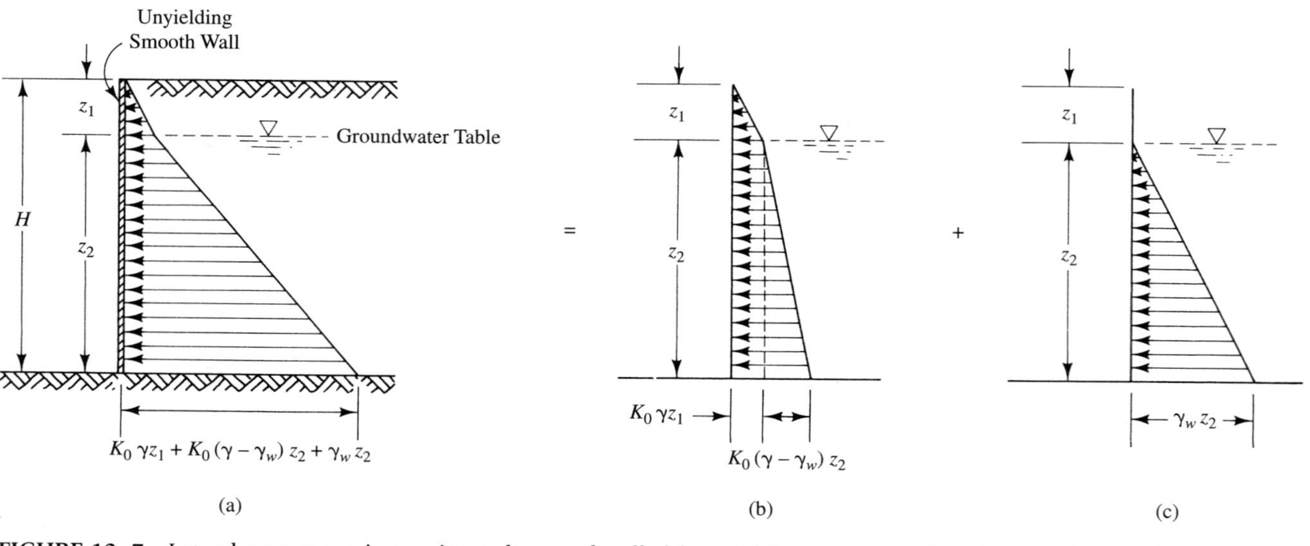

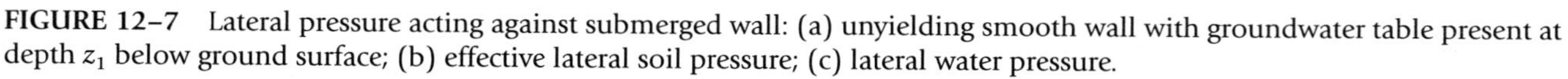

FIGURE 12–7 Lateral pressure acting against submerged wall: (a) unyielding smooth wall with groundwater table present at depth z_1 below ground surface; (b) effective lateral soil pressure; (c) lateral water pressure.

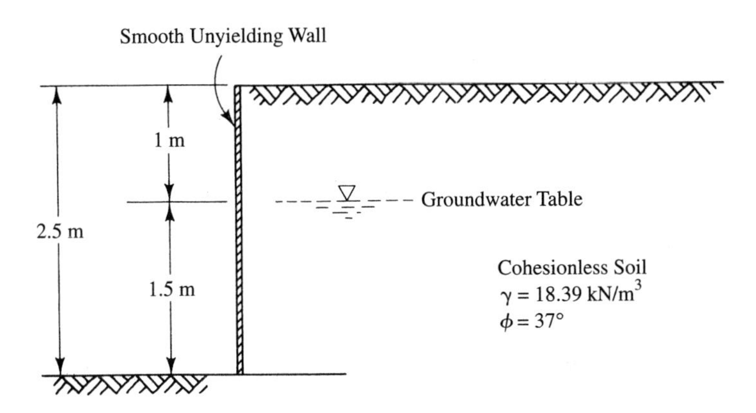

FIGURE 12–8

Required

1. Diagram of lateral earth pressure against the wall.
2. Total lateral force acting on the wall.

Solution

From Eq. (12–5),

$$K_0 = 1 - \sin \phi \qquad (12\text{–}5)$$
$$K_0 = 1 - \sin 37° = 0.398$$

1. *Pressure at 1-m depth (at the water table):* From Eq. (12–7),

$$p_1 = K_0 \gamma z_1 \qquad (12\text{–}7)$$
$$p_1 = (0.398)(18.39\ \text{kN/m}^3)(1.00\ \text{m}) = 7.32\ \text{kN/m}^2$$

Pressure at 2.5-m depth (at the wall base): From Eq. (12–8),

$$p_2 = K_0 \gamma z_1 + K_0 \gamma_{\text{sub}} z_2 + \gamma_w z_2 \qquad (12\text{–}8)$$
$$p_2 = (0.398)(18.39\ \text{kN/m}^3)(1.00\ \text{m}) + (0.398)(18.39\ \text{kN/m}^3 - 9.81\ \text{kN/m}^3)(1.5\ \text{m}) + (9.81\ \text{kN/m}^3)(1.5\ \text{m})$$
$$= 27.16\ \text{kN/m}^2$$

The required diagram of lateral earth pressure against the wall is shown in Figure 12–9.

2. From Eq. (12–9),

$$P_0 = \frac{p_1 z_1}{2} + \frac{p_1 + p_2}{2}(z_2) \qquad (12\text{–}9)$$

$$P_0 = \frac{(7.32\ \text{kN/m}^2)(1.00\ \text{m})}{2} + \frac{7.32\ \text{kN/m}^2 + 27.16\ \text{kN/m}^2}{2}(1.5\ \text{m})$$

$$= 29.52\ \text{kN/m of wall}$$

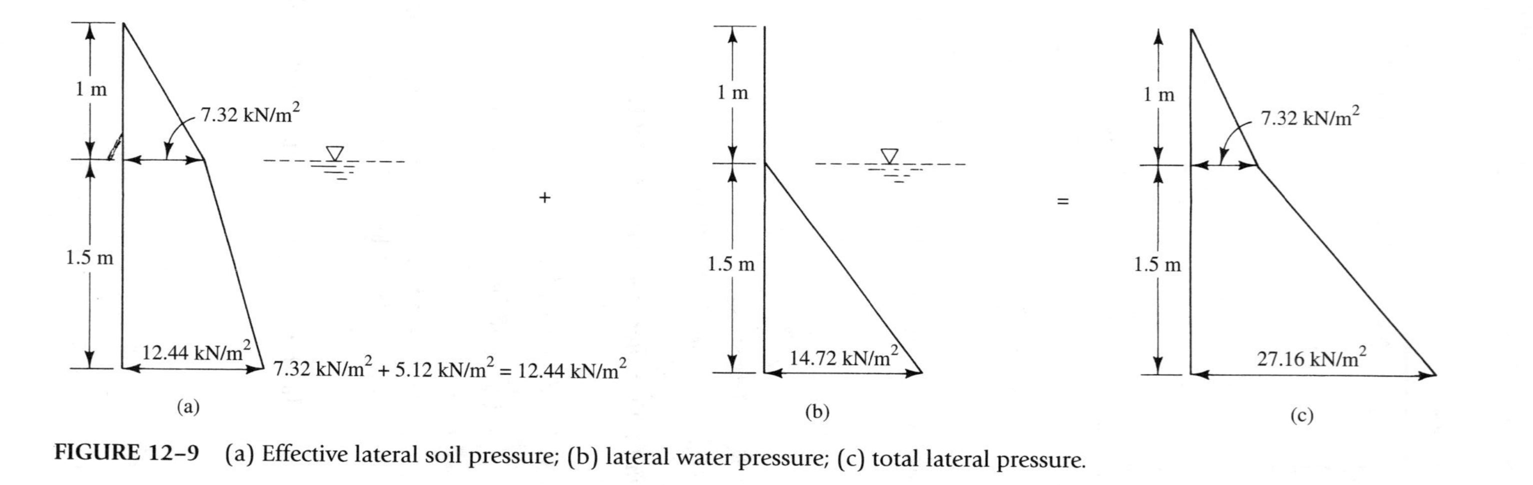

FIGURE 12–9 (a) Effective lateral soil pressure; (b) lateral water pressure; (c) total lateral pressure.

12–3 RANKINE EARTH PRESSURES

The Rankine theory for determining lateral earth pressures is based on several assumptions. The primary one is that there is no adhesion or friction between wall and soil (i.e., the wall is smooth). In addition, lateral pressures computed from Rankine theory are limited to vertical walls. Failure is assumed to occur in the form of a sliding wedge along an assumed failure plane defined as a function of the soil's angle of internal friction (ϕ), as shown in Figure 12–10. Lateral earth pressure varies linearly with depth (see Figure 12–11), and resultant pressures are assumed to act at a distance up from the base of the wall equal to one-third the vertical distance from the heel at the wall's base to the surface of the backfill (Figure 12–11). The direction of resultants is parallel to the backfill surface.

The primary assumption of this theory (i.e., that the wall is smooth) is not valid. Nevertheless, equations derived based on this assumption are widely used for computing lateral earth pressures, and, propitiously, results obtained using these equations may not differ appreciably from results based on more accurate and sophisticated analyses. In fact, results based on Rankine theory generally give slightly larger values, causing a slightly larger wall to be designed, thus giving a small additional safety factor.

The equations for computing lateral earth pressure* based on Rankine theory are as follows [4]:

* P_a and P_p are actually forces per unit length of wall; however, they are commonly referred to as the *lateral earth pressure.*

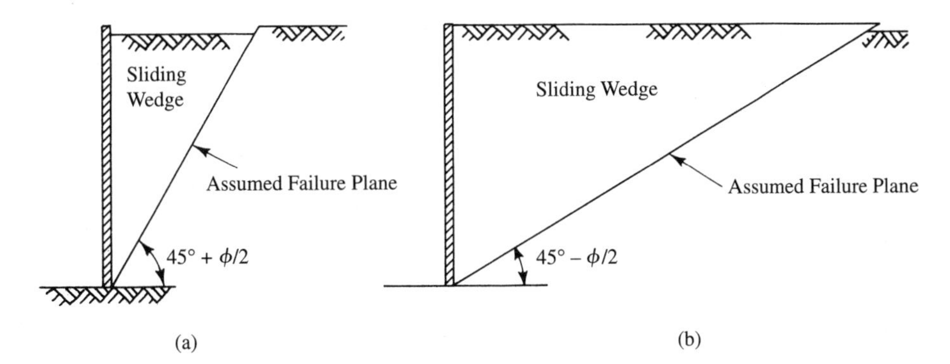

FIGURE 12–10 Assumed failure plane for Rankine theory: (a) Rankine active state; (b) Rankine passive state.

$$P_a = \frac{1}{2}\gamma H^2 K_a \tag{12–10}$$

where

$$K_a = \cos\beta \frac{\cos\beta - \sqrt{\cos^2\beta - \cos^2\phi}}{\cos\beta + \sqrt{\cos^2\beta - \cos^2\phi}} \tag{12–11}$$

$$P_p = \frac{1}{2}\gamma H^2 K_p \tag{12–12}$$

where

$$K_p = \cos\beta \frac{\cos\beta + \sqrt{\cos^2\beta - \cos^2\phi}}{\cos\beta - \sqrt{\cos^2\beta - \cos^2\phi}} \tag{12–13}$$

where P_a = active earth pressure
γ = unit weight of the backfill soil
H = height of the wall (see Figure 12–11)
K_a = coefficient of active earth pressure
β = angle between backfill surface line and a horizontal line (Figure 12–11)
ϕ = angle of internal friction of the backfill soil
P_p = passive earth pressure
K_p = coefficient of passive earth pressure

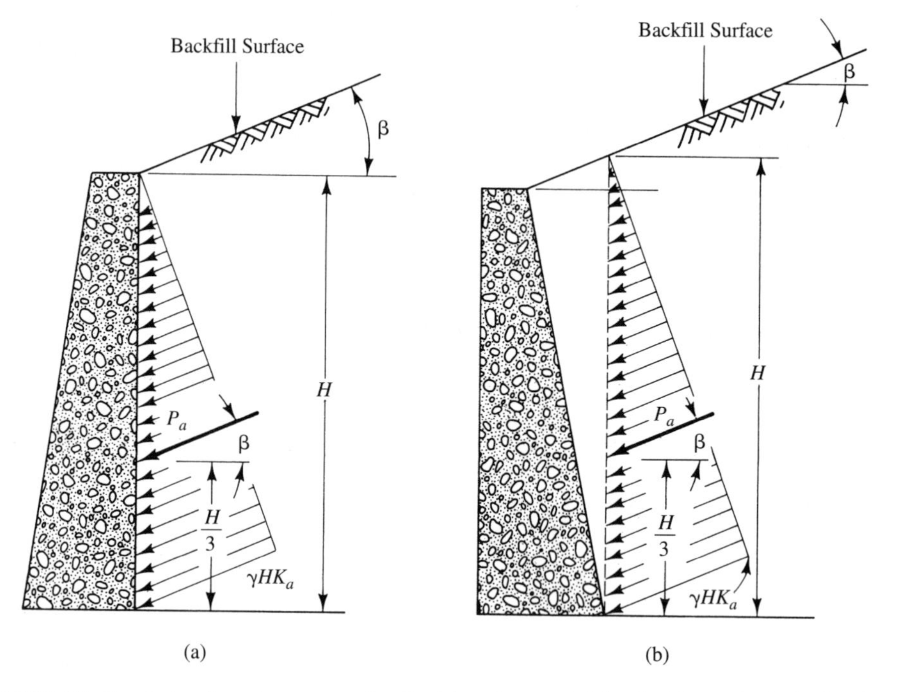

FIGURE 12–11 Lateral earth pressure for Rankine theory: (a) back side vertical; (b) back side inclined.

If the backfill surface is level, angle β is zero, and Eqs. (12–11) and (12–13) revert to

$$K_a = \frac{1 - \sin\phi}{1 + \sin\phi} \tag{12–14}$$

$$K_p = \frac{1 + \sin\phi}{1 - \sin\phi} \tag{12–15}$$

By trigonometric identities,

$$\frac{1 - \sin\phi}{1 + \sin\phi} = \tan^2\left(45° - \frac{\phi}{2}\right)$$

$$\frac{1 + \sin\phi}{1 - \sin\phi} = \tan^2\left(45° + \frac{\phi}{2}\right)$$

Therefore, the equations for determining the coefficients of active and passive earth pressure for level backfill surfaces can also be expressed as follows:

$$K_a = \tan^2\left(45° - \frac{\phi}{2}\right) \tag{12–16}$$

$$K_p = \tan^2\left(45° + \frac{\phi}{2}\right) \tag{12–17}$$

Example 12–2 illustrates the computation of lateral earth pressure for a level backfill surface, and Example 12–3 illustrates the computation for a sloping backfill surface. Example 12–4 gives a technique for computing lateral earth pressure based on Rankine theory for a retaining wall with a back side that is not vertical.

EXAMPLE 12–2

Given

The retaining wall shown in Figure 12–12.

Required

Total active earth pressure per foot of wall and its point of application, by Rankine theory.

Solution

From Eqs. (12–10) and (12–14) (for level backfill),

$$P_a = \tfrac{1}{2}\gamma H^2 K_a \tag{12–10}$$

$$K_a = \frac{1 - \sin\phi}{1 + \sin\phi} \tag{12–14}$$

$$K_a = \frac{1 - \sin 30°}{1 + \sin 30°} = 0.333$$

$$P_a = (\tfrac{1}{2})(110\ \text{lb/ft}^3)(30\ \text{ft})^2(0.333) = 16{,}500\ \text{lb/ft}$$

FIGURE 12–12

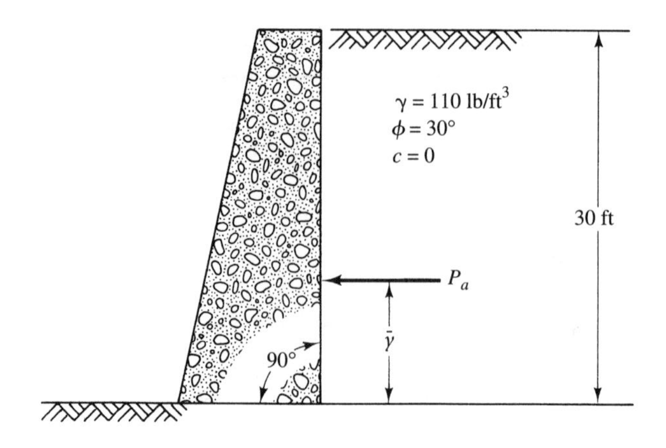

The point of application of the total earth pressure $(\bar{y}) = H/3 = 30\text{ ft}/3 = 10\text{ ft}$ from the base of the wall.

EXAMPLE 12–3

Given

The retaining wall shown in Figure 12–13.

Required

Total active earth pressure per foot of wall and its point of application, by Rankine theory.

Solution

From Eqs. (12–10) and (12–11),

$$P_a = \tfrac{1}{2}\gamma H^2 K_a \tag{12–10}$$

$$K_a = \cos\beta \frac{\cos\beta - \sqrt{\cos^2\beta - \cos^2\phi}}{\cos\beta + \sqrt{\cos^2\beta - \cos^2\phi}} \tag{12–11}$$

$$K_a = (\cos 15^\circ) \frac{\cos 15^\circ - \sqrt{\cos^2 15^\circ - \cos^2 30^\circ}}{\cos 15^\circ + \sqrt{\cos^2 15^\circ - \cos^2 30^\circ}} = 0.373$$

$$P_a = (\tfrac{1}{2})(17.3\text{ kN/m}^3)(9.1\text{ m})^2(0.373) = 267\text{ kN/m}$$

$\bar{y} = H/3 = 9.1\text{ m}/3 = 3.03\text{ m}$ from the base of the wall (see Figure 12–13).

EXAMPLE 12–4

Given

The retaining wall shown in Figure 12–14.

FIGURE 12–13

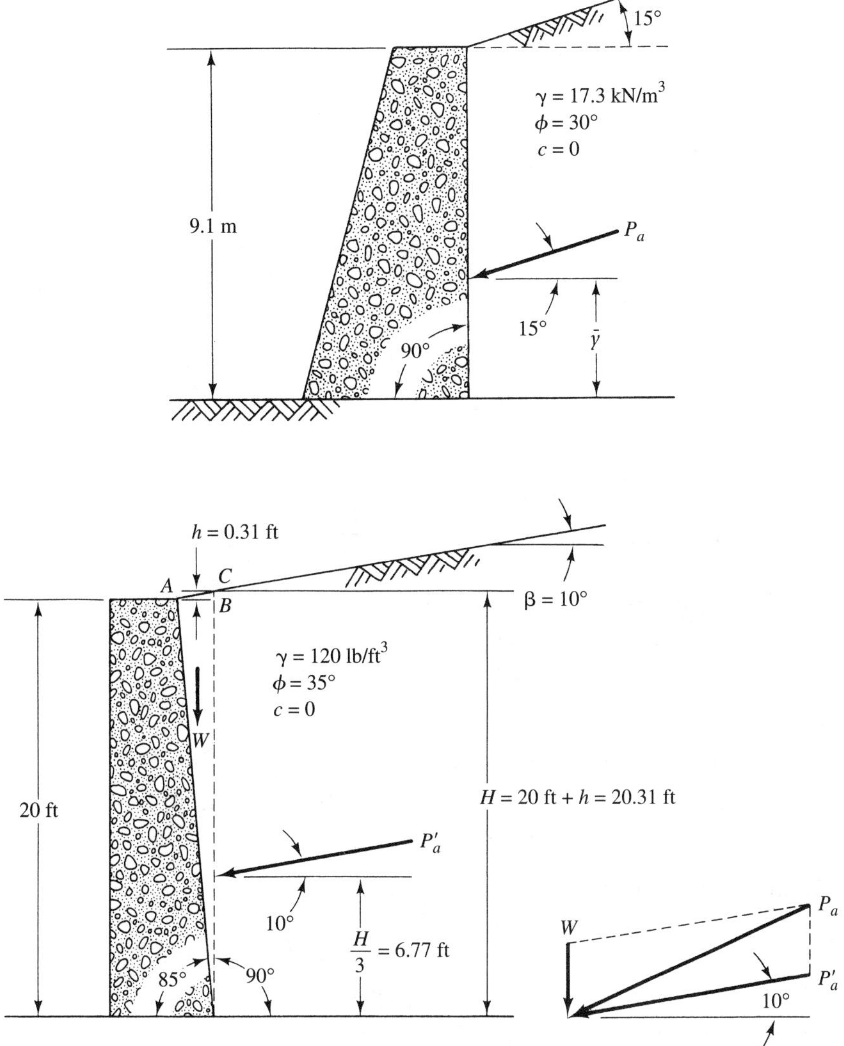

FIGURE 12–14

Required

Total active earth pressure per foot of wall, by Rankine theory.

Solution

As shown in Figure 12–14,

$$\tan 5° = \frac{AB}{20\text{ ft}}$$

$$AB = (20\text{ ft})(\tan 5°) = 1.75\text{ ft}$$

Also,

$$\tan 10° = \frac{BC}{AB} = \frac{h}{1.75 \text{ ft}}$$

$$h = (1.75 \text{ ft})(\tan 10°) = 0.31 \text{ ft}$$

From Eqs. (12–10) and (12–11),

$$P'_a = \tfrac{1}{2}\gamma H^2 K_a \qquad \textbf{(12–10)}$$

$$K_a = \cos\beta \frac{\cos\beta - \sqrt{\cos^2\beta - \cos^2\phi}}{\cos\beta + \sqrt{\cos^2\beta - \cos^2\phi}} \qquad \textbf{(12–11)}$$

$$\gamma = 120 \text{ lb/ft}^3$$
$$H = 20.31 \text{ ft}$$
$$\beta = 10°$$
$$\phi = 35°$$

$$K_a = (\cos 10°)\frac{\cos 10° - \sqrt{\cos^2 10° - \cos^2 35°}}{\cos 10° + \sqrt{\cos^2 10° - \cos^2 35°}} = 0.282$$

$$P'_a = (\tfrac{1}{2})(120 \text{ lb/ft}^3)(20.31 \text{ ft})^2(0.282) = 6979 \text{ lb/ft}$$
$$W = (\tfrac{1}{2})(\gamma)(AB)(H)$$
$$W = (\tfrac{1}{2})(120 \text{ lb/ft}^3)(1.75 \text{ ft})(20.31 \text{ ft}) = 2133 \text{ lb/ft}$$
$$P_h = P'_a \cos\beta = (6979 \text{ lb/ft}) \cos 10° = 6873 \text{ lb/ft}$$
$$P_v = P'_a \sin\beta = (6979 \text{ lb/ft}) \sin 10° = 1212 \text{ lb/ft}$$
$$\Sigma V = W + P_v = 2133 \text{ lb/ft} + 1212 \text{ lb/ft} = 3345 \text{ lb/ft}$$
$$\Sigma H = P_h = 6873 \text{ lb/ft}$$

$$\text{Total active earth pressure } (P_a) = \sqrt{(\Sigma V)^2 + (\Sigma H)^2}$$
$$= \sqrt{(3345 \text{ lb/ft})^2 + (6873 \text{ lb/ft})^2}$$
$$= 7640 \text{ lb/ft}$$

Equations (12–10) through (12–17) are applicable for cohesionless soils. The generalized lateral earth pressure distribution for soils that have both cohesion and friction is, based on Rankine theory, as shown in Figure 12–15. Figure 12–15a gives the pressure distribution for active pressure, and Figure 12–15b gives that for passive pressure. It can be noted that active pressure acts over only the lower part of the wall (Figure 12–15a). The pressure distribution for a particular case can be ascertained by substituting appropriate parameters into the equations indicated in Figure 12–15. Example 12–5 illustrates this method.

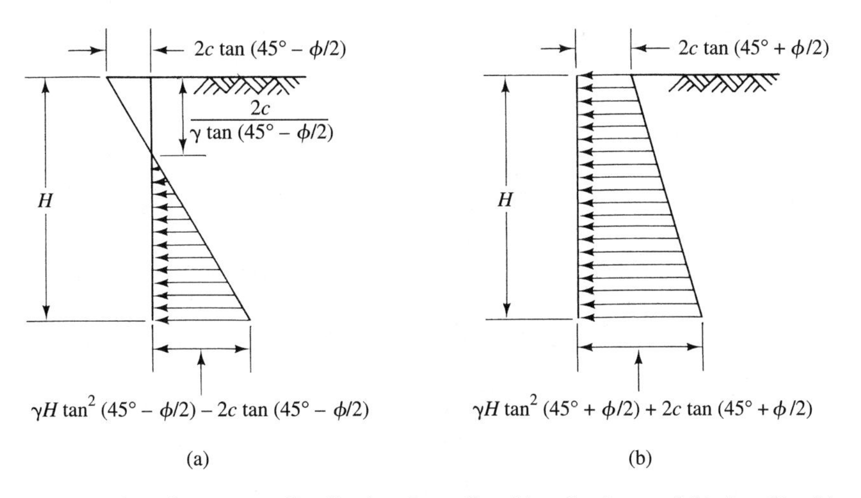

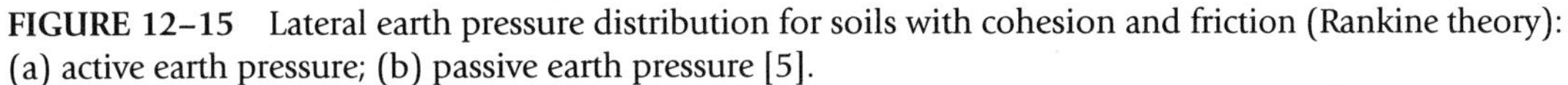

FIGURE 12–15 Lateral earth pressure distribution for soils with cohesion and friction (Rankine theory): (a) active earth pressure; (b) passive earth pressure [5].

EXAMPLE 12–5

Given

The retaining wall shown in Figure 12–16.

Required

Active earth pressure diagram, by Rankine theory.

Solution

From Figure 12–15a,

$$2c \tan\left(45^\circ - \frac{\phi}{2}\right) = (2)(200\ \text{lb/ft}^2)\tan\left(45^\circ - \frac{10^\circ}{2}\right) = 336\ \text{lb/ft}^2$$

$$\gamma H \tan^2\left(45^\circ - \frac{\phi}{2}\right) - 2c\tan\left(45^\circ - \frac{\phi}{2}\right) = (120\ \text{lb/ft}^3)(30\ \text{ft})\tan^2\left(45^\circ - \frac{10^\circ}{2}\right) - (2)(200\ \text{lb/ft}^2)\tan\left(45^\circ - \frac{10^\circ}{2}\right) = 2200\ \text{lb/ft}^2$$

$$\frac{2c}{\gamma \tan\left(45^\circ - \frac{\phi}{2}\right)} = \frac{(2)(200\ \text{lb/ft}^2)}{(120\ \text{lb/ft}^3)\tan\left(45^\circ - \frac{10^\circ}{2}\right)} = 3.97\ \text{ft}$$

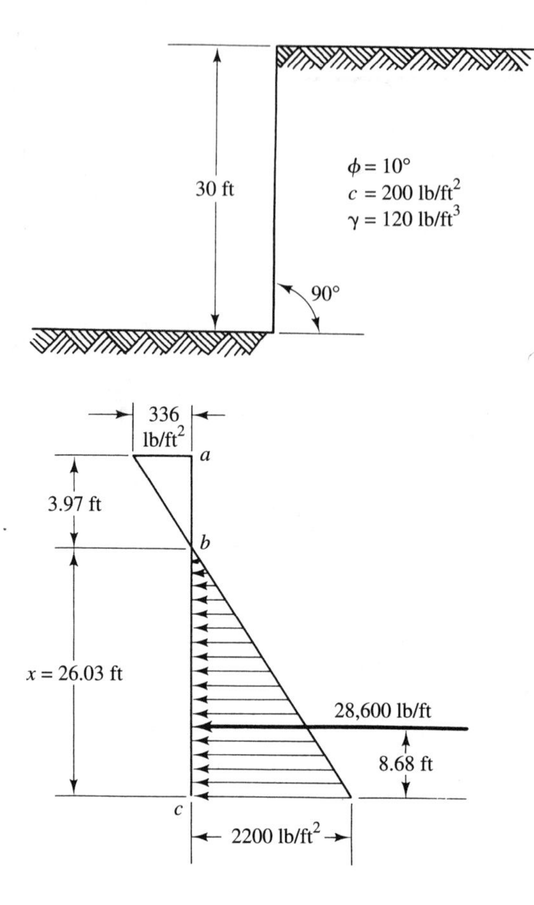

FIGURE 12–16

FIGURE 12–17

$$\text{Resultant} = (\tfrac{1}{2})(2200\ \text{lb/ft}^2)(30\ \text{ft} - 3.97\ \text{ft}) = 28{,}600\ \text{lb/ft}$$

$$\bar{y} = x/3 = (30\ \text{ft} - 3.97\ \text{ft})/3 = 8.68\ \text{ft above the base of the wall}$$

The active earth pressure diagram, based on the preceding computed values, is shown in Figure 12–17.

12–4 COULOMB EARTH PRESSURES

The Coulomb theory for determining lateral earth pressure, developed nearly a century before the Rankine theory, assumes that failure occurs in the form of a wedge and that friction occurs between wall and soil. The sides of the wedge are the earth side of the retaining wall and a failure plane that passes through the heel of the wall (see Figure 12–18). Resultant active earth pressure acts on the wall at a point where a line through the wedge's center of gravity and parallel to the failure plane intersects the wall (see Figure 12–19). The resultant's direction at the wall is along a line that

FIGURE 12–18 Sketch showing failure plane for Coulomb theory.

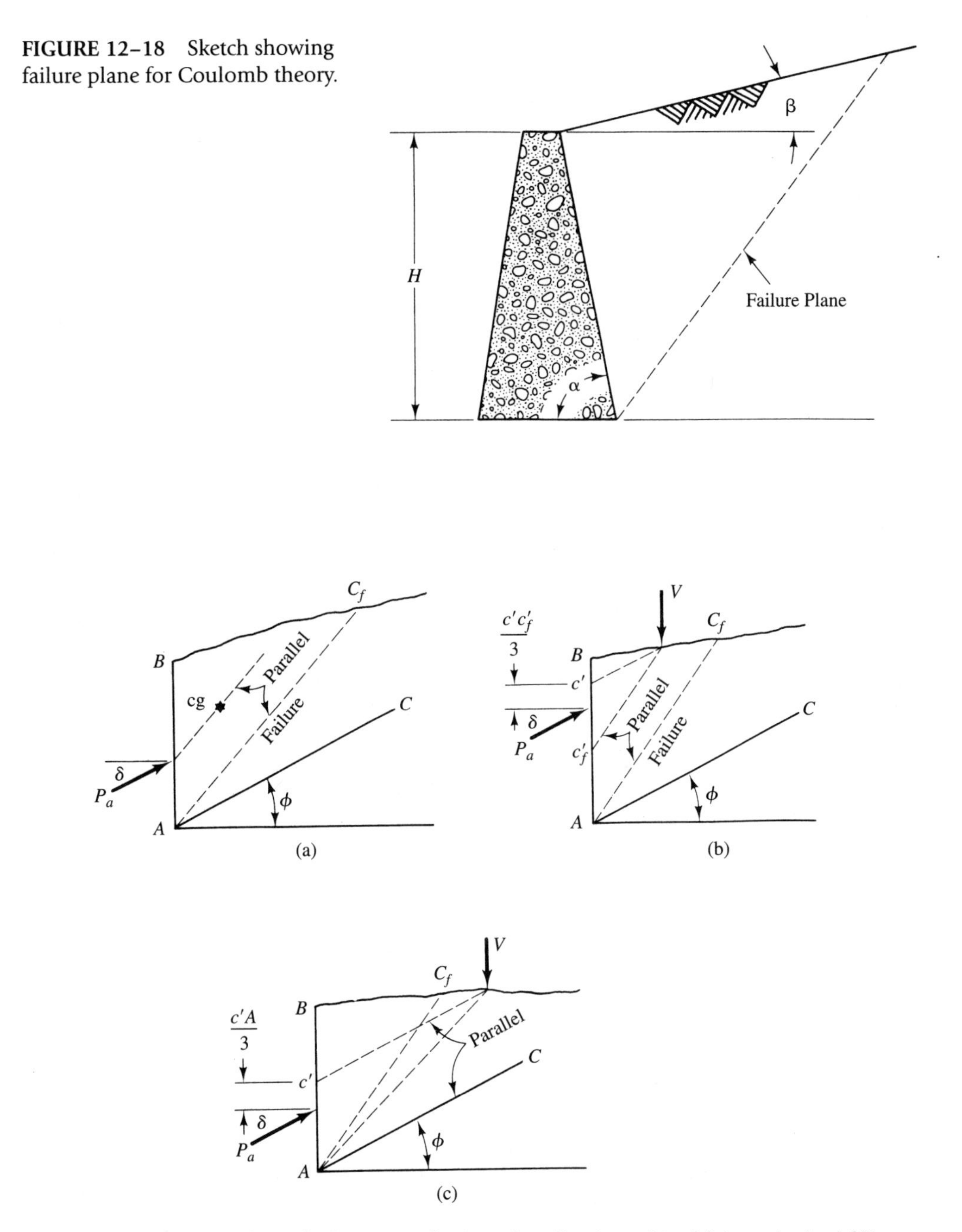

FIGURE 12–19 Procedures for location of point of application of P_a: (a) irregular backfill; (b) concentrated or line load inside failure zone; (c) concentrated or line load outside failure zone (but inside zone *ABC*) [4].

FIGURE 12–20 Sketch showing direction of active pressure resultant for Coulomb theory.

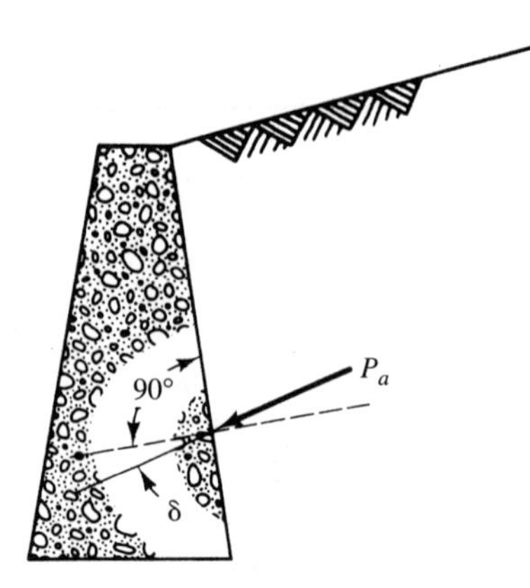

makes an angle δ with a line normal to the back side of the wall, where δ is the angle of wall friction (see Figure 12–20).

Equations for computing lateral earth pressure based on Coulomb theory are as follows [4]:

$$P_a = \frac{1}{2}\gamma H^2 K_a \tag{12–10}$$

where

$$K_a = \frac{\sin^2(\alpha + \phi)}{\sin^2\alpha \sin(\alpha - \delta)\left[1 + \sqrt{\dfrac{\sin(\phi + \delta)\sin(\phi - \beta)}{\sin(\alpha - \delta)\sin(\alpha + \beta)}}\right]^2} \tag{12–18}$$

$$P_p = \frac{1}{2}\gamma H^2 K_p \tag{12–12}$$

where

$$K_p = \frac{\sin^2(\alpha - \phi)}{\sin^2\alpha \sin(\alpha + \delta)\left[1 - \sqrt{\dfrac{\sin(\phi + \delta)\sin(\phi + \beta)}{\sin(\alpha + \delta)\sin(\alpha + \beta)}}\right]^2} \tag{12–19}$$

where P_a = active earth pressure
γ = unit weight of the backfill soil
H = height of the wall (see Figure 12–18)
K_a = coefficient of active earth pressure
α = angle between back side of wall and a horizontal line (Figure 12–18)
ϕ = angle of internal friction of the backfill soil
δ = angle of wall friction
β = angle between backfill surface lines and a horizontal line (Figure 12–18)

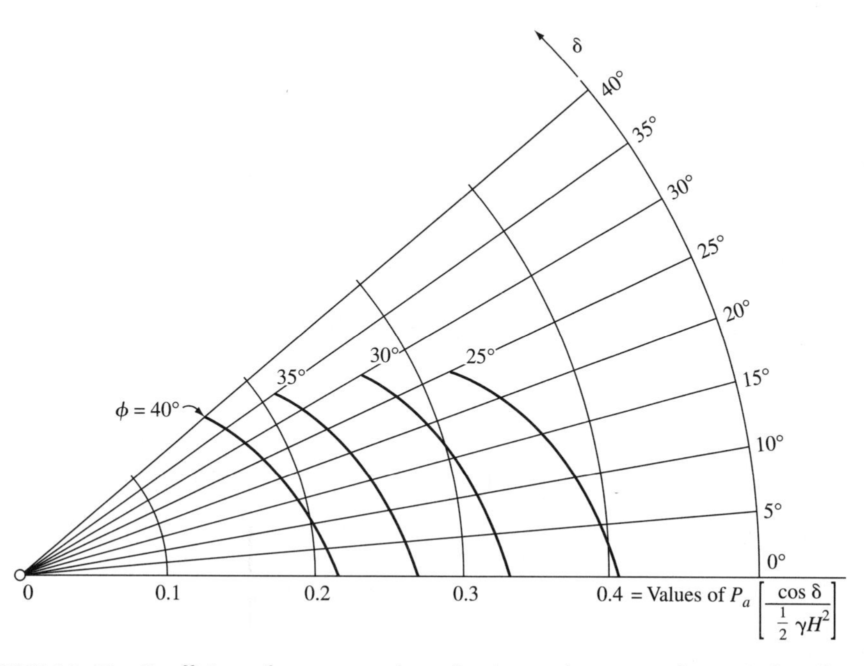

FIGURE 12–21 Coefficients for computation of active earth pressure for vertical walls supporting cohesionless backfill with a horizontal surface [6, 7].
Source: From Karl Terzaghi, Ralph B. Peck, and Gholamreza Mesri, *Soil Mechanics in Engineering Practice,* 3rd ed., John Wiley & Sons, Inc., New York, 1996. Copyright © 1996, by John Wiley & Sons, Inc. Required by permission of John Wiley & Sons, Inc.

P_p = passive earth pressure
K_p = coefficient of passive earth pressure

For vertical walls supporting cohesionless backfill with a horizontal surface (i.e., $\beta = 0$), values of active earth pressure (P_a) can be found from Figure 12–21 in lieu of using Eqs. (12–10) and (12–18).

In the case of a smooth, vertical wall with level backfill, δ and β are each zero and α is 90°; if these values are substituted into Eqs. (12–18) and (12–19), the equations revert to Eqs. (12–14) and (12–15), respectively. The latter two equations are the Rankine equations for the conditions stated (i.e., smooth, vertical wall with level backfill).

Table 12–1 gives some typical values of angles of internal friction, angles of wall friction, and unit weights of common types of backfill soil.

Examples 12–6 and 12–7 illustrate the computation of lateral earth pressure based on Coulomb theory.

TABLE 12–1
Friction Angles and Unit Weights for Backfill Soil [8]

Number	Description of Soil	Angle of Internal Friction, ϕ		Angle of Wall Friction, δ		Unit Weight γ (lb/ft^3)	
		Dry	*Moist*	*Dry*	*Moist*	*Dry*	*Moist*
1	Coarse to medium sand, trace fine gravel	36°00′	27°30′	27°30′	26°10′	—	91
2	Coarse to fine sand, trace + silt (7.5%)	37°40′	27°50′	32°10′	26°20′	101	95
3	Coarse to fine sand, trace + (7.5%) fine gravel	38°40′	30°00′	27°10′	26°20′	106	94
4	Coarse to fine sand	36°30′	30°00′	28°50′	27°10′	95	80
5	Medium to fine sand, some silt (29%), trace fine gravel	35°10′	29°10′	25°10′	21°30′	99	82
6	Fine sand, trace silt	37°50′	29°20′	29°40′	26°20′	94	82
7	Fine sand, some silt	35°00′	30°20′	28°00′	28°00′	103	96
8	Coarse silt, fine sand (45%)	34°50′	26°10′	27°50′	25°40′	94	80
9	Silt, some coarse to fine sand, trace + clay (7%)	—	31°20′	—	28°50′	—	75

EXAMPLE 12–6

Given

Same conditions as in Example 12–2, except that the angle of wall friction between backfill and wall (δ) is 25° (see Figure 12–22).

Required

Total active earth pressure per foot of wall, by Coulomb theory.

Solution

From Eqs. (12–10) and (12–18),

$$P_a = \tfrac{1}{2}\gamma H^2 K_a \tag{12–10}$$

$$K_a = \frac{\sin^2(\alpha + \phi)}{\sin^2\alpha \sin(\alpha - \delta)\left[1 + \sqrt{\dfrac{\sin(\phi + \delta)\sin(\phi - \beta)}{\sin(\alpha - \delta)\sin(\alpha + \beta)}}\right]^2} \tag{12–18}$$

$\gamma = 110\text{ lb/ft}^3$

$H = 30\text{ ft}$

$\alpha = 90°$

$\phi = 30°$

$\delta = 25°$

FIGURE 12–22

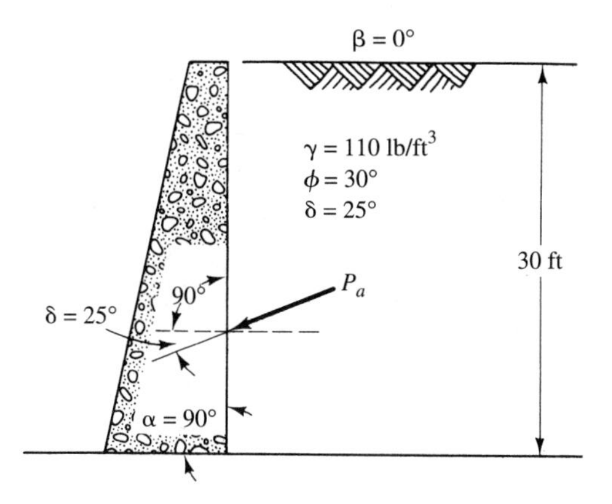

$$\beta = 0° \qquad \text{(level backfill)}$$

$$K_a = \frac{\sin^2(90° + 30°)}{\sin^2(90°)\sin(90° - 25°)\left[1 + \sqrt{\dfrac{\sin(30° + 25°)\sin(30° - 0°)}{\sin(90° - 25°)\sin(90° + 0°)}}\right]^2}$$

$$K_a = 0.296$$

$$P_a = (\tfrac{1}{2})(110\ \text{lb/ft}^3)(30\ \text{ft})^2(0.296) = 14{,}700\ \text{lb/ft}$$

Because this example involves a vertical wall supporting cohesionless backfill with a horizontal surface, an alternative method for finding the solution is to use Figure 12–21. With $\delta = 25°$ and $\phi = 30°$, Figure 12–21 yields the following:

$$P_a\left[\frac{\cos\delta}{\frac{1}{2}\gamma H^2}\right] = 0.27$$

$$P_a\left[\frac{\cos 25°}{(\frac{1}{2})(110\ \text{lb/ft}^3)(30\ \text{ft})^2}\right] = 0.27$$

$$P_a = 14{,}700\ \text{lb/ft}^2$$

EXAMPLE 12–7

Given

Same conditions as Example 12–4, except that the angle of wall friction between backfill and wall (δ) is 20° (see Figure 12–23).

Required

Total active earth pressure per foot of wall, by Coulomb theory.

FIGURE 12–23

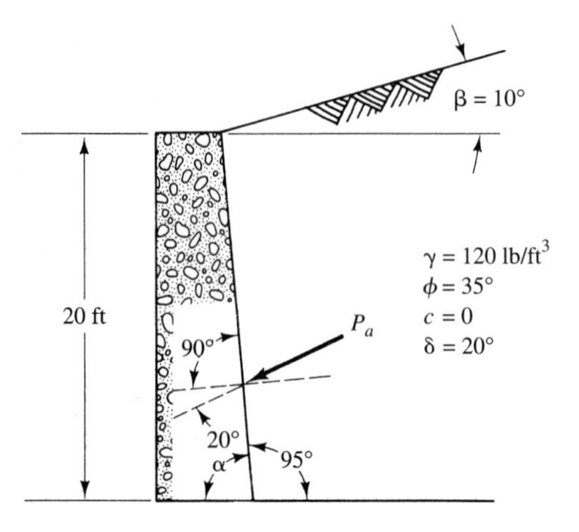

Solution

From Eqs. (12–10) and (12–18),

$$P_a = \tfrac{1}{2}\gamma H^2 K_a \tag{12–10}$$

$$K_a = \frac{\sin^2(\alpha + \phi)}{\sin^2 \alpha \sin(\alpha - \delta)\left[1 + \sqrt{\dfrac{\sin(\phi + \delta)\sin(\phi - \beta)}{\sin(\alpha - \delta)\sin(\alpha + \beta)}}\right]^2} \tag{12–18}$$

$\gamma = 120 \text{ lb/ft}^3$

$H = 20 \text{ ft}$

$\alpha = 180° - 95° = 85°$

$\phi = 35°$

$\delta = 20°$

$\beta = 10°$

$$K_a = \frac{\sin^2(85° + 35°)}{\sin^2(85°)\sin(85° - 20°)\left[1 + \sqrt{\dfrac{\sin(35° + 20°)\sin(35° - 10°)}{\sin(85° - 20°)\sin(85° + 10°)}}\right]^2}$$

$K_a = 0.318$

$P_a = (\tfrac{1}{2})(120 \text{ lb/ft}^3)(20 \text{ ft})^2(0.318) = 7630 \text{ lb/ft}$

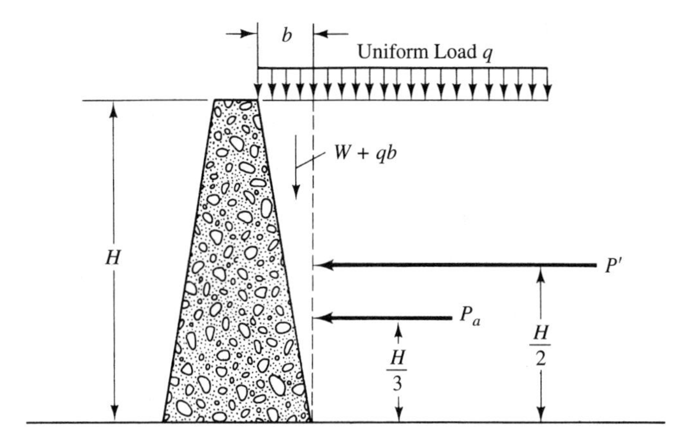

FIGURE 12–24 Sketch showing additional pressure exerted against a retaining wall as a result of a surcharge in the form of a uniform load.
Source: Reprinted with permission of Macmillan Publishing Co., Inc., from *Theory and practice of Foundation Engineering* by Louis J. Goodman and R. H. Karol. Copyright © 1968, Macmillan Publishing Co., Inc.

12–5 EFFECTS OF A SURCHARGE LOAD UPON ACTIVE THRUST

Sometimes backfill resting against a retaining wall is subjected to a surcharge. A surcharge, which is simply a uniform load and/or concentrated load imposed on the soil, adds to the lateral earth pressure exerted against the retaining wall by the backfill. This added pressure must, of course, be considered when the retaining wall is being designed.

Additional pressure exerted against a retaining wall as a result of a surcharge in the form of a uniform load can be computed from the following equation (see Figure 12–24) [9]*:

$$P' = qHK_a \tag{12-20}$$

where P' = additional active earth pressure as a result of uniform load surcharge
q = uniform load (surcharge) on backfill
H = height of wall
K_a = coefficient of active earth pressure [determined from Eq. (12–14)]

Example 12–8, which follows, illustrates the computation of pressure due to a surcharge in the form of a uniform load. Example 12–11 in Section 12–6 illustrates the treatment (graphic solution) of a surcharge in the form of a concentrated load.

* Reprinted with permission of Macmillan Publishing Co., Inc., from *Theory and Practice of Foundation Engineering* by Louis J. Goodman and R. H. Karol. Copyright © 1968, Macmillan Publishing Co., Inc.

EXAMPLE 12–8

Given

1. A smooth vertical wall is 20 ft high and retains a cohesionless soil with $\gamma = 120\ \text{lb/ft}^3$ and $\phi = 28°$.
2. The top of the soil is horizontal and level with the top of the wall.
3. The soil surface carries a uniformly distributed load of $1000\ \text{lb/ft}^2$ (see Figure 12–25).

Required

1. Total active earth pressure on the wall per linear foot of wall.
2. Point of action of the total active earth pressure, by Rankine theory.

Solution

From Eqs. (12–10) and (12–14) (for level backfill),

$$P_a = \tfrac{1}{2}\gamma H^2 K_a \tag{12–10}$$

$$K_a = \frac{1 - \sin\phi}{1 + \sin\phi} \tag{12–14}$$

$$K_a = \frac{1 - \sin 28°}{1 + \sin 28°} = 0.361$$

$$P_a = (\tfrac{1}{2})(120\ \text{lb/ft}^3)(20\ \text{ft})^2(0.361) = 8660\ \text{lb/ft}$$

FIGURE 12–25

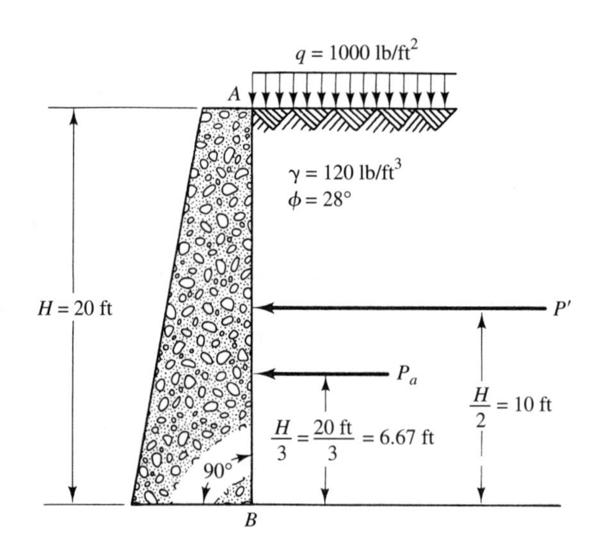

The point of action for $P_a = H/3 = 20$ ft$/3 = 6.67$ ft from the base of the wall. From Eq. (12–20),*

$$P' = qHK_a \quad \textbf{(12–20)}$$

$$P' = (1000 \text{ lb/ft}^2)(20 \text{ ft})(0.361) = 7220 \text{ lb/ft}$$

The point of action for $P' = H/2 = 20$ ft$/2 = 10$ ft from the base of the wall.

1. Total active earth pressure $= P_a + P' = 8660$ lb/ft $+ 7220$ lb/ft $= 15{,}880$ lb/ft.
2. Let the point of application of the total active earth pressure be h ft above the base of the wall. h is obtained by taking moments of forces (i.e., P_a and P') at the base of the wall.

$$(15{,}880 \text{ lb/ft})(h) = (8660 \text{ lb/ft})(6.67 \text{ ft}) + (7220 \text{ lb/ft})(10 \text{ ft})$$

$$h = 8.18 \text{ ft}$$

Hence, total active earth pressure acts at 8.18 ft above the base of the wall.

12–6 CULMANN'S GRAPHIC SOLUTION

Several graphic methods to determine earth pressures are available, one of which is Culmann's graphic solution. The steps in carrying out a Culmann's graphic solution for active earth pressure (P_a) may be summarized as follows [4]:

1. Draw the retaining wall, backfill, and so on, to a convenient scale (see Figure 12–26).
2. From point A (the base of the wall), lay off a line at angle ϕ (angle of internal friction) with a horizontal line. This is line AC in Figure 12–26.
3. From point A, lay off a line at an angle θ with line AC (from step 2). Angle θ is equal to α (the angle between the back side of the wall and a horizontal line, as indicated in Figure 12–26) minus δ (angle of wall friction). This line is AD in Figure 12–26.
4. Draw some possible failure wedges, such as ABC_1, ABC_2, ABC_3, and so on.
5. Compute the weights of the wedges (W_1, W_2, W_3, etc.).
6. Using a convenient weight scale along line AC, lay off the respective weights of the wedges, locating points w_1, w_2, w_3, and so on.
7. Through each point, w_1, w_2, w_3, and so on, draw a line parallel to line AD, intersecting the corresponding line AC_1, AC_2, AC_3, respectively.
8. Draw a smooth curve (*Culmann's line*) through the points of intersection determined in step 7 (i.e., the point of intersection of the line through point w_1 parallel to line AD and of line AC_1, the point of intersection of the line through point w_2 parallel to line AD and of line AC_2, etc.).

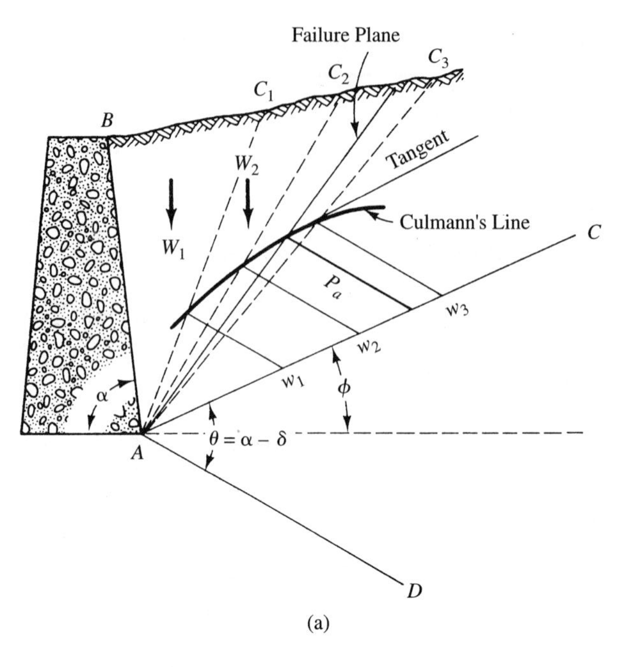

(a)

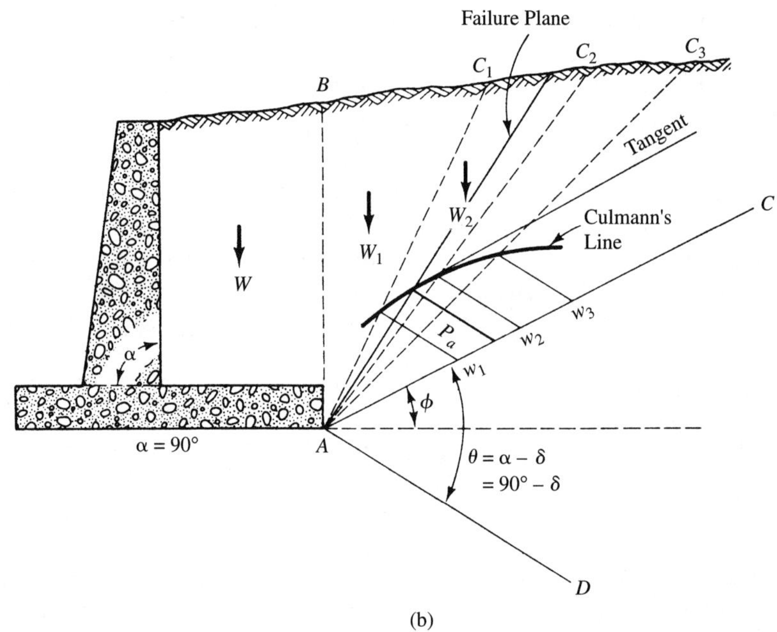

(b)

FIGURE 12–26 Culmann's graphic solution: (a) gravity wall; (b) cantilever wall [4].

9. Draw a line that is both tangent to Culmann's line and parallel to line *AC*.
10. Draw a line through the tangent point (determined in step 9) that is parallel to line *AD* and intersects line *AC*. The length of this line applied to the weight scale gives the value of P_a (Figure 12–26). A line from point *A* through the tangent point defines the failure plane.

As discussed in Section 12–4, the point of application of P_a can be found by drawing a line through the center of gravity of the failure wedge and parallel to the failure plane until it intersects the wall (see Figure 12–19). The direction of P_a is along a line that makes an angle δ (the angle of wall friction) with a line normal to the back side of the wall (see Figure 12–20).

Examples 12–9 through 12–11 illustrate the application of Culmann's graphic solution.

EXAMPLE 12–9

Given

The same conditions as in Example 12–7 (see Figure 12–27).

Required

Total active earth pressure per foot of wall, by Culmann's graphic solution.

Solution

By following the steps outlined previously for Culmann's graphic solution, one first prepares the sketch of Figure 12–28. The weights of the wedges (step 5) are then computed as follows:

$$W_1 = (\tfrac{1}{2})(120 \text{ lb/ft}^3)(4.7 \text{ ft})(21.0 \text{ ft}) = 5920 \text{ lb/ft}$$

$$W_2 = (\tfrac{1}{2})(120 \text{ lb/ft}^3)(4.4 \text{ ft})(22.2 \text{ ft}) = 5860 \text{ lb/ft}$$

$$W_3 = (\tfrac{1}{2})(120 \text{ lb/ft}^3)(5.0 \text{ ft})(27.2 \text{ ft}) = 8160 \text{ lb/ft}$$

$$W_4 = (\tfrac{1}{2})(120 \text{ lb/ft}^3)(3.5 \text{ ft})(31.4 \text{ ft}) = 6590 \text{ lb/ft}$$

From Figure 12–28, the value of P_a is determined to be 7600 lb/ft.

EXAMPLE 12-10

Given

The same conditions as in Example 12–8 (see Figure 12–29).

Required

Total active earth pressure per foot of wall, by Culmann's graphic solution.

Solution

The effect of the surcharge uniform load of 1000 lb/ft^2 is taken into account by superposing an equivalent depth of fill $h = q/\gamma = (1000 \text{ lb/ft}^2)/(120 \text{ lb/ft}^3) = 8.33$ ft on each trial wedge. Then, Culmann's graphic solution is carried out by following the

FIGURE 12–27

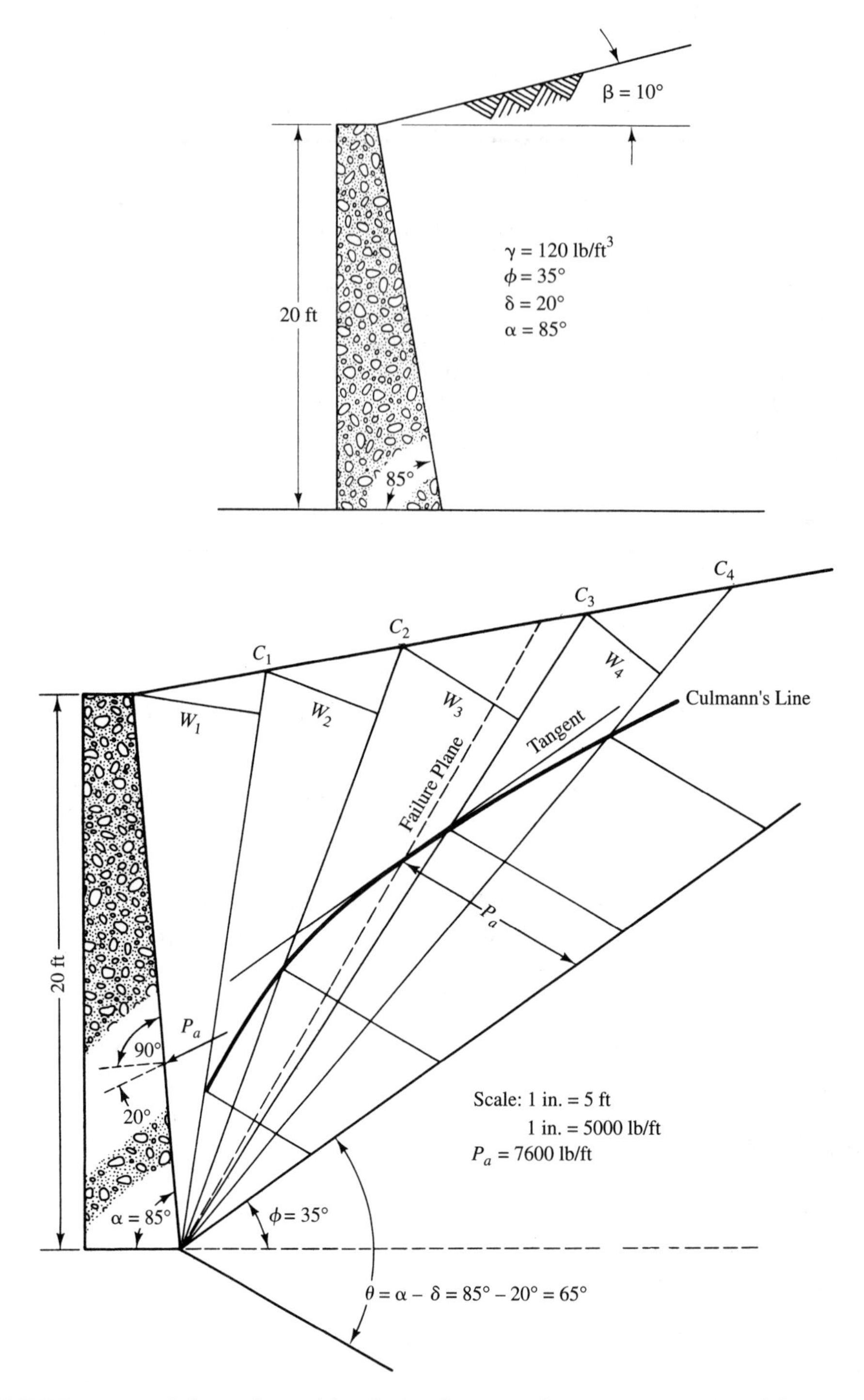

FIGURE 12–28 Culmann's graphic solution for Example 12–9. Note: Original drawing reduced by 25%.

FIGURE 12–29

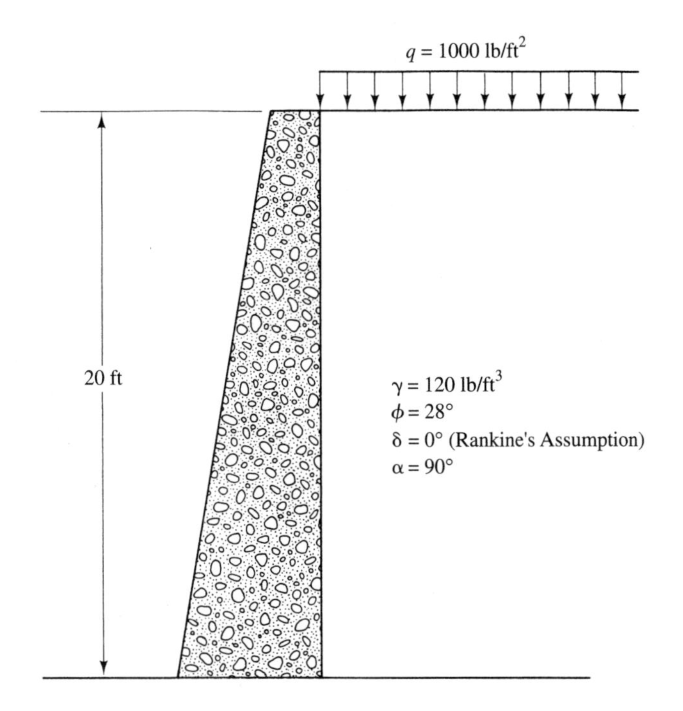

steps outlined previously and preparing the sketch of Figure 12–30. Weights of the wedges (step 5) are computed as follows:

$$W_1 = (\tfrac{1}{2})(120\ \text{lb/ft}^3)(5.0\ \text{ft})(20.0\ \text{ft}) + (120\ \text{lb/ft}^3)(8.33\ \text{ft})(5\ \text{ft}) = 11{,}000\ \text{lb/ft}$$

$$W_2 = (\tfrac{1}{2})(120\ \text{lb/ft}^3)(4.5\ \text{ft})(22.4\ \text{ft}) + (120\ \text{lb/ft}^3)(8.33\ \text{ft})(5\ \text{ft}) = 11{,}050\ \text{lb/ft}$$

$$W_3 = (\tfrac{1}{2})(120\ \text{lb/ft}^3)(4.0\ \text{ft})(25.0\ \text{ft}) + (120\ \text{lb/ft}^3)(8.33\ \text{ft})(5\ \text{ft}) = 11{,}000\ \text{lb/ft}$$

$$W_4 = (\tfrac{1}{2})(120\ \text{lb/ft}^3)(3.5\ \text{ft})(28.3\ \text{ft}) + (120\ \text{lb/ft}^3)(8.33\ \text{ft})(5\ \text{ft}) = 10{,}940\ \text{lb/ft}$$

From Figure 12–30, the value of P_a is determined to be 15,800 lb/ft. As computed in Example 12–8, P_a acts 8.18 ft from the base of the wall (Figure 12–30).

EXAMPLE 12–11

Given

The retaining wall shown in Figure 12–31.

Required

Total active earth pressure, P_a, by Culmann's graphic solution.

Solution

By following the steps outlined previously for Culmann's graphic solution, one first prepares the sketch of Figure 12–32. Weights of the wedges (step 5) are then computed as follows:

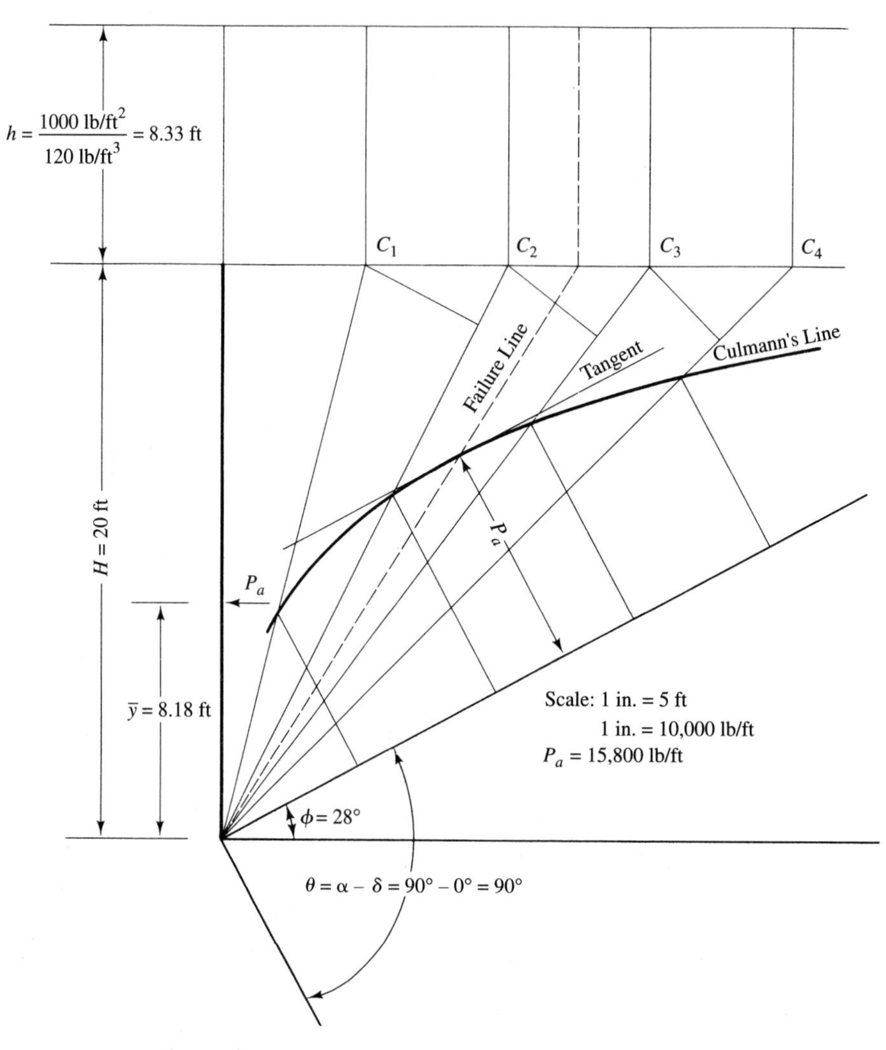

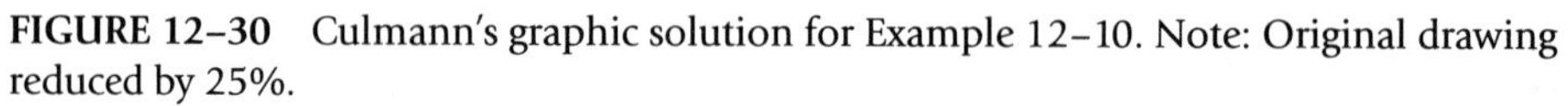

FIGURE 12–30 Culmann's graphic solution for Example 12–10. Note: Original drawing reduced by 25%.

$$W_1 = (\tfrac{1}{2})(0.12 \text{ kip/ft}^3)(5.2 \text{ ft})(25.5 \text{ ft}) = 7.96 \text{ kips/ft}$$

$$W_2 = (\tfrac{1}{2})(0.12 \text{ kip/ft}^3)(2.9 \text{ ft})(26.4 \text{ ft}) = 4.59 \text{ kips/ft}$$

$$W_3 = (\tfrac{1}{2})(0.12 \text{ kip/ft}^3)(2.7 \text{ ft})(27.6 \text{ ft}) = 4.47 \text{ kips/ft}$$

$$W_{3c} = 8 \text{ kips} \quad \text{(concentrated load)}$$

$$W_4 = (\tfrac{1}{2})(0.12 \text{ kip/ft}^3)(2.6 \text{ ft})(29.0 \text{ ft}) = 4.52 \text{ kips/ft}$$

$$W_5 = (\tfrac{1}{2})(0.12 \text{ kip/ft}^3)(3.0 \text{ ft})(31.0 \text{ ft}) = 5.58 \text{ kips/ft}$$

From Figure 12–32, the value of P_a is determined to be 17.0 kips/ft.

FIGURE 12–31

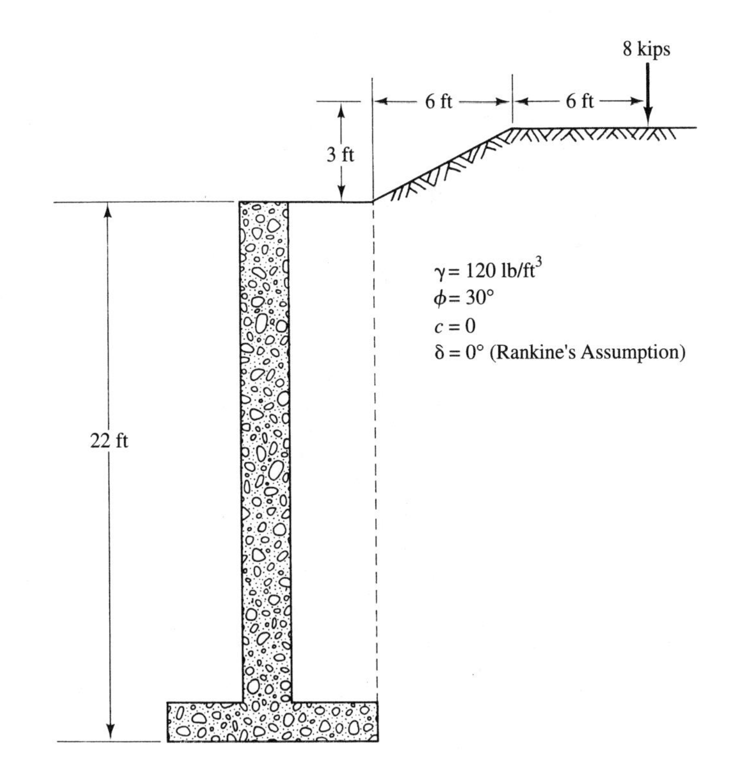

12–7 DESIGN CONSIDERATIONS FOR RETAINING WALLS

In designing retaining walls, the first step is to determine the magnitude and location of the active earth pressures that will be acting on the wall. These determinations can be made by utilizing any of the methods presented previously in this chapter. Active earth pressure is used to design free-standing retaining walls.

The next step is to assume a retaining-wall size. Normally, the required wall height will be known, thus a wall thickness and base width must be estimated. The assumed wall is then checked for three conditions. First, the wall must be safe against sliding horizontally. Second, the wall must be safe against overturning. Third, the wall must not introduce a contact pressure on the foundation soil beneath the wall's base that exceeds the allowable bearing pressure of the foundation soil. If any of these conditions is not safe, the assumed wall size must be modified, and conditions checked again. If (when) the three conditions are met, the assumed size is used for design. If, however, the three conditions are met with plenty to spare, the size might be reduced somewhat, and conditions checked again. Obviously, this is more or less a trial-and-error procedure.

The preceding gives a brief preview of design considerations for retaining walls. This topic is addressed in greater detail in Chapter 13.

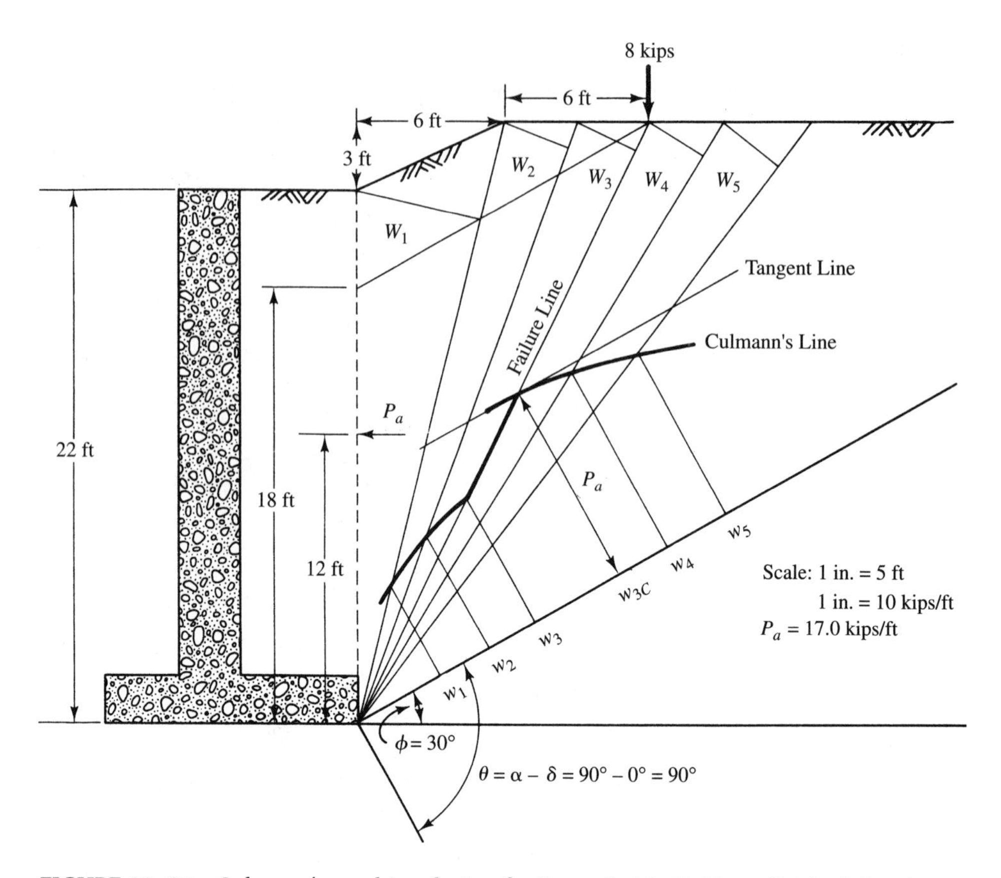

FIGURE 12–32 Culmann's graphic solution for Example 12–11. Note: Original drawing reduced by 37.5%.

12–8 LATERAL EARTH PRESSURE ON BRACED SHEETINGS

Sometimes earth cuts are retained by braced sheetings rather than the rigid walls considered heretofore in this chapter. Commonly made of wood or steel, sheetings are normally driven vertically and often used to retain earth temporarily during a construction project. A sketch of braced sheetings used to retain earth is shown in Figure 12–33. A horizontal brace providing lateral support to resist earth pressure behind the sheeting is known as a *strut.* A continuous horizontal (longitudinal) member extending along a sheeting's face to provide intermediate sheeting support between strut locations is called a *wale.* Examples of struts and wales are shown in Figure 12–33.

Lateral earth pressure on braced sheetings cannot ordinarily be analyzed by the Rankine theory, Coulomb theory, or other theories that are used to analyze pressures on rigid retaining walls. Those theories are based on the condition that the (rigid)

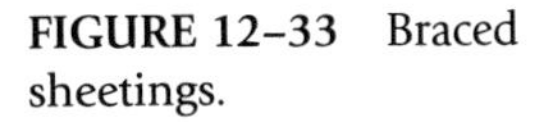

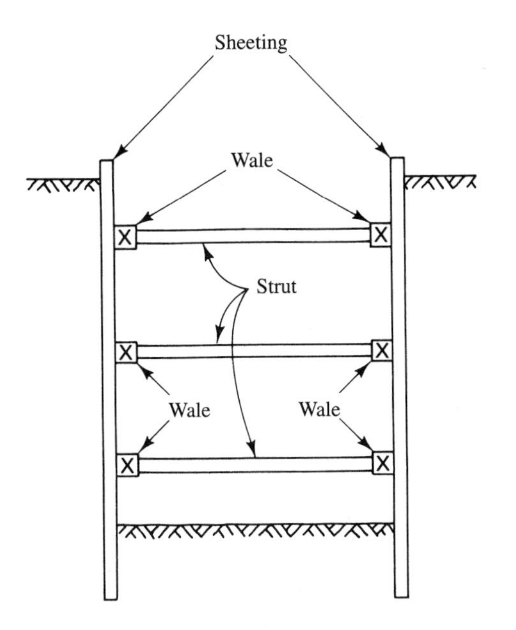

FIGURE 12–33 Braced sheetings.

wall yields laterally, either by sliding sideways or rotating about the bottom of the wall, so that the soil's full shearing resistance can be developed. Braced sheetings are much more flexible; hence, they do not yield in the same manner as rigid walls, thereby giving different shearing patterns.

Braced sheetings can be designed by using empirical diagrams of lateral pressure against braced sheetings. Figure 12–34 gives such diagrams for braced sheetings in sand, soft to medium clays, and stiff-fissured clays. Struts may be designed by assuming that vertical members are hinged at each strut level except those at the top and bottom (see Figure 12–35).

EXAMPLE 12–12

Given

1. A braced sheet pile for an open cut in soft to medium clay is illustrated in Figure 12–36.
2. Struts are spaced longitudinally at 4.0 m center to center.
3. The sheet piles are pinned or hinged at strut levels *B* and *C*.

Required

1. Lateral earth pressure diagram for the braced sheet pile system.
2. Loads on struts *A*, *B*, *C*, and *D*.

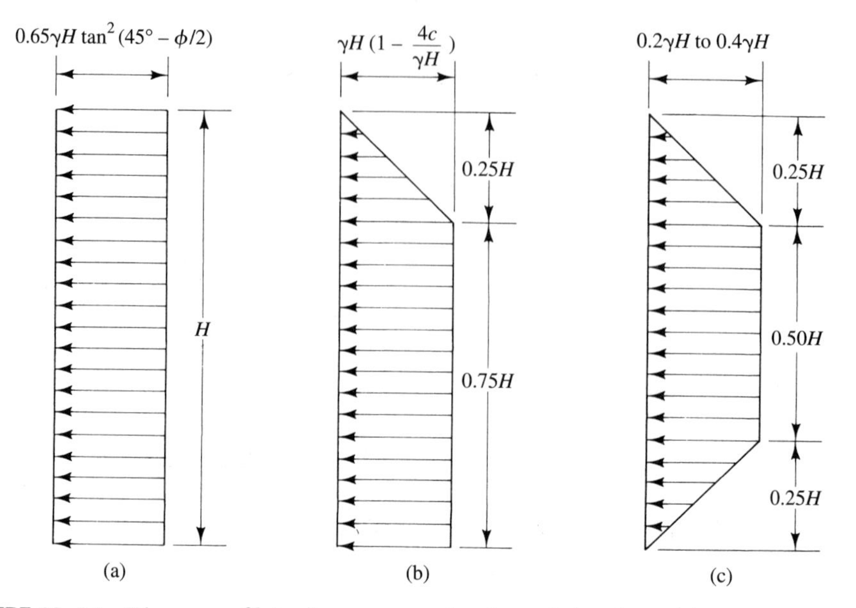

FIGURE 12–34 Diagrams of lateral pressure against braced sheetings: (a) sand; (b) soft to medium clay; (c) stiff-fissured clay [10].

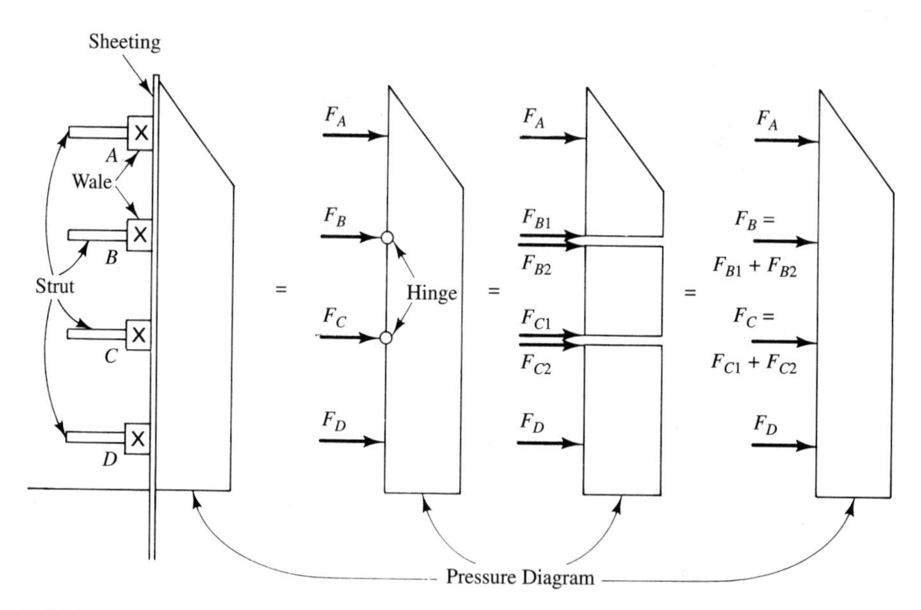

FIGURE 12–35 Forces on struts in braced sheeting.

FIGURE 12–36

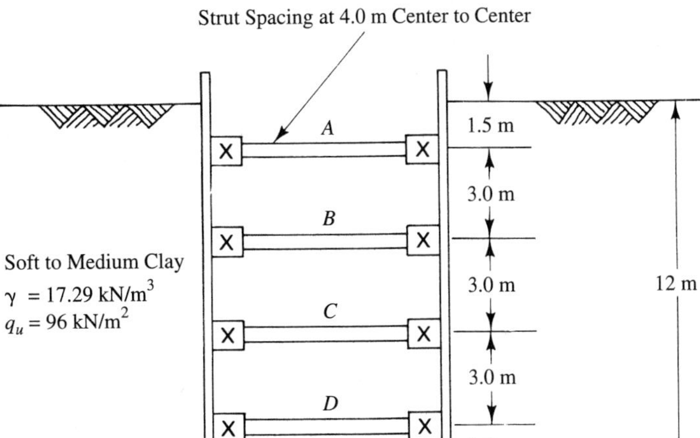

Solution

1. $p = \gamma H\left(1 - \frac{4c}{\gamma H}\right)$ (from Figure 12–34)

$$c = \frac{96\ \text{kN/m}^2}{2} = 48\ \text{kN/m}^2$$

$$p = (17.29\ \text{kN/m}^3)(12\ \text{m})\left[1 - \frac{(4)(48\ \text{kN/m}^2)}{(17.29\ \text{kN/m}^3)(12\ \text{m})}\right] = 15.48\ \text{kN/m}^2$$

The lateral earth pressure diagram for the braced sheet pile system is, therefore, as shown in Figure 12–37.

2. In the free body diagram of Figure 12–38a,

$$\Sigma M_B = 0$$

$$(\tfrac{1}{2})(15.48\ \text{kN/m}^2)(3.0\ \text{m})(4.0\ \text{m})\left(1.5\ \text{m} + \frac{3.0\ \text{m}}{3}\right)$$

$$+ (1.5\ \text{m})(15.48\ \text{kN/m}^2)(4.0\ \text{m})\left(\frac{1.5\ \text{m}}{2}\right) - (F_A)(3.0\ \text{m}) = 0$$

$$F_A = 100.6\ \text{kN}$$

$$\Sigma H = 0$$

$$F_{B1} = (\tfrac{1}{2})(1.5\ \text{m} + 4.5\ \text{m})(15.48\ \text{kN/m}^2)(4.0\ \text{m}) - 100.6\ \text{kN}$$
$$= 85.2\ \text{kN}$$

In the free body diagram of Figure 12–38b,

$$F_{B2} = F_{C1} = (\tfrac{1}{2})(3.0\ \text{m})(15.48\ \text{kN/m}^2)(4.0\ \text{m}) = 92.9\ \text{kN}$$

FIGURE 12–37

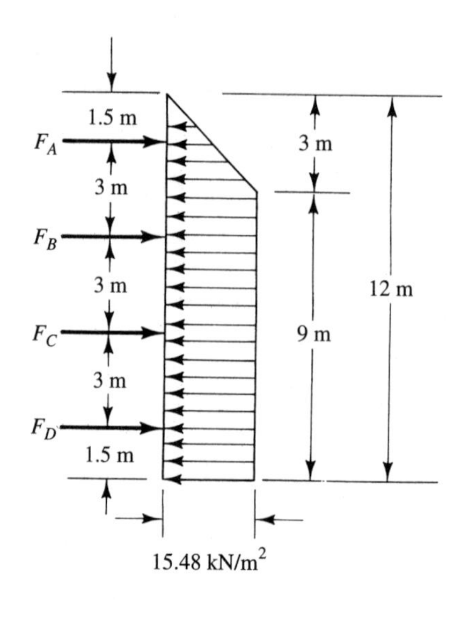

FIGURE 12–38

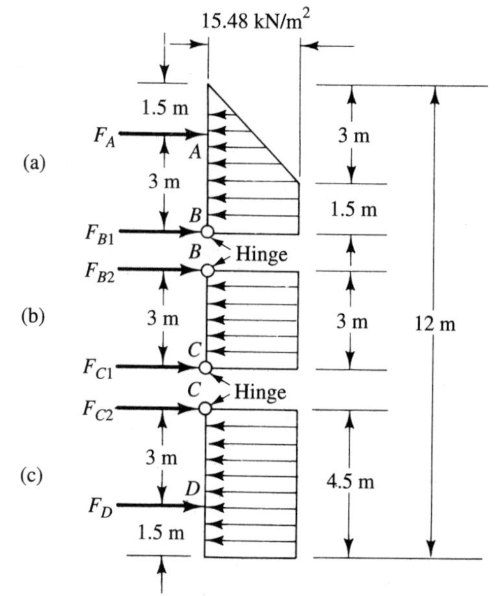

In the free body diagram of Figure 12–38c,

$$\Sigma M_C = 0$$

$$(F_D)(3.0 \text{ m}) - (4.5 \text{ m})(15.48 \text{ kN/m}^2)(4.0 \text{ m})\left(\frac{4.5 \text{ m}}{2}\right) = 0$$

$$F_D = 209.0 \text{ kN}$$

$$\Sigma H = 0$$

FIGURE 12–39

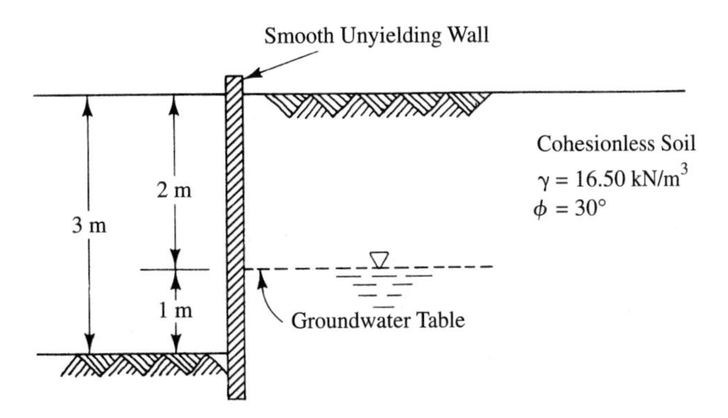

$$F_{C2} + F_D - (4.5\text{ m})(15.48\text{ kN/m}^2)(4.0\text{ m}) = 0$$
$$F_{C2} = (4.5\text{ m})(15.48\text{ kN/m}^2)(4.0\text{ m}) - 209.0\text{ kN} = 69.6\text{ kN}$$

Therefore,

$$F_A = 100.6\text{ kN}$$
$$F_B = 85.2\text{ kN} + 92.9\text{ kN} = 178.1\text{ kN}$$
$$F_C = 92.9\text{ kN} + 69.6\text{ kN} = 162.5\text{ kN}$$
$$F_D = 209.0\text{ kN}$$

12–9 PROBLEMS

12–1. A smooth, unyielding wall retains loose sand (see Figure 12–39). Assume that no lateral movement occurs in the soil mass, and the at-rest condition prevails. Draw the diagram of earth pressure against the wall and find the total lateral force acting on the wall if the groundwater table is located 2 m below the ground surface, as shown in Figure 12–39.

12–2. A vertical retaining wall 25 ft high supports a deposit of sand having a level backfill. Soil properties are as follows:

$$\gamma = 120\text{ lb/ft}^3$$
$$\phi = 35°$$
$$c = 0$$

Calculate the total active earth pressure per foot of wall and its point of application, by Rankine theory.

12–3. A vertical retaining wall 7.62 m high supports a deposit of sand with a sloping backfill. The angle of sloping backfill is 10°. Soil properties are as follows:

$$\gamma = 18.85\text{ kN/m}^3$$
$$\phi = 35°$$
$$c = 0$$

FIGURE 12–40

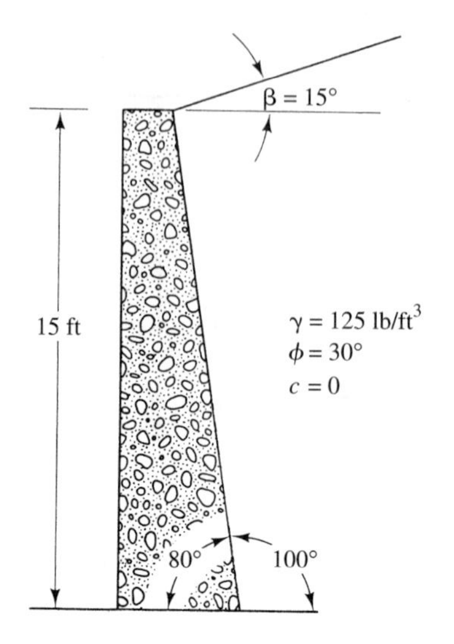

Calculate the total active earth pressure per meter of wall and its point of application, by Rankine theory.

12–4. What is the total active earth pressure per foot of wall for the wall shown in Figure 12–40, using Rankine theory?

12–5. A vertical wall 25 ft high supports a level backfill of clayey sand. The samples of the backfill soil were tested, and the following properties were determined: $\phi = 20°$, $c = 250$ lb/ft^2, and $\gamma = 125$ lb/ft^3. Draw the active earth pressure diagram, using Rankine theory.

12–6. What is the total active earth pressure per foot of wall for the retaining wall in Problem 12–2, with an angle of wall friction between backfill and wall of 20°, using Coulomb theory?

12–7. What is the total active earth pressure per foot of wall for the retaining wall in Problem 12–4, with an angle of wall friction between backfill and wall of 25°, using Coulomb theory?

12–8. A vertical wall 6.0 m high supports a cohesionless backfill with a horizontal surface. The backfill soil's unit weight and angle of internal friction are 17.2 kN/m^3 and 31°, respectively, and the angle of wall friction between backfill and wall is 20°. Using Figure 12–21, find the total active earth pressure against the wall.

12–9. A smooth, vertical wall is 25 ft high and retains a cohesionless soil with $\gamma =$ 115 lb/ft^3 and $\phi = 30°$. The top of the soil is level with the top of the wall, and the soil surface carries a uniformly distributed load of 500 lb/ft^2. Calculate the total active earth pressure on the wall per linear foot of wall, and determine its point of application, by Rankine theory.

12–10. Solve Problem 12–7 by Culmann's graphic solution.

12–11. Solve Problem 12–9 by Culmann's graphic solution.

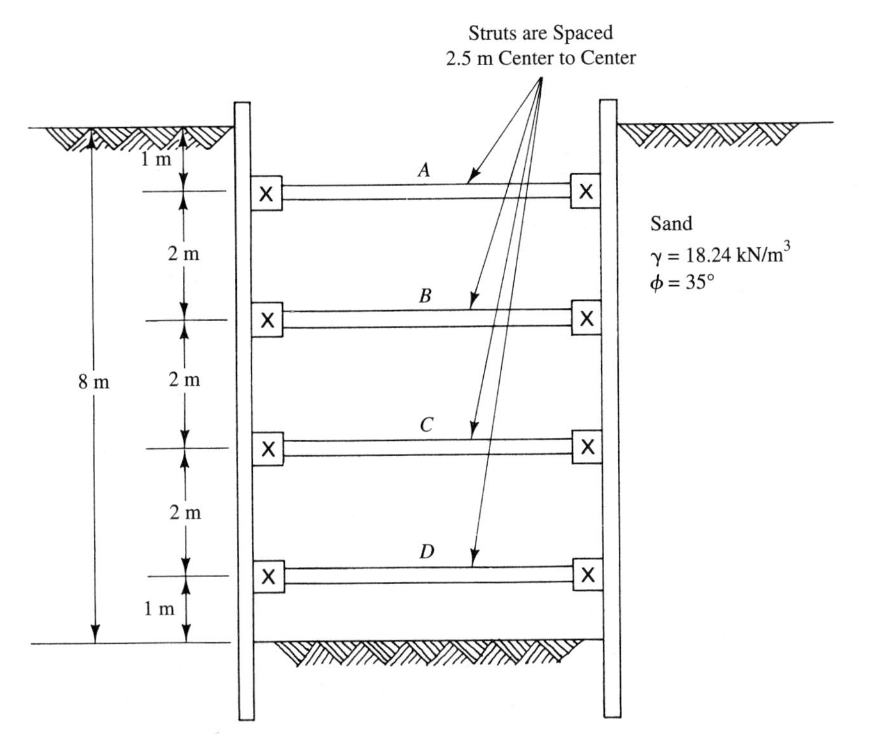

FIGURE 12–41

12–12. A braced sheet pile to be used in an open cut in sand is shown in Figure 12–41. Assume that the sheet piles are hinged at strut levels *B* and *C*. Struts are spaced longitudinally at 2.5-m center-to-center spacing. Draw the lateral earth pressure diagram for the braced sheet pile system and compute the loads on struts *A*, *B*, *C*, and *D*.

References

[1] J. Jaky, "Pressure in Silos," *Proc. 2nd Int. Conf. Soil Mech. Found. Eng.*, **1** (1948).

[2] I.Alpan, "The Empirical Evaluation of the Coefficient K_0 and K_{0r}," *Soils and Foundations*, Japanese Society of Soil Mechanics and Foundations Engineering, Tokyo, **VII**(I) (1967).

[3] Irving S. Dunn, Loren R. Anderson, and Fred W. Kiefer, *Fundamentals of Geotechnical Analysis*, John Wiley & Sons, Inc., New York, 1980. Copyright © 1980, by John Wiley & Sons, Inc. Reprinted by permission of John Wiley & Sons, Inc.

[4] Joseph E. Bowles, *Foundation Analysis and Design*, McGraw-Hill Book Company, New York, 1968.

[5] Wayne C. Teng, *Foundation Design,* Prentice-Hall, Inc., Englewood Cliffs, N.J., 1962.

[6] Karl Terzaghi, Ralph B. Peck, and Gholamreza Mesri, *Soil Mechanics in Engineering Practice,* 3rd ed., John Wiley & Sons, Inc., New York, 1996.

[7] O. Syffert, *Erddrucktafeln* (Earth-pressure tables), J. Springer, Berlin, 1929.

[8] William S. Lalonde, Jr., and Milo F. Janes, eds., *Concrete Engineering Handbook,* McGraw-Hill Book Company, New York, 1961.

[9] Louis J. Goodman and R. H. Karol, *Theory and Practice of Foundation Engineering,* Macmillan Publishing Co., Inc., New York, 1968.

[10] Karl Terzaghi and Ralph B. Peck, *Soil Mechanics in Engineering Practice,* John Wiley & Sons, Inc., New York, 1967. Copyright © 1967, by John Wiley & Sons, Inc. Reprinted by permission of John Wiley & Sons, Inc.

13 RETAINING STRUCTURES

13–1 INTRODUCTION

Retaining structures are built for the purpose of retaining, or holding back, a soil mass (or other material). Probably a majority of retaining structures are concrete walls, which are covered in Sections 13–2 through 13–6. A relatively new type of retaining structure known as *Reinforced Earth* is presented in Section 13–7. Slurry trench walls, specially constructed concrete walls built entirely below ground level, are described in Section 13–8. Anchored bulkheads, covered in Section 13–9, are useful when certain waterfront retaining structures are needed.

13–2 RETAINING WALLS

A simple retaining wall is illustrated in Figure 13–1. This type of wall depends on its weight to achieve stability; hence, it is called a *gravity wall.* In the case of taller walls, large lateral pressure tends to overturn the wall, and for economic reasons *cantilever walls* may be more desirable. As illustrated in Figure 13–2, a cantilever wall has part of its base extending underneath the backfill, and (as is shown subsequently) the weight of the soil above this part of the base helps prevent overturning.

Gravity walls are often built of plain concrete and are bulky. Concrete cantilever walls are generally more slender and must be adequately reinforced with steel. Although there are other types of retaining walls, these two are most common.

Although retaining walls may give the appearance of being unyielding, some wall movement is to be expected. In order that walls may undergo some forward yielding without appearing to tip over, they are often built with an inward slope on the outer face of the wall, as shown in Figures 13–1 and 13–2. This inward slope is called *batter.*

Material placed behind a retaining wall is commonly referred to as *backfill.* It is highly desirable that backfill be a select, free-draining, granular material, such as clean sand, gravel, or broken stones. If necessary, appropriate material should be hauled in from an area outside the construction site. Clayey soils make extremely

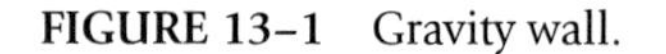

FIGURE 13–1 Gravity wall.

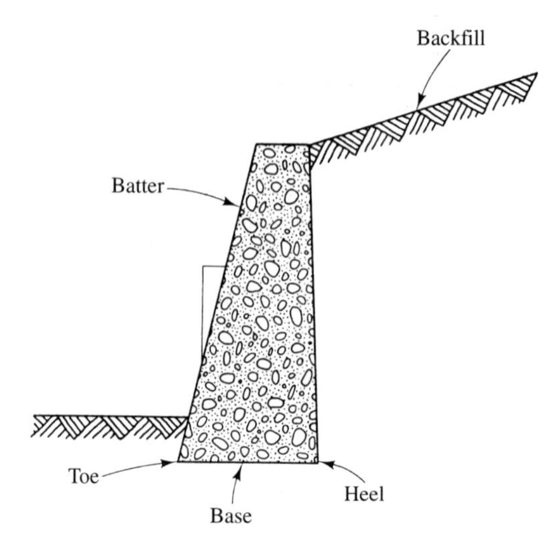

FIGURE 13–2 Cantilever wall.

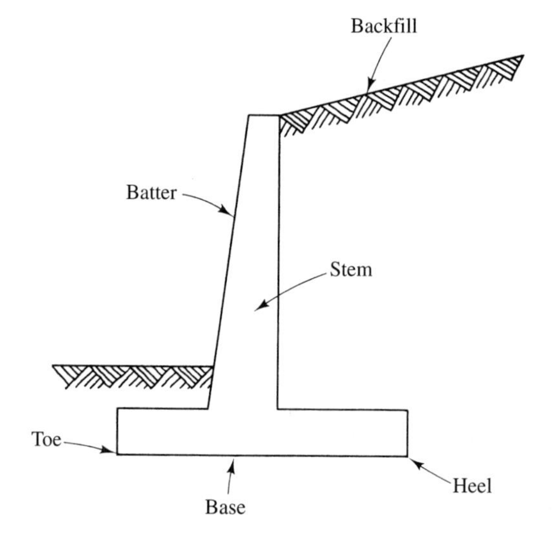

objectionable backfill material because of the excessive lateral pressure they create. The designer of a retaining wall should either (1) write specifications for the backfill and base the design of the wall on the specified backfill, or (2) be given information on the material to be used as backfill and base the design of the wall on the indicated backfill. If it is possible that the water table will rise in the backfill, special designing, construction, and monitoring must go into effect.

In Chapter 12, several methods were presented for analyzing both the magnitude and the location of the lateral earth pressure acting on retaining walls. For economic reasons, retaining walls are commonly designed for active earth pressure, developed by a free-draining, granular backfill acting on the wall. As related in Section 12–7, a retaining wall must (1) be able to resist sliding along the base, (2) be able to resist overturning, and (3) not introduce a contact pressure on the foundation soil beneath the wall's base that exceeds the allowable bearing pressure of the foundation soil. (Walls must also meet structural requirements, such as shear and bending moment; however, such considerations are not covered in this book.) Chapter 13 deals in more detail with retaining-wall design.

13–3 EARTH PRESSURE COMPUTATION

To design a retaining wall, one must, of course, determine the earth pressure acting on the wall. Analytic determinations of earth pressures—including Rankine earth pressure, Coulomb earth pressure, and Culmann's graphic solution—were covered in detail in Chapter 12. Retaining-wall design is normally based on active earth pressure.

In practice, earth pressures for walls less than 20 ft (6 m) high are often obtained from graphs or tables. Almost all such graphs and tables are developed from Rankine theory. One graphic relationship is given in Figure 13–3. Use of this approach to obtain earth pressure should be self-explanatory.

As can be noted by both the analytic methods of Chapter 12 and the graphic method of Figure 13–3, the magnitude of earth pressure on a retaining wall depends in part upon the type of soil backfill.

13–4 STABILITY ANALYSIS

Common procedure in retaining-wall design is to assume a trial wall shape and size and then to check the trial wall for stability. If it does not prove to be stable by conventional standards, the wall's shape and/or size must be revised, and the new wall checked for stability. This procedure is repeated until a satisfactory wall is found.

If a wall is *stable,* it means, of course, that the wall does not move. Essentially, there are three means by which a retaining wall can move—horizontally (by sliding), vertically (by excessive settlement and/or bearing capacity failure of the foundation soil), and by rotation (by overturning). Standard procedure is to check for stability with respect to each of the three means of movement to ensure that an adequate factor of safety (F.S.) is present in each case. Checks for sliding and overturning hark back to the basic laws of statics. Checks for settlement and bearing capacity of foundation soil are done by settlement and bearing capacity analyses, which were presented in Chapters 7 and 9, respectively.

The factor of safety against sliding is found by dividing sliding resistance force by sliding force. The sliding resistance force is the product of the total downward force on the base of the wall and the coefficient of friction (μ) between the base of the retaining wall and the underlying soil. The sliding force is typically the horizontal component of lateral earth pressure exerted against the wall by backfill material.

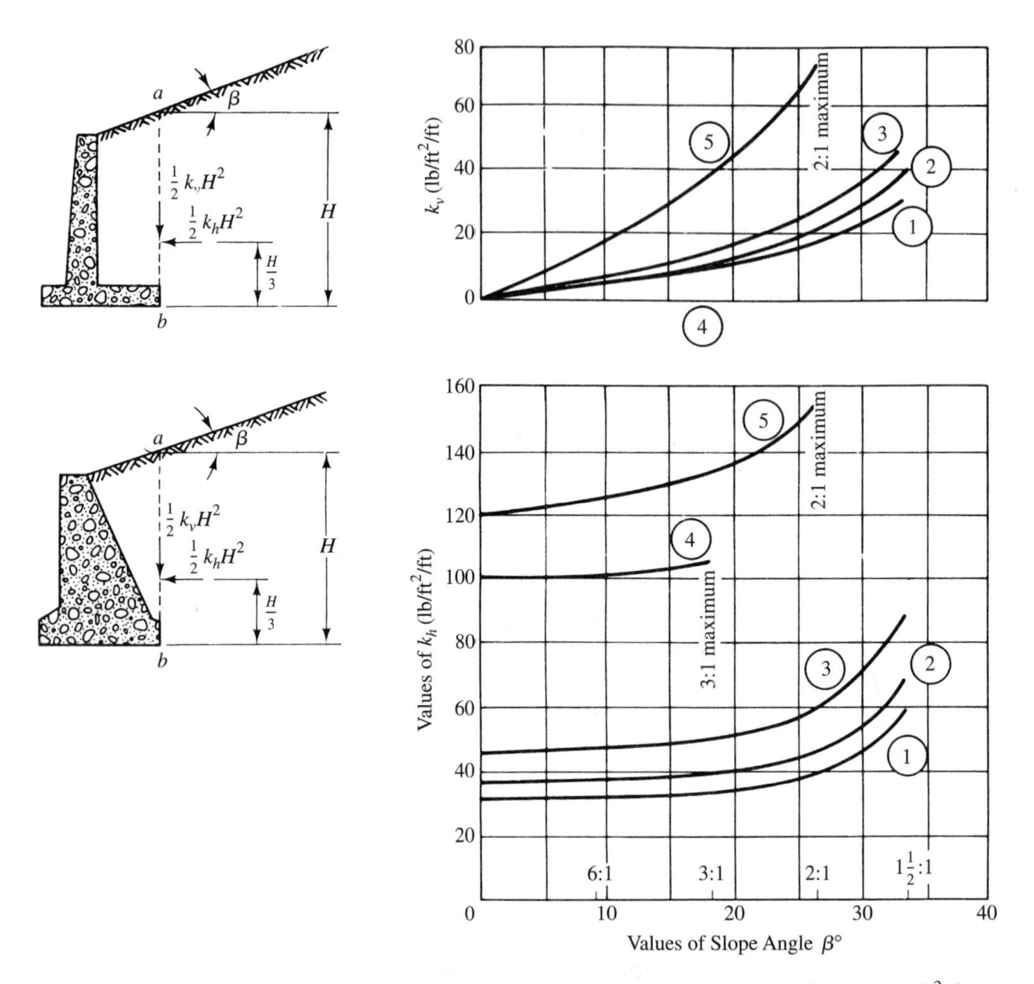

FIGURE 13–3 Earth pressure charts for retaining walls less than 20 ft (6 m) high (1 lb/ft²/ft = 0.1571 kN/m²/m). Notes: Numerals on curves indicate soil types as described here. For material of Type 5, computations should be based on value of H 4 ft less than actual value. Types of backfill for retaining walls: ① Coarse-grained soil without admixture of fine soil particles, very free-draining (clean sand, gravel or broken stone); ② coarse-grained soil of low permeability, owing to admixture of particles of silt size; ③fine silty sand; granular materials with conspicuous clay content, or residual soil with stones; ④ soft or very soft clay, organic silt, or soft silty clay; ⑤ medium or stiff clay that may be placed in such a way that a negligible amount of water will enter the spaces between the chunks during floods or heavy rains [1, 2].

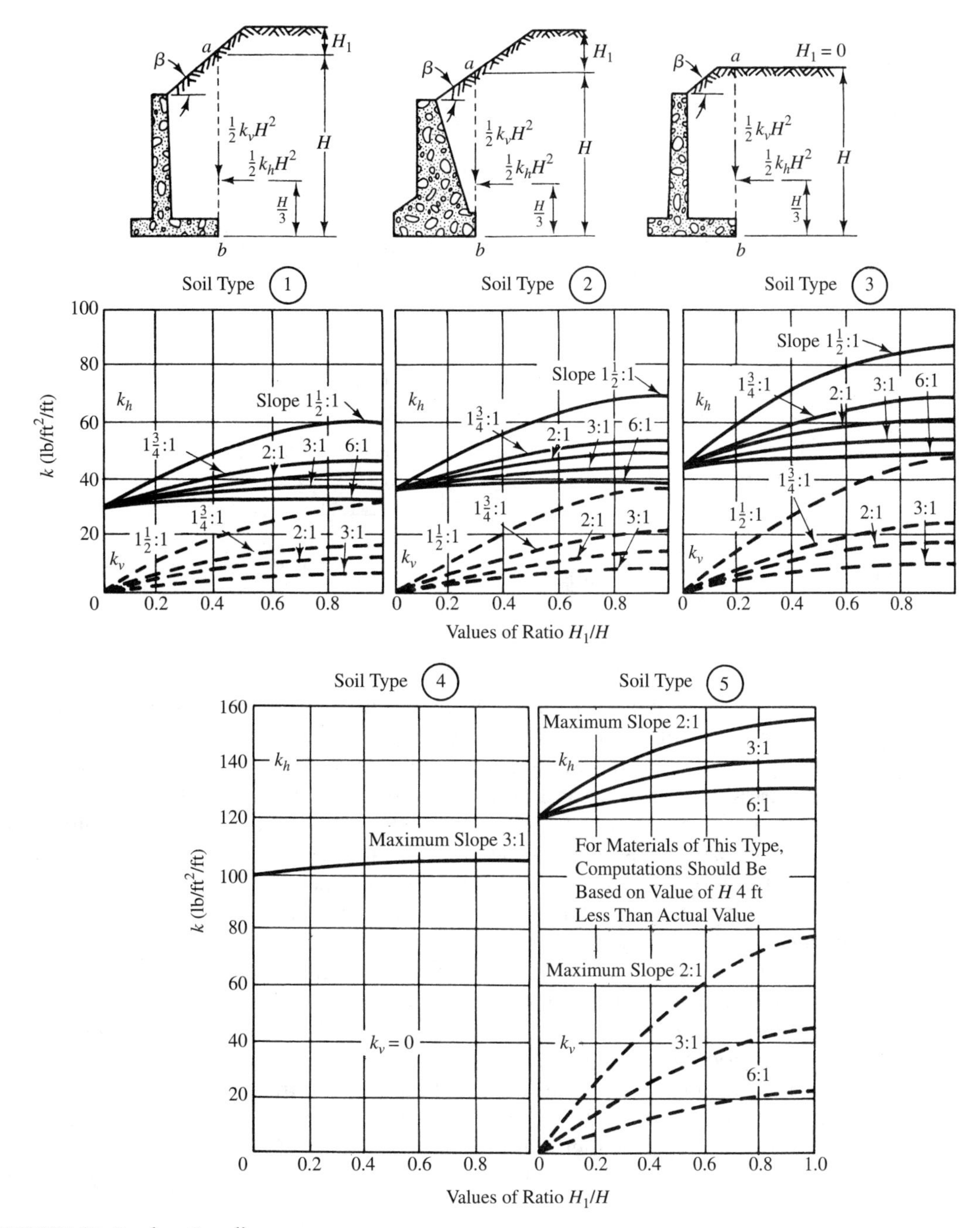

FIGURE 13–3 (*continued*)

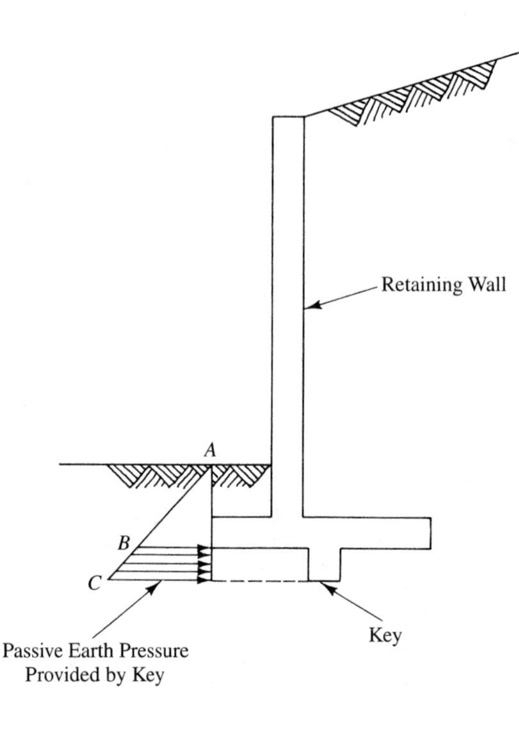

FIGURE 13–4 Sketch showing additional resistance to sliding in the form of passive resistance in front of key.

If an adequate factor of safety against sliding is not obtained with an ordinary flat-bottomed wall, some additional sliding resistance may be achieved by constructing a "key" into the wall's base. As shown in Figure 13–4, soil in front of the key's vertical face provides additional resistance to sliding in the form of passive resistance (i.e., zone *BC* of the earth pressure diagram). Of course, soil in front of the wall and its base furnishes some passive resistance (zone *AB* of the earth pressure diagram of Figure 13–4); however, because this soil may be subsequently removed by erosion, this passive resistance is often ignored in retaining-wall design. Keys are most effective in hard soil or rock.

The factor of safety against overturning is determined by dividing total righting moment by total overturning moment. Because overturning tends to occur about the front base of a wall (at the toe), righting moments and overturning moments are computed about the wall's toe.

The factor of safety against bearing capacity failure is determined by dividing ultimate bearing capacity by actual maximum contact (base) pressure. Contact pressure is computed by the methods presented in Chapter 9.

In summary, the three factors of safety with regard to stability analysis are as follows:

$$(\text{F.S.})_{\text{sliding}} = \frac{\text{Sliding resistance force}}{\text{Sliding force}} \tag{13–1}$$

$$(\text{F.S.})_{\text{overturning}} = \frac{\text{Total righting moment about toe}}{\text{Total overturning moment about toe}} \quad (13\text{–}2)$$

$$(\text{F.S.})_{\text{bearing capacity failure}} = \frac{\text{Soil's ultimate bearing capacity}}{\text{Actual maximum contact (base) pressure}} \quad (13\text{–}3)$$

Some common minimum factors of safety for sufficient stability are as follows:

$(\text{F.S.})_{\text{sliding}} = 1.5$ (if the passive earth pressure of the soil at the toe in front of the wall is neglected) [3]*

$= 2.0$ (if the passive earth pressure of the soil at the toe in front of the wall is included) [3]*

$(\text{F.S.})_{\text{overturning}} = 1.5$ (granular backfill soil)

$= 2.0$ (cohesive backfill soil) [1]

$(\text{F.S.})_{\text{bearing capacity failure}} = 3.0$

The two examples that follow illustrate the investigation of stability analysis for retaining walls. Example 13–1 refers to a gravity wall, and Example 13–2 to a cantilever wall.

EXAMPLE 13–1

Given

1. The retaining wall shown in Figure 13–5 is to be constructed of concrete having a unit weight of 150 lb/ft^3.
2. The retaining wall is to support a deposit of granular soil that has the following properties:

$$\gamma = 115 \text{ lb/ft}^3$$

$$\phi = 30°$$

$$c = 0$$

3. The coefficient of base friction is 0.55.
4. The foundation soil's ultimate bearing capacity is 6.5 tons/ft^2.

* Reprinted with permission of Macmillan Publishing Co., Inc., from *Theory and Practice of Foundation Engineering,* by Louis J. Goodman and R. H. Karol. Copyright © 1968, Macmillan Publishing Co., Inc.

FIGURE 13–5

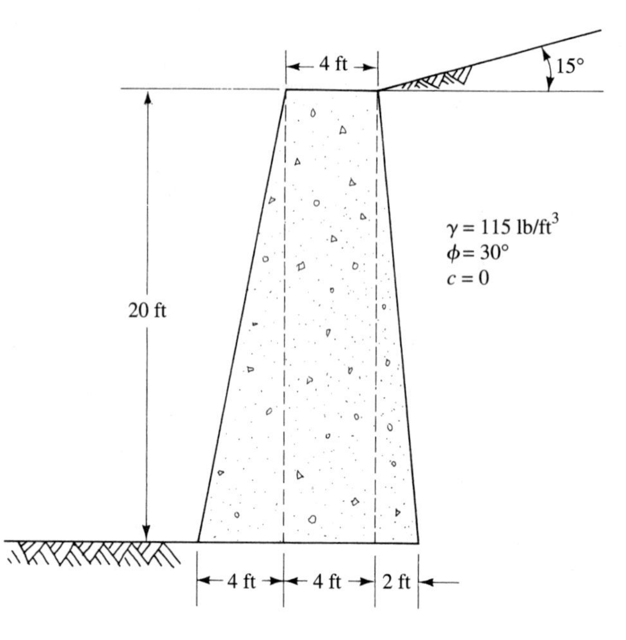

Required

Check the stability of the proposed retaining wall; that is, check the factor of safety against

1. Sliding.
2. Overturning.
3. Bearing capacity failure.

Solution

Calculation of Active Earth Pressure on the Back of the Wall by Rankine Theory

From Eqs. (12–10) and (12–11),

$$P_a = \frac{1}{2}\gamma H^2 \cos\beta \frac{\cos\beta - \sqrt{\cos^2\beta - \cos^2\phi}}{\cos\beta + \sqrt{\cos^2\beta - \cos^2\phi}}$$

Referring to Figure 13–6, one finds that

$$H = \overline{BC} = 20 \text{ ft} + (2 \text{ ft})(\tan 15°) = 20.54 \text{ ft}$$

$$P_a = (\tfrac{1}{2})(0.115 \text{ kip/ft}^3)(20.54 \text{ ft})^2(\cos 15°)\frac{\cos 15° - \sqrt{\cos^2 15° - \cos^2 30°}}{\cos 15° + \sqrt{\cos^2 15° - \cos^2 30°}}$$

$$P_a = 9.05 \text{ kips/ft}$$

FIGURE 13–6

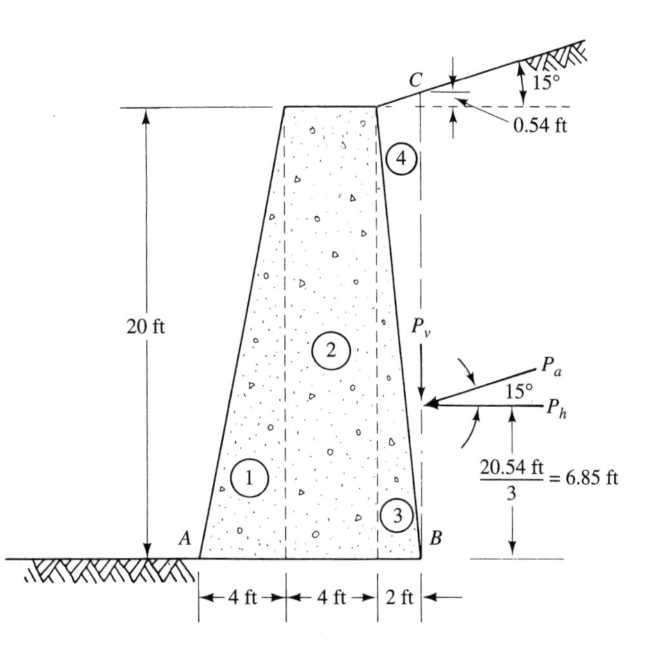

P_a acts parallel to the surface of the backfill; therefore,

$$\text{Horizontal component } (P_h) = P_a \cos 15° = (9.05 \text{ kips/ft}) \cos 15°$$

$$= 8.74 \text{ kips/ft}$$

$$\text{Vertical component } (P_v) = P_a \sin 15° = (9.05 \text{ kips/ft}) \sin 15° = 2.34 \text{ kips/ft}$$

Calculation of Righting Moment (See Figure 13–6)

Component	Weight of Component (kips/ft)	Moment Arm from *A* (ft)	Righting Moment about *A* (ft-kips/ft)
1	$(0.15)(\frac{1}{2})(4)(20) = 6.00$	$(\frac{2}{3})(4) = \frac{8}{3}$	16.0
2	$(0.15)(4)(20) = 12.00$	$4 + \frac{4}{2} = 6$	72.0
3	$(0.15)(\frac{1}{2})(2)(20) = 3.00$	$4 + 4 + (\frac{1}{3})(2) = \frac{26}{3}$	26.0
4	$(0.115)(\frac{1}{2})(20.54)(2) = 2.36$	$4 + 4 + (\frac{2}{3})(2) = \frac{28}{3}$	22.0
P_v	2.34	$4 + 4 + 2 = 10$	23.4
	$\Sigma V = 25.70$		$\Sigma M_r = 159.4$

Calculation of Overturning Moment

Overturning moment $(M_0) = (8.74 \text{ kips/ft})(6.85 \text{ ft}) = 59.9 \text{ ft-kips/ft}$.

1. From Eq. (13–1),

$$(\text{F.S.})_{\text{sliding}} = \frac{\text{Sliding resistance force}}{\text{Sliding force}} \tag{13–1)}$$

$$(\text{F.S.})_{\text{sliding}} = \frac{(\mu)(\Sigma V)}{P_h} = \frac{(0.55)(25.70 \text{ kips/ft})}{8.74 \text{ kips/ft}}$$

$$= 1.62 > 1.5 \quad \therefore \text{ O.K.}$$

2. From Eq. (13–2),

$$(\text{F.S.})_{\text{overturning}} = \frac{\text{Total righting moment about toe}}{\text{Total overturning moment about toe}} \tag{(13–2)}$$

$$(\text{F.S.})_{\text{overturning}} = \frac{\Sigma M_r}{\Sigma M_0} = \frac{159.4 \text{ ft-kips/ft}}{59.9 \text{ ft-kips/ft}}$$

$$= 2.66 > 1.5 \text{ (for granular backfill)} \quad \therefore \text{ O.K.}$$

Base Pressure Calculations

Location of resultant R $(= \Sigma V)$ if R acts at $\bar{x}$ from the toe (point A):

$$\bar{x} = \frac{\Sigma M_A}{\Sigma V} = \frac{\Sigma M_r - \Sigma M_0}{\Sigma V} = \frac{159.4 \text{ ft-kips/ft} - 59.9 \text{ ft-kips/ft}}{25.70 \text{ kips/ft}} = 3.87 \text{ ft}$$

$$e = \frac{4 \text{ ft} + 4 \text{ ft} + 2 \text{ ft}}{2} - 3.87 \text{ ft} = 1.13 \text{ ft} < \frac{L}{6} \text{ (i.e., } {}^{10}\!/_{6} \text{ ft, or } 1.67 \text{ ft)} \quad \therefore \text{ O.K.}$$

(i.e., R acts within the middle third of the base)

Using the flexural formula, from Eq. (9–9) (see Chapter 9), one gets

$$q = \frac{Q}{A} \pm \frac{M_x y}{I_x} \pm \frac{M_y x}{I_y} \tag{9–9}$$

Here,

$$Q = \text{Resultant } (R) = \Sigma V = 25.70 \text{ kips}$$

$$A = (1 \text{ ft})(10 \text{ ft}) = 10 \text{ ft}^2$$

$$M_x = 0 \quad \text{(one-way bending)}$$

$$M_y = Q \times e = (25.70 \text{ kips})(1.13 \text{ ft}) = 29.04 \text{ ft-kips}$$

$$x = \frac{10 \text{ ft}}{2} = 5 \text{ ft}$$

FIGURE 13–7

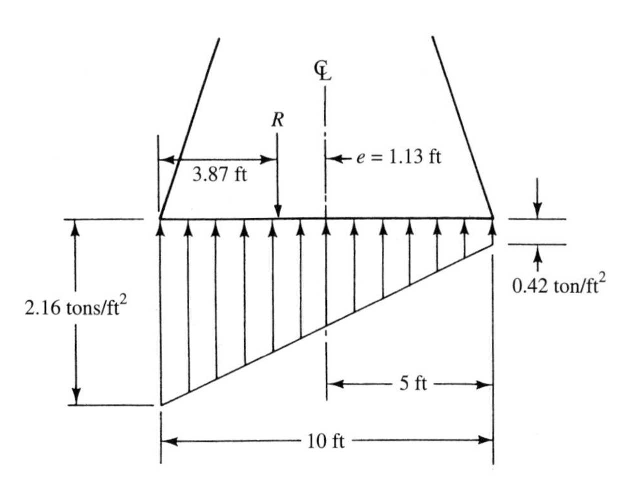

$$I_y = \frac{bh^3}{12} = \frac{(1 \text{ ft})(10 \text{ ft})^3}{12} = 83.33 \text{ ft}^4$$

$$q = \frac{25.70 \text{ kips}}{10 \text{ ft}^2} \pm \frac{(29.04 \text{ ft-kips})(5 \text{ ft})}{83.33 \text{ ft}^4}$$

$$q_L = 2.57 \text{ kips/ft}^2 + 1.74 \text{ kips/ft}^2 = 4.31 \text{ kips/ft}^2 = 2.16 \text{ tons/ft}^2$$

$$q_R = 2.57 \text{ kips/ft}^2 - 1.74 \text{ kips/ft}^2 = 0.83 \text{ kip/ft}^2 = 0.42 \text{ ton/ft}^2$$

The pressure distribution is shown in Figure 13–7.

3. From Eq. (13–3),

$$(\text{F.S.})_{\text{bearing capacity failure}} = \frac{\text{Soil's ultimate bearing capacity}}{\text{Actual maximum contact (base) pressure}} \quad \textbf{(13–3)}$$

$$(\text{F.S.})_{\text{bearing capacity failure}} = \frac{6.5 \text{ tons/ft}^2}{2.16 \text{ tons/ft}^2} = 3.01 > 3 \quad \therefore \text{ O.K.}$$

EXAMPLE 13–2

Given

1. The retaining wall shown in Figure 13–8.
2. The backfill material is Type 1 soil (Figure 13–3).
3. The unit weight and ϕ angle of the backfill material are 120 lb/ft^3 and 37°, respectively.
4. The coefficient of base friction is 0.45.
5. Allowable soil pressure is 3 kips/ft^2.
6. The unit weight of the concrete is 150 lb/ft^3.

FIGURE 13–8

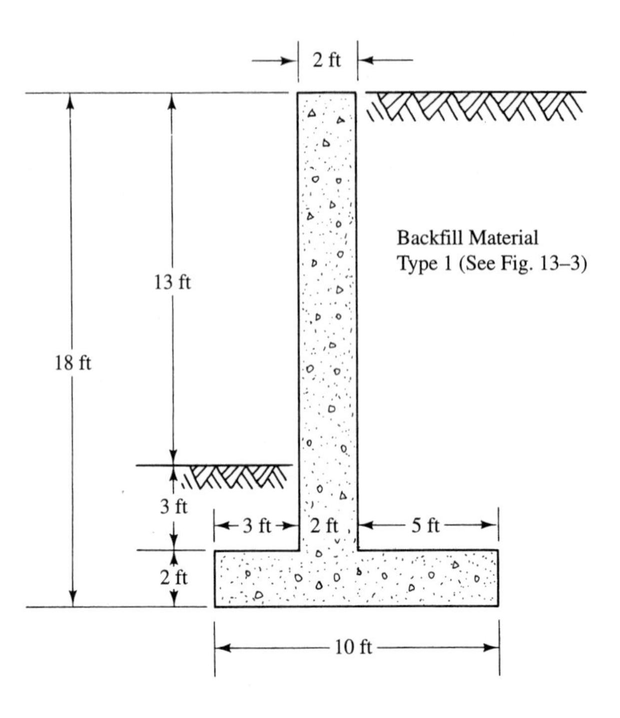

Required

1. The factor of safety against sliding. Analyze both without and with passive earth pressure at the toe.
2. The factor of safety against overturning.
3. The safety against failure of the foundation soil.

Solution

Calculation of the Active Earth Pressure by Figure 13–3

From Figure 13–3,

$$P_h = \tfrac{1}{2}k_h H^2 \text{ with } \beta = 0° \text{ and Type 1 backfill material, } k_h = 30 \text{ lb/ft}^2\text{/ft}$$

$$P_h = (\tfrac{1}{2})(30 \text{ lb/ft}^2\text{/ft})(18 \text{ ft})^2 = 4860 \text{ lb/ft} = 4.86 \text{ kips/ft}$$

$$P_v = \tfrac{1}{2}k_v H^2 \text{ with } \beta = 0°,\ k_v = 0$$

$$P_v = 0$$

FIGURE 13–9*

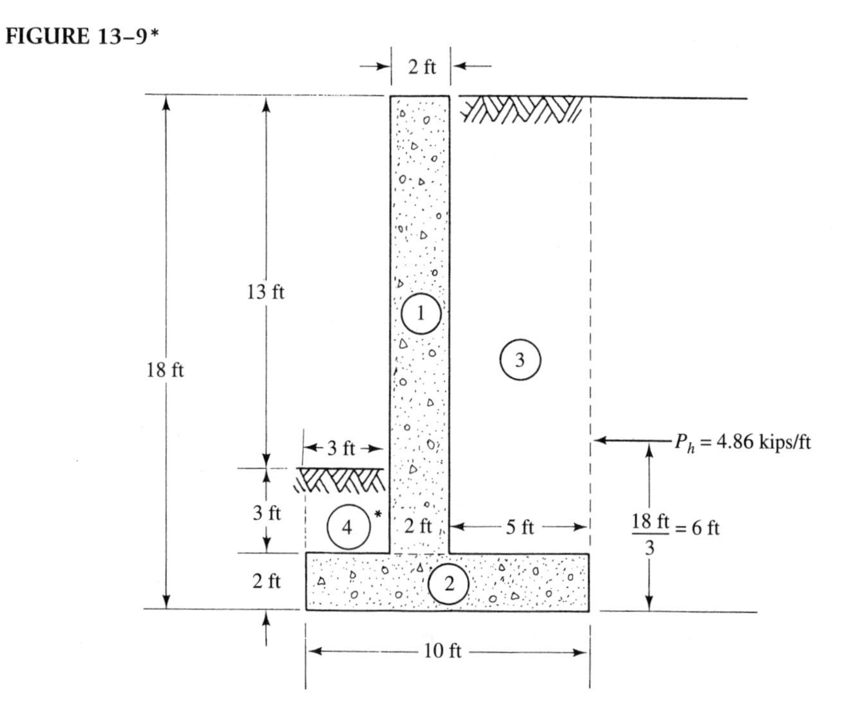

Calculation of Overturning Moment (See Figure 13–9)

Calculation of Righting Moment (See Figure 13–9)

Component	Weight of Component (kips/ft)	Moment Arm from Toe (ft)	Righting Moment about Toe (ft-kips/ft)
1	(0.15)(2)(13 + 3) = 4.8	3 + 2/2 = 4.0	19.2
2	(0.15)(2)(10) = 3.0	10/2 = 5.0	15.0
3	(0.12)(5)(13 + 3) = 9.6	3 + 2 + 5/2 = 7.5	72.0
4	(0.12)(3)(3) = 1.1	3/2 = 1.5	1.6
	ΣV = 18.5		ΣM_r = 107.8

Overturning moment (M_0) = (4.86 kips/ft)(6 ft) = 29.16 ft-kips/ft.

1. *Factor of safety against sliding:*

   a. *Without passive earth pressure analysis (neglect passive earth pressure at the toe):* From Eq. (13–1),

* Assume the soil above the toe is also Type I backfill material.

$$(\text{F.S.})_{\text{sliding}} = \frac{\text{Sliding resistance force}}{\text{Sliding force}} \tag{13–1}$$

$$(\text{F.S.})_{\text{sliding}} = \frac{(\mu)(\Sigma V)}{P_h} = \frac{(0.45)(18.5 \text{ kips/ft})}{4.86 \text{ kips/ft}} = 1.71 > 1.5 \quad \therefore \text{O.K.}$$

b. *With passive earth pressure at the toe:*

Sliding resistance = Passive earth pressure at toe + Friction available along base

According to Rankine theory for level backfill, from Eqs. (12–12) and (12–15),

$$P_p = \tfrac{1}{2}\gamma H^2 \frac{1 + \sin\phi}{1 - \sin\phi}$$

$$P_p = (1/2)(0.12 \text{ kip/ft}^3)(5 \text{ ft})^2 \left(\frac{1 + \sin 37°}{1 - \sin 37°}\right) = 6.03 \text{ kips/ft}$$

$$(\text{F.S.})_{\text{sliding}} = \frac{(\mu)(\Sigma V) + P_p}{P_h}$$

$$(\text{F.S.})_{\text{sliding}} = \frac{(0.45)(18.5 \text{ kips/ft}) + 6.03 \text{ kips/ft}}{4.86 \text{ kips/ft}} = 2.95 > 2.0 \therefore \text{O.K.}$$

2. *Factor of safety against overturning:* From Eq. (13–2),

$$(\text{F.S.})_{\text{overturning}} = \frac{\text{Total righting moment}}{\text{Total overturning moment}} \tag{13–2}$$

$$(\text{F.S.})_{\text{overturning}} = \frac{\Sigma M_r}{\Sigma M_0} = \frac{107.8 \text{ ft-kips/ft}}{29.16 \text{ ft-kips/ft}} = 3.70 > 1.5 \quad \therefore \text{O.K.}$$

Base Pressure Calculations

Location of resultant R ($= \Sigma V$) if R acts at $\bar{x}$ from the toe:

$$\bar{x} = \frac{\Sigma M_{toe}}{\Sigma V} = \frac{\Sigma M_r - \Sigma M_0}{\Sigma V} = \frac{107.8 \text{ ft-kips/ft} - 29.16 \text{ ft-kips/ft}}{18.5 \text{ kips/ft}} = 4.25 \text{ ft}$$

$$e = \frac{10 \text{ ft}}{2} - 4.25 \text{ ft} = 0.75 \text{ ft} < \frac{L}{6} \text{ (i.e., } 10/6 \text{ ft, or } 1.67 \text{ ft)} \quad \therefore \text{O.K.}$$

Using the flexural formula, from Eq. (9–9), one gets

$$q = \frac{Q}{A} \pm \frac{M_x y}{I_x} \pm \frac{M_y x}{I_y} \tag{9–9}$$

Here,

$$Q = \text{Resultant } (R) = \Sigma V = 18.5 \text{ kips}$$
$$A = (1 \text{ ft})(10 \text{ ft}) = 10 \text{ ft}^2$$
$$M_x = 0$$
$$M_y = Q \times e = (18.5 \text{ kips})(0.75 \text{ ft}) = 13.9 \text{ ft-kips}$$
$$x = \frac{10 \text{ ft}}{2} = 5 \text{ ft}$$
$$I_y = \frac{bh^3}{12} = \frac{(1 \text{ ft})(10 \text{ ft})^3}{12} = 83.33 \text{ ft}^4$$
$$q = \frac{18.5 \text{ kips}}{10 \text{ ft}^2} \pm \frac{(13.9 \text{ ft-kips})(5 \text{ ft})}{83.33 \text{ ft}^4}$$

$$q_L = 1.85 \text{ kips/ft}^2 + 0.83 \text{ kip/ft}^2 = 2.68 \text{ kips/ft}^2$$
$$q_R = 1.85 \text{ kips/ft}^2 - 0.83 \text{ kip/ft}^2 = 1.02 \text{ kips/ft}^2$$

3. Because q_L is 2.68 kips/ft^2, which is less than the allowable soil pressure of 3.0 kips/ft^2 (given), the wall is safe against failure of the foundation soil.

13–5 BACKFILL DRAINAGE

If water is allowed to permeate the soil behind a retaining wall, large additional pressure will be applied to the wall. Unless the wall is designed to withstand this large additional pressure (not the usual practice), it is imperative that steps be taken to prevent water that infiltrates the backfill soil from accumulating behind the wall.

One method of preventing water from accumulating behind a wall is to provide an effective means of draining away any water that enters the backfill soil. To accomplish this, it is highly desirable to use as backfill material a highly pervious soil such as sand, gravel, or crushed stone. To remove water from behind the wall, one can place 4- to 6-in. (102- to 152-mm) weep holes, which are pipes extending through the wall (see Figure 13–10a), every 5 to 10 ft (1.5 to 3 m) along the wall. A perforated drainpipe placed longitudinally along the back of the wall (Figure 13–10b) may also be used to remove water from behind the wall. In this case, the pipe is surrounded by filter material, and water drains through the filter material into the pipe and then through the pipe to one end of the wall. In both cases (weep holes and drainpipes), a filter material must be placed adjacent to the pipe to prevent clogging, and the pipes must be kept clear of debris.

If a less pervious material (silt, granular soil containing clay, etc.) has to be used as backfill because a free-draining, granular material is too expensive in the locality, it is highly desirable to place a wedge of pervious material adjacent to the wall, as shown in Figure 13–11. If this is not possible, a "drainage blanket" of pervious material may be placed as shown in Figure 13–12.

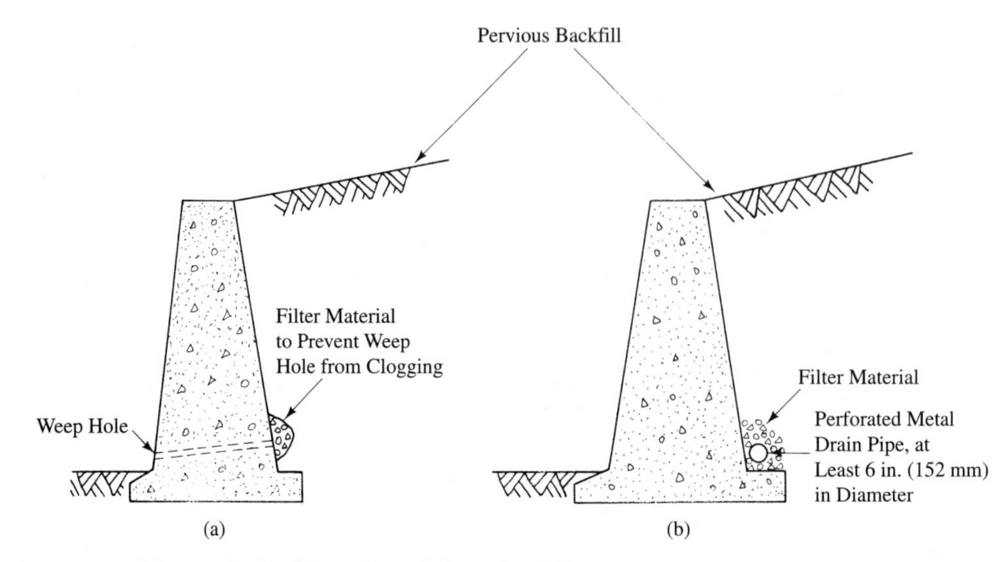

FIGURE 13–10 (a) Weep hole; (b) perforated drainpipe [1].

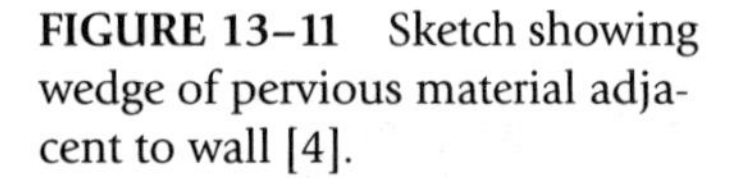

FIGURE 13–11 Sketch showing wedge of pervious material adjacent to wall [4].

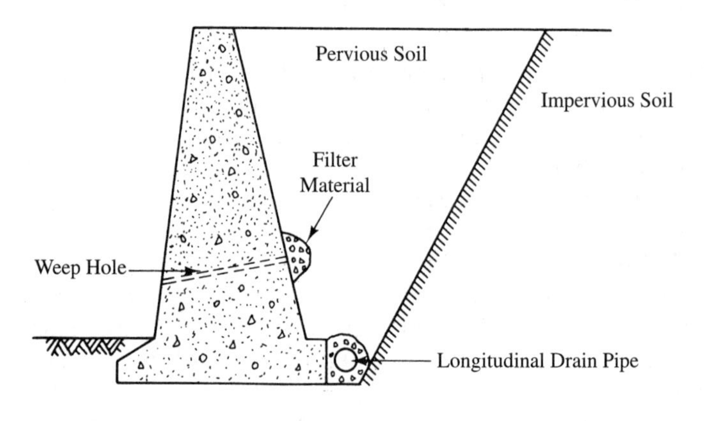

A highly impervious soil (clay) is very undesirable as backfill material because, in addition to the excessive lateral earth pressure it creates, it is difficult to drain and may be subject to frost action. Also, clays are subject to swelling and shrinking. If clayey soil must be used as backfill material, it is advisable to place a wedge of pervious material adjacent to the wall between the wall and clay backfill (as shown in Figure 13–11) [6].

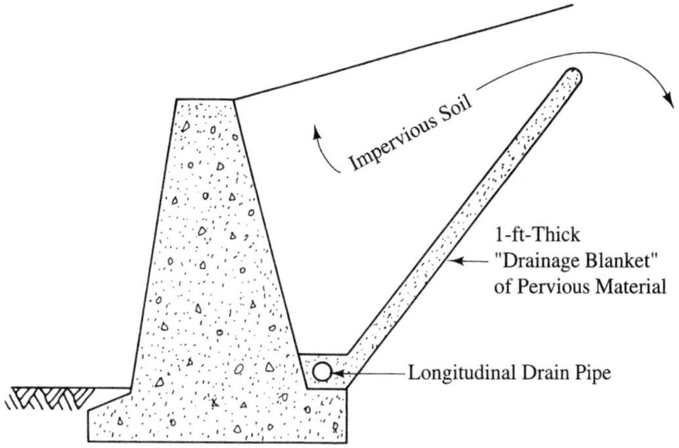

FIGURE 13–12 Sketch showing drainage blanket with longitudinal drain [5].
Source: Reprinted with permission of Macmillan Publishing Co., Inc., from *Introductory Soil Mechanics and Foundations: Geotechnical Engineering*, 4th Edition, by George F. Sowers. Copyright © 1979, Macmillan Publishing Co., Inc.

13–6 SETTLEMENT AND TILTING

A certain amount of settlement by retaining walls is to be expected, just as by any other structures resting on footings or piles. In the case of retaining walls on granular soils, most of the expected settlement will have occurred by the time construction of the wall and placement of backfill have been completed. With retaining walls on cohesive soils, for which consolidation theory is applicable, settlement will occur slowly and for some period of time after construction has been completed.

The amount of settlement for retaining walls resting on spread footings can be determined by using the principles of settlement analysis for footings (see Chapter 7). For walls resting on piles, the amount of settlement can be estimated by using the principles of settlement analysis for pile foundations (see Chapter 10). To keep settlement relatively uniform, one must ensure that the resultant force is kept near the middle of the base.

If the soil upon which a retaining wall rests is not uniform in bearing capacity along the length of the wall, differential settlement may occur along the wall, which could cause the wall to crack vertically. If soil of poor bearing capacity occurs for only a very short distance, differential settlement may not be a problem, as the wall tends to bridge across poor material. If, however, poor bearing capacity of the soil exists for a considerable distance along the length of the wall, differential settlement will likely happen unless the designer takes this into account and implements remedies to correct the situation. Possible remedies include improving the soil (e.g., by replacement, compaction, or stabilization of the soil) and changing the footing's width. If computed settlement is excessive, pile foundations may be used [7].

In addition to settlement, a retaining wall is subject to tilting caused by eccentric pressure on the base of the wall. Tilting can be reduced by keeping the resultant force near the middle of the base. In many cases, walls tilt forward because the resultant force intersects the base at a point between the center and toe.

It is difficult to determine the amount of tilting to be expected, and rough estimates must suffice. If stability requirements are met in accordance with established design procedures (see Section 13–4), the amount of tilting may be expected to be in the order of magnitude one-tenth of 1% of the height of the wall or less. However, if the subsoil consists of a compressive layer, this amount may be exceeded [1].

13–7 REINFORCED EARTH* WALLS

One alternative to the conventional retaining wall for holding soil embankments is known as a *Reinforced Earth wall,* a patented method first developed in France by Vidal. It is particularly useful where high or otherwise-difficult-to-construct retaining structures are needed.

A typical section of a Reinforced Earth structure is shown in Figure 13–13. The wall consists of precast concrete facings resting on an unreinforced concrete leveling pad. However, the "wall" does not have to be particularly strong and can therefore be constructed of thinner materials. It is sometimes referred to as the *skin.* The Reinforced Earth volume is cohesionless soil, spread and compacted in layers. Reinforcing strips, commonly made of ribbed, galvanized steel, are placed atop each layer and bolted to the skin material (i.e., wall element).

Figure 13–14 gives another sketch of a Reinforced Earth wall, which depicts length, width, thickness, and horizontal and vertical spacings (L, w, t, s, and h, respectively) of the reinforcing strips. Design considerations require that (1) the skin (wall element) resist soil pressure from adjacent soil layers, (2) strip length be long enough to support the skin, and (3) the strip be strong enough to resist its internal tension. Typical strip spacings are about 1 ft (0.3 m) vertically and 2 ft (0.6 m) horizontally.

To evaluate the criteria for Reinforced Earth design, consider a strip at depth z below the wall top. Here, the force against the area of wall that must be supported by a strip can be determined by the following equation [8]:

$$T = \gamma z K_a s h \quad \textbf{(13–4)}$$

where T = tensile force per strip
γ = unit weight of backfill soil
z = depth from wall top to strip
K_a = coefficient of active earth pressure (Rankine)
s = horizontal spacing between strips
h = vertical spacing between strips

The frictional resistance of a strip at depth z (F) developed between the strip's top and bottom faces and the backfill soil is as follows [8]:

* Reinforced Earth is a registered trademark of The Reinforced Earth Company, Arlington, Va.

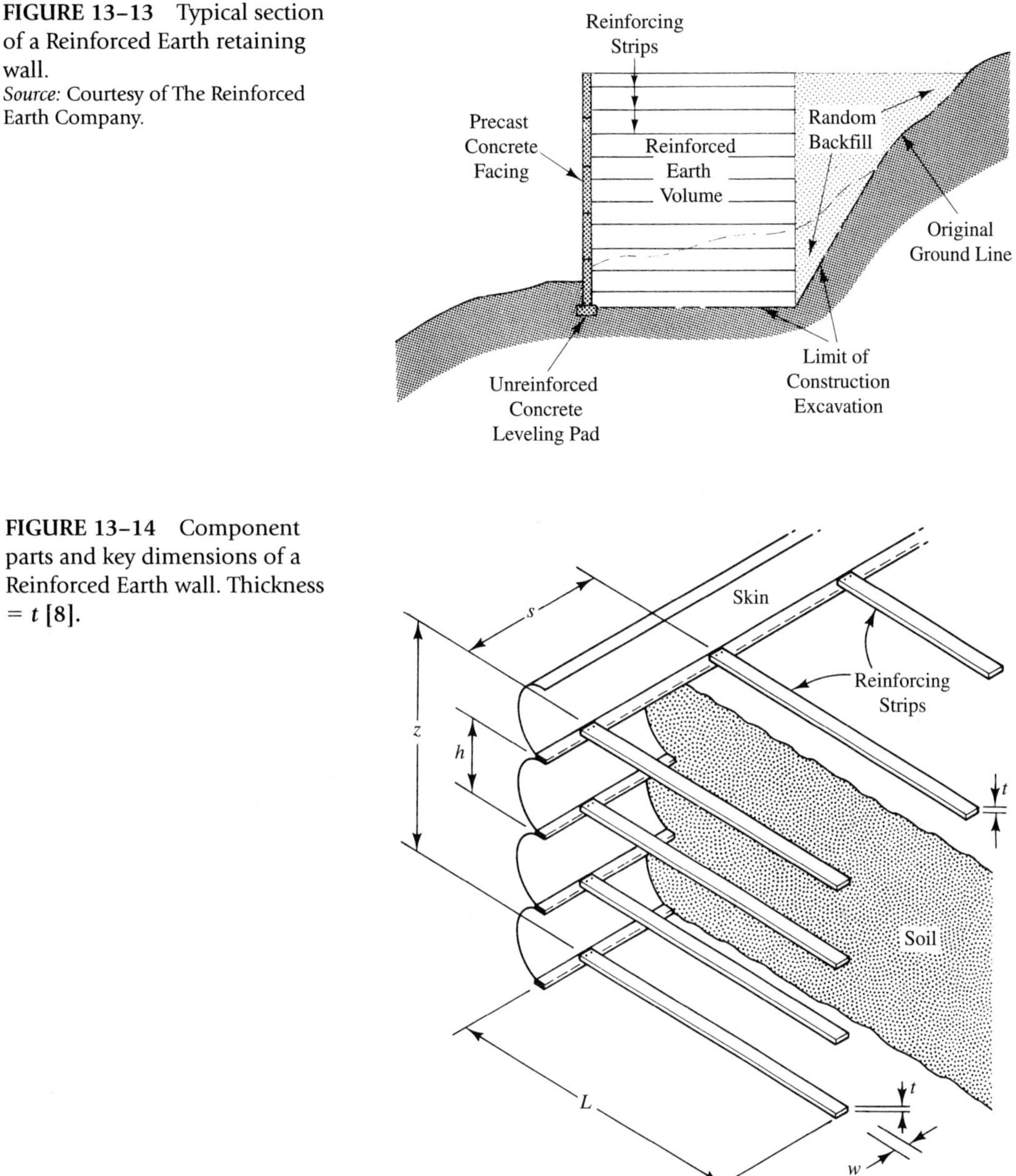

FIGURE 13–13 Typical section of a Reinforced Earth retaining wall.
Source: Courtesy of The Reinforced Earth Company.

FIGURE 13–14 Component parts and key dimensions of a Reinforced Earth wall. Thickness $= t$ [8].

$$F = (\gamma z \tan \delta)(2Lw) \qquad \textbf{(13–5)}$$

where δ is the angle of friction between the strip's surfaces and the backfill soil and other terms are as previously defined. Tan δ can be taken as tan ($\phi/2$), where ϕ is the soil's angle of internal friction.

The required minimum length of the strip (L_{min}) can be evaluated by equating T [Eq. (13–4)] and F [Eq. (13–5)] and including an appropriate factor of safety against pullout (F.S., normally 1.5 to 2.0). Hence,

$$(\text{F.S.})(\gamma z K_a sh) = (\gamma z \tan \delta)(2Lw) \qquad \textbf{(13–6)}$$

Solving for L_{min} (i.e., L) yields the following:

$$L_{min} = \frac{(\text{F.S.})(K_a sh)}{2w \tan \delta} \qquad \textbf{(13–7)}$$

L_{min} is measured beyond the zone of Rankine failure, as shown in Figure 13–15a. That is,

$$L_{total} = L_{Rankine} + L_{min} \qquad \textbf{(13–8)}$$

where

$$L_{Rankine} = H \tan\left(45° - \frac{\phi}{2}\right) \qquad \textbf{(13–9)}$$

For overall stability, a minimum length (L_{total}) of 80% of the wall height, H, is suggested (see Figure 13–15b). That is,

$$(L_{total})_{minimum} = 0.80H \qquad \textbf{(13–10)}$$

Strip thickness can be determined from the basic stress equation

$$t = \frac{T}{wf_s} \qquad \textbf{(13–11)}$$

where t = strip thickness
T = tensile force (per strip) [from Eq. (13–4)]
w = strip width
f_s = allowable stress for strip material

It may be noted from Eq. (13–4) that the tensile force per strip is greatest at the bottom of the wall (i.e., where z is greatest). At lesser values of z, the tensile force is less, but the friction on each strip is also reduced. Accordingly, the total strip area should be constant at all depths to provide the same resistance to strip pullout. Also, from Eq. (13–7) it is clear that the minimum length of the strip is independent of depth. For these reasons as well as simplicity in construction, usually the same size, length (Figure 13–13), and spacing of strips are used throughout a Reinforced Earth structure.

It should be emphasized that Reinforced Earth structures must use cohesionless soils as backfill material because of their needed high friction.

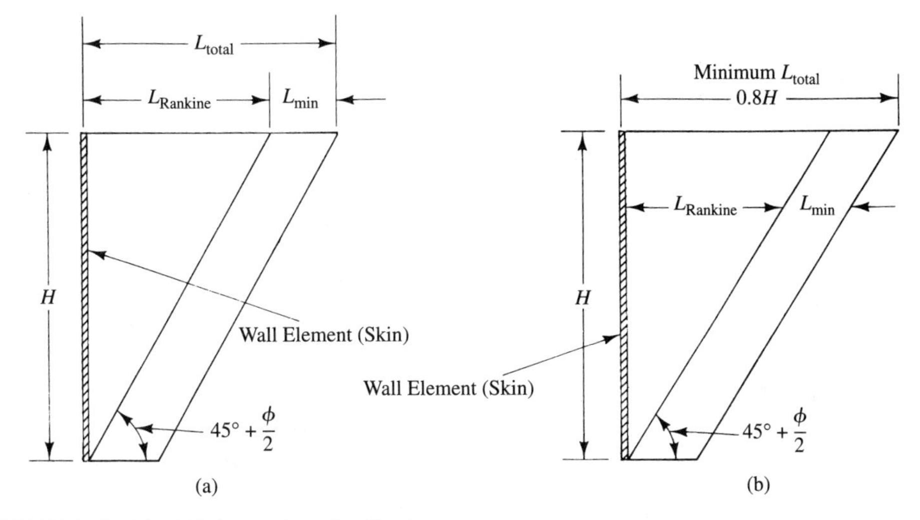

FIGURE 13–15 Minimum length of strip.

EXAMPLE 13–3

Given

A 6-m-high Reinforced Earth wall is to be constructed with level backfill (Figure 13–14) and will have no surcharge on the backfill. A granular soil with a unit weight of 17.12 kN/m^3 and an angle of internal friction of 34° will be used for backfill material. The steel strips' width, vertical spacing, horizontal spacing, and allowable stress are 75 mm, 0.3 m, 1.0 m, and 138,000 kN/m^2, respectively. The factor of safety against pullout is to be 1.5.

Required

1. Total length of strip required.
2. Thickness of strip required.

Solution

1. From Eq. (13–7),

$$L_{min} = \frac{(\text{F.S.})(K_a sh)}{2w \tan \delta} \tag{13–7}$$

$$\text{F.S.} = 1.5 \qquad \text{(given)}$$

From Eq. (12–14),

$$K_a = \frac{1 - \sin \phi}{1 + \sin \phi} \tag{12–14}$$

$$K_a = \frac{1 - \sin 34°}{1 + \sin 34°} = 0.283$$

$$s = 1.0 \text{ m} \qquad \text{(given)}$$

$$h = 0.3 \text{ m} \qquad \text{(given)}$$

$$w = 75 \text{ mm} = 0.075 \text{ m} \qquad \text{(given)}$$

$$\delta = \frac{\phi}{2} = \frac{34°}{2} = 17°$$

Therefore, substituting into Eq. (13–7) yields

$$L_{\text{min}} = \frac{(1.5)(0.283)(1.0 \text{ m})(0.3 \text{ m})}{(2)(0.075 \text{ m})(\tan 17°)} = 2.78 \text{ m}$$

From Eq. (13–9),

$$L_{\text{Rankine}} = H \tan\left(45° - \frac{\phi}{2}\right) \tag{13–9}$$

$$L_{\text{Rankine}} = (6.0 \text{ m}) \tan\left(45° - \frac{34°}{2}\right) = 3.19 \text{ m}$$

From Eq. (13–8),

$$L_{\text{total}} = L_{\text{Rankine}} + L_{\text{min}} \tag{13–8}$$

$$L_{\text{total}} = 3.19 \text{ m} + 2.78 \text{ m} = 5.97 \text{ m}$$

From Eq. (13–10),

$$(L_{\text{total}})_{\text{minimum}} = 0.80H \tag{13–10}$$

$$(L_{\text{total}})_{\text{minimum}} = (0.80)(6.0 \text{ m}) = 4.80 \text{ m}$$

Therefore, use a total length of strip of 5.97 m. (In practice, one would probably specify 6 m.)

2. From Eq. (13–11),

$$t = \frac{T}{wf_s} \tag{13–11}$$

From Eq. (13–4),

$$T = \gamma z K_a s h \tag{13–4}$$

$$\gamma = 17.12 \text{ kN/m}^3 \qquad \text{(given)}$$

$$z = 6.0 \text{ m} \qquad \text{(given)}$$

$$T = (17.12 \text{ kN/m}^3)(6.0 \text{ m})(0.283)(1.0 \text{ m})(0.3 \text{ m})$$

$$= 8.72 \text{ kN}$$

$$f_s = 138{,}000 \text{ kN/m}^2 \qquad \text{(given)}$$

Therefore, substituting into Eq. (13–11) yields

$$t = \frac{8.72 \text{ kN}}{(0.075 \text{ m})(138{,}000 \text{ kN/m}^2)} = 0.00084 \text{ m, or } 0.84 \text{ mm}$$

13–8 SLURRY TRENCH WALLS

The *slurry trench method* of constructing a wall to retain earth is applicable for retaining walls built entirely below ground level. The procedure includes excavating a trench the width of the wall while simultaneously filling the excavation with a viscous bentonite slurry, which exerts a lateral pressure and thereby helps stabilize the excavated wall. When excavation is complete, reinforcing steel is placed in the bentonite-slurry-filled trench, and, with the trench's soil walls acting as forms, concrete is poured from the bottom up. The concrete is delivered by a tremie (a long canvas tube) that is slowly raised as the excavated trench is filled with the concrete. Being displaced by the poured concrete, the lighter bentonite slurry rises to the top, where it is removed and may be saved for later use. Adjacent wall sections may be keyed together by a steel beam that becomes a part of the wall.

After the concrete has hardened, soil is removed from one side to expose the face of the wall and allow installation of a tieback system (see Figure 13–16). Depending on the height of wall, two or more levels of tiebacks may be used.

13–9 ANCHORED BULKHEADS

Anchored bulkheads are often used for various waterfront structures (e.g., wharves and waterfront retaining walls). As illustrated in Figure 13–17, they consist of interlocking sheetpilings driven into the soil and anchored by steel tie-rods or cables attached near the pilings' tops, extended through the ground, and anchored

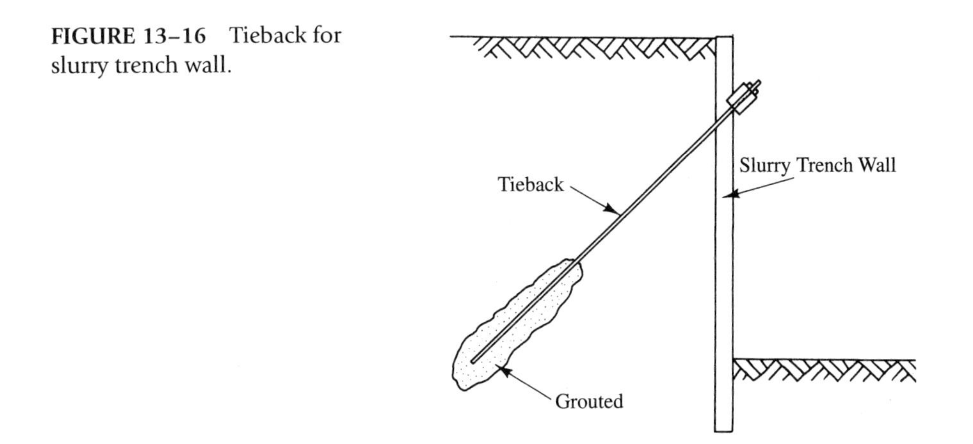

FIGURE 13–16 Tieback for slurry trench wall.

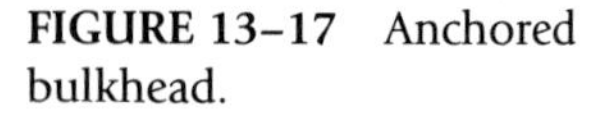

FIGURE 13–17 Anchored bulkhead.

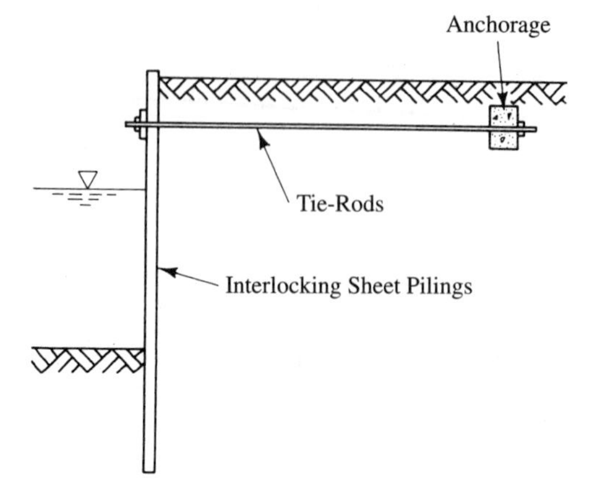

securely somewhere away from the sheet pile wall in firm soil. The distant anchors may be "deadmen" or braced piles.

In general, bulkheads can be constructed by one of two methods. Either the bulkhead can be built (driven) in open water and fill placed behind it, or the bulkhead can be constructed in the natural ground and earth removed from its face. The former type is known as a *fill bulkhead;* the latter, a *dredged bulkhead.*

As related in Figure 13–18, there are several possible anchoring systems available. Concrete and sheet pile deadmen (Figure 13–18a and 13–18b, respectively) are usable in strong soils. They are, however, rather bulky and therefore require sufficient space. Where upper soils are weak or space is severely limited, braced piles (Figure 13–18c) may be used as anchors.

The fill behind a bulkhead applies lateral pressure to it and tends to push it forward. It is restrained at its top by the anchors and at its bottom by the soil in front of the bulkhead. It is also restrained by the standing water, but this tends to be offset by groundwater behind the bulkhead. Anchored bulkheads are usually subjected to fluctuating water levels; hence, engineers must base their design on "worst-case" conditions. (For example, tidal fluctuations produce high pressures behind the wall compared with those in front when the tide is out.)

Figure 13–19 shows the various forces acting on a typical anchored bulkhead. Actuating forces result from active pressure of the soil backfill and from water behind the wall. Resisting forces are composed of tension in the cable or tie-rod to the anchor, water pressure on the wall's front side, and passive resistance pressure of the soil within the penetrating depth of the sheetpiling. The ratio of total resisting force to total actuating force gives the bulkhead's factor of safety. The factor of safety can be increased by driving the sheetpiling deeper into the soil.

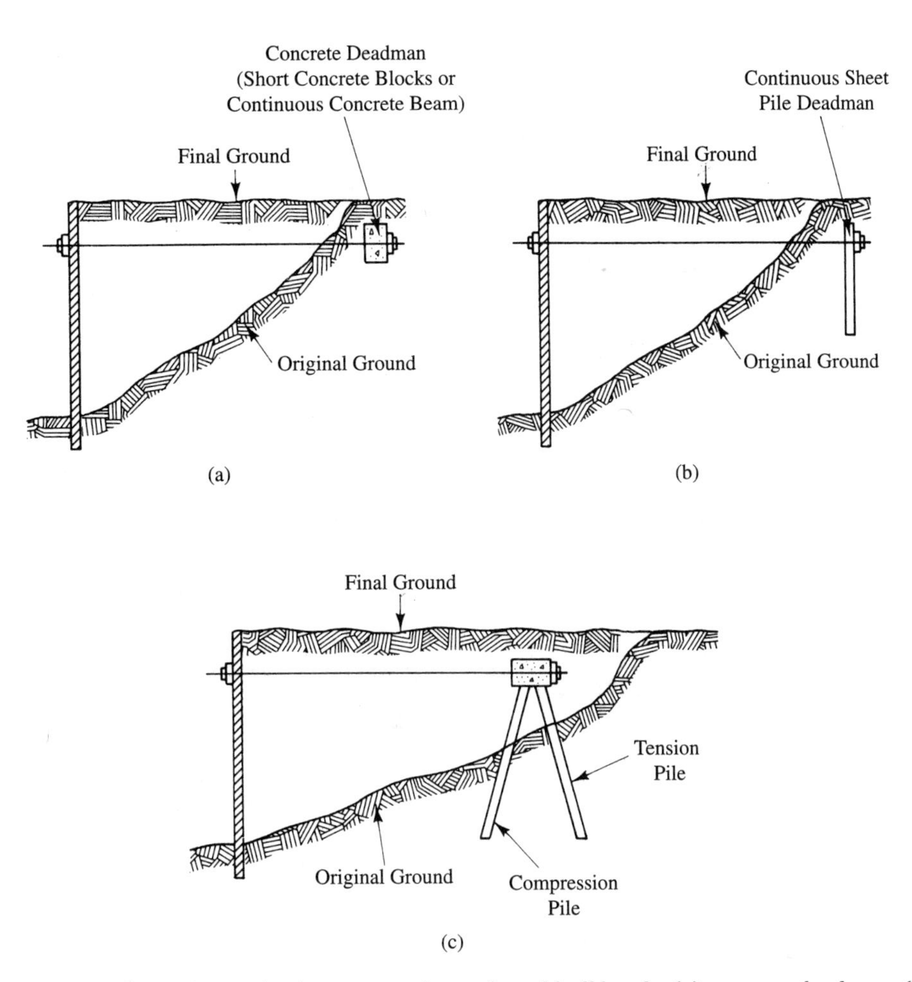

FIGURE 13–18 Alternative anchoring systems for anchored bulkheads: (a) concrete deadman; (b) sheet pile deadman; (c) braced piles.

EXAMPLE 13–4

Given

An anchored bulkhead is to be constructed as shown in Figure 13–20, and a factor of safety of 1.5 is to be used.

Required

Analyze the bulkhead system and determine the tension in the anchor rod (tieback).

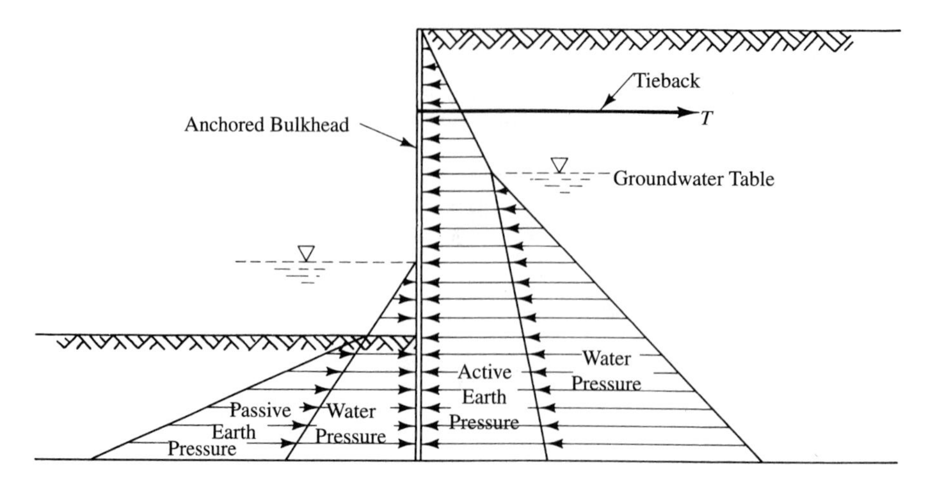

FIGURE 13–19 Forces acting on anchored bulkheads.

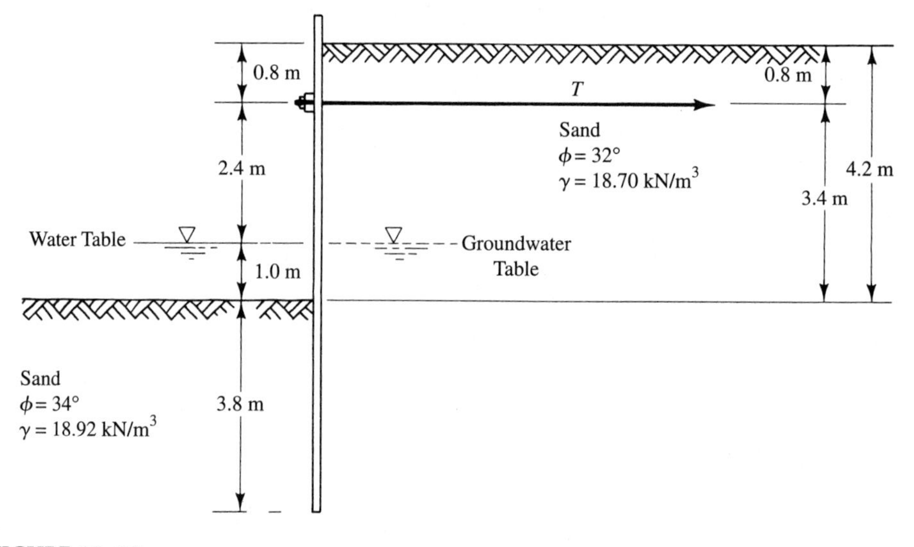

FIGURE 13–20

Solution

From Eqs. (12–10) and (12–12),

$$P_a = \tfrac{1}{2}\gamma H^2 K_a \quad (12\text{–}10)$$

$$P_p = \tfrac{1}{2}\gamma H^2 K_p \quad (12\text{–}12)$$

From Eq. (12–14),

$$K_a = \frac{1 - \sin\phi}{1 + \sin\phi} \quad (12\text{–}14)$$

$$K_a = \frac{1 - \sin 32°}{1 + \sin 32°} = 0.307$$

From Eq. (12–15),

$$K_p = \frac{1 + \sin\phi}{1 - \sin\phi} \quad (12\text{–}15)$$

$$K_p = \frac{1 + \sin 34°}{1 - \sin 34°} = 3.54$$

Component	Active Earth or Water Pressure (kN/m)	Moment about Tie Point (kN · m/m)
1	$(\tfrac{1}{2})(18.7)(3.2)^2(0.307) = 29.39$	$(29.39)[(3.2)(\tfrac{2}{3}) - 0.8] = 39.2$
2	$(3.2)(18.7)(4.8)(0.307) = 88.18$	$(88.18)[(4.8)(\tfrac{1}{2}) + 2.4] = 423.3$
3	$(\tfrac{1}{2})(18.7 - 9.81)(4.8)^2(0.307) = 31.44$	$(31.44)[(4.8)(\tfrac{2}{3}) + 2.4] = 176.1$
4	$(\tfrac{1}{2})(9.81)(4.8)^2 = 113.01$	$(113.01)[(4.8)(\tfrac{2}{3}) + 2.4] = 632.9$
	262.02	1271.5

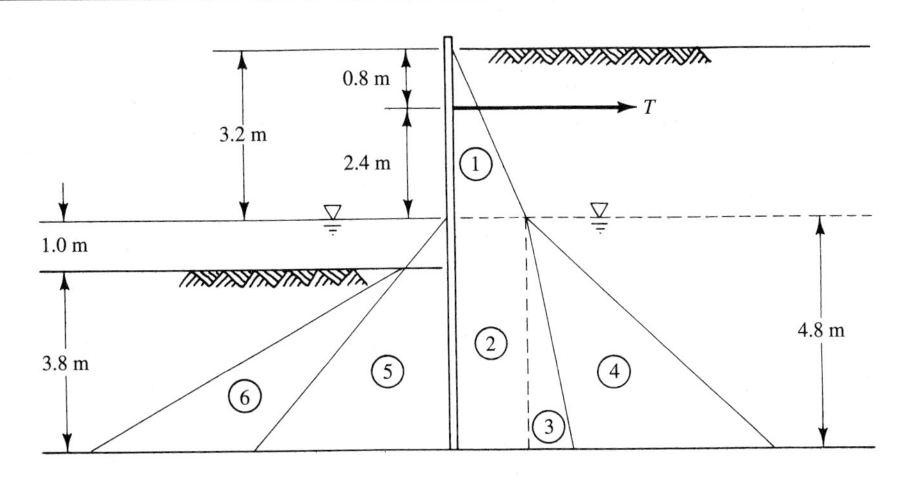

FIGURE 13–21

Referring to Figure 13–21, one may determine applicable pressure and moments as follows:
To simplify the analysis that follows, let

R_{wp} = resistance from water pressure (left of sheetpiling) (i.e., P_5)
$(P_6)_{min}$ = minimum required passive earth pressure in component 6
d_{P_6} = distance from tie point to P_6
$d_{R_{wp}}$ = distance from tie point to R_{wp} (i.e., d_{P_5})
$(P_6)_{mm}$ = maximum mobilizable P_6
$\Sigma\text{Moments}_{\text{tie point}} = 0 \curvearrowleft^{+}$

$$[(P_6)_{min}](d_{P_6}) + (R_{wp})(d_{R_{wp}}) - \text{Sum of moments of } P_1, P_2, P_3, \text{ and } P_4 = 0 \quad \textbf{(A)}$$

$$d_{P_6} = (3.8 \text{ m})(2/3) + 1.0 \text{ m} + 2.4 \text{ m} = 5.933 \text{ m}$$

$$R_{wp} = \text{Force of component 5} = \frac{\frac{1}{2}\gamma H^2}{\text{F.S.}}$$

$$R_{wp} = \frac{(1/2)(9.81 \text{ kN/m}^3)(4.8 \text{ m})^2}{1.5} = 75.34 \text{ kN/m}$$

$$d_{R_{wp}} = (4.8 \text{ m})(2/3) + 2.4 \text{ m} = 5.600 \text{ m}$$

The sum of the moments of P_1, P_2, P_3, and P_4 = 1271.5 kN · m/m (from preceding tabulation). Substituting into Eq. (A) gives the following:

$$[(P_6)_{min}](5.933 \text{ m}) + (75.34 \text{ kN/m})(5.600 \text{ m}) - 1271.5 \text{ kN} \cdot \text{m/m} = 0$$

$$(P_6)_{min} = 143.20 \text{ kN/m}$$

$$(P_6)_{mm} = \frac{1/2(\gamma_{soil} - \gamma_{water})h^2K_p}{\text{F.S.}}$$

$$(P_6)_{mm} = \frac{(1/2)(18.92 \text{ kN/m}^3 - 9.81 \text{ kN/m}^3)(3.8 \text{ m})^2(3.54)}{1.5} = 155.2 \text{ kN/m}$$

Because $[(P_6)_{mm} = 155.2 \text{ kN/m}] > [(P_6)_{min} = 143.20 \text{ kN/m}]$, the design is O.K. because the sheet pile penetration depth is adequate to develop sufficient passive resistance.

To determine the tension (T) in the anchor rod, one must perform the following calculations:

$$\Sigma\text{Forces}_{\text{horizontal}} = 0$$

$$T + R_{wp} + (P_6)_{min} - (P_1 + P_2 + P_3 + P_4) = 0$$

$$P_1 + P_2 + P_3 + P_4 = 262.02 \text{ kN/m} \quad \text{(from previous tabulation)}$$

$$T + 75.34 \text{ kN/m} + 143.20 \text{ kN/m} - 262.02 \text{ kN/m} = 0$$

$$T = 43.48 \text{ kN/m}$$

EXAMPLE 13–5

Given

A continuous deadman is to be designed and installed near the ground surface, as shown in Figure 13–22. Anchor rod (tieback) tension is to be 75 kN/m.

Required

Design the deadman, using a factor of safety against anchor resistance failure of 1.5.

Solution

$$\text{Capacity of deadman} = \text{Tieback tension} = \frac{\frac{1}{2}\gamma H^2(K_p - K_a)}{\text{F.S.}}$$

From Eq. (12–14),

$$K_a = \frac{1 - \sin\phi}{1 + \sin\phi} \tag{12–14}$$

$$K_a = \frac{1 - \sin 32°}{1 + \sin 32°} = 0.307$$

From Eq. (12–15),

$$K_p = \frac{1 + \sin\phi}{1 - \sin\phi} \tag{12–15}$$

$$K_p = \frac{1 + \sin 32°}{1 - \sin 32°} = 3.255$$

$$75\ \text{kN/m} = \frac{(1/2)(18.70\ \text{kN/m}^3)(H)^2(3.255 - 0.307)}{1.5}$$

$$H = 2.02\ \text{m}$$

The deadman should be placed with its bottom 2.02 m below the ground surface.

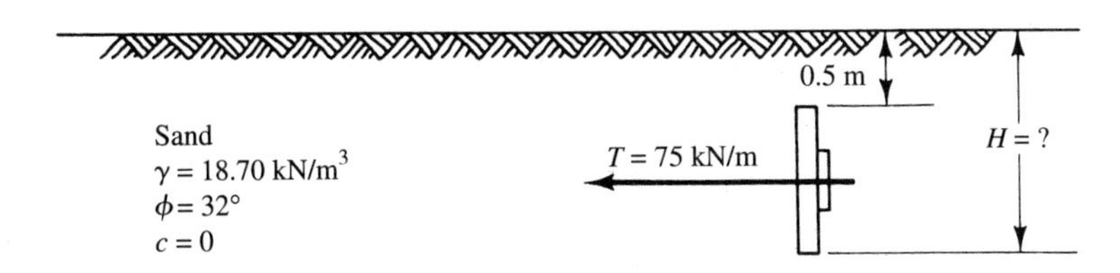

FIGURE 13–22

13–10 PROBLEMS

13–1. A proposed L-shaped reinforced-concrete retaining wall is shown in Figure 13–23. The backfill material will be Type 2 soil (Figure 13–3), and its unit weight is 125 lb/ft^3. The coefficient of base friction is estimated to be 0.48, and allowable soil pressure for the foundation soil is 4 $kips/ft^2$. Determine the (a) factor of safety against overturning, (b) factor of safety against sliding, and (c) safety against failure of the foundation.

13–2. Investigate the stability against overturning, sliding resistance (consider passive earth pressure at the toe), and foundation soil pressure of the retaining wall shown in Figure 13–24. The retaining wall is to support a deposit of granular soil, which has a unit weight of 17.30 kN/m^3 and an angle of internal friction of 32°. The coefficient of base friction is 0.50. Allowable soil pressure for the foundation soil is 144 kN/m^2. Use Rankine theory to calculate both active and passive earth pressures.

FIGURE 13–23

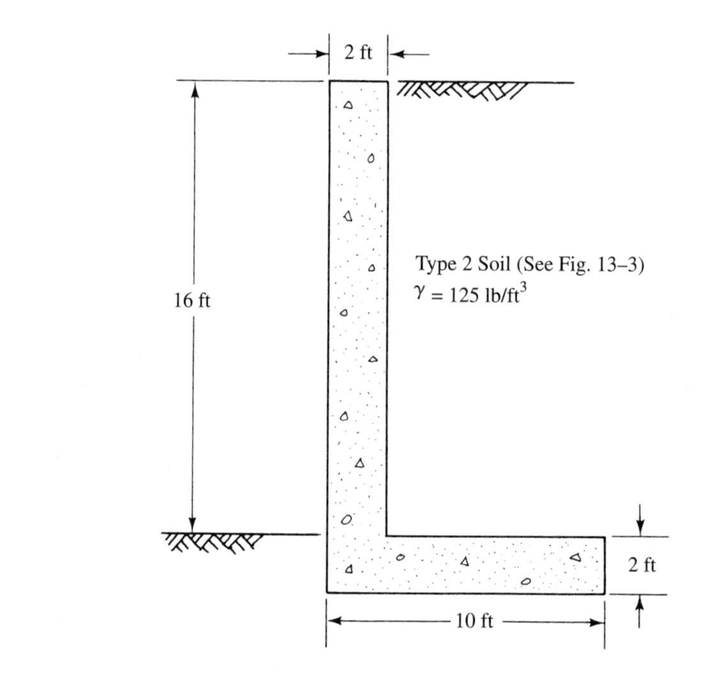

FIGURE 13–24

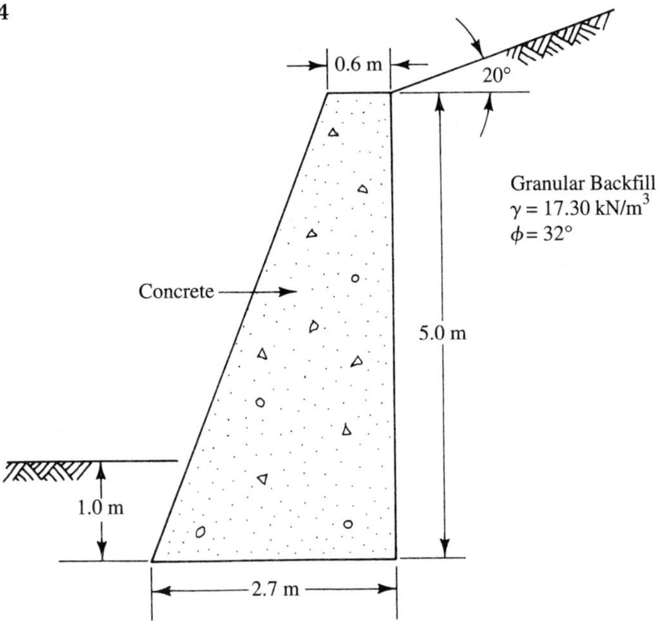

13–3. For the retaining wall shown in Figure 13–25, compute the factors of safety against overturning and sliding (analyze the latter both without and with passive earth pressure at the toe). Also determine the soil pressure at the base of the wall. Use the Rankine equation to compute passive earth pressure.

13–4. An 8-m-high Reinforced Earth wall is to be built with level backfill and without a surcharge on the backfill. Sand with an angle of internal friction of 36° and a unit weight of 17.0 kN/m^3 will be used as backfill material. The steel strips are 90 mm wide and 0.762 mm thick and have an allowable stress of 138,000 kN/m^2. For vertical spacing of 0.4 m, determine the required total length and horizontal spacing of the steel strips.

13–5. Analyze the anchored bulkhead system shown in Figure 13–26, using a factor of safety of 1.5. Determine the anchor rod tension per unit length of sheetpiling.

13–6. A continuous deadman is to be designed and constructed near ground surface, as shown in Figure 13–27. Anchor rod tension is to be 79 kN/m. Using a factor of safety of 1.5, determine how far the bottom of the deadman should be placed below the ground surface.

FIGURE 13–25

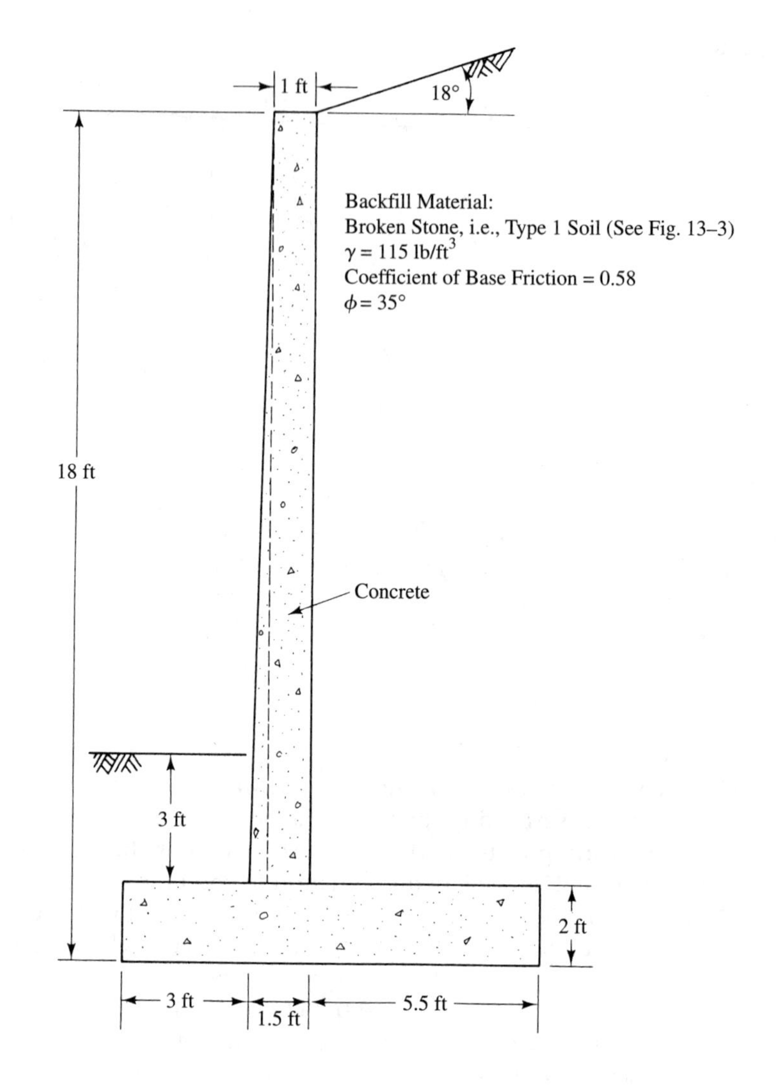

1 ft
18°
18 ft
Backfill Material:
Broken Stone, i.e., Type 1 Soil (See Fig. 13–3)
γ = 115 lb/ft³
Coefficient of Base Friction = 0.58
ϕ = 35°
Concrete
3 ft
2 ft
3 ft
1.5 ft
5.5 ft

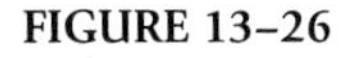

FIGURE 13–26

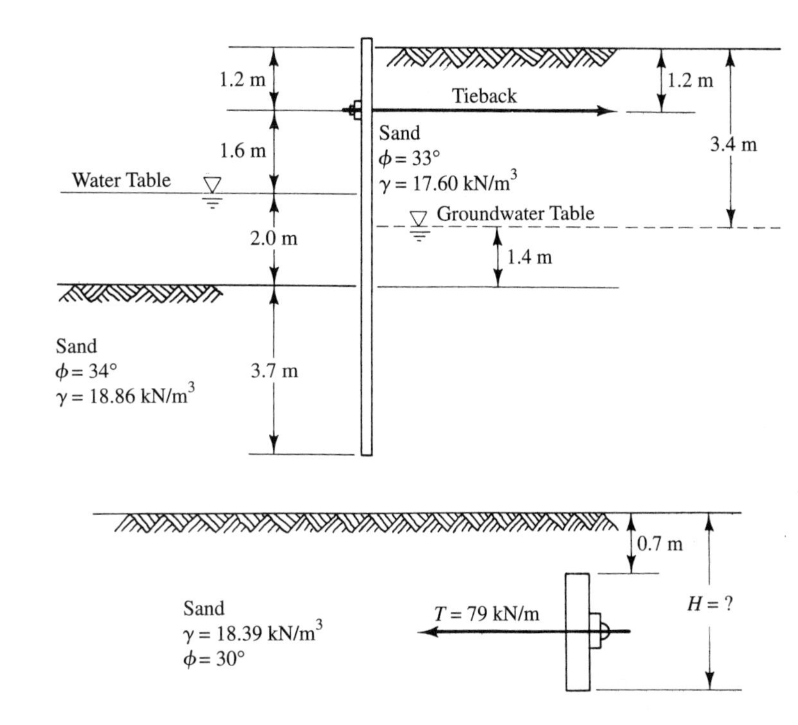

FIGURE 13–27

References

[1] Wayne C. Teng, *Foundation Design,* Prentice-Hall, Inc., Englewood Cliffs, N.J., 1962.

[2] AREA, *Manual of Recommended Practice,* Construction and Maintenance Section, Engineering Division, Association of American Railroads, Chicago, 1958.

[3] Louis J. Goodman and R. H. Karol, *Theory and Practice of Foundation Engineering,* Macmillan Publishing Co., Inc., New York, 1968.

[4] B. K. Hough, *Basic Soils Engineering,* 2nd ed., The Ronald Press Company, New York, 1969. Copyright © 1969, by John Wiley & Sons, Inc. Reprinted by permission of John Wiley & Sons, Inc.

[5] George F. Sowers, *Introductory Soil Mechanics and Foundations: Geotechnical Engineering,* 4th ed., Macmillan Publishing Co., Inc., New York, 1979.

[6] Ralph B. Peck, Walter E. Hansen, and Thomas H. Thornburn, *Foundation Engineering,* 2nd ed., John Wiley & Sons, Inc., New York, 1974. Copyright © 1974, by John Wiley & Sons, Inc. Reprinted by permission of John Wiley & Sons, Inc.

[7] Joseph E. Bowles, *Foundation Analysis and Design,* McGraw-Hill Book Company, New York, 1968.

[8] Kenneth L. Lee, Bobby Dean Adams, and Jean-Marie J. Vagneron, "Reinforced Earth Retaining Walls," *J. Soil Mech. Found. Eng. Div., Proc. ASCE,* **99**(SM 10) 745–764 (1973).

14

STABILITY ANALYSIS OF SLOPES

14–1 INTRODUCTION

Whenever a mass of soil has an inclined surface, the potential always exists for part of the soil mass to slide from a higher location to a lower one. Sliding will occur if shear stresses developed in the soil exceed the corresponding shear strength of the soil. This phenomenon is of importance in the case of highway cuts and fills, embankments, earthen dams, and so on.

This principle—that sliding will occur if shear stresses developed in the soil exceed the corresponding shear strength the soil possesses—is simple in theory; however, certain practical considerations make precise stability analyses of slopes difficult in practice. In the first place, sliding may occur along any of a number of possible surfaces. In the second place, a given soil's shear strength generally varies throughout time, as soil moisture and other factors change. Obviously, stability analysis should be based on the smallest shear strength a soil will ever have in the future. This is difficult, if not impossible, to ascertain. It is, therefore, normal in practice to use appropriate safety factors when one is analyzing slope stability.

There are several techniques available for stability analysis. Section 14–2 covers the analysis of a soil mass resting on an inclined layer of impermeable soil. Section 14–3 discusses slopes in homogeneous cohesionless soils. Section 14–4 gives two methods of analyzing stability for homogeneous soils that have cohesion. The first is known as the *Culmann method*. It is applicable to only vertical, or nearly vertical, slopes. The second might be called the *stability number method*. Section 14–5 presents the *method of slices*.

14–2 ANALYSIS OF A MASS RESTING ON AN INCLINED LAYER OF IMPERMEABLE SOIL

One situation for which slope stability analysis is fairly simple is that of a soil mass resting on an inclined layer of impermeable soil (see Figure 14–1). There exists a tendency for the upper mass to slide downward along its plane of contact with the lower layer of impermeable soil.

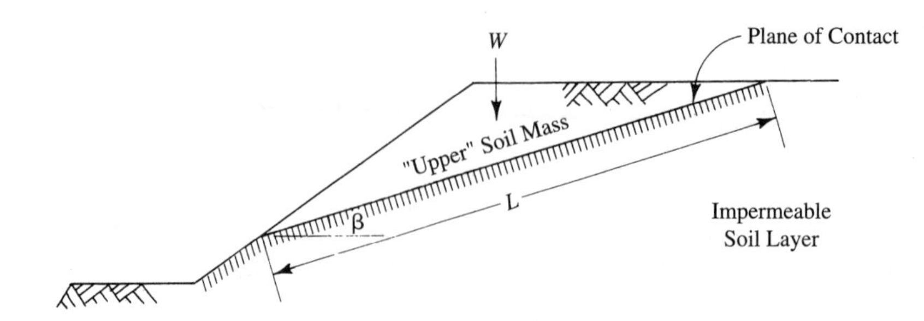

FIGURE 14–1 Sketch showing soil mass resting on an inclined layer of impermeable soil.

The force tending to cause sliding is the component of the upper mass's weight along the plane of contact. By referring to Figure 14–2 and considering a unit width of slope (i.e., perpendicular to wedge *abc*), one can compute the upper mass's weight (W) (i.e., weight of wedge *abc*) by using the following equation:

$$W = \frac{Lh\gamma}{2} \tag{14–1}$$

where γ is the unit weight of the upper mass. Hence, the force tending to cause sliding (F_s) is given by the following equation:

$$F_s = W \sin \alpha \tag{14–2}$$

Forces that resist sliding result from cohesion and friction. In quantitative terms, the cohesion (i.e., adhesion) component is the product of the soil's cohesion (c) times the length of the plane of contact (L in Figure 14–2). The friction component is obtained by multiplying the coefficient of friction between the two strata (tan ϕ) by the component of the upper mass's weight that is perpendicular to the plane of contact ($W \cos \alpha$). Hence, the resistance (to sliding) force, R_s, is given by the following equation:

$$R_s = cL + W \cos \alpha \tan \phi \tag{14–3}$$

where ϕ is the angle of friction between the upper mass and the lower layer of impermeable soil.

The factor of safety (F.S.) against sliding is determined by dividing the resistance (to sliding) force, R_s [Eq. (14–3)], by the sliding force, F_s [Eq. (14–2)]. Hence,

$$\text{F.S.} = \frac{cL + W \cos \alpha \tan \phi}{W \sin \alpha} \tag{14–4}$$

Figure 14–3 gives the formulation required to evaluate L and h, which are needed in applying Eqs. (14–1) and (14–4). Table 14–1 gives the significance of factors of safety against sliding for design. Example 14–1 illustrates the computation of the factor of safety for stability analysis of a soil mass resting on an inclined layer of impermeable soil.

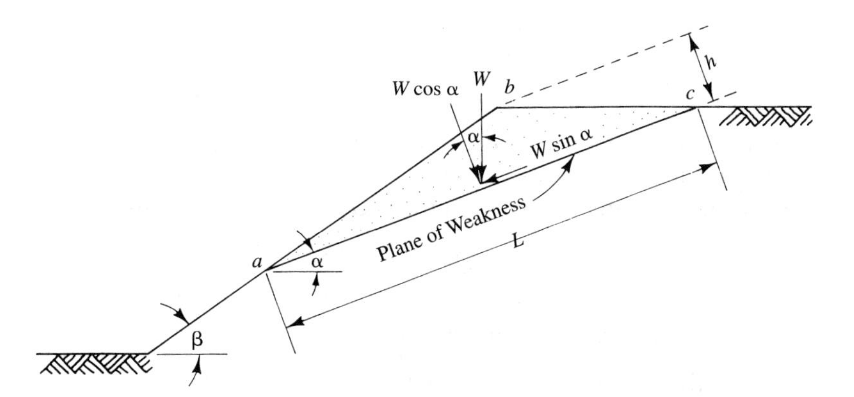

FIGURE 14–2 Sketch showing forces acting on inclined layer of impermeable soil.

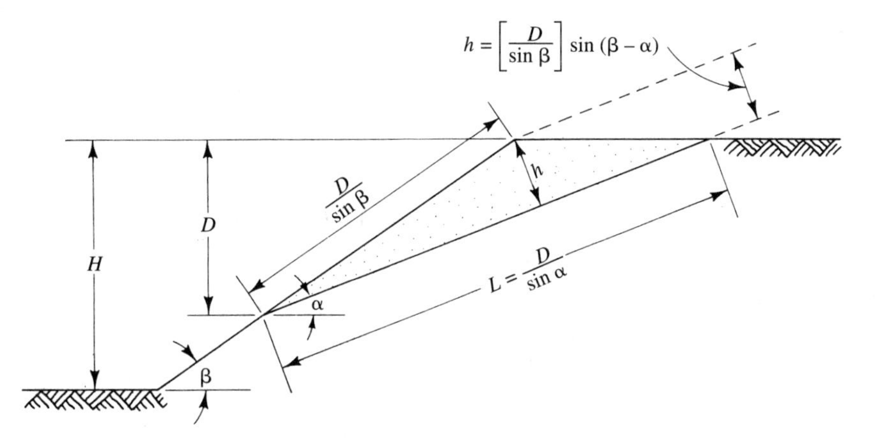

FIGURE 14–3 Sketch showing formulation required to evaluate L and h.

TABLE 14–1
Significance of Factors of Safety for Design [1][1]

Safety Factor	Significance
Less than 1.0	Unsafe
1.0–1.2	Questionable safety
1.3–1.4	Satisfactory for cuts, fills; questionable for dams
1.5–1.75	Safe for dams

[1]Reprinted with permission of Macmillan Publishing Co., Inc., from *Introductory Soil Mechanics and Foundations,* 4th Edition, by George F. Sowers. Copyright © 1979, Macmillan Publishing Co., Inc.

EXAMPLE 14–1

Given

1. Figure 14–4 shows a 15-ft cut through two soil strata. The lower is a highly impermeable cohesive soil.
2. Shearing strength data between the two strata are as follows:

$$\text{Cohesion} = 150 \text{ lb/ft}^2$$
$$\text{Angle of friction} = 25°$$
$$\text{Unit weight of upper layer} = 105 \text{ lb/ft}^3$$

3. Neglect the effects of soil water between the two strata.

Required

Factor of safety against sliding.

Solution

From Figure 14–3,

$$L = \frac{D}{\sin \alpha}$$

$$L = \frac{10 \text{ ft}}{\sin 30°} = 20.0 \text{ ft}$$

Again, from Figure 14–3,

$$h = \left(\frac{D}{\sin \beta}\right) \sin (\beta - \alpha)$$

$$h = \left(\frac{10 \text{ ft}}{\sin 45°}\right) \sin (45° - 30°) = 3.66 \text{ ft}$$

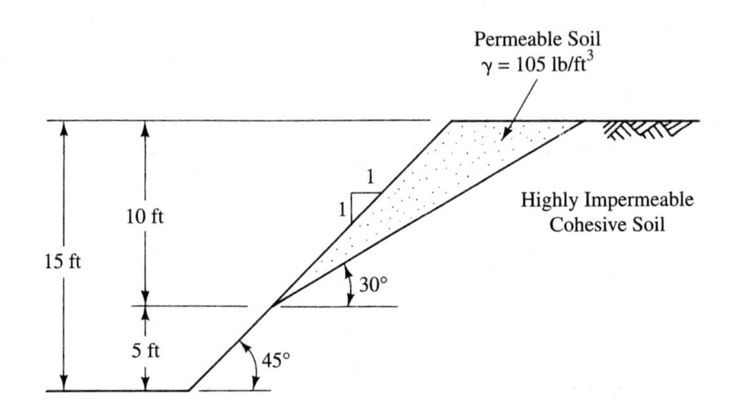

FIGURE 14–4

From Eq. (14–1),

$$W = \frac{Lh\gamma}{2} \tag{14–1}$$

$$W = \frac{(20.0\text{ ft})(3.66\text{ ft})(105\text{ lb/ft}^3)}{2} = 3843\text{ lb/ft}$$

From Eq. (14–4),

$$\text{F.S.} = \frac{cL + W\cos\alpha\tan\phi}{W\sin\alpha} \tag{14–4}$$

$$\text{F.S.} = \frac{(150\text{ lb/ft}^2)(20.0\text{ ft}) + (3843\text{ lb/ft})(\cos 30°)(\tan 25°)}{(3843\text{ lb/ft})(\sin 30°)}$$

$$= 2.37 > 1.5 \quad \therefore \text{O.K.}$$

14–3 SLOPES IN HOMOGENEOUS COHESIONLESS SOILS ($c = 0$, $\Phi > 0$)

When the slope angle (β) of a sand slope exceeds the sand's angle of internal friction (ϕ), the sand slope tends to fail by sliding in a downhill direction parallel to the slope. This phenomenon can be inferred by visualizing individual grains of sand being blocks resting on an inclined plane at the slope angle. If the slope angle is increased, the sand grains will begin to slide down the slope when the slope angle exceeds the sand's ϕ angle. Accordingly, the greatest slope for a free-standing cohesionless soil is at an angle approximately equal to the soil's ϕ angle.

The slope angle at which a loose sand fails may be estimated by its *angle of repose,* the angle formed (with the horizontal) by sand as it forms a pile below a funnel through which it passes. A sand's angle of repose is roughly equal to its angle of internal friction in a loose condition, and sand at or near ground surface is ordinarily in a loose condition and therefore near its maximum value of ϕ.

The factor of safety for slopes in homogeneous cohesionless soils is given by the following equation:

$$\text{F.S.} = \frac{\tan\phi}{\tan\beta} \tag{14–5}$$

Clearly, when slope angle β equals angle of internal friction ϕ, the factor of safety is 1. For slopes with β less than ϕ, the factor of safety is greater than 1.

14–4 SLOPES IN HOMOGENEOUS SOILS POSSESSING COHESION ($c > 0$, $\Phi = 0$, *and* $c > 0$, $\Phi > 0$)

In this section, two methods are presented for analyzing slope stability in homogeneous soils possessing cohesion. One is known as the *Culmann method;* the other might be called the *stability number method.*

Culmann Method

In the Culmann method, the assumption is made that failure (sliding) will occur along a plane that passes through the toe of the fill [2]. Such a plane is indicated in Figure 14–5. As with the analysis of a mass resting on an inclined layer of impermeable soil (Section 14–2), the force tending to cause sliding is given by Eq. (14–2):

$$F_s = W \sin \alpha \tag{14–2}$$

Also similarly, resistance to sliding results from cohesion and friction and is given by Eq. (14–3):

$$R_s = c_d L + W \cos \alpha \tan \phi_d \tag{14–3}$$

where c_d is the developed cohesion ($c/\text{F.S.}_c$), $\tan \phi_d$ is the developed coefficient of friction ($\tan \phi/\text{F.S.}_\phi$), and the other terms are as defined in Figure 14–5. (F.S._c and F.S._ϕ denote factors of safety for cohesion and angle of internal friction, respectively.) As in Section 14–2, the weight of soil in the upper triangle *abc* (W) can be computed by using Eq. (14–1):

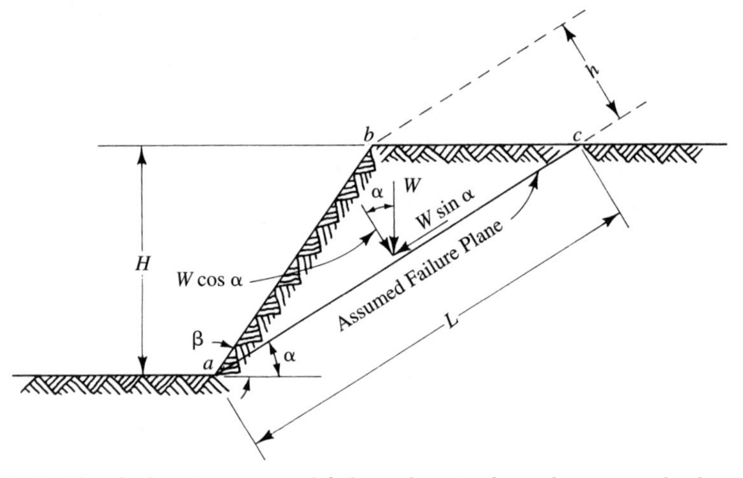

FIGURE 14–5 Sketch showing assumed failure plane in the Culmann method.

$$W = \frac{Lh\gamma}{2} \tag{14-1}$$

but h, the height of the triangle, can be evaluated as follows:

$$h = \left(\frac{H}{\sin\beta}\right)\sin(\beta - \alpha) \tag{14-6}$$

Substituting Eq. (14–6) into Eq. (14–1) gives the following:

$$W = \left(\frac{1}{2}\right)L\left(\frac{H}{\sin\beta}\right)\sin(\beta - \alpha)(\gamma) \tag{14-7}$$

Equating Eqs. (14–2) and (14–3), substituting W from Eq. (14–7) into the new equation, and then solving for c_d gives the following:

$$c_d = \left(\frac{\gamma H}{2\sin\beta}\right)\left[\frac{\sin(\beta - \alpha)\sin(\alpha - \phi_d)}{\cos\phi_d}\right] \tag{14-8}$$

The critical angle for α (i.e., α_c) can be determined by equating the first derivative of c_d with respect to α to zero [i.e., $d(c_d)/d(\alpha) = 0$] and solving for α. The result of this operation is as follows:

$$\alpha_c = \frac{\beta + \phi_d}{2} \tag{14-9}$$

Substituting α_c from Eq. (14–9) into Eq. (14–8) for α gives the following:

$$c_d = \frac{\gamma H[1 - \cos(\beta - \phi_d)]}{4\sin\beta\cos\phi_d} \tag{14-10}$$

Solving for H gives the following [3]:

$$H = \frac{4c_d\sin\beta\cos\phi_d}{\gamma[1 - \cos(\beta - \phi_d)]} \tag{14-11}$$

where H = safe depth of cut
c_d = developed cohesion
β = angle from horizontal to cut surface (Figure 14–5)
ϕ_d = developed angle of internal friction of the soil
γ = unit weight of the soil

In using Eq. (14–11) to compute the safe depth of a cut, one must determine developed cohesion (c_d) and the developed angle of internal friction (ϕ_d) by dividing cohesion and the tangent of the angle of internal friction by their respective safety factors.

The Culmann method gives reasonably accurate results if the slope is vertical, or nearly vertical (i.e., angle β is equal to, or nearly equal to, 90°) [2]. Examples 14–2 and 14–3 illustrate the Culmann method.

EXAMPLE 14–2

Given

1. A vertical cut is to be made through a soil mass.
2. The soil to be cut has the following properties:

$$\text{Unit weight } (\gamma) = 105\ \text{lb/ft}^3$$
$$\text{Cohesion } (c) = 500\ \text{lb/ft}^2$$
$$\text{Angle of internal friction } (\phi) = 21°$$

Required

Safe depth of cut in this soil, by the Culmann method, using a factor of safety of 2.

Solution

From Eq. (14–11),

$$H = \frac{4c_d \sin \beta \cos \phi_d}{\gamma[1 - \cos(\beta - \phi_d)]} \tag{14–11}$$

Here,

$$c_d = \frac{c}{\text{F.S.}_{\cdot c}} = \frac{500\ \text{lb/ft}^2}{2} = 250\ \text{lb/ft}^2$$

($\text{F.S.}_{\cdot c}$ is the factor of safety with respect to cohesion)

$$\tan \phi_d = \frac{\tan \phi}{\text{F.S.}_{\cdot \phi}} = \frac{\tan 21°}{2} = 0.192$$

($\text{F.S.}_{\cdot \phi}$ is the factor of safety with respect to $\tan \phi$)

$$\phi_d = \arctan 0.192 = 10.87°$$

$$\beta = 90° \qquad \text{(vertical cut)}$$

$$H = \frac{(4)(250\ \text{lb/ft}^2) \sin 90° \cos 10.87°}{(105\ \text{lb/ft}^3)[1 - \cos(90° - 10.87°)]} = 11.5\ \text{ft}$$

EXAMPLE 14–3

Given

A 1.8-m-deep vertical-wall trench is to be dug in soil without shoring. The soil's unit weight, angle of internal friction, and cohesion are 19.0 kN/m^3, 28°, and 20.2 kN/m^2, respectively.

Required

Factor of safety of this trench, using the Culmann method.

Solution

From Eq. (14–11),

$$H = \frac{4c_d \sin \beta \cos \phi_d}{\gamma[1 - \cos(\beta - \phi_d)]} \tag{14–11}$$

Try $F.S._{\phi} = 1.0$

$$\tan \phi_d = \frac{\tan \phi}{F.S._{\phi}} = \frac{\tan 28°}{1.0} = \tan 28°$$

Therefore, $\phi_d = 28°$

$$\beta = 90° \qquad \text{(for a vertical wall)}$$

Substituting into Eq. (14–11) yields the following:

$$1.8 \text{ m} = \frac{(4)(c_d) \sin 90° \cos 28°}{(19.0 \text{ kN/m}^3)[1 - \cos(90° - 28°)]}$$

$$c_d = 5.14 \text{ kN/m}^2$$

$$F.S._c = \frac{c}{c_d} = \frac{20.2 \text{ kN/m}^2}{5.14 \text{ kN/m}^2} = 3.93$$

Because $[F.S._c = 3.93] \neq [F.S._{\phi} = 1.0]$, another trial factor of safety must be attempted.

Try $F.S._{\phi} = 2.0$

$$\tan \phi_d = \frac{\tan \phi}{F.S._{\phi}} = \frac{\tan 28°}{2.0} = 0.2659$$

Therefore, $\phi_d = 14.89°$

$$1.8 \text{ m} = \frac{(4)(c_d) \sin 90° \cos 14.89°}{(19.0 \text{ kN/m}^3)[1 - \cos(90° - 14.89°)]}$$

$$c_d = 6.57 \text{ kN/m}^2$$

$$F.S._c = \frac{c}{c_d} = \frac{20.2 \text{ kN/m}^2}{6.57 \text{ kN/m}^2} = 3.07$$

Because $[F.S._c = 3.07] \neq [F.S._{\phi} = 2.0]$, another trial factor of safety must be attempted.

Try $F.S._{\phi} = 3.0$

$$\tan \phi_d = \frac{\tan \phi}{F.S._{\phi}} = \frac{\tan 28°}{3.0} = 0.1772$$

Therefore, $\phi_d = 10.05°$

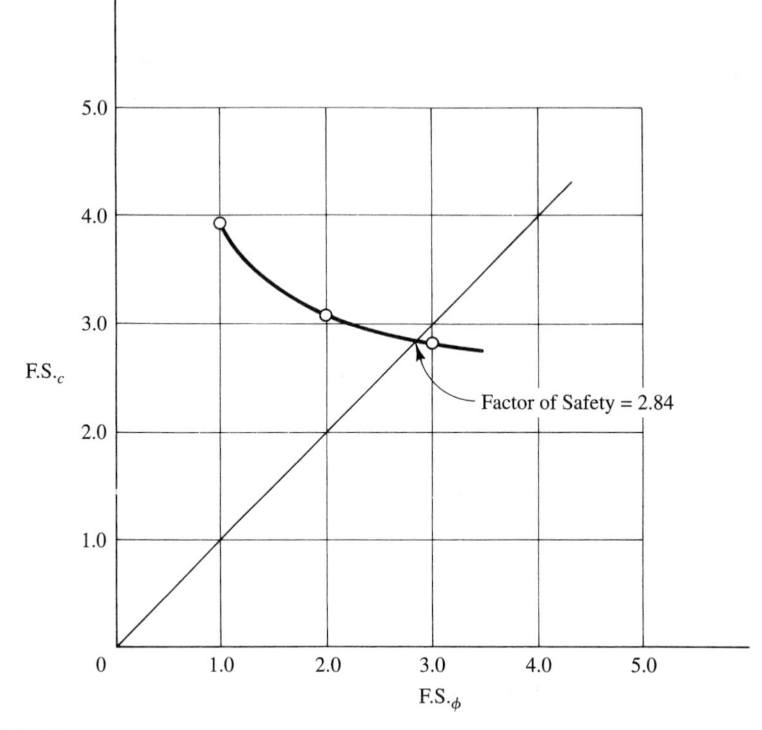

FIGURE 14–6

$$1.8 \text{ m} = \frac{(4)(c_d) \sin 90° \cos 10.05°}{(19.0 \text{ kN/m}^3)[1 - \cos(90° - 10.05°)]}$$

$$c_d = 7.17 \text{ kN/m}^2$$

$$\text{F.S.}_{\cdot c} = \frac{c}{c_d} = \frac{20.2 \text{ kN/m}^2}{7.17 \text{ kN/m}^2} = 2.82$$

Because $[\text{F.S.}_{\cdot c} = 2.82] \neq [\text{F.S.}_{\cdot \phi} = 3.0]$, the correct factor of safety has not yet been found. Rather than continue this trial-and-error procedure, the values of $\text{F.S.}_{\cdot c}$ and $\text{F.S.}_{\cdot \phi}$ are plotted in Figure 14–6, from which the applicable factor of safety of about 2.84 can be read.

Stability Number Method

The stability number method is also based on the premise that resistance of a soil mass to sliding results from cohesion and internal friction of the soil along the failure surface. Unlike the Culmann method, in this method the failure surface is as-

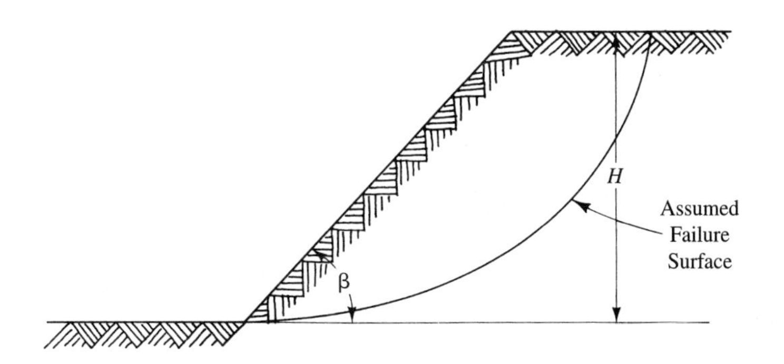

FIGURE 14–7 Sketch showing assumed failure surface as a circular arc.

sumed to be a circular arc (see Figure 14–7). A parameter called the *stability number* is introduced, which groups factors affecting the stability of soil slopes. The stability number (N_s) is defined as follows [4]:

$$N_s = \frac{\gamma H}{c} \tag{14–12}$$

where γ = unit weight of soil
H = height of cut (Figure 14–7)
c = cohesion of soil

For the embankment illustrated in Figure 14–7, three types of failure surfaces are possible. These are shown in Figure 14–8. For the toe circle (Figure 14–8a), the failure surface passes through the toe. In the case of the slope circle (Figure 14–8b), the failure surface intersects the slope above the toe. For the midpoint circle (Figure 14–8c), the center of the failure surface is on a vertical line passing through the midpoint of the slope.

Both the type of failure surface and the stability number can be determined for a specific case based on given values of ϕ (angle of internal friction) and β (slope angle, Figure 14–7). If the value of ϕ is zero, or nearly zero, Figure 14–9 may be used to determine both the type of failure surface and the stability number. One enters along the abscissa at the value of β and moves upward to the line that indicates the appropriate value of n_d. (n_d is a depth factor related to the distance to the underlying layer of stiff material or bedrock and is determined from the relationship indicated in Figure 14–8a.) The type of line for n_d indicates the type of failure surface, and the value of stability number is determined by moving leftward and reading from the ordinate. Observation of Figure 14–9 indicates that if β is greater than 53°, the failure surface is always a toe circle, and if n_d is greater than 4, the failure surface is always a midpoint circle [4].

If the value of ϕ is greater than 3°, the failure surface is always a toe circle [4]. Figure 14–10 may be used to determine the stability number for different values of ϕ [5]. One enters along the abscissa at the value of β, moves upward to the line that

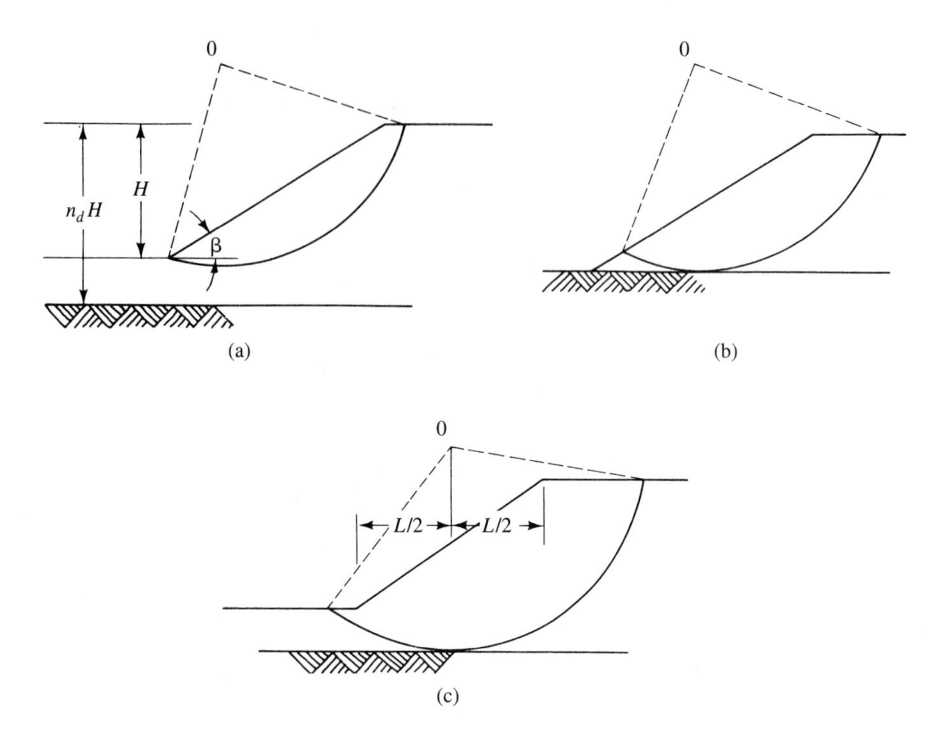

FIGURE 14–8 Types of failure surfaces: (a) toe circle; (b) slope circle; (c) midpoint circle [4]. *Source:* From Tien Hsing Wu, *Soil Mechanics.* Copyright © 1976 by Allyn and Bacon, Inc., Boston. Reprinted with permission.

indicates the ϕ angle, and then leftward to the ordinate where the stability number is read.

The factor of safety for highly cohesive soils (that have $\phi = 0$) can be obtained from Figure 14–9. This is illustrated in Example 14–5. For soils possessing cohesion and having $\phi > 0$, the procedure is more complicated. One procedure is to estimate $F.S._{\phi}$ and determine $\phi_{required}$. Using this value and slope angle β, one can find the stability number from Figure 14–10. With this stability number, $c_{required}$ can be computed from Eq. (14–12). $F.S._{c}$ is the quotient of c_{given} divided by $c_{required}$. If $F.S._{\phi}$ equals $F.S._{c}$, the overall factor of safety is equal to $F.S._{\phi}$ (or $F.S._{c}$). If $F.S._{\phi}$ and $F.S._{c}$ are not equal, additional values of $F.S._{\phi}$ can be estimated and the preceding procedure repeated to determine corresponding values of $F.S._{c}$ until the factor of safety is found where $F.S._{\phi}$ equals $F.S._{c}$. If the correct factor of safety has not been found after several such trials, it may be expedient to plot corresponding values of $F.S._{\phi}$ and $F.S._{c}$ on a graph, from which the overall factor of safety (i.e., where $F.S._{\phi}$ equals $F.S._{c}$) can be read. This procedure is illustrated in Example 14–4.

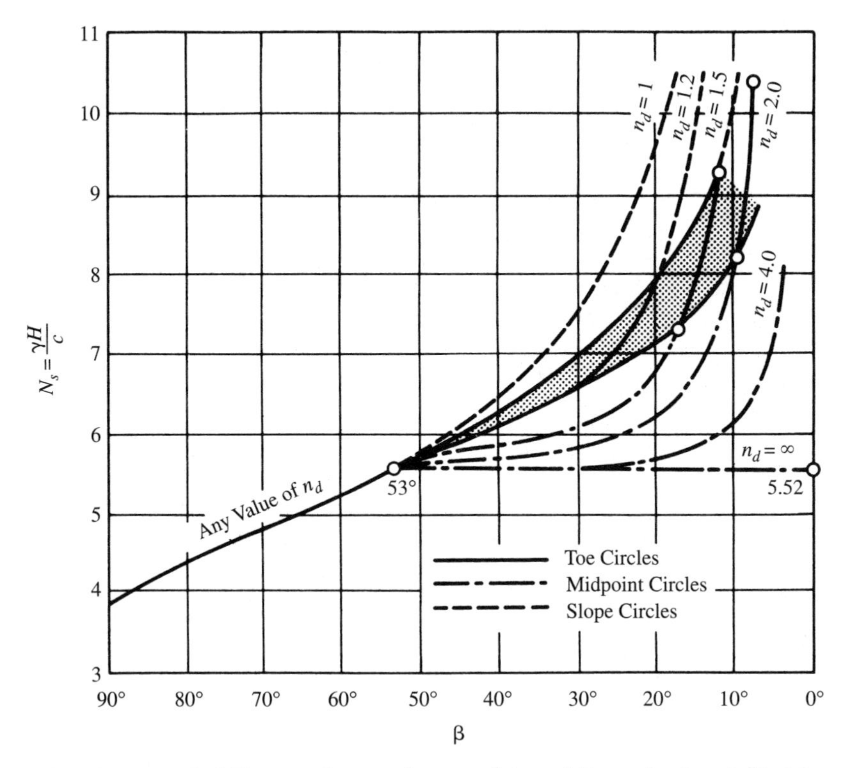

FIGURE 14–9 Stability numbers and types of slope failures for $\phi = 0$ [5, 6].

EXAMPLE 14–4

Given

The slope and data shown in Figure 14–11.

Required

Factor of safety against failure, by the stability number method.

Solution

Because the given angle of internal friction (ϕ) of 10° is greater than 3°, the failure surface will be a toe circle.

Try F.S.$_{\phi} = 1$

$$\tan \phi_{\text{required}} = \frac{\tan \phi_{\text{given}}}{\text{F.S.}_{\phi}} = \frac{\tan 10°}{1}$$

$$\phi_{\text{required}} = 10°$$

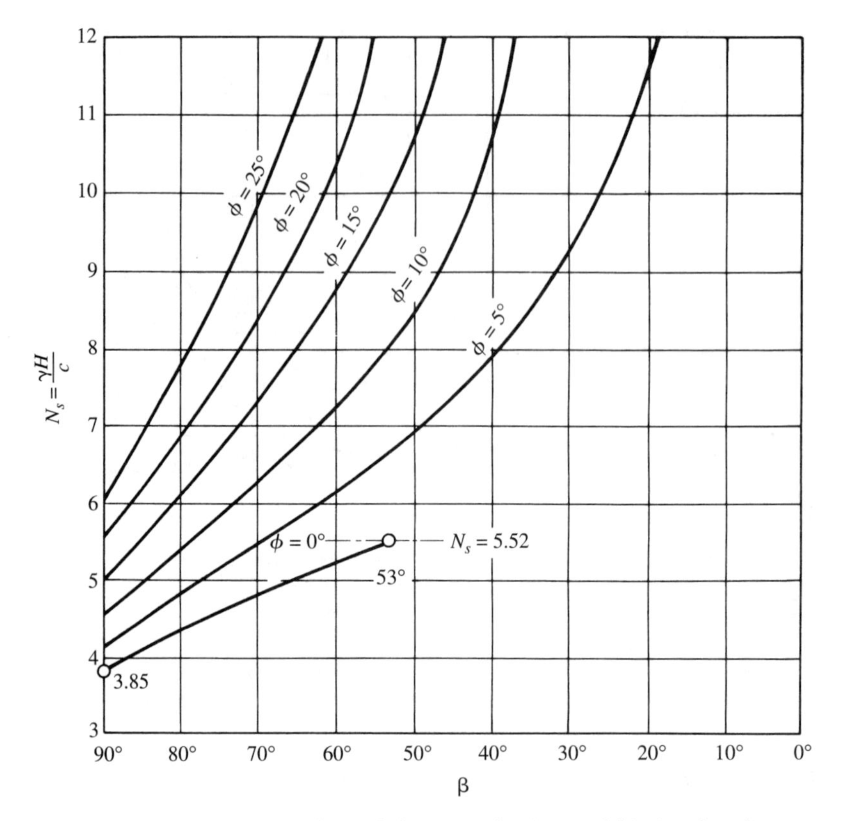

FIGURE 14–10 Stability numbers for soils having cohesion and friction [5, 6].

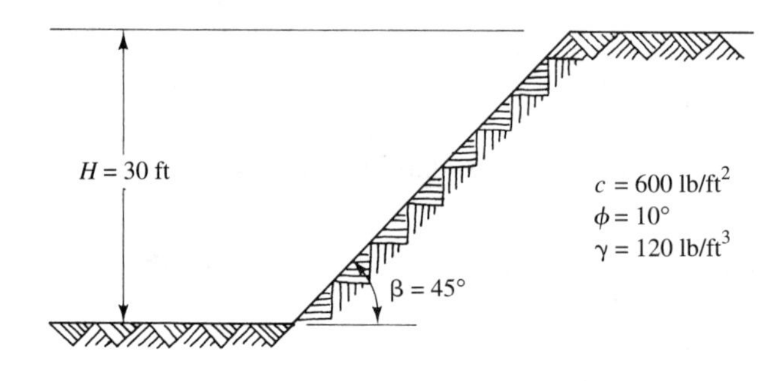

FIGURE 14–11

With $\phi_{\text{required}} = 10°$ and $\beta = 45°$, from Figure 14–10,

$$N_s = 9.2$$

$$N_s = \frac{\gamma H}{c} \tag{14–12}$$

$$\gamma = 120 \text{ lb/ft}^3$$

$$H = 30 \text{ ft}$$

$$9.2 = \frac{(120 \text{ lb/ft}^3)(30 \text{ ft})}{c_{\text{required}}}$$

$$c_{\text{required}} = 391 \text{ lb/ft}^2$$

$$\text{F.S.}_{\cdot c} = \frac{c_{\text{given}}}{c_{\text{required}}} = \frac{600 \text{ lb/ft}^2}{391 \text{ lb/ft}^2} = 1.53$$

Because $\text{F.S.}_{\cdot\phi}$ and $\text{F.S.}_{\cdot c}$ are not the same value, another value of $\text{F.S.}_{\cdot\phi}$ must be tried.

Try $\text{F.S.}_{\cdot\phi} = 1.2$

$$\tan \phi_{\text{required}} = \frac{\tan \phi_{\text{given}}}{\text{F.S.}_{\cdot\phi}} = \frac{\tan 10°}{1.2} = 0.147$$

$$\phi_{\text{required}} = 8.36°$$

With $\phi_{\text{required}} = 8.36°$ and $\beta = 45°$, from Figure 14–10,

$$N_s = 8.6$$

$$c_{\text{required}} = \frac{(120 \text{ lb/ft}^3)(30 \text{ ft})}{8.6} = 419 \text{ lb/ft}^2$$

$$\text{F.S.}_{\cdot c} = \frac{c_{\text{given}}}{c_{\text{required}}} = \frac{600 \text{ lb/ft}^2}{419 \text{ lb/ft}^2} = 1.43$$

Again, $\text{F.S.}_{\cdot\phi}$ and $\text{F.S.}_{\cdot c}$ are not the same value; hence, another value of $\text{F.S.}_{\cdot\phi}$ must be tried.

Try $\text{F.S.}_{\cdot\phi} = 1.5$

$$\tan \phi_{\text{required}} = \frac{\tan \phi_{\text{given}}}{\text{F.S.}_{\cdot\phi}} = \frac{\tan 10°}{1.5} = 0.118$$

$$\phi_{\text{required}} = 6.73°$$

With $\phi_{\text{required}} = 6.73°$ and $\beta = 45°$, from Figure 14-10,

$$N_s = 7.9$$

$$c_{\text{required}} = \frac{(120 \text{ lb/ft}^3)(30 \text{ ft})}{7.9} = 456 \text{ lb/ft}^2$$

$$\text{F.S.}_{\cdot c} = \frac{c_{\text{given}}}{c_{\text{required}}} = \frac{600 \text{ lb/ft}^2}{456 \text{ lb/ft}^2} = 1.32$$

FIGURE 14–12

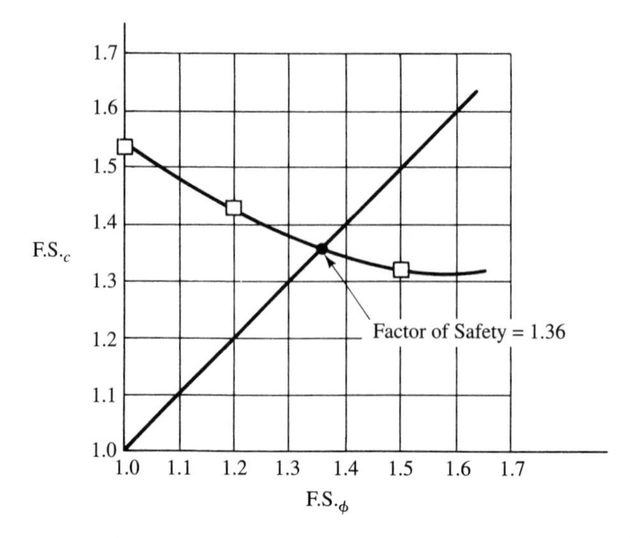

Again, F.S.$_{\cdot\phi}$ and F.S.$_{\cdot c}$ are not the same value. Rather than continue a trial-and-error solution, plot the values computed. From Figure 14–12, the factor of safety of the slope against failure is observed to be 1.36.

EXAMPLE 14–5

Given

1. A cut 25 ft deep is to be made in a stratum of highly cohesive soil (see Figure 14–13).
2. The slope angle β is 30°.
3. Soil exploration indicated that bedrock is located 40 ft below the original ground surface.
4. The soil has a unit weight of 120 lb/ft^3, and its cohesion and angle of internal friction are 650 lb/ft^2 and 0°, respectively.

Required

Factor of safety against slope failure.

Solution

From Figure 14–8a,

$$n_d H = 40 \text{ ft}$$

$$H = 25 \text{ ft}$$

$$n_d = \frac{40 \text{ ft}}{25 \text{ ft}} = 1.60$$

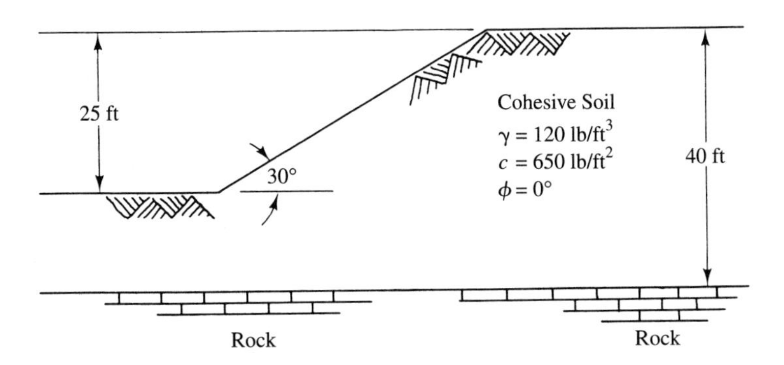

FIGURE 14–13

With $\beta = 30°$ and $n_d = 1.60$, from Figure 14–9,

$$N_s = 6.0$$

$$N_s = \frac{\gamma H}{c_{\text{required}}} \tag{14–12}$$

$$\gamma = 120 \text{ lb/ft}^3$$

$$H = 25 \text{ ft}$$

$$6.0 = \frac{(120 \text{ lb/ft}^3)(25 \text{ ft})}{c_{\text{required}}}$$

$$c_{\text{required}} = 500 \text{ lb/ft}^2$$

$$\text{F.S.} = \frac{c_{\text{given}}}{c_{\text{required}}} = \frac{650 \text{ lb/ft}^2}{500 \text{ lb/ft}^2} = 1.30$$

EXAMPLE 14–6

Given

1. A cut 30 ft deep is to be made in a deposit of highly cohesive soil that is 60 ft thick and underlain by rock (see Figure 14–14).
2. The properties of the soil to be cut are as follows:

$$c = 750 \text{ lb/ft}^2$$

$$\phi = 0°$$

$$\gamma = 120 \text{ lb/ft}^3$$

3. The factor of safety against slope failure must be 1.25.

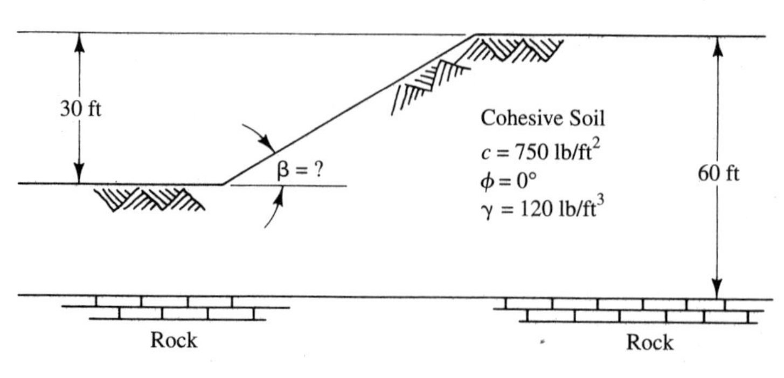

FIGURE 14–14

Required

Estimate the slope angle (β) at which the cut should be made.

Solution

From Figure 14–8a,

$$n_d H = 60 \text{ ft}$$

$$H = 30 \text{ ft}$$

$$n_d = \frac{60 \text{ ft}}{30 \text{ ft}} = 2.0$$

From Eq. (14–12),

$$N_s = \frac{\gamma H}{c_{\text{required}}} \tag{14–12}$$

$$\gamma = 120 \text{ lb/ft}^3$$

$$H = 30 \text{ ft}$$

$$c_{\text{required}} = \frac{c_{\text{given}}}{\text{F.S.}} = \frac{750 \text{ lb/ft}^2}{1.25} = 600 \text{ lb/ft}^2$$

$$N_s = \frac{(120 \text{ lb/ft}^3)(30 \text{ ft})}{600 \text{ lb/ft}^2} = 6.0$$

From Figure 14–9, with $N_s = 6.0$ and $n_d = 2.0$,

$$\beta = 23°$$

EXAMPLE 14–7

Given

A cut 10 m deep is to be made in soil that has the following properties:

$$\gamma = 17.66 \text{ kN/m}^3$$
$$c = 19.2 \text{ kN/m}^2$$
$$\phi = 16°$$

Required

Using a factor of safety of 1.25, estimate the slope angle at which the cut should be made.

Solution

$$c_d = \frac{c}{\text{F.S.}_c} = \frac{19.2 \text{ kN/m}^2}{1.25} = 15.36 \text{ kN/m}^2$$

From Eq. (14–12),

$$N_s = \frac{\gamma H}{c_d} \tag{14–12}$$

$$N_s = \frac{(17.66 \text{ kN/m}^3)(10 \text{ m})}{15.36 \text{ kN/m}^2} = 11.5$$

$$\tan \phi_d = \frac{\tan \phi}{\text{F.S.}_\phi} = \frac{\tan 16°}{1.25} = 0.2294$$

$$\phi_d = 12.9°$$

From Figure 14–10, with $\phi_d = 12.9°$ and $N_s = 11.5$,

$$\beta = 44°$$

14–5 METHOD OF SLICES

In Section 14–4, the assumption was made in the Culmann method that failure (sliding) would occur along a plane that passes through the toe of the slope. It is probably more likely, and observations suggest, that failure will occur along a curved surface (rather than a plane) within the soil. Like the stability number method, the method of slices, which was developed in the 1920s by Swedish engineers, performs slope stability analysis assuming failure occurs along a curved surface.

The first step in applying the method of slices is to draw to scale a cross section of the slope such as that shown in Figure 14–15. A trial curved surface along which sliding is assumed to take place is then drawn. This trial surface is normally approximately circular. Soil contained between the trial surface and the slope is then divided into a number of vertical slices of equal width. The weight of soil within each slice is calculated by multiplying the slice's volume by the soil's unit weight. (This problem is, of course, three-dimensional; however, by assuming a unit thickness throughout the computations, the problem can be treated as two-dimensional.)

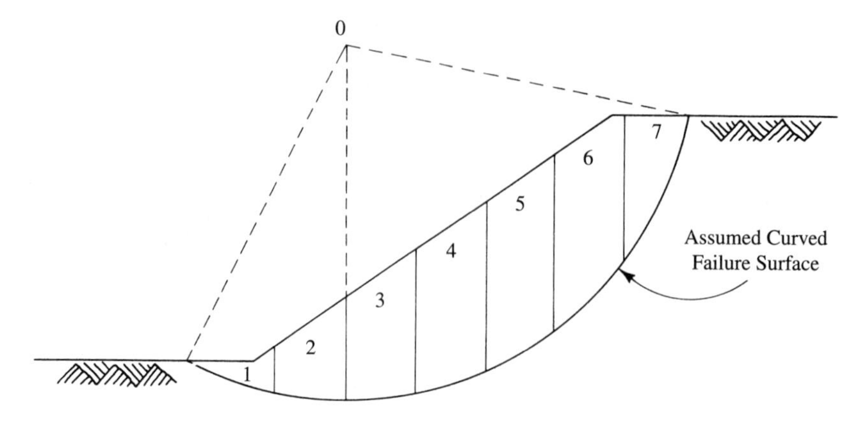

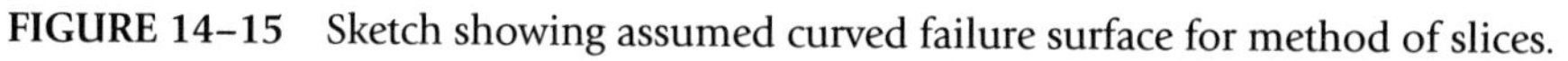
FIGURE 14–15 Sketch showing assumed curved failure surface for method of slices.

FIGURE 14–16 Sketch showing forces on a single slice in method of slices.

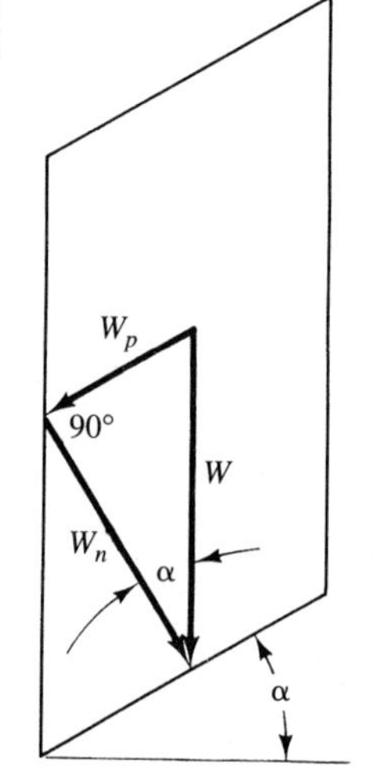

Figure 14–16 shows a sketch of a single slice. The weight of soil within the slice is a vertically downward force (W in Figure 14–16). This force can be resolved into two components—one normal to the base of the slice (W_n) and one parallel to the base of the slice (W_p). It is the parallel component that tends to cause sliding. Resistance to sliding is afforded by the soil's cohesion and internal friction. The cohesion force is equal to the product of the soil's cohesion times the length of the slice's curved base. The friction force is equal to the component of W normal to the base (W_n) multiplied by the friction coefficient (tan ϕ, where ϕ is the angle of internal friction).

Because W_p, the component tending to cause sliding of the slice, is equal to W multiplied by $\sin \alpha$ (Figure 14–16), the *total* force tending to cause sliding of the entire soil mass is the summation of products of the weight of each slice times the respective value of $\sin \alpha$, or $\Sigma\, W \sin \alpha$. Because W_n is equal to W multiplied by $\cos \alpha$, the *total* friction force resisting sliding of the entire soil mass is the summation of products of the weight of each slice times the respective value of $\cos \alpha$ times $\tan \phi$, or $\Sigma\, W \cos \alpha \tan \phi$. The *total* cohesion force resisting sliding of the entire soil mass can be computed simply by multiplying the soil's cohesion by the (total) length of the trial curved surface, or cL. Based on the foregoing, the factor of safety can be computed by using the following equation:

$$\text{F.S.} = \frac{cL + \Sigma\, W \cos \alpha \tan \phi}{\Sigma\, W \sin \alpha} \tag{14–13}$$

(As related subsequently in Example 14–8, the term $W \sin \alpha$ may be negative in certain situations.)

This method gives the factor of safety for the specific assumed failure surface. It is quite possible that the circular surface selected may not be the weakest, or the one along which sliding would occur. It is essential, therefore, that several circular surfaces be analyzed until the designer is satisfied that the worst condition has been considered.

EXAMPLE 14–8

Given

1. The stability of a slope is to be analyzed by the method of slices.
2. On a particular trial curved surface through the soil mass (see Figure 14–17), the shearing component (i.e., sliding force) and the normal component (i.e., normal to the base of each slice) of each slice's weight are tabulated as follows:

Slice Number	Shearing Component ($W \sin \alpha$) (lb/ft)	Normal Component ($W \cos \alpha$) (lb/ft)
1	−63[1]	358
2	−51[1]	1450
3	86	2460
4	722	3060
5	1470	3300
6	1880	3130
7	2200	2270
8	950	91

[1]Because the trial surface curves upward near its lower end, the shearing components of the weights of slices 1 and 2 will act in a direction opposite to those along the remainder of the trial curve, resulting in a negative sign.

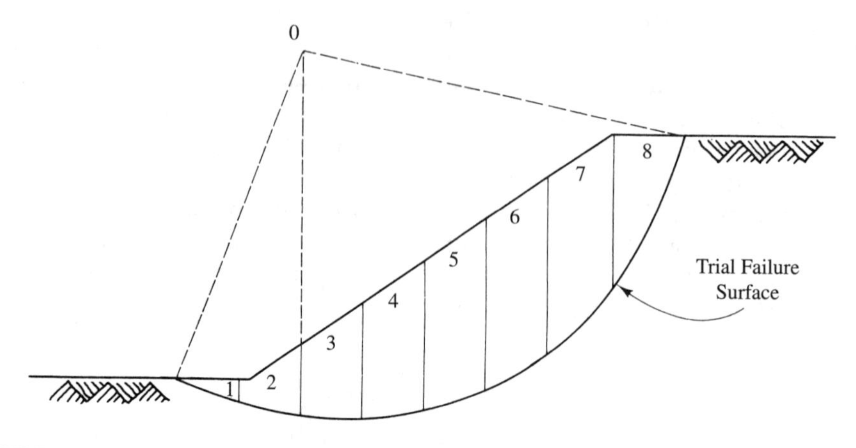

FIGURE 14–17

3. The length of the trial curved surface is 36 ft.
4. The ϕ angle of the soil is 5°, and the cohesion (c) is 400 lb/ft^2.

Required

Factor of safety of the slope along this particular trial surface.

Solution

From Eq. (14–13),

$$\text{F.S.} = \frac{cL + \Sigma W \cos\alpha \tan\phi}{\Sigma W \sin\alpha} \tag{14–13}$$

$$c = 400 \text{ lb/ft}^2$$

$$L = 36 \text{ ft}$$

$$\begin{aligned}\Sigma W \cos\alpha &= 358 \text{ lb/ft} + 1450 \text{ lb/ft} + 2460 \text{ lb/ft} + 3060 \text{ lb/ft} \\ &\quad + 3300 \text{ lb/ft} + 3130 \text{ lb/ft} + 2270 \text{ lb/ft} + 91 \text{ lb/ft} \\ &= 16{,}119 \text{ lb/ft}\end{aligned}$$

$$\phi = 5°$$

$$\begin{aligned}\Sigma W \sin\alpha &= -63 \text{ lb/ft} - 51 \text{ lb/ft} + 86 \text{ lb/ft} + 722 \text{ lb/ft} \\ &\quad + 1470 \text{ lb/ft} + 1880 \text{ lb/ft} + 2200 \text{ lb/ft} + 950 \text{ lb/ft} \\ &= 7194 \text{ lb/ft}\end{aligned}$$

$$\text{F.S.} = \frac{(400 \text{ lb/ft}^2)(36 \text{ ft}) + (16{,}119 \text{ lb/ft}) \tan 5°}{7194 \text{ lb/ft}} = 2.20$$

It should be emphasized that the computed factor of safety of 2.20 is for the given trial surface, which is not necessarily the weakest surface.

Bishop's Simplified Method of Slices

In 1955, Bishop [7] presented a more refined method of analysis. His method uses static equilibrium considerations rather than finding a factor of safety against sliding by computing the ratio of the total force resisting sliding (of the entire soil mass) to the total force tending to cause sliding, as is done in the ordinary method of slices.

To understand Bishop's method, consider the representative slice shown in Figure 14–18. Unlike the slice shown in Figure 14–16, the one in Figure 14–18 shows all forces acting on the slice [i.e., its weight *W*, shear forces *T*, normal forces *H* (on its sides), and a set of forces on its base (shear force *S* and normal force *N*)]. Bishop found that little error would accrue if the side forces are assumed equal and opposite. Equilibrium of the entire sliding mass requires (Figure 14–18) that

$$R \Sigma W \sin \alpha = R \Sigma S \tag{14–14}$$

The shear force on the base of a slice, *S*, is given by the following (Figure 14–18):

$$S = \frac{sl}{\text{F.S.}} = \frac{sb}{\text{F.S.} \cos \alpha} \tag{14–15}$$

where *s* is shear strength; *l*, *b*, and α are as shown in Figure 14–18; and F.S. is the factor of safety. Substituting Eq. (14–15) into Eq. (14–14) yields the following:

$$\frac{R}{\text{F.S.}} \sum \frac{sb}{\cos \alpha} = R \Sigma W \sin \alpha \tag{14–16}$$

from which

$$\text{F.S.} = \frac{\Sigma\,(sb/\cos \alpha)}{\Sigma W \sin \alpha} \tag{14–17}$$

Shear strength *s* can be determined from Eq. (2–16):

$$s = c + \bar{\sigma} \tan \phi \tag{2–16}$$

where c = cohesion

$\bar{\sigma}$ = effective intergranular normal pressure (normal stress across the surface of sliding, *l*)

ϕ = angle of internal friction

$\bar{\sigma}$ can be evaluated by analyzing the vertical equilibrium of the slice shown in Figure 14–18:

$$W = S \sin \alpha + N \cos \alpha \tag{14–18}$$

and

$$\bar{\sigma} = \frac{N}{l} + \frac{N \cos \alpha}{b} = \frac{W}{b} - \frac{S}{b} \sin \alpha \tag{14–19}$$

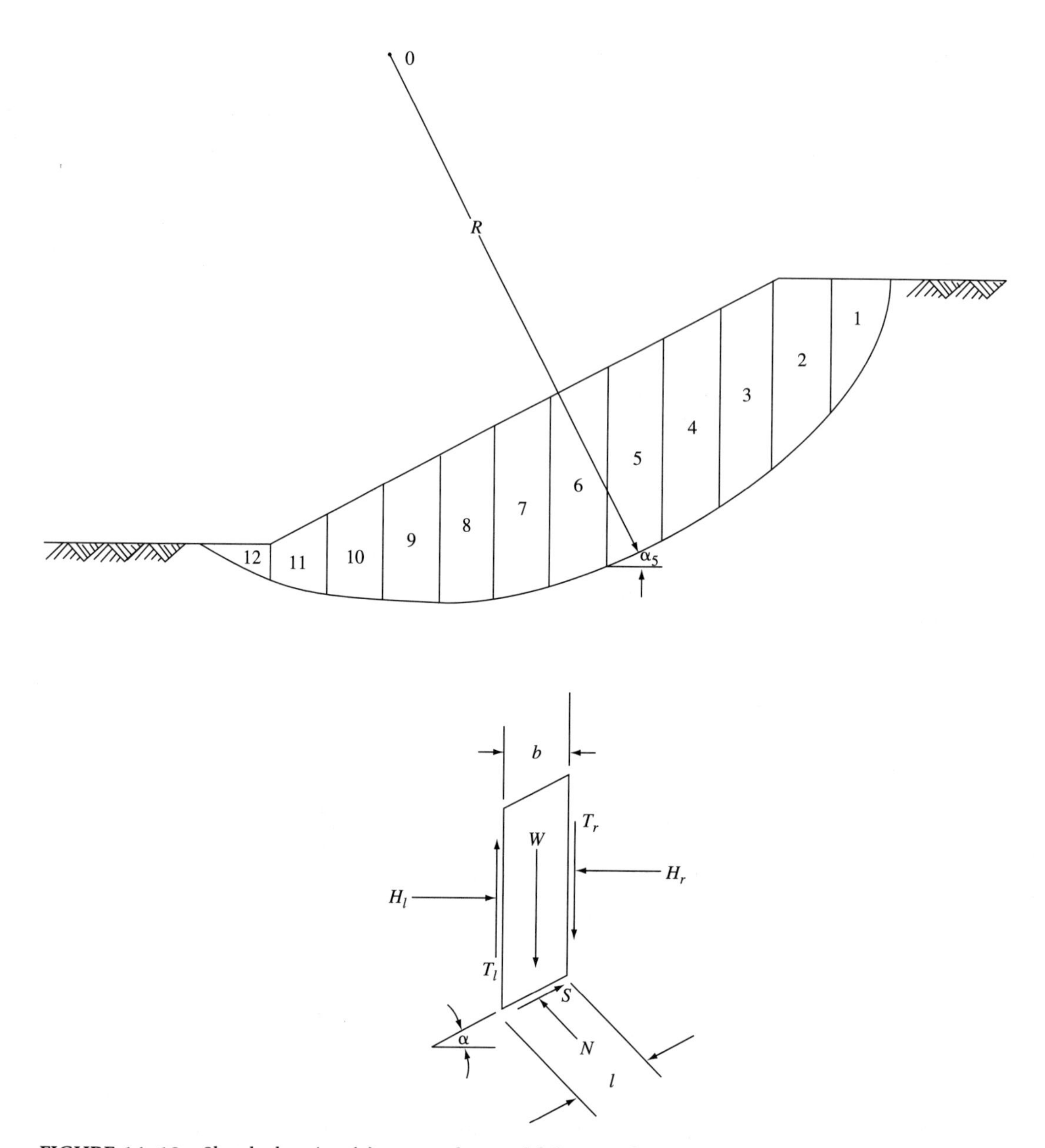

FIGURE 14–18 Sketch showing (a) assumed curved failure surface and (b) forces on a single slice for Bishop's simplified method of slices.

Substituting the latter value of $\bar{\sigma}$ from Eq. (14–19) into Eq. (2–16) gives the following:

$$s = c + \left(\frac{W}{b} - \frac{S}{b}\sin\alpha\right)\tan\phi \qquad (14\text{–}20)$$

But, substituting the value of S from Eq. (14–15) yields the following:

$$s = c + \left(\frac{W}{b} - \frac{s}{\text{F.S.}}\tan\alpha\right)\tan\phi \qquad (14\text{–}21)$$

Solving for s from Eq. (14–21) gives

$$s = \frac{c + (W/b)\tan\phi}{1 + (\tan\alpha\tan\phi)/\text{F.S.}} \qquad (14\text{–}22)$$

To simplify computations, let

$$m_\alpha = \left[1 + \frac{\tan\alpha\tan\phi}{\text{F.S.}}\right]\cos\alpha \qquad (14\text{–}23)$$

Substituting this value of m_α into Eq. (14–22) yields the following:

$$s = \left[\frac{c + (W/b)\tan\phi}{m_\alpha}\right]\cos\alpha \qquad (14\text{–}24)$$

Then substitute Eq. (14–24) into Eq. (14–17):

$$\text{F.S.} = \frac{\sum \dfrac{cb + W\tan\phi}{m_\alpha}}{\sum W\sin\alpha} \qquad (14\text{–}25)$$

Equation (14–25) can be used to find the factor of safety for the given (trial) failure surface. This is complicated somewhat by the fact that the value of m_α to be substituted into Eq. (14–25) to calculate the factor of safety must be determined from Eq. (14–23), which requires the value of the factor of safety (on the right side of the equation). Hence, Eq. (14–25) must be solved by trial and error—i.e., assume a value for the factor of safety, substitute it into Eq. (14–23) to solve for m_α, and substitute that value of m_α into Eq. (14–25) to compute the factor of safety. If the computed value for the factor of safety is the same (or nearly the same) as the assumed value, then that value is the correct one. If not, another value must be assumed and the procedure repeated until the correct value for the factor of safety is found. Figure 14–19 may be used in lieu of Eq. (14–23) to evaluate m_α.

As noted with the ordinary method of slices, it should be emphasized that Bishop's simplified method of slices also gives the factor of safety for the specific assumed failure surface. It is quite possible that the circular surface selected may not be the weakest, or the one along which sliding would occur. It is essential, therefore, that several circular surfaces be studied until the designer is satisfied that the worst condition has been analyzed.

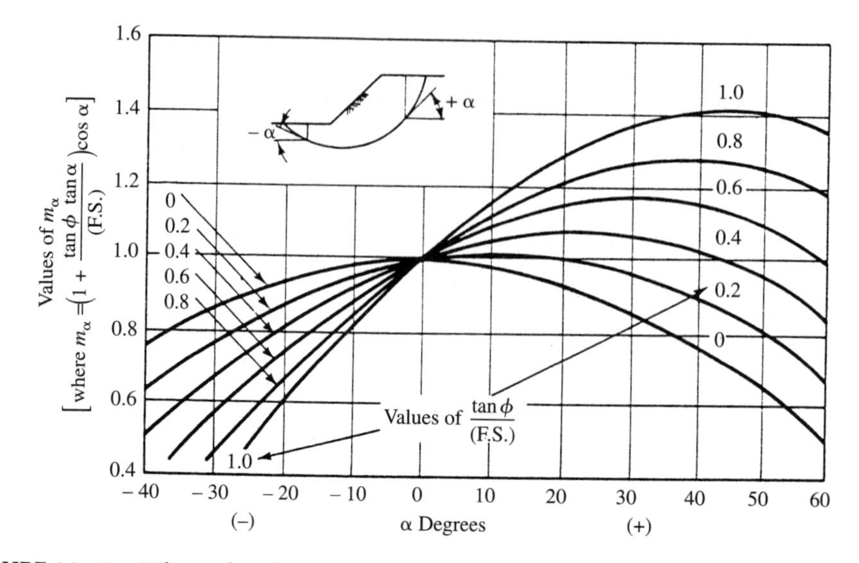

FIGURE 14–19 Values of m_α for Bishop equation [8].

Strictly speaking, both methods of slices apply only if the entire trial circle lies above the water table and no excess pore pressures are present. If these conditions are not met, additional analysis is required.

Generally, the Bishop method gives slightly higher factors of safety than those calculated from the ordinary slice method—hence, the latter is somewhat more conservative. The Bishop method provides too-high factors of safety if the negative alpha angle (see $-\alpha$ in Figure 14–19) approaches 30°. For the same situation, the ordinary method of slices tends to provide too-low values.

14–6 PROBLEMS

14–1. Figure 14–20 shows a 20-ft cut through two soil strata. The lower is a highly impermeable cohesive clay. Shear strength data between the two strata are as follows:

$$c = 220 \text{ lb/ft}^2$$

$$\phi = 12°$$

The unit weight of the upper layer is 110 lb/ft³. Determine if a slide is likely by computing the factor of safety against sliding. Neglect the effects of soil water.

14–2. A vertical cut is to be made in a deposit of homogeneous soil. The soil mass to be cut has the following properties: The soil's unit weight is 120 lb/ft³, cohesion is 350 lb/ft², and the angle of internal friction is 10°. It has been specified that the factor of safety against sliding must be 1.50. Using the Culmann

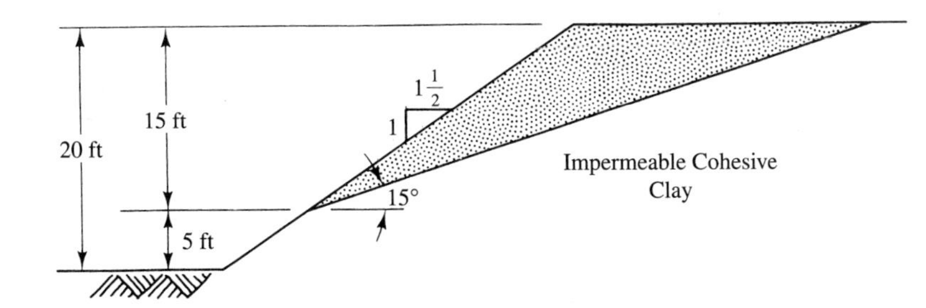

FIGURE 14–20

FIGURE 14–21

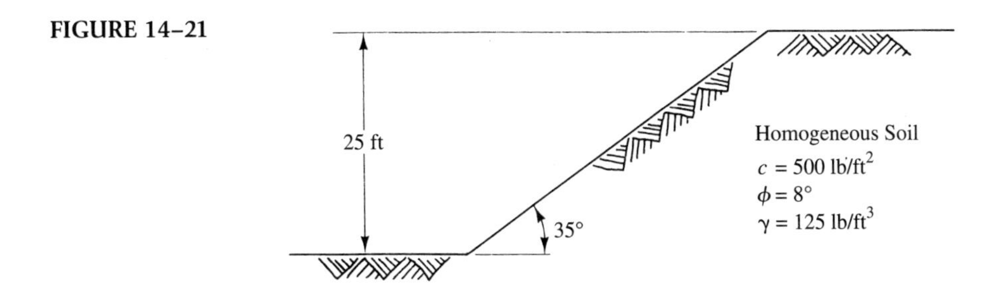

method, determine the safe depth of the cut.

14–3. A 1.5-m-deep vertical-wall trench is to be cut in a soil whose unit weight, angle of internal friction, and cohesion are 17.36 kN/m^3, 25°, and 20.6 kN/m^2, respectively. Determine the factor of safety of this trench, by the Culmann method.

14–4. Determine the factor of safety against slope failure by means of the stability number method for the slope shown in Figure 14–21.

14–5. A cut 20 ft deep is to be made in a stratum of highly cohesive soil that is 80 ft thick and underlain by bedrock. The slope of the cut is 2:1 (i.e., 2 horizontal to 1 vertical). The clay's unit weight is 110 lb/ft^3, and its c and ϕ values are 500 lb/ft^2 and 0°, respectively. Determine the factor of safety against slope failure.

14–6. A cut 25 ft deep is to be made in a deposit of cohesive soil with $c = 700$ lb/ft^2, $\phi = 0°$, and $\gamma = 115$ lb/ft^3. The soil is 30 ft thick and underlain by rock. The factor of safety of the slope against failure must be 1.50. At what slope angle should the cut be made?

14–7. A slope 8 m high is to be made in a soil whose unit weight, angle of internal friction, and cohesion are 16.7 kN/m^3, 10°, and 17.0 kN/m^2, respectively. Using an overall factor of safety of 1.25, estimate the slope angle that should be used.

14–8. The stability of a slope is to be analyzed by the method of slices. On a partic-

Slice Number	Shearing Component ($W \sin \alpha$) (lb/ft)	Normal Component ($W \cos \alpha$) (lb/ft)
1	−38	306
2	−74	1410
3	124	2380
4	429	3050
5	934	3480
6	1570	3540
7	2000	3210
8	2040	2190
9	766	600

ular trial curved surface through the soil mass, the shearing and normal components of each slice's weight are tabulated as shown above. The length of the trial curved surface is 40 ft. The cohesion c and ϕ angle of the soil are 225 lb/ft^2 and $15°$, respectively. Determine the factor of safety along this trial surface.

References

[1] George F. Sowers, *Introductory Soil Mechanics and Foundations: Geotechnical Engineering*, 4th ed., Macmillan Publishing Co., Inc., New York, 1979.

[2] Donald W. Taylor, *Fundamentals of Soil Mechanics*, John Wiley & Sons, Inc., New York, 1948. Copyright © 1948, by John Wiley & Sons, Inc. Reprinted by permission of John Wiley & Sons, Inc.

[3] Merlin G. Spangler and Richard L. Handy, *Soil Engineering*, 3rd ed., Intext Educational Publishers, New York, 1973. Copyright © 1951, 1960, 1973 by Harper & Row, Publishers, Inc. Reprinted by permission of the publisher.

[4] Tien Hsing Wu, *Soil Mechanics*, Allyn and Bacon, Inc., Boston, 1976. Copyright © 1976 by Allyn and Bacon, Inc., Boston. Reprinted with permission.

[5] Karl Terzaghi and Ralph B. Peck, *Soil Mechanics in Engineering Practice*, John Wiley & Sons, Inc., New York, 1967. Copyright © 1967, by John Wiley & Sons, Inc. Reprinted by permission of John Wiley & Sons, Inc.

[6] D. W. Taylor, "Stability of Earth Slopes," *J. Boston Soc. Civil Eng.*, **24** (1937).

[7] A. W. Bishop, "The Use of Slip Circle in the Stability Analysis of Earth Slopes," *Geotechnique*, **5**(1) (1955).

[8] David F. McCarthy, *Essentials of Soil Mechanics and Foundations*, Regents/Prentice Hall, Englewood Cliffs, N.J., 1993.

Answers to Selected Problems

CHAPTER 2

2–2. A-2-4 (0)
SC

2–4. $w = 15.3\%$
$e = 0.56$
$n = 36.0\%$
$S = 72.6\%$
$\gamma = 122.6$ lb/ft^3
$\gamma_d = 106.3$ lb/ft^3

2–6. $\gamma = 121.2$ lb/ft^3
$\gamma_d = 105.4$ lb/ft^3
$n = 36.0\%$
$S = 70.3\%$

2–8. $\gamma = 109.8$ lb/ft^3
$G_s = 2.77$

2–10. $w = 13.04\%$
$\gamma = 104.4$ lb/ft^3

2–12. $\gamma_d = 16.52$ kN/m^3
$e = 0.62$
$S = 54.3\%$

2–16. $e = 0.61$
$\gamma = 111.2$ lb/ft^3

2–18. 6.1%

2–20. 19.9 lb

CHAPTER 3

3–2. 30, 31, 34

3–4. 19, 19, 18

3–6. 382 lb/ft^2

3–8. 63 ft; upper layer: moist to dry silty and sandy soils; lower layer: well-fractured to slightly fractured bedrock with moist soil-filled cracks

CHAPTER 4

4–1. $\gamma = 116.7$ lb/ft^3
$\gamma_d = 107.0$ lb/ft^3

4–3. 9 to 19%

4–5. $\gamma = 124.2$ lb/ft^3
$\gamma_d = 107.1$ lb/ft^3

4–7. 2897 m^3

CHAPTER 5

5–1. 6.56×10^{-4} ft^3/s

5–3. 0.103 cm/s

5–5. 0.0342 cm/s

5–7. 0.0405 cm/s

5–9. 0.485 ft, or 5.8 in.

5–11. 1.57×10^{-6} cm/s

5–13. 0.00328 ft^3/s per ft

CHAPTER 6

6–2. 169 lb/ft^2

6–4. a. $q_{(1m)} = 97.50$ kN/m^2
b. $q_{(3m)} = 39.00$ kN/m^2
c. $q_{(5m)} = 20.89$ kN/m^2

6–6. a. 48.50 kN/m^2
b. 97.64 kN/m^2

6–8. 520 lb/ft^2

6–10. a. 628 lb/ft^2
b. 350 lb/ft^2

6–12. a. 109.5 kN/m of wall
b. 197.9 kN/m of wall

6–14. 2.87 kips/ft^2

CHAPTER 7

7–1. 2.63×10^{-8} cm/s
7–3. 0.79
7–5. 0.062 m
7–7. **b.** 3.66 in.
7–9. 1. 37 yr
 2. 0.69 in.
 3. 1.75 yr
7–11. 0.89 in.
7–13. 3.06 tons/ft^2
7–15. 13.5 mm

CHAPTER 8

8–2. c = 380 lb/ft^2
 $\phi = 19.5°$
8–4. 34°
8–5. 35.4°
8–8. 1925 lb/ft^2
8–10. 96.0 kN/m^2

CHAPTER 9

9–2. 12,100 lb/ft^2
9–4. 695 tons
9–6. 38°
9–8. 228 kN/m^2
9–10. 5.5 ft by 5.5 ft
9–12. 91 kN/m^2
9–15. **1a.** 2.47
 1b. 3.36
 2a. 3.58
 2b. 4.87
9–17. 1. q_R = 3.5 kip/ft^2
 q_L = 0.5 kip/ft^2
 2. 69.6 kips
 3. 148.8 ft-kips
 4. 4.0
 5. 7.2
 6. 3.1

CHAPTER 10

10–2. 21.1 kips
10–4. 83.6 kips
10–6. 265 kN
10–8. 28 ft
10–10. 20 blows/ft
10–12. 37.5 tons
10–14. 382 kips
10–16. **a.** 557 kips
 b. 444 kips
 c. 444 kips
10–18. 675 kN
10–20. Q_{max} = 70 kips
 Q_{min} = 10 kips

CHAPTER 11

11–2. 1.08 m
11–4. 1086 tons

CHAPTER 12

12–1. 39.58 kN/m of wall
12–4. 6900 lb/ft of wall
12–6. 9190 lb/ft of wall
12–8. 89.0 kN/m^2
12–9. 16,150 lb/ft of wall acting at 9.41 ft above the base
12–11. 16,100 lb/ft of wall

CHAPTER 13

13–1. **a.** 3.31
 b. 1.96
 c. q_L = 4.66 kips/ft^2
 q_R = 0.10 kip/ft^2 (*Note:* $q_L > q_a$ of 4 kips/ft^2)
13–4. L_{total} = 6.40 m
 s = 0.669 m
13–6. 2.20 m

CHAPTER 14

14–2. 8.7 ft
14–4. 1.38
14–6. 38°
14–8. 1.86

APPENDIX: CONVERSION FACTORS

1 in. = 25.40 mm	*Length*	1 mm = 0.03937 in.
1 in. = 2.540 cm		1 cm = 0.3937 in.
1 ft. = 0.3048 m		1 m = 3.281 ft
1 mile = 1.609 km		1 km = 0.6214 mile
1 in.^{2} = 645.2 mm^2	*Area*	1 mm^2 = 0.001550 in.^{2}
1 in.^{2} = 6.452 cm^2		1 cm^2 = 0.1550 in.^{2}
1 ft^2 = 0.09290 m^2		1 m^2 = 10.76 ft^2
1 in.^{3} = 16,390 mm^3	*Volume*	1 mm^3 = 0.00006102 in.^{3}
1 in.^{3} = 16.39 cm^3		1 cm^3 = 0.06102 in.^{3}
1 ft^3 = 0.02832 m^3		1 m^3 = 35.31 ft^3
1 lb = 0.4535 kg	*Force*	1 kg = 2.205 lb
1 lb = 4.448 N		1 N = 0.2248 lb
1 kip = 4.448 kN		1 kN = 0.2248 kip
1 ton = 8.896 kN		1 kN = 0.1124 ton
1 lb/in.^{2} = 0.07029 kg/cm^2	*Pressure or stress*	1 kg/cm^2 = 14.23 lb/in.^{2}
1 lb/in.^{2} = 6.894 kN/m^2		1 kN/m^2 = 0.1450 lb/in.^{2}
1 lb/ft^2 = 0.04788 kN/m^2		1 kN/m^2 = 20.89 lb/ft^2
1 kip/ft^2 = 47.88 kN/m^2		1 kN/m^2 = 0.02089 kip/ft^2
1 ton/ft^2 = 95.76 kN/m^2		1 kN/m^2 = 0.01044 ton/ft^2
1 lb/ft^3 = 0.1571 kN/m^3	*Unit weight* 62.4 lb/ft^3 = 9.81 kN/m^3	1 kN/m^3 = 6.366 lb/ft^3

Note: kg is used as a unit of force in the metric system.

INDEX

A

B

C

D

E

F

G

I

L

M

N

O